Space Dynamics

Advances in Astronomy and Astrophysics

A series of books and monographs which aims to supply professionals and serious graduate students with the intellectual tools necessary for the appreciation of the present status of topics at the forefront of current research, and provides a framework upon which future developments may be based.

Series Editors:
V.G. Gurzadyan, Yerevan Physics Institute, Armenia and ICRA, Department of Physics, University of Rome 'La Sapienza', Italy
S. Inagaki, Kyoto University, Japan
G.Meylan, European Southern Observatory, Garching near Munich, Germany

Volume 1
Highly Evolved Close Binary Stars: Catalog
Highly Evolved Close Binary Stars: Finding Charts
A.M. Cherepashchuk, N.A. Katysheva, T.S. Khruzina, S. Yu. Shugarov

Volume 2
Physics of Accretion Disks: Advection, Radiation and Magnetic Fields
Edited by S. Kato, S. Inagaki, S. Mineshige, J. Fukue

Volume 3
Celestial Dynamics at High Eccentricities
V.A. Brumberg and E.V. Brumberg

Volume 4
Polarization of Light and Astronomical Observation
J. Leroy

Volume 5
Modern Celestial Mechanics: Aspects of Solar System Dynamics
A. Morbidelli

Volume 6
Mass Transfer in Close Binary Stars
A.A. Boyarchuk, D.V. Bisikalo, O.A. Kuznetsov, V.M. Chechetkin

Volume 7
Space Dynamics
G.A. Gurzadyan

Space Dynamics

Second edition of
Theory of Interplanetary Flights

G.A. Gurzadyan
Garni Space Astronomy Institute
Yerevan, Armenia

London and New York

Cover: Deep Space 1 spacecraft, credited to NASA, DP1 team

First edition published 1996 as *Theory of Interplanetary Flights*
by Gordon and Breach Publishers

This edition first published 2002 by Taylor & Francis
11 New Fetter Lane, London EC4P 4EE

Simultaneously published in the USA and Canada
by Taylor & Francis Inc,
29 West 35th Street, New York, NY 10001

Taylor & Francis is an imprint of the Taylor & Francis Group

Publisher's note

This book has been prepared from camera-ready copy supplied by the author.

Printed and bound in Great Britain by TJ International Ltd, Padstow, Cornwall

British Library Cataloguing in Publication Data
A catalogue record for this book is available from the British Library

Library of Congress Cataloging in Publication Data
A catalog record for this book has been requested

ISBN 0-415-28202-0

Contents

Preface

Spectacular achievements attained during interplanetary flights in the last several years are the main motivation for the appearance of the present monograph. The author's previous book **Theory of Interplanetary Flights** only in part served as a basis for this monograph.

The biggest Chapter, XV, is dedicated to interplanetary flights opening a new epoch in the understanding and assimilation of the Solar system in all its fantastic diversity. In that chapter we discuss about 20 interplanetary missions, still the minor part of already realized or forthcoming projects, but entirely different one from another and each one attractive because of its performance. Our primary concentration is on flight trajectories, i.e. one of the main subjects of celestial mechanics, the result of profound theoretical searches and, at the same time, demonstrating the possibilities of computational techniques, control and communication systems, etc. In certain cases these flights exploit theoretical results of celestial mechanics which not long ago still have been considered as quite abstract.

In Chapters XII, XIII and XIV new types of trajectories for interplanetary flights are presented and analyzed, among them trajectories not only from the Earth but, say, from Mars to Earth or to Venus.

Chapter X is devoted to the application, for the first time, of the idea of equipotential surfaces (Chapter IX) to binary stars, which lead to the formation of so-called roundchroms around those binaries with a spatial volume of emission of enhanced chromospheric-type emission. Another unexpected area of application of the same idea of equipotential surfaces are the binary globular clusters and galaxies with binary nuclei. In essence, we have examples of the invasion of celestial mechanics into astrophysics.

The opening chapters are devoted to the basics of celestial mechanics, including the two-body problem, ephemerides, determination of orbits, the N-body problem, the restricted three-body problem, perturbation theory, etc.

In various sections, a number of novel problems, such as the problem of extrasolar planets, binary asteroids, the determination of the mass of a celestial body by perturbations in the spacecraft's orbital motion, creation of space observatories in libration points etc. are discussed, occasionally with unexpected, non-ordinary and far-reaching conclusions. Certain new insights on the chaos problem are discussed as well.

Throughout the book, while dealing both with classical issues of celestial mechanics and with novel ideas, clarity of representation has been our priority.

Chapter I

The Universal Law of Gravitation

1   The Antique Period of Celestial Mechanics

By Plato's definition the motion of the celestial bodies is a problem beyond human understanding. Apparently the persistence revealed by humanity while deciphering the secrets of the celestial bodies and their mystic motions is comparable only with the desire to understand the secret of life itself.

The historical data, rather obscured and not always reliable, attribute an astronomical activity to Egyptian priests already in the fourth millennium BC. Certain features of the construction of the pyramids and similar facts seem to indicate the capacity of the ancient Egyptians to perform rather accurate astronomical observations in those olden days. Chinese astronomical records are attributed up to the XXVth century BC, and various astronomical events are mentioned in ancient Mesopotamian texts. Cuneiform Babylonian inscriptions of the second and especially of the first millennium BC contain numerous records of astronomical events which are used for dating the events and obtaining the absolute chronology of the epoch, such as the Solar eclipse of June 15, 763 BC linked with eponyms of Assur. Among the Babylonian astronomical sources of the second millennium BC are the famous Venus Tablet attributed to the reign of King Ammisaduqa, and records of a pair of lunar eclipses connected with the Third Dynasty of Ur, which dominated Babylonia around the XXth century BC.

However, the first reliable recordings of astronomical observations appear only in the VIIIth century BC and belong to the period of the rise and prosperity of the Greek civilization. Thus, to **Thales of Miletus** in the VIIth century BC is attributed the initiation of implementing Egyptian astronomical knowledge within Greece. In the VIth century BC the **Pythagorean** theory stated that the Earth is a globe, as are all other celestial bodies, and is hanging freely in space without any support. Later **Philolaus**, a Pythagorean student, was the first to claim the possibility of axial rotation of the Earth.

Presumably the first attempts to bring a logical basis to the problem of the motion of celestial bodies are due to **Eudoxus** and **Calyppus**, students of Plato. They proposed the idea of concentric spheres on which the motion of the celestial bodies is occurring, including the sphere of motionless stars. They can be considered as the initiators of the transition from purely descriptive images to the scientific stage of the formation of dynamical astronomy. This approach found its highest development due to **Aristotle** in the mid-IVth century BC. Arguing for the geocentric system, Aristotle was aiming to reach a better understanding of the observed motion of the celestial bodies, e.g. The claimed the number of concentric crystal spheres to be 56, but even in that case the resulting picture of the motion of the celestial bodies was rather far from reality.

The contribution of Aristotle to the proof of the sphericity of the Earth is, however, not in doubt. He was the author of the idea that lunar eclipses are caused by the Earth's shadow, so that the circular boundary of the shadow on the Moon is indicating the spherical form of the Earth. His other argument for a spherical Earth was the variation of the positions of stars while moving to the North and South. Rather impressive is his brave idea that if one celestial body is spherical, so can the others be; his comprehending mind was uncovering the various phenomena of Nature and the processes that occur.

As for the lunar and Solar eclipses, not only Aristotle but all the Greek scholars of that epoch were familiar with their real nature, namely, with their being the results of the mutual positions of the Sun and Moon. To Aristotle's epoch belongs the first appearance of the idea of a heliocentric world. Thus, according to **Heracleides of Pontus**, Aristotle's contemporary, one can describe celestial events by assuming the sky to be motionless while the Earth is rotating.

After Aristotle, the center of Greek thought, including astronomy, moved to Alexandria, the city founded in 332 BC by Alexander the Great. The most eminent representative of the Alexandrian epoch is **Aristarchus of Samos**, in the first half of the IIIth century BC, who was a contemporary of Euclid. To Aristarchus is attributed the first attempt at measuring the distances to the Moon and Sun, and the first mathematical concepts to explain the character of motion of the celestial bodies. The latter includes the idea of the theory of "moving eccentrics" according to which any planet is moving in an eccentric circle, i.e. a circle with a center not coinciding with the Earth's center, though all the centers are situated along the line joining the centers of the Sun and Earth. It is certain that the idea of the heliocentric system was clearly and fully formulated by Aristarchus. Two of his contemporaries, **Thimocharis** and **Aristillus**, were presumably the authors of the first attempts at mathematical constructions of the consequences of the rotation of the celestial sphere and several simple celestial motions. They also obtained the positions of stars and therefore created the first star catalogs, in the modern sense of that word.

While evaluating the intellectual outburst of ancient thinkers, one must, nevertheless, confess that only a few of their purely scientific results have survived to the present day. Among these few discoveries one should outline the contribution of **Eratosthenes** (late IIIrd to early IInd century BC), a brilliant representative of the Alexandrian school. He performed the first scientific measurement of the size of the Earth, by means of an exceptionally simple but mathematically strict and convincing method. He noticed that in Alexandria on the day of the summer Solar equinox the Sun is situated $7°.2$ from the zenith, whereas in Siena in Upper Egypt, the Sun is exactly at its zenith. Knowing the distance to Siena, he estimated the length of the Earth's circumference and therefore the Earth's radius. Eratosthenes's result for the Earth's circumference was 39 250 km, which differs from the modern value by less than 2%.

Remarkable progress in antique astronomy is associated with the name of Hipparchus (IInd century BC), one of the foremost figures in astronomy. He was the author of the first truly scientific theories to explain the motion of celestial bodies not only qualitatively but also quantitatively. He was the creator of a fundamental catalog of 1 000 stars with given coordinates, latitudes and longitudes, and with magnitudes on a six-grade scale. Hipparchus carried out the first systematic comparisons with the previously obtained observational data, to discover discrepancies, and regularities occurring within time periods exceeding the human lifespan. In this way, he made one of the essential discoveries in astronomy, namely that of a systematic variation in the latitudes of stars, later called *precession*. He estimated its rate to be 48" per year, which is quite close to the present estimation, 50".3. Hipparchus also determined the Earth's obliquity, i.e. the angle between the celestial equator and the ecliptic, with surprising accuracy, $23°.5$, the modern value being $23°44'$.

Hipparchus, with ideas a few centuries ahead of his contemporaries, correctly guessed the advantages of geometrical analogs of visible motions of celestial bodies, and proposed the idea of a principally new geometrical system, of eccentrics, in order to explain the regularities in the motions of the Sun and the Moon. He was also successful in an area with a direct practical outcome, predictions of Solar and Lunar eclipses.

Hipparchus introduced the concept of geographic latitude and longitude and offered to use Rhodes, the island on which he was performing his observations, as a center of coordinates. Much later, the "null" meridian was moved to the Canary islands, then to the Paris Observatory and only in 1884 to the Greenwich Observatory.

To Hipparchus are also attributed important improvements in trigonometry, closely related to his astronomical achievements. His contribution in astronomy is not limited to the previously mentioned points; all of his results are amazing in their bravery and comprehensiveness. It is natural that d'Alembert, an expert on the history of science, considered Hipparchus among the "amazing and great thinkers of antiquity".

For approximately three centuries after Hipparchus's death, a silent "foggy" period in astronomy elapsed before there arose in the early IInd century BC a brilliant new scholar – Ptolemy.

2   Ptolemaic Cosmography

The apogee of Greek astronomy is undoubtedly related to the name of **Claudius Ptolemy** (70–140 AD). The extraordinary fame of Ptolemy is mainly based on his great astronomical book known us its shortened Arabic title "Almagest", which had tremendous influence for an unprecedentedly long period. That book is based largely on the works of previous astronomers, first of all that of Hipparchus, who had inspired Ptolemy in an essential way. "Almagest" in fact became a kind of encyclopedia for generations of astronomers during the entire Middle Ages.

"Almagest" consists of 13 books, in which a number of astronomical facts and events are considered successively: the daily motion of the celestial sphere, visual motions of the Sun, Moon and planets, the durations of days, the times of rising and outing of stars at various points on the Earth, arguments and proofs of the spherical form of the Earth, etc.

Ptolemy is describing his cosmographical postulates, the basic ones being as follows: the sky is spherical and is rotating like a sphere; the Earth is situated in the center of this sphere; the Earth is motionless and is a point compared with the distances to stars. "Almagest" contains the highly original ideas and constructions of Ptolemy himself concerning the structure of the surrounding world and first of all, the motion of planets. His aim was the explanation of a number of irregularities in the visible motion of the planets, a task beyond the capabilities of his predecessors. It was known that the planets are not always moving from west to east, but sometimes stop and move backwards, then again stopping and continuing their motion.

Already, before Ptolemy, there had been attempts to create a theory of planetary motion, in particular, to describe the retrograde motion of planets, for example, proposing an oscillatory motion relative to a fictitious planet smoothly moving along the celestial sphere (Eudoxus) or via rotating spheres or epicycles (Apollonius). Ptolemy developed the idea of epicycles and ideal spheres to a perfect level. As a result, for the first time the laws of planetary motions were formulated, enabling the prediction of planetary positions, the creation of corresponding tables, etc. The practical importance of this novelty was rather large for navigation and the composition of geographical maps; this fact was crucial in preserving the authority of "Almagest" and its author during subsequent epochs.

Ptolemy brilliantly solved the problem of motion of all five of the then-known planets. By means of the proper choice of ideal circles of various sizes, **deferents** and **epicycles,** i.e. fitting the visible anomalies of planetary motion via a sequence of circular motions, Ptolemy reached an amazing coincidence with the observed motions of planets.

Fig. 1.1 reflects Ptolemy's main idea on constructing planetary motions. At the center of the ideal circle, the deferent, the Earth is situated. No planet is moving in that circle as one might have expected; it is in another circle, the epicycle, that the observed planet is performing its motion. It is easy to see that the visible motion of the planet from the Earth should be either from west to east or vice versa. In such a simple and natural way he explained phenomena that had puzzled earlier astronomers – the retrograde motion of planets. Obviously, via proper choice of corresponding sizes of the epicycles and deferents one can reach complete coincidence with the observed periods of retrograde motions, as well as with the observed frequency of such events during one complete revolution of the planet in its deferent.

Ptolemy went even further: to eliminate the minor inaccuracies in the observed motions, and therefore to reach absolute accordance with the observations, he superposed another, smaller epicycle on the main epicycle, and in certain cases even more epicycles; a converging sequence of epicycles appeared. Finally, he introduced a deferent eccentrically situated with respect to the Earth, the so-called deferent eccentric, as shown in Fig. 1.2. This enabled him to explain the most complicated and mysterious planetary motions. By means of the idea of n circles, epicycles of decreasing size, Ptolemy in fact predicted the idea of converging and infinite Fourier series.

Within the described scheme naturally there was explained the simplest of all motions – the motion of the Sun, as had been noticed already by Hipparchus; therefore Ptolemy included his theory of Solar motion in "Almagest" without any changes. As shown in Fig. 1.2, Solar motion is performed via the deferent only; it has no epicycle.

The main defect of Ptolemaic cosmography is not in the central position of the Earth, as usually is considered, while other bodies including the Sun are rotating around the Earth. From a philosophical point of view it even has some attractiveness, since it contains the idea that certain observations can be explained equally both by motion of stars or of the Earth. Moreover, one might recall also the preferred idea of his time, to attribute motion to the stars, i.e. objects with the nature of fire, rather than to the solid Earth.

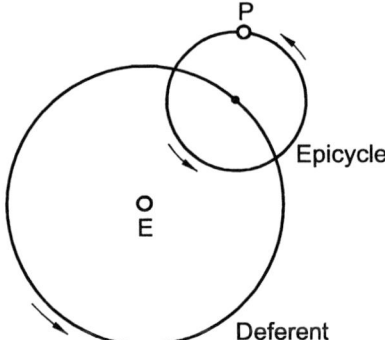

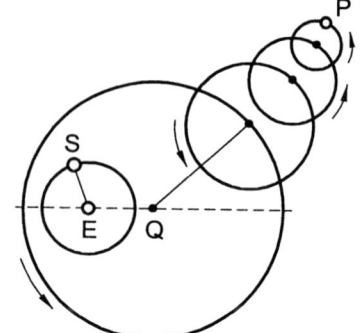

Fig. 1.1. *The Ptolemaic geocentric system: planet P is rotating in an epicycle, while the epicycle itself is rotating in a deferent around the Earth, E.*

Fig. 1.2. *A system with three epicycles and eccentrically situated deferent explaining the complicated observed motion of planets.*

The main weak point of Ptolemaic cosmography is the absence of the concept of *force* controlling the motion of planets. The thinkers of antiquity, Aristotle, Hipparchus, and Ptolemy, had only the concept of motion, while the question concerning its reason did not exist. The concept of force in celestial mechanics, and in mechanics in general, was introduced surprisingly late, only in the XVIIth century in the time of Descartes.

Thus, the Ptolemaic cosmology was based on geometrical abstraction in the complete absence of causality, hence its apparent absurdity from the modern point of view. His arguments nowadays seem rather synthetic and lacking physical background. It is hard to imagine in terms of the present epoch's scientific mentality the possibility of explaining something within known physical laws using the idea of deferents and epicycles. One should note, however, that Ptolemy himself seems not to attribute too much importance to his own postulates and constructions; he considers them as "non-important truths", presumably requiring deeper background and further mathematical treatment. His only aim was to explain *the observations*, nothing else.

Yes, the genius of Ptolemy is unique, since he succeeded in keeping mankind aware of his ideas for 14 centuries, up to the time of Copernicus. It is impossible to find another similar example, either before or after Ptolemy, in any area of science.

3 Copernican Heliocentrism

Among the most important discoveries of XIXth–century mathematics is the proof of the possibility of expansion of any function into series, including functions that have no physical content. For the function satisfying, say, the equation of motion of a certain planet, one can always find a sequence of discrete quantities, the members of the series, such that their sum gives the positions of the planet $x(t)$, $y(t)$:

$$x(t) = f_1(t) + f_2(t) + \ldots + f_n(t) = \sum^n f_i(t),$$

$$y(t) = \varphi_1(t) + \varphi_2(t) + \ldots + \varphi_n(t) = \sum^n \varphi_i(t).$$

Obviously, by retaining a large enough number of members of these series, one can reach the required coincidence with the observed position of the planet, if of course the series is converging.

In fact this is Ptolemy's approach, the only difference being that he performed the expansion of functions by a geometrical method, replacing the pairs of functions f_1, φ_1 and f_2, φ_2, etc., on corresponding epicycles of decreasing size, by which both the imaginary and real planets are moving with constant velocities. The basic idea, however, is the same.

Ptolemy, in fact, is solving an inverse problem, i.e. without knowing the functions $x(t)$ and $y(t)$ by means of arbitrary and purely geometrical constructions with no physical content he is finding each of the functions f_i and φ_i. This is the idea of estimation of the parameters of epicycles that ensure the best fit with the observed position of the planet. It is important that he performs this procedure anew for each planet. In other words, Ptolemy had no uniform principle and parameters, that could be used for the estimation of the position of any planet. In fact Ptolemy is finding an empirical formula, *ad hoc* and *ad libitum*, with only one aim, namely, to obtain the position of the planet. The content of the expansion terms, i.e. of epicycles, did not bother him.

Ptolemy's constructions did not reveal any underlying relation between the observed motion of the planets and Sun. His geocentric scheme cannot answer the question of why the Sun has no epicycles, though it is also rotating around the Earth. This fact should have prompted Ptolemy and his successors to assign a unique role to the Sun; however, it did not happen. Obviously, just due to the absence of a basic idea on the relation of the planets and the Sun, one can explain the fact that for 14 centuries nobody succeeded in deriving any general law, as was done later for the heliocentric system in the form of Kepler's laws. Therefore the necessity of a drastic reformation of astronomy, the transition from purely imaginary geometrical constructions to physical discipline, was absolutely clear.

The aim formulated by **Nicholaus Copernicus** (1473–1543) was to find an underlying relation between the motions of the planets. This was the aim of his life, and he finally demonstrated that:

1. all planets are moving around the Sun;

2. the Earth is an ordinary planet and is rotating around the Sun;

3. the Earth is rotating around its axis, with one revolution per day.

The heliocentric Copernican system was created in 1543. Copernicus claimed that the planets are moving around the Sun in one direction without any stops and retrograde motions, and that the observed loops in planetary motions are due to the observations being from the moving Earth. The resulting effects are seen as shifts of planetary projections, later called **parallaxes**. Thus for the observer on the Earth E_1 the planet P_1 (Fig. 1.3) is seen at projection P_1', so that the subsequent motion of the Earth and the planet will give an illusion of retrograde motion. As a convincing argument Copernicus noticed that the retrograde motions of planets have periods equal to the period of revolution of the Earth around the Sun; this fact had been known earlier but had not received proper attention. Thus the mystery of the retrograde planetary motions and hence of epicycles was solved.

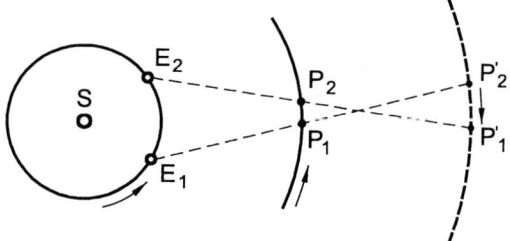

Fig. 1.3. *A scheme explaining the retrograde motion of a planet P relative to an observer on the Earth E in the Copernican heliocentric system.*

Copernicus destroyed the idea of a motionless Earth, which had been held for many centuries. Ptolemy, following Aristotle, was arguing the impossibility of the Earth's motion since it would lead to the disruption of the Earth. Copernicus, considering this point from a physical point of view, mentioned that in the case of a motionless Earth, the "sky" should rotate at such high velocity that it would itself be destroyed.

The basic point of Copernican philosophy is the unification of the world, the Earth, the Sun, and all the planets. His world would have to be governed by universal laws that were to be discovered soon thereafter.

4 Tycho Brahe and Kepler

It might be surprising, but for the creation of the Copernican heliocentric system the high inaccuracy of observations was crucial. Thus, he could not distinguish the elliptical orbits of the planets, which were obtained later, from their circular motions, which greatly simplified the deciphering of the macrostructure of the planetary system.

The next landmark in the evolution of astronomy is associated with the name of Tycho Brahe (1546–1601). Having at his disposal an observatory with the best facilities of the time, Tycho Brahe observed the positions of celestial bodies, including the planets, for 20 years, with an unprecedented accuracy, up to 1 arc min (!). Note that this was the epoch of observations by the unaided eye; the first telescope was operated in 1609 by Galileo. Much later it became evident that 1 arc min is the theoretical limit of the eye as an optical instrument (Rayleigh criterion).

After the death of Tycho Brahe, all his rich observational data were inherited by Johannes Kepler (1571–1630). Based on them, Kepler tried to decipher the underlying laws of the Copernican heliocentric system. After huge efforts he came to the astonishing conclusion that the orbit of Mars is not a circle as he had previously believed, but an ellipse. Immediately he derived the first crucial regularities of that motion and formulated his first two laws in 1607:

I. Every body is moving in an ellipse, so that the Sun is located at one of the foci.

II. The radius-vectors of the planets sweep over equal areas in equal intervals of time.

The second law is stating, first, the nonuniform motion of the planets, and second, that this nonuniformity nevertheless has its regularity, i.e. the faster the motion, the closer the planet is to the Sun. Since the areas can be estimated easily, the second law therefore enables the prediction of the position of planets at any moment of time.

Kepler's first two laws deciphered the deep physical content of the heliocentric system. However, both laws, though they enable predictions, only quantitatively characterize the motion of each orbit; they do not reflect the interrelation of various orbits, besides the fact that the Sun is situated at the foci of all ellipses.

Twelve years later, in 1619, Kepler announced his third law: the squares of the periods of orbital motions P are related to the cubes of their semimajor axes a:

$$\frac{p_1^2}{a_1^3} = \frac{p_2^2}{a_2^3} = \frac{p_3^2}{a_3^3} = constant\,.$$

It is important to outline that the third law is the first **quantitative** law in the history of celestial mechanics to relate different parameters of the motion.

It is hard to evaluate the whole role of Kepler's laws in celestial mechanics. Still having no idea of the forces governing the motion of the planets, Kepler, first, proved that a force the same for all the planets is associated with the Sun; second, purely empirically, he succeeded in deriving certain regularities of the motions. If many people were already doubting the idea of a motionless Earth before Copernicus, nevertheless nobody, including Copernicus himself, doubted the circular orbits. While Kepler proved that planets can move only by ellipses, Kepler's laws reveal the existence of a certain harmony in the planetary motions with deep underlying regularities that were unreachable for his contemporaries.

Kepler's laws are the culmination before the final step – the discovery of the law of universal gravitation. That step awaited the genius of Newton.

5   The Newtonian Law of Gravitation

Isaac Newton (1643–1727) discovered the law of universal gravitation by proceeding from Kepler's three laws. Of course, he had already discovered the laws of motion, consolidating the work by **Galileo** (1564–1642), **Descartes** (1596–1650) and **Huygens**(1629–1695).

Kepler's laws are as follows:

The First Law. The motion of the planets around the Sun is performed in a such way that the area swept in some interval of time is proportional to that interval, or in other words, the sectorial velocity is constant. This fact is expressed in polar coordinates as follows:

$$r^2\frac{d\theta}{dt} = constant \tag{1}$$

or

$$x\frac{dy}{dt} - y\frac{dx}{dt} = constant \tag{2}$$

in Cartesian coordinates.

The Second Law. The motion of planets occurs in ellipses, which can be represented via the expression

$$r = \frac{p}{1 + e\,\cos(\theta - \omega)}\,, \tag{3}$$

where $p = a(1 - e^2)$, a and e are the semimajor axis and eccentricity of the ellipse, and ω is the angular distance of the semimajor axis from some chosen direction.

The Third Law. The squares of periods of revolutions P are proportional to the cubes of the semimajor axes, or:

$$\frac{P^2}{a^3} = constant\,.\tag{4}$$

This law, as we shall see in next chapter, should be rewritten in a more precise form:

$$\left(1 + \frac{m}{M}\right)\frac{P^2}{a^3} = constant\,,\tag{5}$$

where M and m are the masses of the Sun and the planet, respectively. The largest m/M that we have is for Jupiter, about 0.01, i.e. 1.001 in (4), instead of 1.000. However, first, this discrepancy could not have been noticed by Kepler; second, it did not prevent Newton discovering the true law of gravitation. In any case Newton was proceeding from Kepler's third law, i.e. from Eq. (4) when he deduced the law of gravitation.

Denoting by F_x and F_y the projections of the force F acting between the Sun and planet, one can write the equations of motion:

$$m\frac{d^2x}{dt^2} = F_x\,,$$

$$m\frac{d^2y}{dt^2} = F_y\,.$$

Multiplying the first equation by dx/dt, the second by dy/dt and summing, we have

$$\frac{m}{2}\frac{d}{dt}\left[\left(\frac{dx}{dt}\right)^2 + \left(\frac{dy}{dt}\right)^2\right] = F_x\frac{dx}{dt} + F_y\frac{dy}{dt}\,,\tag{6}$$

and using the polar coordinates

$$\begin{aligned}x &= r\cos\theta; & F_x &= F\cos\theta;\\ y &= r\sin\theta; & F_y &= F\sin\theta;\end{aligned}\tag{7}$$

we have

$$\frac{m}{2}\frac{d}{dt}\left[\left(\frac{dr}{dt}\right)^2 + r^2\left(\frac{d\theta}{dt}\right)^2\right] = F\frac{dr}{dt}\,.\tag{8}$$

Writing

$$\frac{dr}{dt} = \frac{dr}{d\theta}\frac{d\theta}{dt}$$

and in view of Kepler's first law, we obtain

$$\frac{mc^2}{2}\frac{d}{dt}\left[\left(\frac{d\,(1/r)}{d\theta}\right)^2 + \frac{1}{r^2}\right] = F\frac{dr}{dt}\,. \tag{9}$$

Substituting for $1/r$ from Kepler's second law, we have

$$F = \frac{mc^2}{pr^2}\,. \tag{10}$$

The left-hand side in (1) is the doubled area of the ellipse swept by the radius-vector in one revolution, so that

$$2\pi ab = cP\,, \tag{11}$$

where $b = a(1 - e^2)^{1/2}$. From this we have for c

$$c = \frac{2\pi a^2\,(1 - e^2)^{1/2}}{P} \tag{12}$$

and substituting into (10) using also $p = a\,(1 - e^2)$ we obtain

$$F = 4\pi^2\frac{a^3}{P^2}\frac{m}{r^2}\,. \tag{13}$$

Insofar as Kepler's third law that P^2/a^3 is the same for all planets applies, we have from (13)

$$F = \mu\frac{m}{r^2}\,, \tag{14}$$

where

$$\mu = 4\pi^2\frac{a^3}{P^2}\,. \tag{15}$$

This result means that the motion of planets is performed under the action of a force directed to the center of Sun and equal to $\mu m/r^2$, where m is the mass of the planet, and r its distance from the Sun's center. On the other hand, it is obvious that the planet is also attracting the Sun with the force $\mu_1 M/r^2$. In accord with the law of equality of action and counteraction discovered by Newton earlier, these forces should be equal:

$$\mu\frac{m}{r^2} = \mu_1\frac{M}{r^2}$$

or

$$\frac{\mu}{M} = \frac{\mu_1}{m}\,.$$

We denote this ratio by f, which as we shall see, will be the same for the Sun and the planets. Substituting this into (14), we finally obtain

$$F = f\frac{mM}{r^2}.$$ (16)

This formula, derived to describe the interaction between the Sun and a planet, can naturally be generalized for any two masses. Thus the force of interaction between two masses is proportional to their masses and inversely proportional to the square of their mutual separation. This is the law of universal gravitation, published by Newton in 1686.

6   Gravitation and Potential

In a number of problems of celestial mechanics there is an important role for the **force function**, i.e. a function $V(r)$, the derivative of which is the force F:

$$F = \frac{dV}{dr}.$$ (17)

When $V = V(x, y, z)$ and X, Y, and Z are the components of the force F in coordinate axes x, y, and z, we have

$$X = \frac{dV}{dx}, \quad Y = \frac{dV}{dy}, \quad Z = \frac{dV}{dz}.$$

Therefore the function V defines the force acting on the point P. From Eq. (17) we have

$$V = \int F \, dr.$$ (18)

and when F is represented by Newton's law, after integration in Eq. (18) we arrive at:

$$V = -f\frac{mM}{r}.$$ (19)

Let us consider the work which should be performed in order to increase the distance between two masses m and M from r up to R_* when the masses are interacting by Newton's law (16) and $r_* > r_1$. The work performed when the distance is increased by dr will be

$$f\frac{mM}{r^2} \, dr,$$

and at the increase of the distance between the masses from r to r_* the work performed will be

$$\int_r^{r_*} f\frac{mM}{r^2} \, dr = -f \, mM \left(\frac{1}{r_*} - \frac{1}{r} \right).$$

In the case when $r_* = \infty$, the work performed will be

$$f \frac{mM}{r} .$$

Comparing this result with Eq. (19), we conclude that the latter, i.e. Eq. (19), is nothing other than the amount of work required to move the point mass M from infinity to a distance r from the central body. This amount of work with negative sign is called the **potential**.

In other words, the potential of a mass is the force function with a negative sign. Thus Newtonian gravity belongs to the type of force possessing a potential given by Eq. (19).

The attraction of two point masses, one of an elementary mass dm, the other with unit mass $M = 1$, from Eq. (19), will be

$$dV = \frac{dm}{r} , \tag{20}$$

where the sign is changed and the distance r is chosen in such a way as to have the coefficient $f = 1$ (Fig. 1.4); the mass dm belongs to an arbitrary point of a finite object $\mathcal{M}$. Obviously the total potential corresponding to this interaction can be obtained via integration over the whole mass $\mathcal{M}$:

$$V = \int_{\mathcal{M}} \frac{dm}{r} . \tag{21}$$

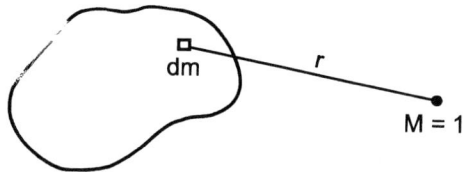

Fig. 1.4. *The problem of determination of the potential of an arbitrary mass configuration, in units $M = 1$.*

In the general case when $m = m(x, y, z)$ or $m = m(r, \varphi, \theta)$ we will have

$$V = \int \int \int \frac{dm(x, y, z)}{(x^2 + y^2 + z^2)^{1/2}} \tag{22}$$

in Cartesian coordinates, or

$$V = \int \int \int \frac{dm(x, y, z)}{r} \tag{23}$$

in polar coordinates.

In the particular case in which the body with finite configuration coincides with the point of a mass $\mathcal{M}$, we will have from Eq. (21)

$$V = \frac{\mathcal{M}}{x}.\qquad(24)$$

This result can be formulated as follows: in all cases when a finite configuration is attracting a mathematical point and the potential of their interaction can be represented by Eq. (24), these bodies are interacting as **material points.**

The concept of the potential, which was later to become the basic one in mathematical physics, in the theory of potential, was introduced by **Greene** in 1828 concerning theories of electricity and magnetism.

7 The Attraction of Two Spherical Bodies

The Newtonian law of gravitation is valid for material points of masses m and M situated at a distance r from another. The planets are not material points, but have finite sizes, so that the problem is to obtain the law of interaction between them.

Consider the simplest case: the planet is spherical of a radius R_0 but not with constant density and consisting of concentric layers of various density, $\rho(r)$.

The first problem is to find out the law of attraction between the material point P of mass $m = 1$ and the sphere of mass $\mathcal{M}$, so that

$$\mathcal{M} = 4\pi \int_0^{R_0} R^2 \rho(R)\,dR.\qquad(25)$$

The material point is located outside the sphere at a distance x from its center (Fig. 1.5).

Consider a spherical layer of radius R and thickness dR containing the center of our sphere. Let $ds = Rd\varphi dR$ be an area element of this layer as shown in Fig. 1.5 (dashed line), and the distance r be that between the layer ds and point P. Obviously all points of the torus with cross-section ds and radius $R\sin\varphi$ are situated at the same distance r from P. Therefore, denoting the mass element of the layer, dm, we will have for the potential between dm and P

$$dV = \frac{dm}{r},\qquad(26)$$

where

$$dm = 2\pi R^2 \rho(R)\,\sin\varphi\,d\varphi\,dR.\qquad(27)$$

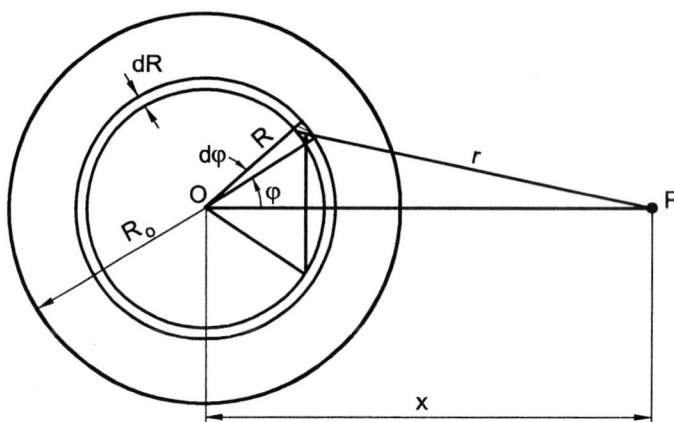

Fig. 1.5. *The problem of the law of attraction between the material point P of mass m and the concentric sphere of mass M.*

For the total potential between the sphere $\mathcal{M}$ and the point P we will have

$$V = \int_{\mathcal{M}} \frac{dm}{r} = 2\pi \int_0^{R_0} \int_0^\pi \frac{R^2 \rho(R) \sin\varphi}{r} \, dR \, d\varphi =$$

$$2\pi \int_0^{R_0} R^2 \rho(R) dR \int_0^\pi \frac{\sin\varphi \, d\varphi}{r} \, . \tag{28}$$

Insofar as

$$r^2 = x^2 + R^2 - 2xR \cos\varphi \tag{29}$$

and therefore

$$r\,dr = xR \sin\varphi \, d\varphi \, , \tag{30}$$

we have

$$\frac{\sin\varphi \, d\varphi}{r} = \frac{dr}{xR} \, . \tag{31}$$

Substituting Eq. (31) into Eq. (28), we obtain

$$V = 2\pi \int_0^{R_0} R^2 \rho(R) \, dR \int_{x-R}^{x+R} \frac{dr}{xR} = \frac{4\pi}{x} \int_0^{R_0} R^2 \rho(R) \, dR \tag{32}$$

or comparing with Eq. (25), finally

$$V = \frac{\mathcal{M}}{x} \, . \tag{33}$$

However, this is nothing other than the potential (24) between two material points, of mass $\mathcal{M}$ and $m = 1$. In our case $\mathcal{M}$ is the mass of the sphere, i.e. of a non-point object.

The conclusion is rather important: the attraction between a spherically concentric body of finite size and a material point is performed as if the mass of the sphere were located at its center.

Almost all bodies of the Solar system are far from spherical, so that strictly speaking this conclusion is not applicable for them. However, due to the fact that mutual distances between the planets and the Sun greatly exceed their sizes, the Sun and the planets can be considered as material points.

The situation, however, is different in the case of artificial satellites at smaller distances from the planets, the Earth or the Moon. In those cases, accounting for the non-sphericity of the planets is necessary.

Eq. (33) is valid at all distances $x > R_0$ and also at $x = R_0$, i.e. when the body is situated on the surface of the sphere. In the latter case the potential is:

$$V = \frac{\mathcal{M}}{R_0}$$

and the attraction force between P and $\mathcal{M}$:

$$F = \left| \frac{dV}{dx} \right|_{x=R_0} = f \frac{m\mathcal{M}}{R_0^2} . \tag{34}$$

8   The Attraction of a Point within the Sphere

Consider the case in which the point P is situated inside the sphere at distance $x < R_0$ from its center. Divide the sphere into two parts, inner (1) of radius x and outer (2), with radius from x to R_0, with masses M_1 and M_2, respectively, so that $M = M_1 + M_2$, where

$$M_1 = 4\pi \int_0^x R^2 \rho(R)\, dR ,$$
$$M_1 = 4\pi \int_x^{R_0} R^2 \rho(R)\, dR . \tag{35}$$

Correspondingly we will have two potentials: V_1 of the inner part, and V_2 of the outer. Then the attraction to the point P will be

$$F = F_1 + F_2 = \frac{dV_1}{dx} + \frac{dV_2}{dx} . \tag{36}$$

For the inner part, obviously the considerations of the previous section are valid:

$$V_1 = \frac{M_1}{x} \tag{37}$$

and therefore

$$F_1 = f\frac{M_1}{x^2}\,. \tag{38}$$

Let us find the potential of the outer part by considering again a layer of radius $R > x$ and thickness dR (Fig. 1.6). The elementary mass of the layer is obviously again given by Eq. (27), and Eqs. (28)–(32) apply. Only the limits of integration in Eq. (32) should be changed:

$$V_2 = 2\pi \int_0^{R_0} R^2 \rho\left(R\right) dR \int_{R-x}^{R+x} \frac{dr}{xR} = 4\pi \int_0^{R_0} R\rho\left(R\right) dR\,. \tag{39}$$

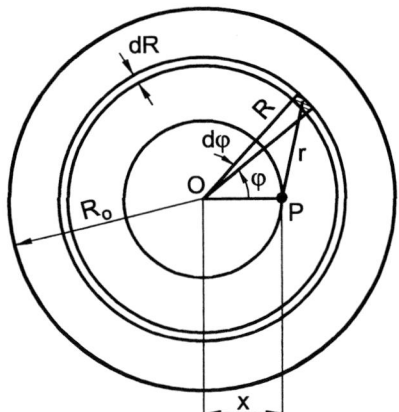

Fig. 1.6. *The problem of the law of attraction between of a point P of mass m situated within the sphere, and the sphere of mass M.*

This integral has no relation to the mass of the outer part M_2, as in the first case, and is constant:

$$V_2 = constant\,. \tag{40}$$

so that

$$F_2 = \frac{dV_2}{dx} = 0\,. \tag{41}$$

We see that the attraction between a hollow sphere and its internal point is zero, irrespective of the location of the point, and that this is valid also in the case in which the point is located on the inner boundary of the layer. Therefore, the attraction within the sphere is determined only by the amount of mass situated further towards the center than the localization of the point, so that the outer mass has no role.

This conclusion is rather interesting and has a number of consequences. For example, while going inside the Earth the weight of a body should decrease, it will decrease to half its weight at the half radius (considering the

Earth to be homogeneous). At the Earth's center $x = 0$ the attraction must vanish; the condition $x = 0$ does not lead to infinities as in Eq. (28), since the mass decreases faster, as x^2, so that $F \sim M/x^2 \sim x \to 0$ when $x \to 0$.

The problem of the mutual attraction of two spheres was originally considered by Newton in a purely geometrical manner. Newton also solved quite elegantly the problem of the attraction of the point by a spherical layer. The idea of his method is seen from Fig. 1.7, where the spherical layer of radius R and thickness Δx is shown, while the point P is located arbitrarily. Consider the lines symmetrical to the one passing through P and the center of the sphere, the angle between those lines being $\Delta\varphi$. The point P is attracted by force F_A from a mass element A, and there is a corresponding force F_B, so that F_A and F_B are directed opposite to each other.

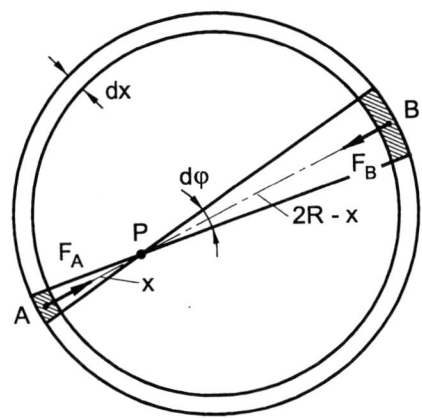

Fig. 1.7. *The problem of the law of attraction of the point P by a spherical layer (Newton's method).*

The volume element of A is $(\pi/4)(x \ \Delta\varphi)^2 \ \Delta x$ and the mass is $(\pi/4)(x \ \Delta\varphi)^2 \ \Delta x \rho$ so that we have

$$F_A = f \frac{(\pi/4)(x \ \Delta\varphi)^2 \ \Delta x \ \rho}{x^2} = f \, (\pi/4)(\Delta\varphi)^2 \ \Delta x \, \rho \, .$$

The volume element of B will be $(\pi/4)\left[(2R - x) \ \Delta\varphi \right]^2 \ \Delta x$, so that analogously

$$F_B = f \, \frac{(\pi/4)\left[(2R - x) \ \Delta\varphi \right]^2 \ \Delta x \ \rho}{(2R - x)^2} = f \, (\pi/4) \, (\Delta\varphi)^2 \ \Delta x \, \rho \, ,$$

i.e. $F_A = F_B$ and hence the point P is not influenced by any attraction force.

9   The Proof of the Newtonian Law of Gravitation

From the mathematical point of view the law discovered by Newton should be considered as a consequence of Kepler's three laws. However, Kepler's laws, first, were obtained based on the observations of only two planets, Mars and Venus; and second, concern a single type of orbit, the ellipse. The question of the possibility of, say, parabolic orbits still remained open, since nobody had observed them before Newton. Non could anything be said regarding the limits on the application of this law, on its long-range character, etc. This list could be continued.

Thus the law derived by Newton required confirmation before it could be claimed to be universal. Newton himself started to seek the necessary confirmation by means of more generally formulated problems. He considered that his law should explain not only Kepler's three laws, but all observational discrepancies from those laws and all "strange" events in the behaviors of celestial bodies. Newton succeeded in finding confirmations. The basic ones are as follows.

1. After solving the inverse problem and the derivation of the law of gravitation, Newton returned to the direct problem, i.e. to the derivation of Kepler's three laws in the two-body problem. Part of the results, Kepler's first two laws, absolutely coincided with the empirical laws. The discrepancies began with the third law; it appears not in the form of Eq. (4) as Kepler had derived, but of Eq. (5):

$$\left(1 + \frac{m}{M}\right) \frac{P^2}{a^3} = constant.$$ (42)

The difference between the precise law (42) and Kepler's law (4) is qualitative; in the latter case the constant depends only on the combination of two quantities, the period P and semimajor axis a, whereas in Eq. (42) the combination of three parameters is constant, including the ratio of the masses of the two bodies, m/M.

By means of the corrected Kepler law (42) Newton estimated the masses of Jupiter and Saturn. He therefore could conclude that the law $F \sim r^{-2}$ is really universal for all planets.

2. By solving the two-body problem Newton derived an equation for the motion of the planets around the central body, the Sun. From that he had the possibility of predicting the position of a planet moving in an elliptical orbit at any moment of time. The comparison with observational data ensured their satisfactory coincidence with the predictions, unless the time periods were not large enough.

3. During the solution of the two-body problem, Newton came to the conclusion that the motion of celestial bodies around the Sun should be performed via conic sections, the ellipse being a particular case only. Therefore two other types of orbits should be possible, the parabola and the hyperbola, their being conic sections as well. No celestial bodies moving in parabolic or hyperbolic orbits were known at the time of Newton. Newton guessed

intuitively that comets could be such objects. During that time, in 1680, when Newton was thinking about these problems, a bright comet was observed. Newton, by means of his method of determination of the parabolic orbit by three observations, immediately estimated its future trajectory. In particular, he predicted the position of the comet relative to bright stars for certain periods of time, if its orbit were parabolic. The first observations after the comet passed its perihelion confirmed the parabolic orbit of the comet. Thus the existence of parabolic orbits was proved, in which at least some comets are moving. The discovery of the parabolic orbits should evidently be considered as one of the crucial confirmations of the universality of the Newtonian law.

4. Newton attributed a special role to the observational discrepancies with Kepler's laws, considering that those facts could also be described by his gravitational law. The most important discrepancies were related to the motion of the Moon and were already well known at the time of Newton. An incomplete list of them is as follows:

a) The orbital plane of the Moon is not motionless; its ascending node on the ecliptic performs a retrograde motion with a velocity $19°21'$ per year. This motion is superposed upon oscillations of amplitude $1°26'$.

b) The inclination of the orbit to the ecliptic is also not constant but is oscillating within the limits $5°0'$ and $5°18'$.

c) The lunar elliptical orbit is rotating in its plane, so that the semimajor axis is shifting by $40°41'$ per year; this shift is superposed upon oscillations of amplitude $8°41'$.

d) The orbital motion of the Moon cannot be completely described by the law of areas, i.e. its radius-vector does not always cover equal areas during the same interval of time.

All these anomalies should be attributed to the same reason, i.e. the divergence from Kepler's laws. Newton, having seriously considered this problem, proved that the reason for all the discrepancies is the same – the Sun. Therefore the motion of the Moon should be considered in terms of a three-body problem. Newton discovered some properties of that problem using the fact that the Moon–Earth distance is much less than the Moon's and Earth's distance from the Sun. Based on his law of gravity he estimated the theoretical values for each of those discrepancies; in the majority of cases the coincidence was quite satisfactory.

Thus the rather complicated motion of the Moon not only did not contradict the Newtonian law of gravity but could be explained with all its peculiarities in terms of that law.

5. The next step in the confirmation of the Newtonian law of gravity was related to the interpretation of the Earth's form. Before Newton, the scientific world already had a definite viewpoint on this problem, i.e. that the Earth is exactly spherical and the problem is to obtain its exact radius.

Newton, however, proved that the Earth cannot be spherical but should be oblate as a rotating spheroid, as follows from his theory of gravitation. Newton argued that precisely spherical bodies can only be nonrotating whereas the rotating ones should be oblate in their equatorial areas due to centrifugal forces acting oppositely to the gravitational ones. This will lead to an increase in the Earth's radius near the equator and a decrease at the poles.

Newton even estimated the degree of oblateness, using the model of a homogeneous fluid, to be

$$\alpha = \frac{R_e - R_p}{R_e},$$

where R_e and R_p are the equatorial and polar radii of the Earth, respectively. Numerically he obtained $\alpha = 1/230$, which does not differ so much from its modern value $\alpha = 1/294$.

Newton's results concerning the Earth's form and the inevitable flattening of the planets were essential for recognition of the law of gravity, since these arguments were qualitatively different from those of the planetary motions. The necessity of the development of the theory of the forms of the planets, including the Earth, therefore became clear. Many years later **Clairaut** (1713–1765), on the basis of the Newtonian gravitation law, created his famous "Theory of the Earth's Figure" (1742), applying for the first time the principles of hydrostatics; this theory has survived up to now almost unchanged.

6.   One of the important conclusions of Newton was connected with the tides. He showed that this phenomenon is completely determined by the relative motions of the Earth and the Moon. Newton found theoretical confirmation of a long-known association between the variation of the oceanic level and lunar motion, which revealed the relation between the propagation of tidal waves and the motions of Moon and Sun. One hundred years later **Laplace** (1749–1827), on the basis of Newton's ideas and the differential equations of hydrodynamics, developed the dynamical theory of tides.

7.   The phenomenon of **precession** that had been discovered by Hipparchus, i.e. the conic motion of the Earth's axis by 50".2 per year with one revolution every 26 000 years, was explained by Newton based completely on the law of gravitation. It appears that precession is caused by the action of the Sun and Moon upon the equatorial regions of the terrestrial ellipsoid, when neither the Sun nor the Moon is situated in that equatorial plane. For almost 20 centuries, from Hipparchus to Newton, nobody knew about the equatorial bulge of the Earth and therefore it is not surprising that nobody could guess the reason for the precession. Newton estimated the value of the precession and concluded that the contribution of the Moon is twice as large (34") as that of the Sun (16").

10   The Universal Law of Gravitation

It seems that Newton did enough to provide convincing confirmation of his law. Nevertheless, the history of astronomy ascribes the final recognition of the Newtonian law within the rank of universality, as a "law of universal gravitation", to a later period. It was associated with events that highly impressed contemporaries and served to bring about the final recognition of the Newtonian law.

The first event is related to Halley's comet. Newton's student **Halley** (1656–1742) devoted his whole life to the study of the motion of comets. After Newton's discovery of the first parabolic orbit in 1680, Halley estimated parabolic orbits for 24 comets. He noticed almost equal periods between the appearance of the comets in 1456, 1531, 1607 and 1682. This fact pushed him to assume that he was dealing with the appearance of one and the same comet, but not one having a parabolic orbit but rather a strongly elongated elliptic one, with a period of 75 or 76 years. He predicted that the next appearance of this comet would be in 1758. Those years coincide with the period of activity of the young Clairaut, who besides geometry and the problem of the Earth's form, was also interested in the problem of the perturbation of planetary orbits, the motion of the Moon, etc. Thus he doubted that the return of Halley's comet should happen in the year mentioned by Halley, since Halley had not taken into account the inevitable perturbations in the motion of the comet which would be caused by the influence of the large planets, Jupiter and Saturn. Clairaut performed extremely complicated and tedious calculations, first developing the corresponding technique – the first serious sketches of future perturbation theory. He concluded that Halley's comet should be delayed by 518 days due to the perturbation of Jupiter and by 100 days due to Saturn. In his report in a publication of the Observatoire de Paris he announced the probable date of the return of the comet to be April 13, 1759 with a possible error of a month on either side.

Recall that the last return of Halley's comet in March 1986 was marked by successful space missions enabling close study of the comet's nucleus for the first time: Vega, space mission, Giotto, etc.

Clairaut's calculations were based on the Newtonian law of gravity, of which he had been an admirer for many years. The situation was unprecedented; for the first time, a strict mathematical technique was predicting a unique event – the return of a comet. One can understand the impatience of the astronomical community, since the point was not only the serious testing of the Newtonian theory for a new type of celestial body – a returning comet, but its testing for distances beyond Saturn, the farthest known planet of the Solar system at that time.

Halley's comet really did return, and passed the perihelion on May 14, 1759, one month and one day later than the date mentioned by Clairaut.

This event was a triumph of Newton's law; nobody doubted its reality and universality. Simultaneously, Halley's prediction of the existence of comets with elliptical orbits, besides the known parabolic ones, was confirmed. Moreover, due to the brilliant work of Clairaut, the first convincing proof of the possibility of the development of the perturbation theory, enabling one to explain the anomalies in planetary motions, was demonstrated.

The next remarkable event to confirm the Newtonian law is related to the discovery of Neptune.

From ancient times up to the XVIIIth century the last known planet of the Solar system was Saturn, and no reason existed to suggest the possibility of there being another planet beyond Saturn. Therefore one can imagine the world's surprise when **William Herschel** (1738–1822) announced in 1781 the discovery of Uranus by means of his mirror telescope. The active efforts to estimate the orbit of Uranus unexpectedly showed the impossibility of fitting the observed motion by the predicted trajectory. The discrepancies were increasing from year to year. Among the possible explanations was the prediction of the existence of an unknown planet perturbing the motion of Uranus.

Thus a principally new problem arose: to estimate the mass and the position of a hypothetical planet, by means of the data of Uranus' motion, assuming that the perturbations were caused by Newtonian gravity.

The first serious attempt was by Adams who reported in October 1845 to the Royal Astronomical Society the results of his rather complicated calculations, mentioning the position of the hypothetical planet. Unfortunately, his report did not receive proper attention.

Almost at the same time Le Verrier, an eminent astronomer and member of the Paris Academy of Sciences, independently of Adams, performed the same calculations and also estimated the position of that planet. On September 23, 1846, Galle, an astronomer at the Berlin Observatory, received a letter from Le Verrier mentioning the coordinates of the planet. On the same evening Galle discovered, close to the position mentioned by Le Verrier, an object, different from a star and most probably the planet predicted by Le Verrier. Later it was called Neptune.

The perturbation of Uranus caused by Neptune was about $2'$, i.e. almost undetectable by the unaided eye, and in any case small enough not to cause serious searches. This tiny value led to the discovery of the planet Neptune on a sheet of paper. The story of Neptune's discovery became a real celebration, no less impressive than that of Halley's comet. This completed about a one and a half century period of testing of the Newtonian law of gravity. Since then, the latter could obviously be considered as the Newtonian law of universal gravity.

11 The Gravitational Constant

Physicists write the Newtonian law of gravity in the form

$$F = G \frac{mM}{r^2}, \tag{43}$$

whereas astronomers prefer the form

$$F = k^2 \frac{mM}{r^2}. \tag{44}$$

Why k^2 and not k is explained simply; in the majority of formulas in celestial mechanics that coefficient is entered under a square root, hence k^2 is preferable.

The coefficient in Eqs. (43) and (44), G or k^2, is called the "gravitational constant". Its physical content is clear; it corresponds to the force of attraction between two bodies with parameters equal to unity, $M = 1$, $m = 1$, and $r = 1$. Its numerical value has been obtained experimentally: in CGS units it yields:

$$G = 6.668 \times 10^{-8} \, dyn \, cm^2 \, g^{-2}$$

Astronomers preferred not to use the CGS system but rather the one proposed by Gauss, which has been used unchanged up to now. Gauss had proposed to use the Solar mass as the mass unit, the semimajor axis of the Earth's orbit as the distance unit, and the average day for the time unit. As for the concrete value of k, as the basis for its estimation the formula relating the period and the semimajor axis of an elliptical orbit was taken:

$$k = \frac{2\pi a^{3/2}}{P\sqrt{M}}. \tag{45}$$

On substituting the Earth's values, i.e. $a = 1$, $M = m_\odot + m_\oplus$, where $m_\odot = 1$, $m_\oplus = 1/354\,710$ and $P = 365.2563835$, we find

$$k = 0.01720209895,$$

i.e. with a precision up to the eleventh digit. This magnitude was obtained by Gauss in 1809. Later on, with the increase of the accuracy of input parameters and in view of the needs of space research, the precision of Gauss's constant was improved. The following value of Gauss's constant is currently adopted:

$$k = 0.017\,202\,098\,950\,000.$$

Such precision of the Gaussian constant is the guarantee of the reliability of space flights, including the landing of manned spacecraft on the Moon, close flybys of the planets and their satellites, perturbation maneuvers within planets' gravitational field, rapprochement and docking of spaceships, etc.

12   Celestial Mechanics after Newton

The development of celestial mechanics after Newton occurred with spectacular rapidity. Its methodological aspect was seriously reconstructed. In particular, the geometric methods widely used by Newton himself were replaced by **analytical methods**, including the study of diverging series, derivation and analysis of differential equations. The initiators and inspirers of this activity were **Clairaut, d'Alembert** (1717–1783) and **Euler** (1707–1783), who almost simultaneously and independently started to study the three-body problem and perturbation theory. The concrete subject of studies became the Moon – one of the most difficult problems of celestial mechanics to be evaluated by Newton himself.

To be objective, one should mention that the mathematical difficulties that appeared were so serious that sometimes Clairaut and d'Alembert even doubted the accuracy of Newton's law and proposed additional terms, such as $\alpha M r^{-3}$ and $\alpha M r^{-4}$. The latter, however, were withdrawn, especially after the understanding of the necessity of more accurate solutions of the differential equations.

The theory of lunar motion pioneered by Clairaut, d'Alembert and Euler was highly developed in fundamental work by **Laplace** (1749–1827), who succeeded in revealing some essential irregularities in the motion of the Moon.

The method of variation of the elements had exceptional importance for the study of the perturbed planetary motions. In spite of the complicated form of the differential equations of the osculation elements, their solution was nevertheless possible by the method of successive approximations. In the case of planets, the method of variation of the elements led to relatively quick (in a few cycles) approximate solutions that were acceptable for practical purposes. At the same time, its inapplicability to lunar motion became clear, in spite of some efforts by Poisson. The problem of lunar motion was solved later, when **Delaunay** (1816–1872) specified a method of integration of the equations of the perturbed motion.

All these achievements served as the bases for the fundamental studies on the development of the theory of lunar motion, the last stage of which was outlined by the work of Hanzen (1795–1874), Hill (1838–1914) and Brown (1866–1938). These works enabled the creation of tables of high precision for lunar positions – ephemerides. Thus lunar motion became one of the first (and independent) directions of celestial mechanics.

The system of differential equations derived by Euler was used to obtain the perturbed elements of the planets. Lagrange had made an essential contribution to the improvement of the method of variation of the elements by means of the representation of Euler's differential equations via a potential function, which was identified with the perturbation function. Remarkable work on the creation of a more precise theory of planetary motion, based again on the method of variation of elements, was done by **Le Verrier** (1811–1877). The story of the discovery of Neptune is an eloquent outcome of that work. Later, Le Verrier created the theories of motion of Mercury, Venus, the Earth, and Mars; his tables (ephemerides) fit the observations so well that they leave no doubt concerning the validity of the theory.

The time came also for the creation of theories of motion of the remaining four planets – Jupiter, Saturn, Uranus and Neptune. The main contributions here came from **Hill**, **Newcomb**, and **Lowell**. Systematic searches for a trans-Neptunian planet were performed; they led much later to the discovery of Pluto, again on a sheet of paper as for Neptune, due to the efforts of **Pickering** and others.

The middle of the XIXth century was typified by interest in the special problems of celestial mechanics, first of all, in the restricted three-body problem. The important results here are associated with the name of **Jacobi**, who derived the so-called integral of energy for an infinitely small mass in the system. Later Hill performed a detailed analysis of the null velocity curves of the infinitely small mass in rotating coordinates.

Hamilton's principle had an outstanding role for the general theory because it unified the basic laws of dynamics. One of the consequences of that principle was the derivation of generalized Lagrange equations and the transformations known as **canonical** ones enabling the convenient and symmetrical representation of equations of motion.

The end of the XIXth century was marked by the study of problems of celestial mechanics that were of global importance. They concerned particularly the problem of the stability of motion, i.e. the problem of the cosmogonical background. The initiators of this direction were **Poincaré** and **Lyapunov**. Parallel to the main problem of celestial mechanics, the theory of planetary motion, intense studies of interdisciplinary problems were performed, including the theory of the forms of celestial bodies, of the Earth in particular. Based on the works of Clairaut, the use of the potential function was crucial for further work. New independent directions of celestial mechanics had appeared, the **theory of attraction and the theory of potential**, studying mainly the gravitational fields of various static configurations. This problem is closely associated with another direction of celestial mechanics – the theory of tides, considering not only the role of the tides on the behavior of the oceanic surface but also the effect on the Earth's crust.

All planets except Mercury and Venus have natural satellites, i.e. miniature solar systems do exist, the study of which is another direction of celestial mechanics and the perturbation theory.

With the appearance of large telescopes, photographic and other facilities, the avalanche of discovery of minor planets, asteroids, started. Their masses are not large so that they are easily influenced by the perturbations of large planets. The motion of minor planets is another direction of celestial mechanics.

A tremendous role in the study of minor planets was played by **Gauss's** (1777–1855) elegant theory of determination of the planetary orbits by three observations. It is interesting that the very reason for developing that method was the minor planet, Ceres, first discovered in 1801. The latter had been lost after its discovery by **Piazzi**, so that Gauss tried to estimate the orbit of Ceres from Piazzi's few but reliable observations. Gauss's method later became the pride of celestial mechanics, and it is still used nowadays almost in its original form.

The problem of cometary motions was a bit isolated. They are the only objects in the Solar system that can move not only in elliptical but also in infinite – parabolic and hyperbolic–orbits. In particular, their ability to radically change their theoretical orbits, say from an elliptical orbit to a parabolic one has already been noted, and vice versa a comet captured by the Sun can be transformed from its initial parabolic orbit to an elliptical orbit. The problem of the origin and cosmogonical role of comets is, therefore, closely associated with the study of the past and future evolution of their orbits.

Thus, the Newtonian law was shown to be universal on the scale of the Solar system. Was it also universal for systems of other categories, in particular for stellar systems; and what are the indications of the Newtonian law for those systems? The first attempts to consider these questions presumably date to the early XVIIIth century, to Herschel's first star counts. However, the fundamental work on stellar motion in our Galaxy (Kapteyn and Oort), leading to the creation of an independent discipline, stellar dynamics, was performed in the early XXth century by Schwarzschild, Eddington, and Jeans. Important steps here were the consideration of the origin of the spiral structure of galaxies (Lindblad), relaxation driving effects (Chandrasekhar, 1960), and the discovery of the effect of evaporation of stars from the star clusters (Ambartzumian, 1938; Spitzer, 1940). All this and later activity, though still leaving open many problems, in general constituted the demonstration of the universal role of the Newtonian law of gravity in governing the structure and evolution of galaxies and presumably also of objects of even larger scale – clusters of galaxies.

Finally, the youngest direction in celestial mechanics – the theory of motion of spacecraft, artificial satellites and interplanetary flights in general, has become of particular importance. The basis of this theory is the Euler differential equations that enable one to obtain the perturbed orbital elements. The theory of interplanetary flights is notable for its essential and very wide use of computational techniques, in terms both of software and of hardware. The role of computers is not limited to the estimation of the traditional parameters of the orbits, velocities, etc., but occasionally involves the necessity of immediate consideration of divergences in the parameters of the flight, and their correction. In other words, computer techniques are needed to control the flight of the spacecraft in the given trajectory, taking into account not only the laws of celestial mechanics but also the resources and facilities of the ground-based control systems and those available on board the spacecraft.

13 The Nature of Gravity

Newton discovered only the law of universal gravitation. Regarding the nature of this phenomenon, he refrained from answering this question. However, he could hardly avoid thinking about it; he clearly understood that this force should be determined by some phenomenon. This problem interested Newton, but presumably he considered the events and facts that were available in his time insufficient for the construction of a deeper theory of gravity, as he confessed in his works. He considered remarkable the very existence of gravitation, its regulation by the law discovered by him, and its description of all types of motions of the planets.

The mysterious nature of gravitation was seriously reconsidered two centuries after Newton in the General Relativity of **Einstein** (1916). By means of a deep synthesis of the Special Theory of Relativity and geometrical ideas developed earlier by a group of outstanding mathematicians, Einstein formulated the concept of gravitation as a property of a 4D space–time manifold curved due to the presence of matter. If in Newtonian gravity the action of one body on the other takes place immediately, in the Einsteinian theory any interaction, including gravitation, cannot propagate faster than the speed of light. Like relativistic mechanics, which becomes identical to classical Galilean–Newtonian mechanics at velocities much smaller than the speed of light, General Relativity also has its classical limit – Newtonian gravitation – for small masses and fields.

Einstein also looked for astronomical consequences for his theory. To avoid mathematical difficulties, he considered the motion of a particle in a spherically symmetrical gravitational field. He showed that the motion should be performed by an ellipse as in the Newtonian case; however, he obtained a qualitatively new effect, which is absent in the Newtonian limit: the rotation of the semimajor axis of the ellipse. Einstein evaluated its rotational velocity:

$$\dot{\Omega} = \frac{24\pi^3}{c^2(1-e^2)}\frac{a^2}{P^2},\qquad(46)$$

where a, e, and P are the semimajor axis, the eccentricity and the rotation period of the elliptical orbit, respectively, and c is the speed of light.

The relation between period P and semimajor axis a is given by

$$P = \frac{2\pi a^{3/2}}{k\sqrt{M}}, \tag{47}$$

where M is the total mass, e.g. of the Sun and the planet. From Eq. (46) we have

$$\dot{\Omega} \sim \frac{1}{a}, \tag{48}$$

i.e. the angular shift of the semimajor axis or of the perihelion of an elliptical orbit is larger, the closer the planet is to the Sun.

In view of the relation between the velocity V of the planet and the semimajor axis,

$$V^2 = \frac{k^2 M}{a}. \tag{49}$$

From Eq. (48) we obtain

$$\dot{\Omega} \sim V^2, \tag{50}$$

i.e. the perihelion will move faster the larger V is.

Among the planets of the Solar system, Mercury has the maximal V and hence the smallest a, so that this effect of General Relativity should be the largest for it: for Mercury one has $\Omega = 8".847$ while, say, for the Earth it yields $0".064$ per year. The motion of the perihelion of Mercury was known to astronomers long before, being one of the unexplained puzzles. Owing to General Relativity, it has found not only a qualitative but also a quantitative explanation.

Einstein predicted two other effects, not associated with celestial mechanics but still requiring astronomical confirmation. The first one concerns the bending of light rays in the gravitation field of a star; this effect was triumphantly confirmed as early as during the 1919 Solar eclipse. The second effect predicts reddening of light, i.e. a decrease in the frequency of electromagnetic waves, in a gravitational field. Both effects have been confirmed by numerous and highly accurate measurements and observations. Among the recent impressive confirmations of General Relativity one should mention the discovery and study of the extraordinary properties of the binary pulsar PSR 1913+16 and the gravitational lenses which appear to be common features of accurate observations of any kind of object, from stars to quasars.

The further development of gravity theories is being continued along the lines outlined by Einstein, i.e. towards unification with the theories of other types of interaction.

Chapter II

The Two-Body Problem

1 Statement of the Problem

The present chapter is devoted to the problem of two celestial bodies. Both bodies are considered to be spherical in form and to possess spherical symmetry, i.e. they consist of homogeneous spherical layers with an arbitrary law of distribution of density of matter along the radius. The radii of both bodies are arbitrary but finite, and both masses, m_1 and m_2, are also finite and constant. Then, in accord with the conclusion of the previous chapter, we can replace both these spherical bodies by mathematical points with masses m_1 and m_2 concentrated at their centers of inertia, i.e. the geometrical centers of these spheres. Let the distance between the centers of these bodies, m_1 and m_2, be r.

We shall proceed from the assumption that both these bodies are absolutely isolated from any external influence; here we have in mind first of all the absence of other celestial bodies. Influences or forces stipulated by the surrounding medium in which the motion of both bodies takes place, for example light pressure, electromagnetic forces, hydrodynamical actions etc., are also considered to be absent. The only force acting between the bodies m_1 and m_2 is Newtonian gravity:

$$F = k^2 \, \frac{m_1 \, m_2}{r^2} \,, \tag{1}$$

where r is the distance between the components of the system, the material points m_1 and m_2. The magnitude of the strength of gravitation F is the same in the case of the action of body m_1 on m_2 as for the action of m_1 on m_2, but the directions are opposite. The motion of each of these bodies, both in space and relative to each other, is determined and controlled by the force F only.

According to the statement of the problem, the masses of both components are constant, i.e. they do not change during motion. However, the distance r between components is changing smoothly which in its turn adversely affects the magnitude of F. Our problem is to obtain the equations of motion for each of these bodies under such conditions.

By the equation of motion we understand every mathematical dependence between coordinates and time, i.e. we have a relationship of the type:

$$
\begin{aligned}
x &= f_1(t)\ , \\
y &= f_2(t)\ , \\
z &= f_3(t)\ .
\end{aligned}
\tag{2}
$$

To find the equations of motion in the form of (2) is our final aim although, as we shall see below, achieving it is possible only by means of certain simplifications in the statement of the problem. Because the equations of motion are known, the following problems may be solved.

a) Determination of the position of the body in space, i.e. its coordinates at a given moment of the time.

b) Determination of the parameters of the motion – velocity, acceleration, trajectory etc. of every point of the motion.

The essence of the two–body problem is to find out the functions (2) in the case in which two absolutely isolated bodies m_1 and m_2 are interacting only by the Newtonian force of gravitation. The full solution of the two–body problem in such a statement, as well as a partial solution of the three–body one, is studied in one of the extensive branches of celestial mechanics that is occasionally called *Theoretical Astronomy.*

2   Differential Equations of Motion

Let us choose an arbitrary system of Cartesian coordinates in space with fixed directions of axes ξ, η and ζ with an arbitrary fixed initial point O. Positions of the bodies m_1 and m_2 relative to this system are determined by coordinates ξ_1, η_1, ζ_1 and ξ_2, η_2, ζ_2 (Fig. 2.1). Denote the distance between m_1 and m_2 by r, then

$$
r = [(\xi_1 - \xi_2)^2 + (\eta_1 - \eta_2)^2 + (\zeta_1 - \zeta_2)^2]^{1/2}\ .
\tag{3}
$$

The force F between these bodies is given by the relationship (1), and its direction coincides with the line connecting the positions of the bodies. Vector F has arbitrary angles with respect to coordinate axes, therefore it

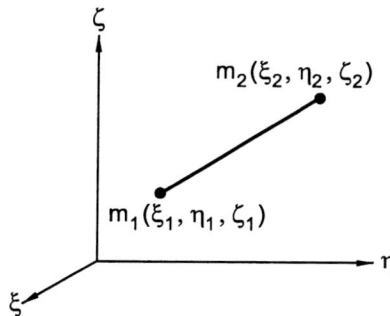

Fig. 2.1. *The derivation of differential equations of motion of mass m_1 and m_2 in an arbitrary coordinate system.*

will be expedient to decompose the spatial motion into three components along axes $O\xi$, $O\eta$ and $O\zeta$. Then we can write the differential equations of motion for body m_1 in the following form:

$$m_1 \frac{d^2\xi_1}{dt^2} = -F_\xi \ ,$$

$$m_1 \frac{d^2\eta_1}{dt^2} = -F_\eta \ , \tag{4}$$

$$m_1 \frac{d^2\zeta_1}{dt^2} = -F_\zeta \ ,$$

where F_ξ, F_η and F_ζ are components of the vector F along corresponding axes. They are connected with F through direction cosines:

$$F_\xi = F \cos{(\xi\, O\, m_1)} = F\, \frac{\xi_1 - \xi_2}{r} \ ,$$

$$F_\eta = F \cos{(\eta\, O\, m_1)} = F\, \frac{\eta_1 - \eta_2}{r} \ , \tag{5}$$

$$F_\zeta = F \cos{(\zeta\, O\, m_1)} = F\, \frac{\zeta_1 - \zeta_2}{r} \ .$$

Substituting these expressions into Eq. (4), in view of Eq. (2), we find finally for the system of differential equations of motion of the body m_1 with respect to the center of the coordinate frame O:

$$m_1 \frac{d^2\xi_1}{dt^2} = -k^2\, m_1\, m_2\, \frac{\xi_1 - \xi_2}{r^3} \ ,$$

$$m_1 \frac{d^2\eta_1}{dt^2} = -k^2\, m_1\, m_2\, \frac{\eta_1 - \eta_2}{r^3} \ , \tag{6}$$

$$m_1 \frac{d^2\zeta_1}{dt^2} = -k^2\, m_1\, m_2\, \frac{\zeta_1 - \zeta_2}{r^3} \ .$$

Analogously we can write the differential equations for the motion of m_2 with respect to the same point O considering that in this case the force F acts not in the direction from m_1 to m_2 as before, but in the opposite one, from m_2 to m_1:

$$m_2 \frac{d^2\xi_2}{dt^2} = -k^2 m_1 m_2 \frac{\xi_2 - \xi_1}{r^3} \,,$$

$$m_2 \frac{d^2\eta_2}{dt^2} = -k^2 m_1 m_2 \frac{\eta_2 - \eta_1}{r^3} \,, \tag{7}$$

$$m_2 \frac{d^2\zeta_2}{dt^2} = -k^2 m_1 m_2 \frac{\zeta_2 - \zeta_1}{r^3} \,.$$

In both systems, Eq. (6) and Eq. (7), r is presented, according to Eq. (2), through the same coordinates ξ_1, η_1, ζ_1 and ξ_2, η_2 and ζ_2.

An interesting property of the system written above needs particular attention. Namely, in Eq. (6) m_1 is absent, whereas in Eq. (7) – m_2 is absent. This means that, in the case of Newtonian gravity, the motion of the body (m_1) does not depend on its mass (m_1); the motion is determined only by the mass of the body (m_2) influencing the motion of body m_1.

In Eq. (6), written for the body m_1, there are present – both explicitly and indirectly, through r – the coordinates of the body m_2 (ξ_2, η_2, and ζ_2). Also, in system (7), written for the body m_2, one has the coordinates of m_1 (ξ_1, η_1, and ζ_1). Therefore, systems (6) and (7) cannot be solved separately – they should be solved as one system consisting of six equations.

All these equations are of second order. Therefore we must find twelve integrals or, which is the same, twelve constants of integration. The final aim of integration is the determination of six equations of type (2), by means of three equations for m_1 and m_2, including these twelve constants.

Let us turn to the choice of these integrals.

3   The Motion of the Center of Mass

By summing the first, second and third equations in system (6) correspondingly with the first, second and third equations of system (7), we obtain:

$$m_1 \frac{d^2\xi_1}{dt^2} + m_2 \frac{d^2\xi_2}{dt^2} = 0 \,,$$

$$m_1 \frac{d^2\eta_1}{dt^2} + m_2 \frac{d^2\eta_2}{dt^2} = 0 \,, \tag{8}$$

$$m_1 \frac{d^2\zeta_1}{dt^2} + m_2 \frac{d^2\zeta_2}{dt^2} = 0 \,.$$

These equations may be integrated to give

$$m_1 \frac{d\xi_1}{dt} + m_2 \frac{d\xi_2}{dt} = \alpha_1 \,,$$

$$m_1 \frac{d\eta_1}{dt} + m_2 \frac{d\eta_2}{dt} = \alpha_2 \,, \tag{9}$$

$$m_1 \frac{d\zeta_1}{dt} + m_2 \frac{d\zeta_2}{dt} = \alpha_3 \,,$$

where α_1, α_2 and α_3 are the first three constants of integration. Integrating (9) once more we obtain

$$m_1 \xi_1 + m_2 \xi_2 = \alpha_1 t + \beta_1 \,,$$

$$m_1 \eta_1 + m_2 \eta_2 = \alpha_2 t + \beta_2 \,, \tag{10}$$

$$m_1 \zeta_1 + m_2 \zeta_2 = \alpha_3 t + \beta_3 \,.$$

Here β_1, β_2 and β_3 are three new constants of integration. Thus, the first six integrals have been obtained with six constants of integration from twelve.

Let M be the total mass of the system: $M = m_1 + m_2$, and ξ, η and ζ be the coordinates of the center of mass of this system. Then, according to the definition of the center of mass, we can write the following relationship:

$$M\xi = m_1 \xi_1 + m_2 \xi_2 \,,$$

$$M\eta = m_1 \eta_1 + m_2 \eta_2 \,, \tag{11}$$

$$M\zeta = m_1 \zeta_1 + m_2 \zeta_2 \,.$$

Comparing this with Eq. (10), we obtain for the equation of motion of the center of mass

$$\xi = \frac{1}{M} (\alpha_1 t + \beta_1) \,,$$

$$\eta = \frac{1}{M} (\alpha_2 t + \beta_2) \,, \tag{12}$$

$$\zeta = \frac{1}{M} (\alpha_3 t + \beta_3) \,.$$

By their structure these equations are analogous to Eq. (2), i.e. they represent a mathematical dependence of coordinates on time with the only difference being that these equations are now related to the center of mass. The character of the motion of the center of mass is determined by the dependence of the coordinates ξ, η, and ζ on time t. In this particular case this dependence is of first order, i.e. it is linear and, hence, the coordinates of the center of mass change smoothly with respect to time.

Though we still know nothing about the motion and behavior of bodies m_1 and m_2, instead we know the law of motion of the center of the mass of the system; it is given by Eq. (12). However, we can find the parameters of this motion, in particular, the velocity, acceleration and trajectory of the center of mass.

So we have for the velocity of the center of mass

$$V = \left[\left(\frac{d\xi}{dt}\right)^2 + \left(\frac{d\eta}{dt}\right)^2 + \left(\frac{d\zeta}{dt}\right)^2\right]^{1/2} ,$$

or, substituting the magnitudes of ξ, η and ζ from (12):

$$V = \frac{1}{M} \left(\alpha_1^2 + \alpha_2^2 + \alpha_3^2\right)^{1/2} = constant , \tag{13}$$

i.e. the velocity of the center of mass does not depend on time, independent of the mutual position of components of the system and the character of their motion.

Naturally, for the acceleration a of the center of mass we have

$$a = \frac{dV}{dt} = 0 . \tag{14}$$

The imaginary path of the center of mass in three-dimensional space during its motion is called its trajectory. The trajectory is a curve that can be described by an equation not including the time – but only the coordinates. Therefore, excluding the time from Eq. (12) we obtain

$$\frac{M\xi - \beta_1}{\alpha_1} = \frac{M\eta - \beta_2}{\alpha_2} = \frac{M\zeta - \beta_3}{\alpha_3} \tag{15}$$

or in a generalized view

$$\varphi(\xi, \eta, \zeta) = C . \tag{16}$$

As we see, all three coordinates of the center of mass are related to each other linearly, hence, the trajectory of the center of mass itself represents a straight line in space, which, however, does not pass through the center of the coordinate frame ($C \neq 0$).

Thus, the center of mass of a two-body system moves in space in a straight line and with a constant velocity.

As a matter of fact, this is all that can be "extracted" from the system of differential equations (6) and (7): we have discovered only six integrals from twelve. Importantly, these six integrals give us information only about the character of the motion of the center of mass. As for the motion of bodies m_1 and m_2 themselves, the obtained six integrals do not give any information about their behavior. Furthermore, they all exist in an arbitrarily chosen system of coordinates. Moreover, celestial mechanics was able to prove, even while it was still being developed, the hopelessness of a search for new integrals and, hence, the possibility of the solution of the two–body problem in the form of Eq. (2) for each body, m_1 or m_2, insofar as their motion is examined in an **arbitrary** system of coordinates.

The solution of the two–body problem may be reached only by rejection of the concept of absolute motion.

4 Differential Equations of Relative Motion

The situation is quite different when the center of the coordinate system is occupied by only one of the bodies, say, m_1 (Fig. 2.2). In this case the problem of the motion of m_1 – let it be the central body – is absent; its co-ordinates now and permanently are $0, 0, 0$. Hence, the first three differential equations should be omitted in (6), as well as in (7). The only problem now will be the determination of the motion of the second body, m_2, in this new coordinate frame. Insofar as m_1 is located at the center of the frame, one can speak of the **relative** motion of m_2 with respect to m_1.

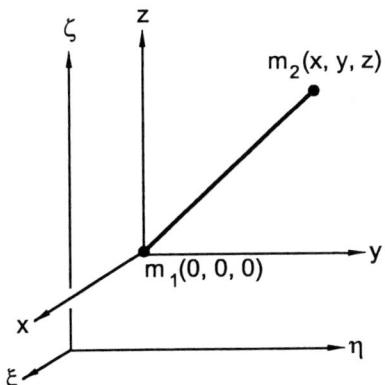

Fig. 2.2. *The derivation of differential equations of motion of a body m_2 with respect to the body m_1.*

The relative motion represents a definite simplification not only from a mathematical point of view; it has an obvious practical interest. In fact, if the observations are carried out from a planet, the problem becomes to obtain the behavior of the other body with respect to the observer. In principle, the observer could also be situated on the Sun, so that one could take the Sun as one of these bodies ($m_1 = m_0$), while the second could be a planet ($m_2 = m$). Other possibilities would be Jupiter, for example, as the central body and one of its satellites as the second one etc.

Denote the new coordinate system x, y, z. For its parallel motion with respect to the ξ, η, ζ frame, the coordinates of $m_2 = m$ will now be x, y, z. The new frame is related to the old one in the following manner:

$$x = \xi_1 - \xi_2 , \quad y = \eta_1 - \eta_2 , \quad z = \zeta_1 - \zeta_2 . \tag{17}$$

By substituting, from Eqs. (6) and (7) we obtain

$$\frac{d^2(\xi_1 - \xi_2)}{dt^2} = -k^2 (m_1 + m_2) \frac{\xi_1 - \xi_2}{r^3} ,$$

$$\frac{d^2(\eta_1 - \eta_2)}{dt^2} = -k^2 (m_1 + m_2) \frac{\eta_1 - \eta_2}{r^3} , \tag{18}$$

$$\frac{d^2(\zeta_1 - \zeta_2)}{dt^2} = -k^2 (m_1 + m_2) \frac{\zeta_1 - \zeta_2}{r^3} ,$$

or, in view of Eq. (17) and also using $M = m_1 + m_2 = m_0 + m$ for the total mass of the system, we find

$$\frac{d^2x}{dt^2} = -k^2 M \frac{x}{r^3} ,$$

$$\frac{d^2y}{dt^2} = -k^2 M \frac{y}{r^3} , \tag{19}$$

$$\frac{d^2z}{dt^2} = -k^2 M \frac{z}{r^3} .$$

During these modifications of the coordinate system the distance between bodies m_0 and m_1, evidently, remains unchanged. So we shall have in the new system

$$r = (x^2 + y^2 + z^2)^{1/2} . \tag{20}$$

The relationships (19) are just the differential equations of relative motion for the body m with respect to the central one m_0. In contrast to the absolute motion, the position of the body m in the case of relative motion depends on the total mass of the system.

The differential equations of relative motion are also of second order; however, the number of equations is less by a factor of two – only three. Therefore the solution of the two–body problem in this case leads to the discovery of six integrals or six constants of integration.

Chapter III

Derivation of Kepler's Laws from Newton's Law

1 Statement of the Problem

Kepler derived his three laws in an exceptionally empirical way, on the basis of analysis of observations of the positions of planets. In Kepler's epoch, the law of gravitation had not yet been discovered. However, Kepler did realize that such a law must exist, and that his three laws should be consequences of a more general law determining the essence of the interaction between celestial bodies. The latter was discovered later by Newton on the basis of careful examination of Kepler's laws. A unique circle does yield: Newton's law of gravitation and Kepler's three laws characterizing the motion of celestial bodies are consequences of each other. The pure mathematical demonstration of this fact must undoubtedly be of interest. Historically, first the transition from Kepler's laws to the Newtonian one was realized. However, even when Newton's law of gravitation is known, one needs to make definite efforts to find its consequences for an isolated system of two bodies. Kepler's three laws may be among the consequences. However, in principle there may be others as well, and no less important ones, either. Given such a statement, the problem evidently gains an independent interest.

So, two isolated celestial bodies m_0 and m are interacting with each other according to the Newtonian law. The problem is to derive the consequences of such an interaction. More definitely it reads: to find the properties and peculiarities in the motion of m with respect to the central body m_0.

2   The Integral of Areas. Kepler's First Law

The system of differential equations of relative motion (19), derived in the previous chapter, should be our initial position.

Multiply the first equation in (19) by $-y$, the second one by $+x$ and take the sum of both, then multiply the third equation by $+y$, the second by $-z$ and again take the sum of both, and finally multiply the first equation by $+z$, the third by $-x$ and take the sum again. As a result we have

$$
\begin{aligned}
x\frac{d^2y}{dt^2} - y\frac{d^2x}{dt^2} &= 0 \ , \\
y\frac{d^2z}{dt^2} - z\frac{d^2y}{dt^2} &= 0 \ , \\
z\frac{d^2x}{dt^2} - x\frac{d^2z}{dt^2} &= 0 \ .
\end{aligned}
\tag{1}
$$

After integration we obtain

$$
\begin{aligned}
x\frac{dy}{dt} - y\frac{dx}{dt} &= a_1 \ , \\
y\frac{dz}{dt} - z\frac{dy}{dt} &= a_2 \ , \\
z\frac{dx}{dt} - x\frac{dz}{dt} &= a_3 \ .
\end{aligned}
\tag{2}
$$

Here a_1, a_2 and a_3 are constants of integration.

Each equation in (2) represents a combination of coordinates and of their first differentials, and these combinations are at the same time constants in every separate coordinate plane. To understand the physical essence, for example, of the combination

$$
x\frac{dy}{dt} - y\frac{dx}{dt} = a_1 ,
\tag{3}
$$

we introduce a polar coordinate system r_{xy} and θ_{xy} for the projection of body m on the XYZ plane as shown in Fig. 3.1. We have

$$
\begin{aligned}
x &= r_{xy} \ \cos\theta_{xy} \ , \\
y &= r_{xy} \ \sin\theta_{xy} \ .
\end{aligned}
\tag{4}
$$

Differentiating by time t and combining correspondingly, we obtain

$$
x\frac{dy}{dt} - y\frac{dx}{dt} = r_{xy}^2 \ \frac{d\theta_{xy}}{dt} \ .
\tag{5}
$$

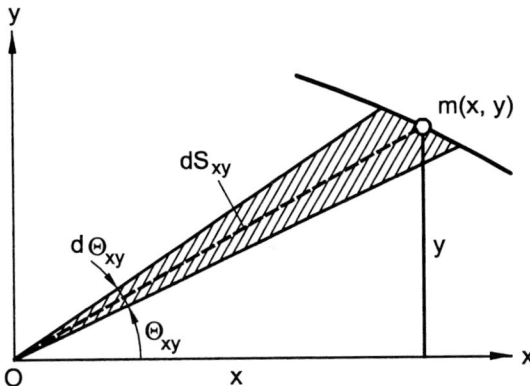

Fig. 3.1. *A scheme explaining the geometrical essence of Eq. (5).*

It is not difficult to see that the expression $r_{xy}^2 \, d\theta_{xy}$ represents twice the elementary area drawn by the projection r_{xy} of radius-vector r on the plane XOY in the time interval dt (the shaded area in Fig. 3.1). On combining Eq. (5) with the first equation in (2) we obtain

$$r_{xy}^2 \, \frac{d\theta_{xy}}{dt} = a_1 = constant \qquad (6)$$

i.e. the elementary area drawn by radius-vector r_{xy} on the plane XOY in unit time is now a constant equal to a_1.

Thus, the constants of integration a_1, a_2 and a_3 in Eq. (2) correspond to an area and are equal to those areas covered by the radius-vector in unit time. That is why the constants in Eq. (2) are often called "integrals of areas".

However, the elementary area covered by the radius-vector in unit time has another name – **sectorial velocity**. Therefore the expression (6) represents nothing other than the condition of the constancy of sectorial velocity on the plane XOY of the motion of body m with respect to m_0 situated at the center of the coordinate system.

Analogously we obtain for the other two planes

$$r_{xy}^2 \, \frac{d\theta_{xy}}{dt} = a_2 \, ,$$

$$r_{zx}^2 \, \frac{d\theta_{zx}}{dt} = a_3 \, . \qquad (7)$$

Thus, all three components of sectorial velocity are constants. For the total sectorial velocity we have

$$r^2 \, \frac{d\theta}{dt} = \left[\left(r_{xy}^2 \, \frac{d\theta_{xy}}{dt} \right)^2 + \left(r_{yz}^2 \, \frac{d\theta_{yz}}{dt} \right)^2 + \left(r_{zx}^2 \, \frac{d\theta_{zx}}{dt} \right)^2 \right]^{1/2} \qquad (8)$$

or

$$r^2 \frac{d\theta}{dt} = c_1 , \tag{9}$$

where

$$c_1 = \left(a_1^2 + a_2^2 + a_3^2 \right)^{1/2} . \tag{10}$$

Obviously $c_1/2$ is the area drawn in unit time by the radius-vector r of celestial body m in the plane of its motion with respect to the central one m_0.

Thus up to now we still do not know the law of motion of body m with respect to m_0, we do not know the equation of this motion either, we know nothing about the trajectory, orbit and other parameters of this motion, but one of its properties is already known, namely, the motion of m with respect to the central body m_0 takes place with a constant sectorial velocity. This is **Kepler's first law** mathematically represented by expression (9).

If we denote by dS the elementary area covered in space by the radius-vector r during a time dt (dS_{xy} in Fig. 3.1 is the projection of dS on the plane XOY), then the formula (9) may be written as follows:

$$\frac{dS}{dt} = \frac{1}{2} c_1 \tag{11}$$

or, after integration:

$$S = \frac{1}{2} c_1 (t - t_0) , \tag{12}$$

i.e. the area covered by the radius-vector of the body m on the plane of the motion *orbital plane* is proportional to time t.

The constancy of sectorial velocity ($dS/dt = constant$ in Fig. 3.2) automatically reflects the various magnitudes of linear velocity of the body: it is small at larger distances from the central body and larger at small distances. In the scheme shown in Fig. 3.2 we have $V_2 > V_1$ at $r_2 < r_1$.

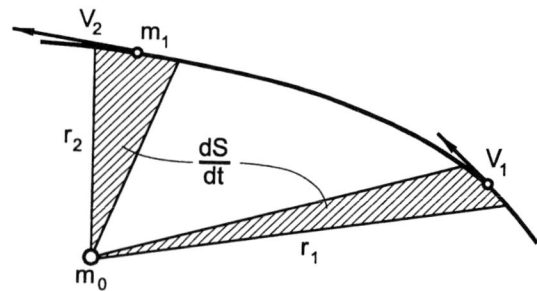

Fig. 3.2. *An illustration of the constancy of sectorial velocity dS/dt (Kepler's first law) and the inconstancy of the linear velocity V during relative motion of a body m around the body m_0.*

3 The Orbital Plane

Multiplying Eq. (2) by x, y and z, correspondingly, and taking the sum we obtain

$$a_1 z + a_2 x + a_3 y = 0 . \tag{13}$$

This is the equation of a plane passing through the center of the coordinate system. However, just there we have the central body m, so that the coordinates x, y and z pertain to it. Therefore the final formulation will be as follows: the expression (13) is the equation of the plane passing through both bodies, m and m_0, so that m, moving relative to m_0, is permanently situated on this plane. Since Eq. (13) describes the plane of the orbit, hence, the orbit of the relative motion represents a plane curve as shown in Fig. 3.3 by the shaded area.

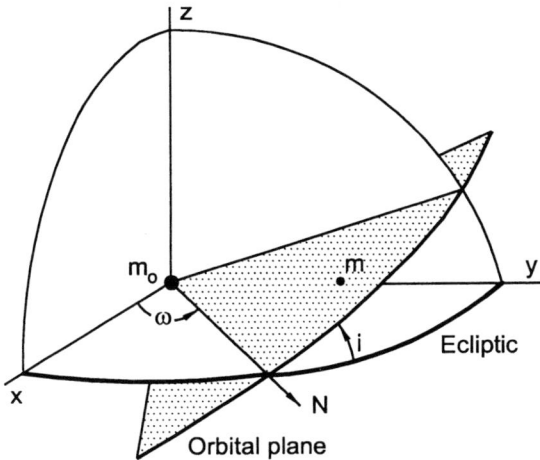

Fig. 3.3. *The spatial position of the orbit's plane in the case of relative motion.*

The last conclusion essentially changes the situation: now the motion of m relative to the central body m_0 is already not in a three-dimensional (3D) space but rather in a two-dimensional (2D) one. Then, by matching the coordinate plane XOY with the plane of the orbit (motion) we obtain a system of differential equations consisting not of three equations as before (Eq. (19) in Chapter II), but only of two:

$$\frac{d^2 x}{dt^2} = -k^2 M \frac{x}{r^3} ,$$

$$\frac{d^2 y}{dt^2} = -k^2 M \frac{y}{r^3} . \tag{14}$$

Owing to this, the problem of the relative motion is reduced from six equations to four. From the formal point of view, for the description of the relative motion it is now enough to have only four integrals, i.e. four constants of integration. As for the remaining two constants, they of course do not disappear, though their presence is in hidden form; we have in mind the two parameters, Ω and i, determining the orientation of the orbital plane in 3D space.

The integration constants a_1, a_2 and a_3 of the equation of the orbital plane (13) may be represented via the parameters just introduced, Ω and i.

We have for the projection of the sectorial velocity, dS/dt, on the coordinate plane

$$\frac{dS_{xy}}{dt} = \frac{dS}{dt}\,\cos(XOY)\,,$$

$$\frac{dS_{yz}}{dt} = \frac{dS}{dt}\,\cos(YOZ)\,, \tag{15}$$

$$\frac{dS_{zx}}{dt} = \frac{dS}{dt}\,\cos(ZOX)\,,$$

where the corresponding directing cosines are equal (see Fig. 3.3):

$$\cos(XOY) = \cos i\,,$$

$$\cos(YOZ) = \sin i\,\sin\Omega\,, \tag{16}$$

$$\cos(ZOX) = -\sin i\,\cos\Omega\,.$$

These relationships may be derived easily with the help of formulas of spherical trigonometry (see Fig. 3.3).

Substituting Eq. (16) into Eq. (15) and in view of Eqs. (6), (7) and (11), we finally obtain

$$a_1 = c_1\,\cos i\,,$$

$$a_2 = c_1\,\sin i\,\sin\Omega\,, \tag{17}$$

$$a_3 = -c_1\,\sin i\,\cos\Omega\,.$$

Then the integral of areas (2) has the following form:

$$x\,\frac{dy}{dt} - y\,\frac{dx}{dt} = c_1\,\cos i\,,$$

$$y\,\frac{dz}{dt} - z\,\frac{dy}{dt} = c_1\,\sin i\,\sin\Omega\,, \tag{18}$$

$$z\,\frac{dx}{dt} - x\,\frac{dz}{dt} = -c_1\,\sin i\,\cos\Omega\,.$$

These relationships will be used, in particular, for the determination of the perturbation elements of the orbit (see Chapter X).

4 The Integral of Kinetic Energy. Kepler's Second Law

On multiplying the first equation in (14) by $2dx/dt$, and the second one by $2dy/dt$ and taking the sum, we obtain

$$\frac{d}{dt}\left[\left(\frac{dx}{dt}\right)^2 + \left(\frac{dy}{dt}\right)^2\right] = -2\frac{k^2 M}{r^3}\left(x\frac{dx}{dt} + y\frac{dy}{dt}\right). \qquad (19)$$

Passing to the polar coordinates, r and θ,

$$x = r\cos\theta\,, \quad y = r\sin\theta\,, \quad r^2 = x^2 + y^2\,, \qquad (20)$$

and modifying the right-hand side using the third equation in (20), i.e.

$$x\frac{dx}{dt} + y\frac{dy}{dt} = r\frac{dr}{dt}\,, \qquad (21)$$

we have

$$\frac{d}{dt}\left[\left(\frac{dx}{dt}\right)^2 + \left(\frac{dy}{dt}\right)^2\right] = 2k^2 M\frac{d}{dt}\left(\frac{1}{r}\right). \qquad (22)$$

After integration we obtain

$$\left(\frac{dx}{dt}\right)^2 + \left(\frac{dy}{dt}\right)^2 = \frac{2k^2 M}{r} + c_2\,, \qquad (23)$$

where c_2 is a new constant of integration.

The left-hand side of Eq. (22) is nothing other than the square of the velocity V. Therefore we can write

$$V^2 = \frac{2k^2 M}{r} + c_2\,. \qquad (24)$$

Insofar as the square of the velocity is twice the kinetic energy of a body with unit mass, relationship (24) is also called the **integral of kinetic energy**. We shall return to it later.

Note that on the left-hand side of Eq. (23) Cartesian coordinates (x, y) are present but on its right-hand side there are polar coordinates. Using the first two relationships of (20) we can present both sides of Eq. (23) in polar coordinates. Then we obtain

$$\left(\frac{dr}{dt}\right)^2 + r^2\left(\frac{d\theta}{dt}\right)^2 = \frac{2k^2 M}{r} + c_2\,. \qquad (25)$$

In this equation, three variables are present simultaneously: coordinates r and θ and time t, the last one as a parameter of the problem. In order to find the equation of relative motion, i.e. the functions

$$r = \varphi_1 (t),$$
$$\theta = \varphi_2 (t),$$

(26)

it is necessary to derive one more equation containing r, θ and t. For that we can use Kepler's first law (9), in which all three variables r, θ and t are present. Solving both these equations, (9) and (25), jointly, we can find, at least in principle, the solution in the form of Eq. (26). Then, the parameters of the motion can be obtained, in particular, the trajectory, i.e. the orbit of the body m with respect to m_0. However, it appears that joint integration of the system consisting of Eqs. (9) and (25) to obtain the equation of motion in the form Eq. (26) is impossible.

Therefore, one has to find another solution of the problem, namely, by obtaining the equation of the trajectory by exclusion of the time from the system (9) and (25). Here, first, it is necessary to derive the differential equation of the trajectory, then to integrate it to deduce the equation of the trajectory itself as a geometric curve.

We have from Kepler's first law (9)

$$\frac{d\theta}{dt} = \frac{c_1}{r^2},$$

(27)

therefore

$$\frac{dr}{dt} = \frac{dr}{d\theta} \frac{d\theta}{dt} = \frac{c_1}{r^2} \frac{dr}{d\theta}.$$

(28)

By substituting Eqs. (27) and (28) into Eq. (25) we exclude the time t from the equations:

$$\frac{c_1^2}{r^4} \left(\frac{dr}{d\theta} \right)^2 + \frac{c_1^2}{r^2} = \frac{2k^2 M}{r} + c_2.$$

(29)

This is the differential equation of the trajectory represented as a relation between coordinates r and θ, only – the time is absent. After integration of this equation we eventually obtain the mathematical expression of the curve or of a trajectory of the body m moving within the gravitational field of the central body m_0.

From Eq. (29) we obtain

$$d\theta = \frac{d \left(c_1/r \right)}{\left(\dfrac{2k^2 M}{r} - \dfrac{c_1^2}{r^2} + c_2 \right)^{1/2}}.$$

(30)

In the denominator, let us add and subtract $k^4 M^2/c_1^2$ inside the square root. Then we obtain, after some modifications,

$$d\theta = \frac{d\,(c_1/r)}{\left[c_2 + \dfrac{k^4\,M^2}{c_1^2} - \left(\dfrac{k^2\,M}{c_1} - \dfrac{c_1}{r}\right)^2\right]^{1/2}} \cdot \tag{31}$$

Introducing the notations

$$B^2 = c_2 + \frac{k^4\,M^2}{c_1^2}\,, \tag{32}$$

$$-u = \frac{k^2\,M}{c_1} - \frac{c_1}{r}\,, \tag{33}$$

$$du = d\left(\frac{c_1}{r}\right). \tag{34}$$

we can write Eq. (31) in the form

$$d\theta = \frac{\pm du}{(B^2 - u^2)^{1/2}} \cdot \tag{35}$$

After integration we obtain for the upper sign

$$\frac{u}{B} = \cos(\theta - c_3)\,,$$

and for the lower sign

$$\frac{u}{B} = \sin(\theta - c_3')\,,$$

where c_3 and c_3' are the integration constants. It is not difficult to notice, however, that in the given case these constants differ from one an other by the magnitude $\pi/2$. Therefore the general solution of (31) may be represented in the form

$$u = B\,\cos(\theta - c_3). \tag{36}$$

Having in mind Eqs. (32) and (33), we derive the equation of the trajectory

$$r = \frac{c_1}{\dfrac{k^2\,M}{c_1} + \left(c_2 + \dfrac{k^4\,M^2}{c_1^2}\right)^{1/2}\cos(\theta - c_3)} \tag{37}$$

or finally

$$r = \frac{\dfrac{c_1^2}{k^2 M}}{1 + \left(1 + \dfrac{c_1^2 c_2}{k^4 M^2}\right)^{1/2} \cos(\theta - c_3)} . \tag{38}$$

This is the definitive form of the trajectory represented in polar coordinates r and θ.

It is not difficult to be convinced that the equation just obtained represents the equation of the conical section with respect to one of its focuses. Indeed, from analytical geometry one has the equation of a conical section in general, i.e. parametric, form:

$$r = \frac{p}{1 + e \, \cos(\theta - \omega)} , \tag{39}$$

where p is the parameter of the conical section, e is its eccentricity, and ω is the angle between polar and semimajor axis of conical section.

On comparing (39) with (38) we find

$$p = \frac{c_1^2}{k^2 M} ,$$

$$e^2 = 1 + \frac{c_1^2 c_2}{k^4 M^2} , \tag{40}$$

$$\omega = c_3 .$$

From here we obtain the integration constants c_1, c_2 and c_3 expressed through elements of the conical section:

$$c_1 = k \, (p M)^{1/2} ,$$

$$c_2 = -\frac{k^2 M (1 - e^2)}{p} , \tag{41}$$

$$c_3 = \omega .$$

The type of conical section and its properties are determined by the magnitudes of e and p. Therefore before going ahead it is useful to have a look first at the various types of conical sections.

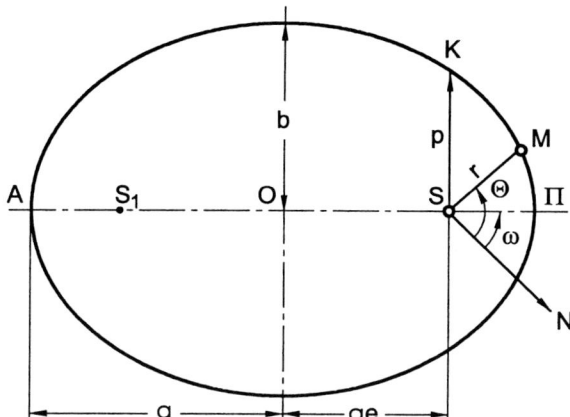

Fig. 3.4. *The parameters deter-mining a conical section, for an ellipse.*

5 Conical Sections. Types of Trajectories of Relative Motions

The Ellipse. In this case the eccentricity is less than unity: $e < 1$. According to definition, the center of the coordinate system in Fig. 3.3 (body m_0) must be localized at one of the focuses of the ellipse, i.e., in S (Fig. 3.4).

For the ellipse one has $p = a(1 - e^2)$ where a is a semimajor axis, in Fig. 3.4 this corresponds to the section SK. The semiminor axis $b = AB$ is related to the eccentricity by the following expression:

$$b = a\,(1 - e^2)^{1/2}, \qquad e = \frac{(a^2 - b^2)^{1/2}}{a}. \tag{42}$$

The distance of the focus S from the center of the ellipse A is the linear eccentricity c and is equal to $AO = c = ae$. In the case of an ellipse $e < 1$ and therefore $c < a$.

In Fig. 3.4 the direction SN, not defined yet, is shown as well, by which the angles ω and θ are defined. Here ω is the angular distance between the semimajor axis and the direction SN. In Fig. 3.4 the position M of the body m is also shown for the given moment of time t, i.e. the radius-vector r and angle θ in accord with Eq. (39). The point Π nearest to the focus of the ellipse is called the **perigee**, and the most distant point A is the **apogee**. The distance of the perigee from focus S is equal to $S\Pi = a - ae = a(1 - e)$; the distance of the apogee from focus S is $SA = a + ae = a(1 + e)$.

Therefore, the integration constants c_1, c_2 and c_3 for an ellipse are

$$c_1 = k\,[\,a\,(1-e^2)\,M\,]^{1/2}\,,$$

$$c_2 = -\frac{k^2\,M}{a}\,, \tag{43}$$

$$c_3 = \omega\,,$$

i.e. c_2 is negative. A circle or circular orbit is a particular case of an ellipse for which $e = 0$, $p = a$, $b = a$, and $c = 0$.

Then we shall have for the radius-vector r:

$$r = \frac{a(1-e^2)}{1+e\,\cos(\theta-\omega)}\,. \tag{44}$$

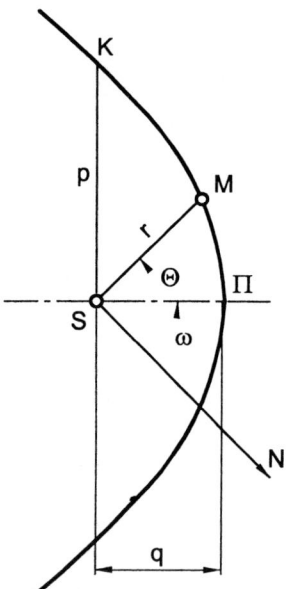

Fig. 3.5. *The parameters determining a conical section, for a parabola.*

The Parabola. For the parabola $e = 1$ and hence $c_2 = 0$. The second focus lies at infinity so that $a = \infty$. In this case $p = 2q$ (SK in Fig. 3.5), where q is the closest distance from the focus (the only one): $q = S\Pi$. Thus, in the case of a parabola, instead of two constants, a and e, we have only one, q. A parabola has only perigee Π, but no apogee – it lies at infinity. The angular notations are the same as in the case of an ellipse. Constants

c_1, c_2 and c_3 for a parabola are equal to

$$c_1 = k\,(\,2q\,M\,)^{1/2}\,,$$
$$c_2 = 0\,, \tag{45}$$
$$c_3 = \omega\,.$$

Then we shall have for the radius-vector r

$$r = \frac{q}{\cos^2(\theta - \omega)/2}\,. \tag{46}$$

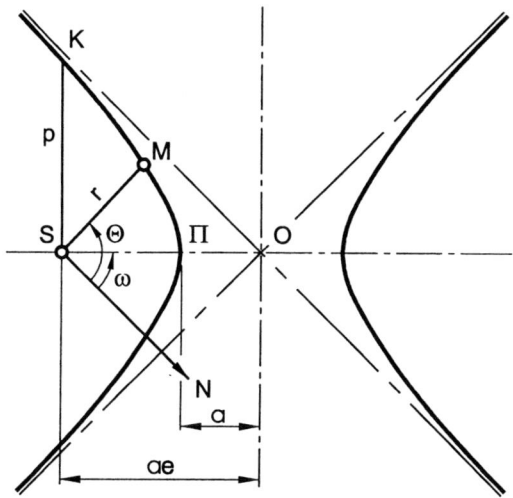

Fig. 3.6. *The parameters determining a conical section, for a hyperbola.*

The Hyperbola. For a hyperbola the eccentricity is larger than unity, $e > 1$, and the parameter $p = a\,(e^2 - 1)$ where a is a large half-axis – $a = O\Pi$ and $p = SK$ in Fig. 3.6. The distance of the focus from the center O is also equal to $c = SO = ae$; however, bearing in mind that $e > 1$, the distance between focuses turns out to be larger than the half-axis, i.e. $c > a$. A hyperbola has perigee Π; the apogee lies at infinity.

The integration constants c_1, c_2 and c_3 in the case of a hyperbola are equal to

$$c_1 = k\,[\,a\,(e^2 - 1)\,M\,]^{1/2}\,,$$
$$c_2 = +\frac{k^2\,M}{a}\,, \tag{47}$$
$$c_3 = \omega\,.$$

Then we shall have for the radius-vector r:

$$r = \frac{a(e^2 - 1)}{1 + e\,\cos(\theta - \omega)}\,. \tag{48}$$

In comparison with the case of an ellipse, we see that the main difference between them is the sign of c_2: it is negative for an ellipse and positive for a hyperbola.

6   Velocity of Motion Depending on the Type of the Orbit

Let us return to Eq. (24) which represents a universal expression for the velocity of relative motion on the orbit. Universal, because it describes the various types of orbits through the integration constant c_2. By substituting the value of c_2 for one or another type of orbit, we obtain the desired formulas for the velocity of motion for all three types of orbits.

For the parabolic orbit, as was shown above, $c_2 = 0$. For the other two orbits $c^2 = \pm k^2 M/a$, where the *minus* sign concerns an ellipse, and *plus* concerns a hyperbola. Therefore we shall have from (24)

$$V^2 = k^2 M \left(\frac{2}{r} \pm \frac{1}{a} \right), \tag{49}$$

where *minus* relates to an elliptic and *plus* to a hyperbolic orbit. In the case of a parabola one has $a = \infty$ in Eq. (49).

It is interesting that the eccentricity e is absent in Eq. (49). This means that the velocity of the motion is independent of the degree of elongation of the ellipse (hyperbola), it depends only on the large half-axis (at given distance r). The vector of the velocity is directed along the tangent to any given point of the orbit.

According to Eq. (49), the velocity of the motion decreases with increasing distance r. That decrease occurs too slowly, to a first approximation according to the law

$$V \sim r^{-1/2}. \tag{50}$$

Eq. (49) is interesting as well since by its help and with known initial conditions the type of the conical section, i.e. the type of the trajectory, can be determined.

Thus, the motion will be elliptic if

$$V^2 < \frac{2k^2 M}{r}, \tag{51}$$

and hyperbolic if

$$V^2 > \frac{2k^2 M}{r}. \tag{52}$$

One can obtain the velocity in the case of a parabolic orbit V_p just by substituting $a = \infty$ into Eq. (49):

$$V_p^2 = \frac{2k^2 M}{r}. \tag{53}$$

If applied to the Solar system, we have $M = 1$ for all planets, especially for small planets, as well as for interplanetary spacecraft. On the other hand, modern radio locating facilities enable measurement with high precision and for large distances both of the magnitude of the velocity V of a given object and of its distance r. Now, it is enough to compare the magnitude of V obtained from direct observations with the magnitude of $2k^2/r$: of course, one of the conditions described above, i.e. Eqs. (51), (52) or (53), should be satisfied, and, as a result, one can obtain the type of the orbit of the object – elliptic, parabolic or hyperbolic. Meanwhile, as we shall see below (Chapter VII), preliminary knowledge of the type of orbit is extremely important for preliminary calculations of a position of a celestial body.

We shall now examine a particular case of the motion of a celestial body; we have in mind motion in a circular orbit as an extreme case of an elliptical orbit, i.e. when $e = 0$. In this case we have $r = a = constant$, and the *minus* sign should be used in the parentheses in Eq. (49). Then we obtain for the velocity of motion V in a **circular orbit**

$$V_0^2 = \frac{k^2 M}{r}. \tag{54}$$

The velocity V_p of a body at distance r moving through a parabolic orbit is given by Eq. (53). However, if the motion takes place in a circular orbit, the velocity V_0 of the same body and at the same distance r will be different and given by Eq. (54). Then we obtain the following rather remarkable and highly simple relationship between the two velocities, V_p and V_0, for a given distance r of the body from the Sun:

$$\frac{V_p}{V_0} = \sqrt{2}, \tag{55}$$

i.e. the ratio of the parabolic velocity to the circular one, V_p/V_0, is constant and is universal for all distances from the central body; V_p is larger than V_0 by 40%.

7 Parabolic Velocity at Different Distances from the Sun

The eccentricities of elliptic orbits for almost all planets of the Solar system are very small, and, hence, to a first approximation these orbits may be considered to be circular. On the other hand, the appearance of comets moving in parabolic orbits and intersecting the circular orbits of the majority of the planets is not so rare. Halley's comet may be a typical example; its extremely elongated elliptical orbit ($e = 0.967$) intersects even Pluto's orbit. Eq. (55) give us an interesting possibility for the determination of the velocity of a comet moving in a parabolic orbit at the points of its intersections with each of the circular orbits of the planets. For example, in the case of the intersection with the orbit of the Earth ($V_0 = 29.8 \approx 30 \ km \ s^{-1}$) the parabolic velocity V_p of a comet will be $V_p = 30 \sqrt{2} = 42 \ km \ s^{-1}$ with respect to the center of the coordinate system, i.e. the Sun. For the rest of the planets the picture is given in Table 3.1. These results are represented more exactly in Fig. 3.7.

T a b l e   3.1

Orbital velocity V_0 and parabolic velocity of escape from the orbit V_p for planets

Planet	V_0 $km \ s^{-1}$	V_p $km \ s^{-1}$
Mercury	47.9	67.7
Venus	35.0	49.5
Earth	29.8	42.1
Mars	24.1	34.1
Jupiter	13.1	18.5
Saturn	9.64	13.6
Uranus	6.81	9.63
Neptune	5.43	7.68
Pluto	4.74	6.70

These results are of definite interest also from the following point of view. A parabola represents an infinite orbit, i.e. every body moving with such an orbit should leave the solar system, sooner or later. The question is whether it is possible to launch an interplanetary spacecraft from the Earth so that it can leave the Solar system. Yes, it is possible. To do that, one must launch that spacecraft, at the moment of its departure from the Earth, with a velocity equal to $42 \ km \ s^{-1}$. This is a rather large velocity, and it is not so easy to realize. Fortunately, this spacecraft can be launched directly from the Earth with a velocity much less than the parabolic velocity but in such a manner that it can reach, eventually (10–12 years), the region of Pluto. The parabolic velocity for Pluto's distance is $6.6 \ km \ s^{-1}$, so that it is enough to increase the velocity of the spacecraft by only $6.6 \ km \ s^{-1}$; then it will assume a parabolic orbit and sooner or later has to leave the Solar system.

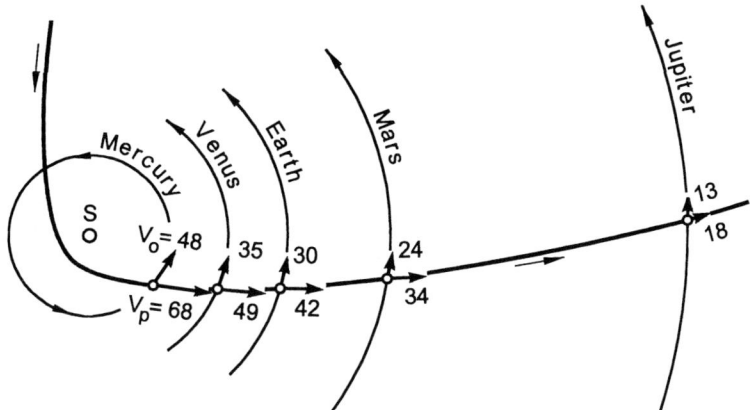

Fig. 3.7. *Circular velocities in km s^{-1} (upper line) and parabolic velocities (lower line) of planets at various distances from the Sun.*

8   Factors Determining the Form and Orientation of an Elliptic Orbit

The size of an elliptic orbit is determined by the semimajor axis a which, in its turn depends, as was shown above, on the initial velocity V_0 according to the relationship (49), which we rewrite in the following form:

$$a = \frac{k^2 M}{\dfrac{2k^2 M}{r_0} - V_0^2}.$$

As follows from this relationship, the larger the initial velocity V_0 (at given distance r) the larger will be the size of the ellipse and vice versa. It is important to notice that the major half-axis a depends only on the absolute magnitude of the velocity V_0, but not on its direction.

However, the elliptic orbit is characterized also by its form, i.e. by the degree of its elongation – by the eccentricity e. It can be shown, however, that the eccentricity depends on the vector V_0, i.e. not only on the absolute magnitude of velocity but also on its direction. Turn to Fig. 3.8 in which an element of the trace ds of a celestial body is shown; S is the Sun and r is the radius-vector.

The components of the element ds both along and perpendicular to the radius-vector are equal: $AP = ds \sin \eta$ and $AN = ds \cos \eta$. Having in view

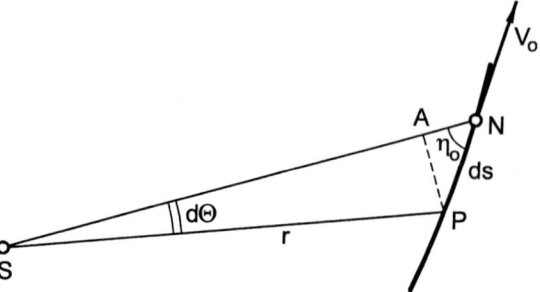

Fig. 3.8. *The problem of the dependence of the eccentricity of the orbit on the direction η_0 of the initial velocity V_0.*

that $PA = r\,d\theta$ and $AN = dr$, we can write

$$dr = ds\,\cos\eta,$$
$$r\,d\theta = ds\,\sin\eta$$

or

$$\frac{dr}{dt} = \frac{ds}{dt}\,\cos\eta,$$
$$r\frac{d\theta}{dt} = \frac{ds}{dt}\,\sin\eta.$$

However, ds/dt is the velocity of celestial body V, and the value of $d\theta/dt$ is known from Kepler's first law. Then we find

$$\frac{c_1}{r^2}\frac{dr}{d\theta} = V\,\cos\eta,$$

$$\frac{c_1}{r} = V\,\sin\eta.$$

On the other hand, for the elliptic orbit we have

$$\frac{1}{r} = \frac{1}{p} + \frac{e}{p}\,\cos(\theta - \omega).$$

Therefore, the above equations may be written, after some modifications, in the following way:

$$e\sin(\theta_0 - \omega) = V_0^2\,\frac{r_0}{k^2\,M}\,\cos\eta_0\,\sin\eta_0,$$

$$e\cos(\theta_0 - \omega) = V_0^2\,\frac{r_0}{k^2\,M}\,\sin^2\eta_0 - 1.$$

As follows from these relationships, the form of the ellipse, i.e. its eccentricity depends not only on the absolute magnitude of the initial velocity V_0, but on its direction η_0 as well.

However, in these equations we have one more element of the elliptic orbit – w, i.e. the angular distance of the perigee from the so called "line of knot"; this element determines the orientation, i.e. the position of the ellipse itself on the orbital plane. Judging by the above formula, the orientation of the ellipse is also determined by the vector of initial velocity V_0. We shall use the results obtained in this section while deriving the trajectories of interplanetary flights.

9   Departure of the Spacecraft from the Earth

Eq. (55) can obviously be applied also for the Earth as a central body (m_0) with an artificial satellite (m) moving around it. In this case, orbits both circular and parabolic are possible.

The magnitude of the circular velocity around the Earth is known; it equals $V_0 = 7.9 \ km \ s^{-1}$. To launch a spacecraft from the Earth, it is necessary to pass the limit of a circular orbit, i.e. to take the parabolic one. Departure from the Earth implies passage to a parabolic orbit with the Earth at its focus. In accord with the universal law (55), the parabolic velocity V_p of departure from the Earth must be $\sqrt{2}$ times larger than the circular velocity:

$$V_p = \sqrt{2} \, V_0 = 1.41 \times 7.9 = 11.2 \ km \ s^{-1}.$$

A body leaving the Earth with such a velocity should theoretically move away as far as possible. A body arriving at the Earth from infinity should fall onto it with the same velocity of $11.2 \ km \ s^{-1}$. However, a body with the velocity $11.2 \ km \ s^{-1}$ can never leave the Solar system (in the framework of the two–body problem).

10   Limiting Velocities of Meteors Falling onto the Earth

There are clouds of meteors moving in parabolic orbits. On meeting the Earth, these meteorites naturally fall onto it. However, depending on the direction of motion of meteor clouds with respect to the vector of orbital motion of the Earth, the relative velocity of falling may be quite different. Hence, the problem of the determination of limiting velocities of meteors falling onto the Earth arises.

The parabolic velocity of a meteor cloud V_p when it meets the Earth is obtained by means of the formula (55): $V_p = \sqrt{2} \, V_0 = 1.41 \times 30 = 42 \ km \ s^{-1}$, where $V_0 = 30 \ km \ s^{-1}$ is used for the heliocentric circular (orbital) velocity of the Earth. Below we consider three limiting cases.

I. Both the meteors and the Earth are moving in one and the same direction. In this case the velocity of falling will be minimum and equal to

$$V_{min} = V_p - V_0 = 42 - 30 = 12 \ km \ s^{-1}. \tag{56}$$

II. Meteors moving perpendicularly to the vector V_0. In this case the falling velocity will be

$$V = V_p = 42 \ km \ s^{-1}.$$

III. Both the meteors and the Earth move towards each other. In this case the velocity of falling is maximal and equals

$$V_{max} = V_p + V_0 = 42 + 30 = 72 \ km \ s^{-1} \tag{57}$$

Thus, the limiting velocities of falling meteors can differ from each other by up to a factor of six, and the kinetic energies by a factor of 36. That is why the consequences of falling meteors in case III can be much more catastrophic, especially when their masses are not small. The Tunguska (1908) and Sichote-Aline (1947) events could well be such events, though the polemics on their real nature still continue; in particular, their cometary nature should be considered highly probable, especially in view of the Shoemaker-Levy 9 comet's impact with Jupiter in July 1994. In general, the understanding of the consequences of the impact of meteorites, comets or asteroids with the Earth remains a problem of constant interest (Hills, Leonard, 1995).

11   Launching a Spacecraft in the Direction of Sun

Can one send a spacecraft towards the Sun? It seems not, bearing in mind the modern possibilities of rocket technique. Anyway, it seems an awkward problem, to throw a stone onto the Sun.... .

In order to send a body in the direction of the Sun, it is necessary first of all to depart from the Earth, i.e. to get the second cosmic velocity $V_2 = 11.2 \ km \ s^{-1}$. This is an easy task. Much more difficult is the second problem: it is necessary to eliminate the orbital velocity of the Earth V_0, which, naturally, passes to the body after its launch. V_0 is very large – $30 \ km \ s^{-1}$. Present space technology still cannot allow us the compensation of such a velocity.

The gravitational field, while moving away from the Earth, falls according to the square law. Correspondingly, the law of motion of the spacecraft will be derived from the following differential equation:

$$\frac{d^2 r}{dt^2} = -g_0 \left(\frac{R}{r} \right)^2, \tag{58}$$

where g_0 is the acceleration at the Earth's surface, R is the radius of the Earth, and r is the distance from the Earth.

Substituting $dV = dr/dt$ into (58), after integration, we obtain

$$V^2(r) = 2g_0 \frac{R^2}{r} + C, \qquad (59)$$

i.e. the same integral of energy as (24) derived above. The constant of integration C one can find from the condition $V(R) = V_0$ at the Earth's surface ($r = R$), where V_0 is the initial velocity often called the **firing velocity**. Then we shall have for the exact expression of $V(r)$

$$V^2(r) = V_0^2 - \left(1 - \frac{R}{r}\right) V_n^2, \qquad (60)$$

where the parabolic velocity of departure, V_n, is used from Eq. (53) as well as the fact that $g_0 = k^2 M/R^2$:

$$V_n^2 = 2g_0 R. \qquad (61)$$

While moving away from the Earth the term in parentheses in Eq. (60) tends to unity. Therefore we shall have for the velocity of the spacecraft at large distances (formally, at infinity, i.e. $r \to \infty$)

$$V^2(\infty) = V_0^2 - V_n^2. \qquad (62)$$

In fact $V(\infty)$ is the velocity V_{out} which we have at the moment of the departure from the sphere of the Earth's gravitation. Therefore, one may well write Eq. (62) in the following form:

$$V_{out}^2 = V_0^2 - V_n^2. \qquad (63)$$

Quite often we have a problem in finding out the initial velocity of launch V_0 from the Earth's surface in order to reach the given object. For that velocity one has

$$V_0 = (V_{out}^2 + V_n^2)^{1/2}. \qquad (64)$$

Let us apply these formulas to the problem formulated above – to send a spacecraft towards the Sun. We have $V_{out} = 30 \ km \ s^{-1}$, which corresponds to the orbital heliocentric velocity of the Earth, and $V_n = 11.2 \ km \ s^{-1}$ is the heliocentric velocity of departure from the Earth. From that we find

$$V_0 = V_4 = 31.8 \ km \ s^{-1}.$$

At such a velocity – often called the **fourth** cosmic velocity, the probe will tear off from the Earth and simultaneously its orbital motion will be compensated; the spacecraft, for a moment, will remain in absolute rest relative to the Sun. In the next moment, influenced by the gravity of the Sun it will move towards the latter – it simply will fall onto the Sun moving strongly in a straight line directed towards its center. In particular, if the velocity of the launch is smaller than 31.8 $km\ s^{-1}$, the spacecraft will go past the Sun. Correspondingly, by choice of V_0, we can organize a flight around the Sun at an initially fixed distance from its center, for example, through various layers of the Solar corona.

The problem discussed here – the launching of a spacecraft towards the Sun, is far from being realized in the near future, though it is of particular interest for astrophysics. We have been particularly interested for many years in the idea of various kinds of observations of the Sun's polar regions by means of the flight of special spacecraft provided with special measuring equipment. However, for that one has to launch the spacecraft with a velocity near the fourth cosmic velocity, i.e. of the order of 30 $km\ s^{-1}$, which can hardly be reached in the foreseeable future. However, the importance of the problem of flight over the Solar poles and the layers of the corona is so large that it is worth seeking other solutions to this problem. A solution has been found, namely, to use the massive planets, and most of all Jupiter, as an accelerator of a spacecraft. We will deal with this problem in Chapter XIII.

12   Away from the Solar System

A space probe, after leaving the Earth with the second cosmic velocity (11.2 $km\ s^{-1}$), becomes some kind of "property" of the Solar system – its subsequent motion is completely controlled by the Sun. The motion itself will take place, if we ignore the perturbations by the Moon and planets, in an elliptic, i.e. finite, closed orbit with the Sun at its focus. The spacecraft, launched from the Earth with that velocity, will, however, never leave the solar system because its velocity at any given point will always be smaller than the parabolic velocity of departure from the Solar system. In this connection a problem arises, namely, to find the conditions under which a spacecraft launched from the Earth can leave the Solar system.

Insofar as the launching of the probe is performed directly from the Earth but not from some planet, Pluto, for example, these conditions are quite definite: one should ensure that the parabolic velocity corresponds to the position of the Earth relative to the Sun. This velocity is known: $V_p = \sqrt{2}\ V_0 = 42.1\ km\ s^{-1}$. At such an initial velocity the spacecraft will continue its motion in a parabolic orbit (Fig. 3.7) and at least should leave the Solar system. However, the velocity 42 $km\ s^{-1}$ is of a colossal magnitude and is not so easy to gain, bearing in mind the principal difficulties of energetic character. A way out of the situation may be to launch in the direction of the vector of orbital velocity of the Earth V_0, at a tangent to the Earth's orbit, when, as a result, the orbital velocity of the Earth will automatically be passed to the probe. Therefore, it should be enough to create an additional velocity $V_{out} = V_p - V_0 = 42.1 - 29.8 = 12.3 \ km \ s^{-1}$.

Together with this, consider the usual escape from the Earth with the "Earth's" parabolic velocity $V_n = 11.2 \ km \ s^{-1}$: one should have an initial velocity in accordance with the formula (64)

$$V_0 = (\ 12.3^2 + 11.2^2 \ )^{1/2} = 16.6 \ km \ s^{-1} \ .$$

With such a velocity directed, however, parallel to the vector of the Earth's orbital velocity, the probe can be passed on the parabolic orbit around the Sun and, as a result, sooner or later should leave the Solar system. Such a scheme was realized in 1973–1974 while launching the *Pioneer 1* and *Pioneer 2* automatic space stations carrying some kind of message to the neighboring stars – a gold slab with the engraved images of a man and a woman, as well as coded information on the structure of Solar system, etc. These spacecraft, having reached the zone of Pluto's orbit in 1987, have to leave the Solar system forever.

The velocity $V_0 = 16.6 \ km \ s^{-1}$ is known as the **third** cosmic velocity.

Thus, depending on the given problem associated with the particular space mission from the Earth, we must manipulate at least four types of "cosmic velocities", as summarized below

First cosmic velocity:		
circular motion around the Earth	7.9	$km \ s^{-1}$
Second cosmic velocity:		
breaking away from the Earth	11.2	$km \ s^{-1}$
Third cosmic velocity:		
leaving the solar system	16.6	$km \ s^{-1}$
Fourth cosmic velocity:		
flight in the direction of the Sun	31.8	$km \ s^{-1}$

It is hard to imagine a practical problem for whose solution a velocity exceeding the fourth cosmic velocity would be required.

13   The Orbital Period of a Body in a Closed Orbit

The ellipse is a closed curve. Therefore the problem of finding the period of one complete revolution P – the orbital period – of celestial body m_0 around the central body m situated at its focus arises.

Turn to Eq. (12) with $t - t_0 = P$. Indeed, during a full rotation P the radius-vector of the body m will cover an area equal to the surface S of the ellipse with the major half-axis a and eccentricity e, i.e.

$$S = \pi ab = \pi a^2 \ (\ 1 - e^2 \ )^{1/2} \ . \tag{65}$$

By substituting S as well as the value of the constant c_1 for an ellipse from Eq. (46) into Eq. (12), we obtain

$$P = \frac{2\pi a^{3/2}}{k\sqrt{M}}.$$ (66)

This is the formula for the determination of the orbital period of a body moving in an elliptic orbit.

The peculiarity of Eq. (66) is, besides its simplicity, the complete independence of the orbital period from the eccentricity of the orbit – the parameter e is absent in Eq. (66). The orbital period depends on one parameter only – the orbit's semimajor axis a. This means, in particular, that the orbital period must be one and the same in the case of a purely circular orbit ($e = 0$) as well as for a strongly elongated ellipse ($e \sim 1$) with the same half-axis a (see Fig. 3.9).

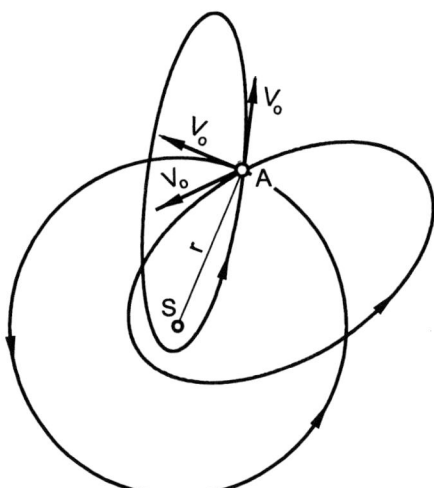

Fig. 3.9. *An illustration of the independence of the velocity of motion from the orbit's eccentricity at one and the same value of semimajor axes of the ellipses.*

The velocity of the motion through the elliptic orbit given by Eq. (49) depends on the semimajor axis a and on the distance r. Therefore, at the point of intersection A in Fig. 3.9, all three elliptic orbits with various eccentricities must have one and the same magnitudes of orbital velocity V_0 in all three cases. The influence of the eccentricity affects only the direction of the velocity vector – it is different in all three cases.

14   Kepler's Third Law

Eq. (66) is universal; it may be used for all elliptic orbits with given magnitude of major half-axis a. Let us examine now the motion of two celestial bodies of masses m_1 and m_2 rotating around one and the same central body, for example, the Sun, in elliptic orbits with the semimajor axes a_1 and a_2. Then we can write for the periods of rotation P_1 and P_2 in accord with Eq. (66):

$$P_1 = \frac{2\pi a_1^{3/2}}{k\sqrt{M_1}}, \qquad (67)$$

$$P_2 = \frac{2\pi a_2^{3/2}}{k\sqrt{M_2}}, \qquad (68)$$

where $M_1 = m_1 + m_\odot$ and $M_2 = m_2 + m_\odot$.

We have from Eqs. (67) and (68)

$$\left(\frac{P_1}{P_2}\right)^2 = \left(\frac{a_1}{a_2}\right)^3 \frac{M_2}{M_1}. \qquad (69)$$

If by the term celestial bodies we understand the planets of the Solar system, then to a first approximation their masses could be easily ignored compared with the Solar mass, so that $M_1 \approx M_2 = 1$, since that even Jupiter's mass is still $1\,000$ times smaller than the Solar one. Then, instead of Eq. (69) we obtain:

$$\left(\frac{P_1}{P_2}\right)^2 = \left(\frac{a_1}{a_2}\right)^3. \qquad (70)$$

This is Kepler's **third** law which runs: the squares of orbital periods of planets are functions of the cubes of the semimajor axes of their orbits.

The exact formulation of Kepler's third law is given, however, by Eq. (69). Kepler discovered his third law based on the analysis of Tycho Brahe's observations. Those data are associated with the epoch of the history of astronomy before the discovery of the telescope (the first telescope was operated by Galileo in 1609, nearly half a century after Tycho Brahe's observations). Naturally their accuracy could not be higher than the angular resolution of the naked eye, i.e. one arc minute. Obviously Kepler could not have noticed a one-thousandth part of the ratio of M_1/M_2, and therefore discovered his law in the form of Eq. (70). The exact form of this law (69) was discovered much later, and exceptionally in a mathematical way, as described above, as one of the consequences of the solution of the problem of the two bodies interacting according to the Newtonian law of gravitation.

15   Weightlessness in the Orbit

Consider now the motion in a circular orbit around the Earth. The velocity
of such a motion is determined by Eq. (54):

$$V_0^2 = \frac{k^2 M}{r},$$

where r is the radius of the orbit. Multiplying both sides of this relationship
by m/r, we obtain

$$\frac{m V_0^2}{r} = k^2 \frac{m M}{r^2}. \tag{71}$$

The left-hand side of this expression is nothing other than the centrifugal
force: $R_c = m V_0^2/r$. The right-hand side is the force of gravitation: $R_g = k^2 m M/r^2$. The condition (71) runs: both forces should be in balance (Fig.
3.10) if the motion in this orbit is uniform, i.e. $R_c = R_g$. Hence, any body,
independently of its mass, should lose its weight[1]. This is the reason for
the phenomenon of weightlessness in spaceships. All items in the ship – the
ship itself, the astronaut, all the ship's contents, including the particles of its
"atmosphere", gain the velocity vector V_0 and, hence, must permanently be
under the action of two opposite acting forces – centrifugal and gravitational.
Also, due to the strong balance of the two forces, weightlessness occurs,
which can be disturbed only partially by local actions of various kinds (air
circulation, jet engines, etc.)

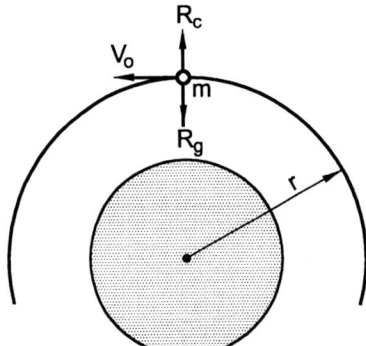

Fig. 3.10. *The origin of weightlessness in a circular orbit around the Earth.*

However, weightlessness can also occur in the course of motion in a nonclosed trajectory, for example, during the motion of a spacecraft towards the
Moon. However, in that case the "motion" is nothing other than free fall
towards the Moon. Strictly speaking, motion in a circular orbit around the

---

[1]The weight is mg, where m is the mass and g is the acceleration due to gravity.

Earth is also a free-fall; however, it is a fall in a circular trajectory, infinite and with constant velocity. Therefore it would be true to formulate the effect of the weightlessness as being a consequence of the inertial motion of a body.

16   Stationary Orbit

An artificial satellite in a circular or almost circular orbit around the Earth complete one rotation in about one and a half hours. The rotation period increases, in accordance with Eq. (66), with increasing distance or radius of the orbit. The Moon, for example, completes a rotation around the Earth in 27 days. Evidently, a circular orbit must exist at some distance from the Earth with a rotation period equal to one day. Any body, including an artificial satellite in such an orbit should complete a full rotation exactly in a day. It is quite clear that the position of such a satellite with respect to the observer on the Earth's surface should remain absolutely unchanged – the satellite will be "suspended" above the head of the observer – insofar as both the satellite and the observer accomplish one rotation on the same time scale – one day. Such an orbit is called **stationary**. Our problem is to find its radius a_0.

From Eq. (70) we have

$$\left(\frac{P_0}{P}\right)^2 = \left(\frac{a_0}{a}\right)^3 .$$

Then we can write for the radius a_0 of a stationary orbit

$$a_0 = a \left(\frac{P_0}{P}\right)^{2/3} , \tag{72}$$

where $P_0 = 1$ day $= 24$ h for a stationary orbit, a and P are the parameters for another well-known orbit.

For artificial satellites in ordinary orbits around the Earth we have $P \approx 1.5$ h, $a = 6\,370 + 300 = 6\,670$ km. Then we obtain from Eq. (72) $a_0 = 42\,340$ km. For the Moon, i.e. $P = 27^d.31$ and $a = 384\,000$ km, we have $a_0 = 42\,380$ km. On average we can take $a_0 = 42\,000$ km. This is 6.3 times farther out than ordinary orbits around the Earth.

The practical significance of a stationary orbit is extremely large. Every space station, located in this orbit, will stay permanently within the zone of 24-hour ground-based radio communications, and three such stations located triangularly in the orbit will provide communications between every point on the Earth's surface. As envisaged, in the future a whole complex of space stations, cosmic colonies, industrial centers, powerful Solar electric plants (with consequent transmission of energy to the Earth), etc., should be created on the stationary orbit. The only defect of the stationary orbit will be its expensiveness – a kilogram of mass delivered to that orbit will cost much more than for the nearest orbits. Really, the energy E_* spent for the delivery of a mass m to the nearest orbit around the Earth with a radius a is

$$E_* = \int_{r_*}^{a} F \, dr = k^2 \frac{m}{r_*} \left(\frac{a}{r_*} - 1 \right) , \qquad (73)$$

where r_* is the radius of the Earth. For the same mass m and a stationary orbit of radius a_0 one has

$$E_0 = k^2 \frac{m}{r_*} \left(\frac{a_0}{r_*} - 1 \right) . \qquad (74)$$

From Eqs. (73) and (74) we find

$$\frac{E_0}{E_*} = \frac{\dfrac{a_0}{r_*} - 1}{\dfrac{a}{r_*} - 1} .$$

For $r_* = 6\,370 \; km$, $a = 6\,670 \; km$, $a_0 = 42\,000 \; km$, we find for the ratio $E_0/E_* \approx 100$, i.e. for one and the same power of the rocket the given mass delivered to the stationary orbit will be 100 times less than that delivered to the nearest geocentric orbits.

17   Period of Revolution of Earth's Satellite

The circular velocity of an artificial satellite around the Earth decreases with moving away from the Earth's center. As a result, the period P of revolution of the satellite should be increased, i.e. the farther away the satellite, the larger the period of its revolution.

A typical artificial satellite performs a Keplerian circular motion around the Earth, namely, the Earth's mass $M_\oplus$ is concentrated in its center as a point, while the satellite is at a distance $R_\oplus + H$ from the Earth's center, where $R_\oplus$ is the Earth's radius, H the altitude of the satellite from the Earth's surface (sphere). Then the period P of circulation of the satellite around the Earth is

$$P = 2\pi \frac{(R_\oplus + H)^{3/2}}{(G \, M_\oplus)^{1/2}} . \qquad (75)$$

In Table 3.2 the periods P of the revolution of an artificial satellite around the Earth at various distances from its surface H are given. In the last column the linear velocities V_0 of the satellite on various H and on various circular orbits are given as well; that velocity is determined by Eq. (53)

$$V_0 = \left(\frac{G M_\oplus}{R_\oplus + H} \right)^{1/2}. \tag{76}$$

The last line in this table is related to the stationary orbit around the Earth, on which the rotation period of the satellite is exactly a day, i.e. 24 hours. That orbit lies at a distance $35\,800 + 6\,380 = 42\,180\,km$ from the Earth's center.

T a b l e   3.2

Period of the revolution P and circular velocity V_0 of an artificial satellite around the Earth at various altitudes

Altitude from Earth's surface km	Period P hours	Circular velocity $km\ s^{-1}$
0	1.41	7.909
250	1.49	7.759
500	1.58	7.617
750	1.66	7.482
1000	1.75	7.354
1500	1.93	7.116
2000	2.12	6.901
5000	3.35	5.921
10000	5.78	4.935
35800	23.935	3.072

The stationary orbit chosen is not always exactly circular. For various purposes, for example, when the satellite is for astrophysical research, the elliptic orbit turns out to be more efficient. As an example, in Fig. 3.11 the orbit of the astrophysical satellite IUE (International Ultraviolet Explorer) is shown (from the left). It had an elliptic orbit with the parameters

Semimajor axis	$35\,778\,km$
Perigee	$25\,669\,km$
Apogee	$45\,887\,km$
Eccentricity	$e = 0.240$
P e r i o d	23 hours 55 min 33 s

At such an orbit the IUE operated efficiently for 16 years, i.e. four times the originally proposed duration.

Fig. 3.11 also shows the elliptic orbit, strongly stretched, of a communication satellite Molniya 10, with an eccentricity $e = 0.975$. In this case the

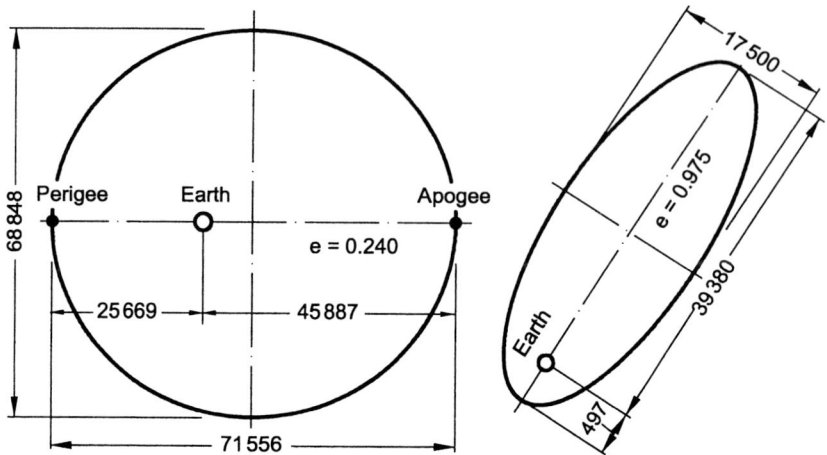

Fig. 3.11. *The stationary orbit of an astrophysical satellite, IUE (left), and semi-stationary orbit of a communication satellite, Molniya 10, both around the Earth.*

apogee of the semi-stationary orbit was $500km$, the perigee nearly $40\,000\ km$ with an orbital period of 11 hours 48 min.

Stationary orbits are convenient for communication and intelligence satellites.

18   Limiting Velocities of Motion in an Elliptic Orbit

It is interesting to know the ratio of the limiting velocities of motion of a celestial body in an elliptic orbit. Recall Eq. (49) written with the minus sign:

$$V^2 = k^2 M \left(\frac{2}{r} - \frac{1}{a} \right) . \tag{77}$$

One has the largest velocity V_{max} at the perigee, i.e. at the smallest distance from the Sun, where

$$r_p = a - ae = a\,(1 - e) \tag{78}$$

and

$$V^2_{max} = \frac{k^2 M}{a} \frac{1 + e}{1 - e} \tag{79}$$

and the smallest at the apogee, i.e. at the largest distance from the Sun, where

$$r_a = a + ae = a\,(1 + e) \tag{80}$$

and

$$V_{min}^2 = \frac{k^2\,M}{a}\frac{1 - e}{1 + e}. \tag{81}$$

From Eqs. (79) and (81) we obtain the following simple relationship for the ratio of both velocities:

$$\frac{V_{max}}{V_{min}} = \frac{1 + e}{1 - e}, \tag{82}$$

i.e. the ratio of the maximum orbital velocity to the minimum one for an elliptic orbit depends only on the orbit's eccentricity.

It is interesting to note that we shall have exactly the same expression for the ratio of the two distances, maximum in apogee $r_{max} = r_a$ and minimum in perihelion $r_{min} = r_p$, from Eqs. (78) and (80):

$$\frac{r_{max}}{r_{min}} = \frac{r_a}{r_p} = \frac{1 + e}{1 - e}. \tag{83}$$

For planets the eccentricities are not large, therefore the ratio V_{max}/V_{min} does not differ much from unity. However, for the majority of comets the eccentricities are very large, and correspondingly the scatter between maximum and minimum velocities should be larger. For three planets and three comets, the values of the ratios V_{max}/V_{min} and r_{max}/r_{min} are given in Table 3.3.

T a b l e   3.3

The ratios V_{max}/V_{min} and a_{max}/a_{min} for three planets and three comets

O b j e c t	e	V_{max}/V_{min}	r_{max}/r_{min}
E a r t h	0.0167	1.034	1.034
M a r s	0.0933	1.20	1.20
P l u t o	0.247	1.65	1.65
Comet Wolf-I	0.400	2.3	2.3
Comet Enke	0.848	12.2	12.2
Comet Halley	0.967	59.6	59.6

19   Determination of the Mass of Binary Stars

The coefficient of proportionality $k^2 \equiv G$ in the Newtonian law of gravitation, called the "gravitation constant", is one of the fundamental physical constants. Evidently, Eq. (69) is applicable for stellar systems, in particular the binary systems, as well. In contrast to the Solar system, in binary stellar systems the masses of companions are comparable. So Eq. (69) may be used for determination of the ratio of the masses of components of binary systems. Combined with spectral observations, this method has been successfully applied to obtain masses of stars of various spectral and luminosity classes.

Consider the motion around the Sun of a planet of mass m, major semi-axis a and period P:

$$\frac{a^3}{P^2\left(M_\odot + m\right)} = \frac{k^2}{4\pi^2}. \tag{84}$$

Consider now a binary system of stars with masses $\mathbf{M}_1$ and $\mathbf{M}_2$. The type of binary system – visual, photometric or spectroscopic – is not of importance, it is enough to carry out spectral observations to obtain the period P of rotation of the components around their center of mass, which is not difficult to realize, as well as the determination of the absolute magnitude of the major half-axis a_*, which is not so easy to realize, for it is necessary to know the distance away of the binary system as well as the angular separation between its components. Then, for such a system we can write

$$\frac{a_*^3}{P_*^2\left(\mathbf{M}_1 + \mathbf{M}_2\right)} = \frac{k^2}{4\pi^2}. \tag{85}$$

Owing to the universal character of the Newtonian law of gravitation, i.e. the constancy of k^2 for any mass scales, from artificial satellites up to stars and galaxies, we have from (84) and (85):

$$\frac{\mathbf{M}_1 + \mathbf{M}_2}{M_\odot + m} = \frac{a_*^3}{a^3}\frac{P^2}{P_*^2}. \tag{86}$$

From that we obtain for the total mass of a binary system $\mathbf{M}_* = \mathbf{M}_1 + \mathbf{M}_2$ in units of the solar mass

$$\frac{\mathbf{M}_*}{M_\odot} = \left(1 + \frac{m}{M_\odot}\right)\frac{a_*^3}{a^3}\frac{P^2}{P_*^2}. \tag{87}$$

Assuming that $1 + m/M_\odot \approx 1$, we obtain

$$\frac{\mathbf{M}_*}{M_\odot} = \frac{a_*^3}{a^3}\frac{P^2}{P_*^2}. \tag{88}$$

By means of this formula astrophysicists determine the total masses of the binary stars.

The ratio of the masses of the components in binary systems, $\mathbf{M_2}/\mathbf{M_1}$, can be obtained only under specific conditions; if, say, the spectral class of one of the companions and hence the absolute magnitude of its mass $\mathbf{M_1}$ is known. Then, for the ratio of the masses of the components $\mathbf{M_2}/\mathbf{M_1}$, one has

$$\frac{\mathbf{M_2}}{\mathbf{M_1}} = \frac{\mathbf{M_*}}{\mathbf{M_1}} - 1 \,. \tag{89}$$

Eqs. (88) and (89) have wide application in stellar astronomy.

Chapter IV

Motion of Celestial Bodies

1 Statement of the Problem

Our final aim is to obtain the position of the celestial body at a given point of the orbit. For that it is necessary to have the equation of motion of the celestial body, i.e. the dependence of its coordinates on time t. For determination of the position of the body in the orbit it is enough to know the dependence of **two** coordinates, r and θ (in the polar system of coordinates) on time t, i.e. to find the explicit form of the functions

$$r = f_1(t),$$
$$\theta = f_2(t). \tag{1}$$

The starting point for the solution of this problem is Kepler's first and second laws, namely (for an arbitrary orbit)

$$r^2 \frac{d\theta}{dt} = c_1, \tag{2}$$

$$r = \frac{p}{1 + e \, \cos(\theta - \omega)}. \tag{3}$$

Here r is the radius-vector of the body, i.e. its linear distance from the focus of the conical section where the central body is situated, θ is the angle between the radius-vector and some conditional direction named the **line of nodes** (see below) and ω is the **longitude of perigee**, i.e. the angular separation between the major axis of the conical section and the line of nodes.

For our purposes it is convenient to determine the position of the celestial body in the orbit not via θ but by angle v between the radius-vector and the major axis of the orbit, i.e.

$$v = \theta - \omega. \tag{4}$$

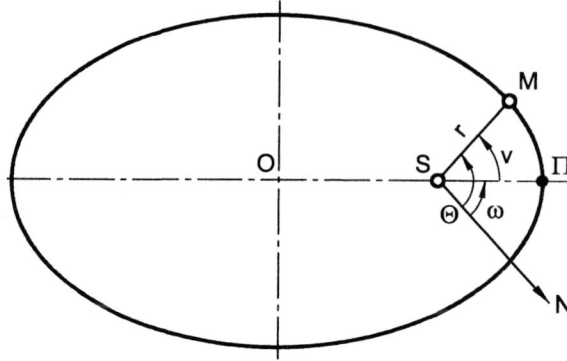

Fig. 4.1. *The derivation of the equations of motion in an elliptic orbit.*

The angle v is called the **true anomaly** of the celestial body. In the case of an elliptic orbit, for example, the angles θ, v and ω are shown in Fig. 4.1. Taking into account Eq. (4), Eqs. (2) and (3) may be rewritten in the following form:

$$r^2 \frac{dv}{dt} = c_1 , \qquad (5)$$

$$r = \frac{p}{1 + e \cos v} . \qquad (6)$$

Thus, in order to determine the position of a body in the orbit it is enough to know, instead of (1), the form of the functions $r = f_1(t)$ and $v = f_3(t)$. These two functions, $f_2(t)$ and $f_3(t)$, differ from each other by a constant $-\omega$.

Eqs. (5) and (6) form a system for the sought parameters r and v. On substituting r from Eq. (6) into Eq. (5) we obtain

$$\int \frac{dv}{(1 + e \cos v)^2} = \frac{c_1}{p^2} t . \qquad (7)$$

If we can realize the integration of the left-hand side of this relationship, i.e. find the explicit form of the function $v = f_3(t)$, then we can also obtain the function $r = f_1(t)$ by substituting v into (6).

The entire problem, however, comes down to the impossibility of integrating it for typical cases, i.e. when $e \neq 0$ and $e \neq 1$. In order to perform the integration, it is necessary to realize a transformation of coordinates, i.e. to find a new variable of integration. The search, however, does not lead to a uniform transformation acceptable for all three types of orbit – elliptic, parabolic and hyperbolic. The integration could be realized only for each case separately. Precisely in this peculiarity there is based the method of solution of the problem of the determination of the motion of a celestial body separately for all three types of orbits. In the present chapter we start with the problem of the motion of a celestial body in an elliptic orbit.

2   Motion in an Elliptic Orbit

The initial equations in the case of an elliptic orbit are, instead of Eqs. (5) and (6), rather

$$r^2 \frac{dv}{dt} = k \left[a \left(1 - e^2 \right) M \right]^{1/2} , \tag{8}$$

$$r = \frac{a \left(1 - e^2 \right)}{1 + e \cos v} . \tag{9}$$

The new variable of integration – denote it by E – is found from the following construction (see Fig. 4.2). Taking the major axis of the ellipse as a diameter, we draw a circle of radius a. From point M – the position of the celestial body in the elliptic orbit, we draw a perpendicular up to the intersection with the major axis at the point G, and up to the intersection with the circle at the point K. Connect now the point K with the center of the ellipse O. The angle $KOS = E$ is precisely the sought variable, which is called the **eccentric anomaly**.

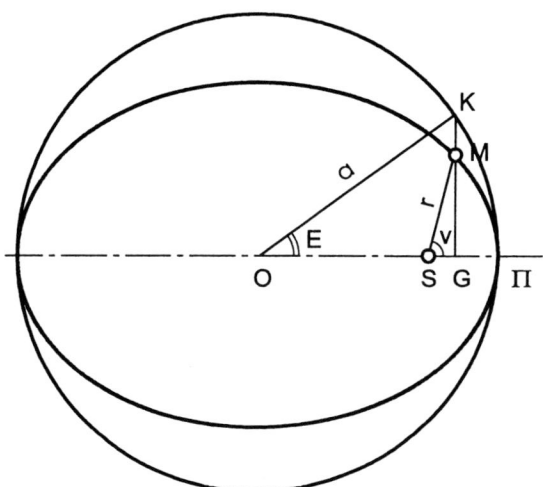

Fig. 4.2. *The geometrical construction connecting the eccentric anomaly E with the true anomaly v.*

The angle v will be replaced by E. It is necessary, therefore, to represent both the radius-vector r and true anomaly v via E. From triangulation in Fig. 4.2 we have

$$r \cos v = a \cos E - ae\,, \tag{10}$$

where $ae = OS$ is the semifocal distance. On substituting the magnitude of $\cos v$ into (9), we obtain

$$r = a\,(1 - e \cos E)\,. \tag{11}$$

This is the formula for the dependence of the radius-vector r on the eccentric anomaly E. It is not difficult to be assured that the relationship (11) is satisfied for all positions of the body M in the orbit.

Turn now to the determination of the dependence of v on E. It may be realized easily by equating Eqs. (9) and (11):

$$\cos v = \frac{\cos E - e}{1 - e \cos E}\,. \tag{12}$$

Based on this relation, the expression for the functions $\cos(v/2)$ and $\sin(v/2)$ may be derived:

$$\cos \frac{v}{2} = \left(\frac{1 - e}{1 - e \cos E}\right)^{1/2} \cos \frac{E}{2}\,, \tag{13}$$

$$\sin \frac{v}{2} = \left(\frac{1 + e}{1 - e \cos E}\right)^{1/2} \sin \frac{E}{2} \tag{14}$$

or

$$\tan \frac{v}{2} = \left(\frac{1 + e}{1 - e}\right)^{1/2} \tan \frac{E}{2}\,, \tag{15}$$

thus completing the representation of the true anomaly v via the eccentric anomaly E.

3   Kepler's Equation

The forms of the dependences both of r and of v on E are known. Now it is necessary to find the dependence of the eccentric anomaly E on time t.

By differentiating Eq. (15), we obtain, after some transformations,

$$dv = \left(\frac{1 + e}{1 - e}\right)^{1/2} \frac{\cos^2 \dfrac{v}{2}}{\cos^2 \dfrac{E}{2}}\, dE \tag{16}$$

or, bearing in mind (13) and (11),

$$\frac{dv}{dt} = \frac{a}{r} (1 - e^2)^{1/2} \frac{dE}{dt}. \qquad (17)$$

On substituting in that equation the magnitude of dv/dt, as well as that of v from Eq. (11), we obtain

$$(1 - e \cos E)\, dE = n\, dt\,, \qquad (18)$$

where n is the mean daily motion of the celestial body (see Eq. (66) in Chapter III):

$$n = \frac{2\pi}{P} = \frac{k \sqrt{M}}{a^{3/2}}. \qquad (19)$$

After integration we find

$$E - e \sin E = nt + C\,, \qquad (20)$$

where C is the constant of integration, which we obtain from the condition $E = 0$ at the moment of passage of the celestial body through the perigee, i.e. at the moment $t = T$. Then we will have finally

$$E - e \sin E = n\,(t - T)\,. \qquad (21)$$

This relationship is known as **Kepler's equation** though Kepler had no part in its derivation. This equation, however, is transcendent to the relation of E.

As a rule, the parameter n is measured in degree/day units, while $t - T$ is the number of days counting from the moment of passage through the perigee, and $n(t - T)$ is the angle in the case of uniform motion of the planet in a circle with mean velocity n. Usually this angle is denoted by M and is called the **average anomaly**:

$$M = n\,(t - T)\,, \qquad (22)$$

where T is the moment of passage of the celestial body through the perigee. Then Eq. (21) may be written in the form

$$E - e \sin E = M\,. \qquad (23)$$

For given E, the magnitude of t may be found from that quite easily. However the main purpose of Kepler's equation is for just the opposite problem, namely, to find E for given t, which in the epoch of computers does not represent a serious problem. However, previously various approximation sequences were used for the solution of this problem – graphical, the method of differential corrections, Lagrange series, etc. Though all these methods are now mostly of historical interest, for methodical reasons it is interesting to briefly consider one of them – the method of **differential corrections**.

Rewrite Kepler's equation (23) in the form

$$E = M + e \sin E.$$

On the right-hand side we have the sought magnitude E. In view of the smallness of e, we can to a first approximation assume that $E = M$, i.e.

$$E_1 = M + e \sin M.$$

With the help of this relationship and for a given value of M we obtain the first approximation E_1, which, indeed, will differ from its true value by some amount $\pm\Delta E$; our aim is to find this "differential" correction. Putting the value of E_1 just obtained into Kepler's equation (23) we get for M_1 some value, which, of course, will differ from M by an amount $\Delta M = M - M_1$, where

$$M_1 = E_1 - e \sin E_1.$$

On differentiating Kepler's equation, we find

$$\Delta E = \frac{\Delta M}{1 - e \cos E}.$$

Then we shall have for ΔE_1

$$\Delta E_1 = \frac{M - M_1}{1 - e \cos E_1}$$

and in the second approximation

$$E_2 = E_1 + \Delta E_1.$$

Of course this process should be repeated up to the condition $M_i = M$, i.e $\Delta M_i = 0$. Note that the method of differential correction is nothing other than the method of series approximation.

Having the magnitude of E for the given moment of time t, we can find the radius-vector r with the help of Eq. (11), and then the true anomaly v – by means of Eq. (15), thus completing the solution of our problem, namely, to find the position of a celestial body in an elliptic orbit at a given time t. As for Eqs. (11) and (15), they are exactly the functions that we sought: $r = f_1(t)$ and $\theta = f_2(t)$, represented, however, via an auxiliary function $E = E(t)$.

Let us enumerate the formulas necessary for the calculation of polar coordinates $r(t)$ and $\theta(t)$ at a given moment of time t for a celestial body moving through an elliptic orbit. The calculations with the help of these formulas may be realized only in the case in which the elements of the elliptic orbit as a geometric curve on a plane are known. Those elements are four in number: the semimajor axis a, the eccentricity e, the moment of passage of the body through the perigee T (or the rotation period P) and the longitude of perigee ω. The mass of the planet (or satellite) m is in units of Solar mass, hence, the magnitude $M = 1 + m$ under the square root in Eq. (19) is also considered known.

First of all we obtain n from

$$n = \frac{k\sqrt{M}}{a^{3/2}}, \tag{24}$$

or

$$n = \frac{2\pi}{P}. \tag{25}$$

Then we determine M for the given moment of time t:

$$M = n(t - T). \tag{26}$$

The next step is connected with the solution of Kepler's equation, i.e. with the determination of the eccentric anomaly E by computers. In the case, for example, of a Lagrange series one has

$$E = M + e \sin M + \frac{e^2}{2} \sin 2M. \tag{27}$$

Having E, we find then the true anomaly v from the formula

$$\tan \frac{v}{2} = \left(\frac{1+e}{1-e}\right)^{1/2} \tan \frac{E}{2}, \tag{28}$$

after which we find one of the two polar coordinates, $\theta(t)$, from the relationship

$$\theta(t) = v + \omega. \tag{29}$$

This step is finished by determining the second coordinate – radius-vector $r(t)$, by means of the formula

$$r(t) = a(1 - e\cos E). \tag{30}$$

Note that these polar coordinates r and θ are related to the system of coordinates with its center coinciding with the "central" body, in this particular case the Sun.

In the determination of E by the Lagrange method, we gave an expression up to second order in e. If higher accuracy is required for determination of

E, the higher orders of expansion should be added too. In particular, the formula for the determination of E to fourth order has the form

$$E = M + e \sin M + \frac{e^2}{2} \sin(2M) + \frac{e^3}{3! \, 2^2} [3^2 \sin(3M) - 3 \sin(2M)]$$

$$+ \frac{e^4}{4! \, 2^3} [4^3 \sin(4M) - 4 \times 2^3 \sin(2M)].$$

$$(31)$$

This series converges if $e \leq 0.6627434...$. This "puzzling" number led Cauchy to the creation of a new direction in mathematics – complex calculus.

Kepler's equation had an essential role not only in celestial mechanics but in mathematics in general. Such fundamental concepts and powerful methods of mathematical calculus such as the Bessel functions, the Fourier series, the topological number of vector fields, and the "argument principle" in the theory of complex calculus appeared during the study of Kepler's equation (Arnold, 1989).

There are a number of special expressions representing, for example, the true anomaly through the eccentric anomaly, average anomaly through the true anomaly, expressions for the solution of Kepler's equation etc., in the form of Bessel functions, Fourier series etc. (for more details see Brauwer, Clemence, 1961).

4   The Position of a Celestial Body in Three-Dimensional Space

The coordinates of a celestial body $r(t)$ and $\theta(t)$ obtained above concern its position in the orbit. The latter, being a two-dimensional plane curve, lies within the plane passing through the central body (the Sun). Meanwhile, the orientation of the plane – the orbit's plane – in three-dimensional space has not yet been determined. Various planets and satellites move in orbits lying on different planes that are diversely oriented in space. Obviously it is necessary to have some unified method of description in order to determine uniquely the spatial orientation of the orbital plane.

For the determination of a plane in space it suffices to introduce two parameters, both angular in essence. Correspondingly it is necessary to create some system of notation. We can create such a system by combination of some chosen plane with one conditionally defined direction in three-dimension space. In the first case, it is convenient to take the ecliptic – the plane of the Earth's motion containing evidently also the Sun (there is another definition: the ecliptic is the plane of annual motion of the Sun with respect to the observer on the Earth). In the second case, the direction of the vernal equinox of the Earth was taken as a conditional direction.

Then we act in the following manner. We choose a Cartesian coordinate system XYZ in such a way that the plane XOY coincides with the ecliptic as shown in Fig. 4.3, and the axis OX is directed towards the point of the vernal equinox. In the center of this system is the Sun.

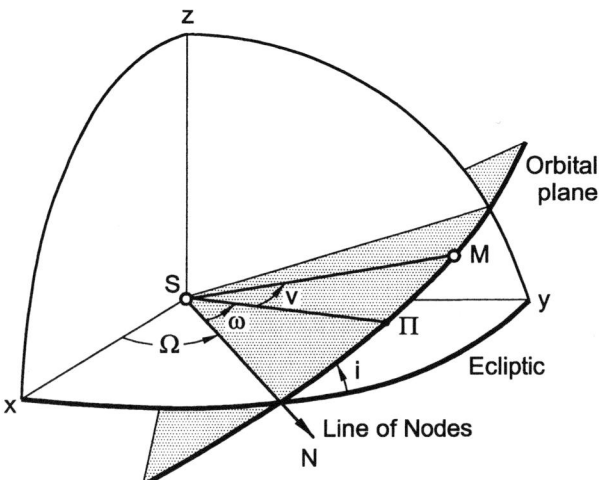

Fig. 4.3. *The determination of the position of a planet M in space.*

Of course, an analogous system may be created for any configuration of celestial bodies, say, for Jupiter's or Saturn's natural satellites: in this case the Sun (the origin of the coordinate system) will be replaced by Jupiter or Saturn, the ecliptic by the plane of orbital motion of any satellite, and the axis OX will be directed in the direction of any star in the sky or even towards the same point, that of the vernal equinox of the Earth.

In general, the orbital plane of a celestial body under examination forms some angle i with the ecliptic, leading to the appearance of the fifth element i – angle of inclination between the orbital plane and the ecliptic. Numerically i may be within limits $+90°$ on both sides of the ecliptic – north $(+)$ and south $(-)$.

In the majority of cases the inclination i differs from zero and, hence, the intersection of the orbital plane with the ecliptic is inevitable; the line of this intersection ON is called the **line of nodes**. The position (direction) of the line of nodes on the ecliptic is determined by the angular distance from the point of vernal equinox; this angular distance is denoted by Ω, which is at the same time the sixth element of the elliptical orbit. Numerically Ω may be within the range $360° > \Omega > 0°$.

The other angular parameters relating to the orbit itself are also exhibited in Fig. 4.3. In particular, the direction of major axis SP, the angular distance ω of perigee P from the line of nodes SP from the line of nodes SN, is shown (SN in Fig. 4.1), as are the position of the planet M and its angular distance

v (true anomaly) from the semimajor axis (perigee). Thus the position of any planet is determined in polar coordinates by the five parameters $\Omega, i, \omega, v(t)$ and $r(t)$.

5   Transition to Cartesian Coordinates

Up to now we have been dealing with the question of the determination of the position of a planet in polar coordinates with respect to the Sun. Meanwhile, the observer is on the Earth and, hence, it is necessary to perform a transition from the Sun to the Earth, i.e. a transition from one system of coordinates to another. For this purpose it is necessary to move from r and θ to the Cartesian coordinates XYZ with the Sun at the center.

Draw a sphere of unit radius around the origin of coordinates O, where the Sun is located. We also take the coordinate axes X, Y, Z in the same manner as shown in Fig. 4.3, by matching the plane XOY with the ecliptic. We also denote the position of the planet M and the orbital plane and the corresponding angles; as a result we have Fig. 4.3. Now the orientation in space of the planet M is determined by coordinates x, y, z, which we shall call **rectangular heliocentric ecliptic coordinates** or **Descartes heliocentric ecliptic coordinates of the celestial body.**

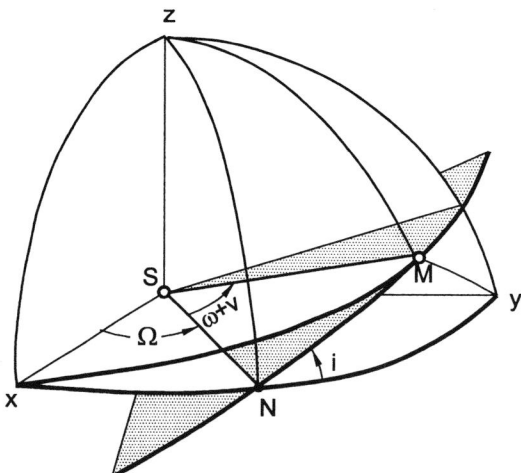

Fig. 4.4. *The angular elements determining the position of a celestial body in space.*

Coordinates x, y, z can be obtained from the following relationships:

$$
\begin{aligned}
x &= r\,\cos(MOX)\,, \\
y &= r\,\cos(MOY)\,, \\
z &= r\,\cos(MOZ)\,,
\end{aligned}
\qquad (32)
$$

where $r = r(t)$ is the radius-vector of the point M and is known. If we are able to determine the direction cosines in these formulas then we will also know the coordinates x, y and z. We can solve this problem using the formulae of spherical trigonometry as well as the scheme shown in Fig. 4.4.

Let us start from $\cos(MOX)$. Take a plane passing through OX and OM; its intersection with the sphere represents an arc of great circle XM. As a result a spherical triangle MXN is formed in which $MX = \angle MOX$, $XN = \Omega$, $NM = \omega + v$, and $XNM = 180 - i$. Applying the formula of cosines from spherical trigonometry for this triangle, we obtain

$$\cos(MOX) = \cos(v + \omega) \, \cos \Omega - \sin(v + \omega) \, \sin \Omega \, \cos i. \qquad (33)$$

The intersection of the plane passing through the lines OM and OY with the sphere will result in an arc; a new spherical triangle will be formed with $MY = \angle MOY$, $NM = \omega + v$, $NY = 90 - \Omega$, and $\angle MNY = i$. Then we obtain

$$\cos(MOY) = \cos(v + \omega) \, \sin \Omega + \sin (v + \omega) \, \cos \Omega \, \cos i. \qquad (34)$$

At least, for the determination of $\cos(MOZ)$ we have the spherical triangle NZM in which $MZ = \angle MOZ$, $NM = \omega + v$, $NZ = 90°$, and $\angle(MNZ) = 90 - i$:

$$\cos(MOZ) = \sin(v + \omega) \, \sin i. \qquad (35)$$

Substituting the obtained expressions for direction cosines from Eqs. (33), (34) and (35) into Eq. (33), we obtain finally

$$x = r \, [\, \cos(v + \omega) \, \cos \Omega - \sin(v + \omega) \, \sin \Omega \, \cos i \,],$$
$$y = r \, [\, \cos(v + \omega) \, \sin \Omega + \sin(v + \omega) \, \cos \Omega \, \cos i \,], \qquad (36)$$
$$z = r \, \sin(v + \omega) \, \sin i.$$

These formulas are universal in the sense that they are suitable for all types of orbit – elliptic, parabolic and hyperbolic. Simply, in every case it is necessary to substitute into Eq. (36) the magnitudes of r and v for the corresponding orbit. In the case of an elliptic orbit r and v are given by Eqs. (11) and (15).

We can compare now all the constants determining the motion of a celestial body through its orbit. The six constants that we saw above are called **elements of the elliptic orbit**. These elements are the following.

i – the inclination of the orbital plane to the plane of the ecliptic. Numerically i is within $\pm 90°$.

Ω – the longitude of the line of nodes or the longitude of the ascending node of the orbital plane. It is counted in the direction from west to east and lies within the limits $0°$–$360°$.

ω – the longitude of the perigee from the line of nodes. It is also counted in the direction from west to east, and lies within the limits $0°$–$360°$.

a – the semimajor axis of the elliptic orbit.

e – the eccentricity of the elliptic orbit: $e < 1$.

T – the time of transit of the celestial body through its perigee.

The elements i and Ω determine the position of the orbital 2D plane in 3D space. The element ω determines the position of the orbit itself in its plane. The element a determines the orbit's linear size; it may be replaced by the mean daily motion n dependent on a by the relation

$$n = \frac{k \sqrt{M}}{a^{3/2}} .$$

The element a may be also replaced by the orbital period P of an elliptic orbit according to the relationship

$$P = \frac{2\pi \, a^{3/2}}{k \sqrt{M}} .$$

Finally, the element T determines the time of transition of a celestial body through the perigee and, hence, determines the position of the body in the orbit in time.

In the framework of the problem of two isolated bodies, all six elements have constant magnitudes in all stages of the motion. Owing to this, the elements enumerated serve as constants in all calculations connected with the determination of the position of a celestial body in its orbit. This peculiarity account for the simplicity and at the same time the vast advantage of the two-body problem. In the presence of outer perturbations, i.e. in the case of the two bodies not being isolated, these six elements will fail to be constants – they become dependent on time. We shall see in Chapter X in now one should act in that case.

The mathematical treatment of the two–body problem results in the discovery of six integrals, i.e. six constants of integration: this is the conclusion at which we arrived in the first sections of the present chapter. So, all six elements of an elliptic orbit mentioned above are those constants of integration.

6 Motion of a Celestial Body in a Parabolic Orbit

As an initial position for the solution of the problem of motion of a celestial body in a parabolic orbit we have, as in the case of an elliptic orbit, the integral of the areas – Kepler's first law, as well as the equation of the conical section in polar coordinates, i.e.

$$r^2 \frac{dv}{dt} = k \, (p M)^{1/2}, \tag{37}$$

$$r = \frac{p}{1 + e \, \cos v}, \tag{38}$$

where $v = \theta - \omega$.

Certain comets and meteorites are moving in parabolic orbits. Their masses are very small compared with the Solar mass, and therefore we can put $M = 1$ in Eq. (37).

A parabola is characterized by two peculiarities: $e = 1$ and $p = 2q$, where q is the distance of the perigee from the Sun. Then we shall have from Eqs. (37) and (38)

$$r^2 \frac{dv}{dt} = k \, (2q)^{1/2}, \tag{39}$$

$$r = \frac{q}{\cos^2 (v/2)}. \tag{40}$$

Note that in the case of a parabola the angle v between the radius-vector of the celestial body and the parabola axis is also called the **true anomaly**. The polar coordinates $r(t)$ and $v = v(t)$ completely determine the position of the body in the orbit. Our problem consists in the discovery of the functions $r(t)$ and $v(t)$.

The problem of motion in a parabolic orbit is solved incomparably more easily and in a more elementary manner owing to the fact that, for the parabola, $e = 1$. In particular, in this case it is not necessary to introduce an auxiliary variable such as E without which we cannot solve the problem in the case of an elliptic orbit.

On substituting the expression for r from Eq. (40) into Eq. (39), we obtain

$$\frac{k \sqrt{2}}{q^{3/2}} \, dt = \frac{dv}{\cos^4 (v/2)}. \tag{41}$$

Because

$$\frac{1}{\cos^4 \dfrac{v}{2}} = \frac{1}{\cos^4 \dfrac{v}{2}} + \frac{\tan^2 \dfrac{v}{2}}{\cos^2 \dfrac{v}{2}},$$

we can write, instead of Eq. (41)

$$\frac{k\sqrt{2}}{q^{3/2}}\, dt = \left(\frac{1}{\cos^2 \dfrac{v}{2}} + \frac{\tan^2 \dfrac{v}{2}}{\cos^2 \dfrac{v}{2}} \right) dv ,\tag{42}$$

and after integration we obtain finally

$$\frac{k}{q^{3/2}\sqrt{2}}\,(t - T) = \tan \frac{v}{2} + \frac{1}{3}\,\tan^3 \frac{v}{2} .\tag{43}$$

Here T is the constant of integration and has the same physical content as in the case of the elliptic orbit, i.e. T is the time of passing of the body through the perigee as well as at $v = 0$ which is, in the case of the perigee, when $t = T$ according to Eq. (43).

In fact Eq. (43) – known as Barker's equation – replaces Kepler's equation (21) in the case of the elliptic orbit with the only difference being that, in the case of a parabola, the solution $v = v(t)$ is obtained directly without intermediate manipulations. In one aspect, however, Eq. (43) does not differ from Kepler's equation: it is again of transcendent type with respect to v and, hence, in this case one cannot avoid, at least in principle, the sequence approximation method. For representation of the solution of Eq. (43), special tables were created for the function $M(t)$:

$$M(t) = \frac{k\,(t - T)}{q^{3/2}\sqrt{2}} .\tag{44}$$

However, now, in the epoch of computers, there is no longer any necessity for the approximation methods and those tables for the solution of Eq. (43).

Having the value of v for a given moment of time t, we then obtain $r = r(t)$ with the help of (40). So, the solution of the problem concerning the determination of the position of a celestial body in a parabolic orbit is completed.

The obtained polar coordinates $r(t)$ and $v(t)$, however, determine the position of the comet in the orbital plane. As for the problem of orientation of the orbital plane in space, it is solved exactly in the same manner as in the case of the elliptic orbit, namely, by introduction of two additional parameters, i and Ω, where i is the inclination of the orbit's plane relative to the ecliptic, as before, and Ω is the longitude of the line of nodes, i.e. the angular distance of the line of nodes (the line of intersection of the orbit's plane with the ecliptic) from the ascending node.

The transition from the polar system of coordinates r and v to the Cartesian heliocentric system x, y, z in the case of a parabolic orbit is realized in exactly in the same manner as in the case of an elliptic orbit, i.e., using the same Eq. (36) without any change. In contrast to the elliptic orbit, in the case of a parabolic orbit the functions $r(t)$ and $v(t)$ used in Eq. (36) are determined by means of Eqs. (40) and (43).

The elements that determine the motion of a celestial body through a parabolic orbit are as follows.

i – the inclination of the orbital plane to that of the ecliptic. Numerically i is within the limits $\pm 90°$.

Ω – the longitude of the line of nodes or the longitude of the ascending node of the orbital plane. Numerically it is within $0°$–$360°$.

ω – the longitude of the perigee from the line of nodes. It is within the limits $0°$–$360°$.

q – the linear distance of the perigee from the focus of the parabola or the nearest distance from the Sun.

T – the time of transit of the celestial body through perigee.

Thus, in contrast to the elliptic orbit, the motion of a celestial body through a parabolic orbit and its position in space are determined not by six but by five elements – one element less.

7   Euler's Equation

Consider the position of a comet in a parabolic orbit at the times t_1 and t_2. The corresponding radius-vectors are r_1 and r_2 and the chord, connecting their ends, we denote by s (Fig. 4.5). Let the corresponding true anomalies be v_1 and v_2.

We introduce some more notation:

$$z = \tan \frac{v}{2}. \tag{45}$$

Then Eq. (43) may be written in the form

$$\frac{k\,(t - T)}{q^{3/2}\,\sqrt{2}} = z + \frac{1}{3}\,z^3. \tag{46}$$

We write this equation for t_1 and t_2:

$$\frac{k\,(t_1 - T)}{q^{3/2}\,\sqrt{2}} = z_1 + \frac{1}{3}\,z_1^3, \tag{47}$$

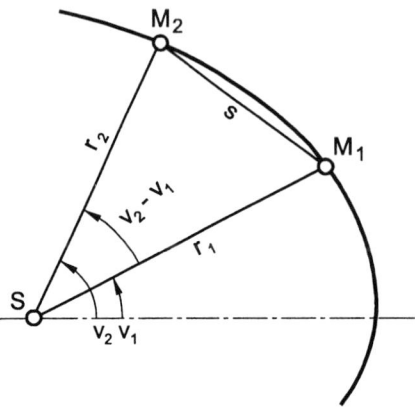

Fig. 4.5. *The derivation of Euler's equation.*

$$\frac{k\,(t_2 - T)}{q^{3/2}\,\sqrt{2}} = z_2 + \frac{1}{3}\,z_2^3\,. \tag{48}$$

Subtracting Eq. (47) from Eq. (48), we obtain

$$\frac{k\,(t_2 - t_1)}{q^{3/2}\,\sqrt{2}} = z_2 - z_1 + \frac{1}{3}\,(z_2^3 - z_1^3)\,,$$

or

$$\frac{3k\,(t_2 - t_1)}{q^{3/2}\,\sqrt{2}} = (z_2 - z_1)(3 + z_2^2 + z_2 z_1 + z_1^2)\,.$$

On the right-hand side, in the second set of parentheses, we add $2z_2 z_1$ and deduct $2z_2 z_1$. Then we have

$$\frac{3k\,(t_2 - t_1)}{q^{3/2}\,\sqrt{2}} = (z_2 - z_1)\,[\,3\,(1 + z_2 z_1) + (z_2 - z_1)^2\,]\,. \tag{49}$$

As we see, on the right-hand side of this equation there are two combinations of z_2 and z_1, namely, $(1 + z_2 z_1)$ and $(z_2 - z_1)$. The problem is to find these two functions in terms of their dependences on r_1, r_2 and s.

From a triangle in Fig. 4.5 we have

$$s^2 = r_1^2 + r_2^2 - 2r_1 r_2\,\cos(v_2 - v_1)\,.$$

On making the substitution

$$\cos(v_2 - v_1) = 2\cos^2\frac{v_2 - v_1}{2} - 1\,,$$

we obtain

$$s^2 = (r_1 + r_2)^2 - 4r_1 r_2 \cos^2 \frac{v_2 - v_1}{2},$$

which results in

$$2 (r_1 r_2)^{1/2} \cos \frac{v_2 - v_1}{2} = \pm(AB)^{1/2}, \tag{50}$$

where the notations

$$\begin{aligned} A &= r_1 + r_2 + s, \\ B &= r_1 + r_2 - s \end{aligned} \tag{51}$$

are used.

The upper sign before the square root in Eq. (50) should be taken when $v_2 - v_1 < 180°$, and the lower sign when $v_2 - v_1 > 180°$. Usually the moments of observations t_1 and t_2 in the case of a comet do not differ from each other by very much, therefore $v_2 - v_1$ must be smaller than $180°$. Hence, in Eq. (50) in practice we should have the upper sign.

Substituting into Eq. (50) the values of r_1 and r_2 from Eq. (40), i.e.

$$\begin{aligned} r_1 &= \frac{q}{\cos^2 \dfrac{v_1}{2}}, \\ r_2 &= \frac{q}{\cos^2 \dfrac{v_2}{2}}, \end{aligned} \tag{52}$$

we obtain

$$2q \left(1 + \tan \frac{v_1}{2} \tan \frac{v_2}{2}\right) = \pm(AB)^{1/2}$$

or, considering (45), we have

$$1 + z_1 z_2 = \pm \frac{(AB)^{1/2}}{2q}, \tag{53}$$

and the determination of one of the combinations of z_1 and z_2 is completed. In order to find the second combination, $z_2 - z_1$, we take the sum of Eq. (52):

$$r_1 + r_2 = q \left(\frac{1}{\cos^2 \dfrac{v_1}{2}} + \frac{1}{\cos^2 \dfrac{v_2}{2}} \right).$$

However, because

$$\frac{1}{\cos^2 \dfrac{v}{2}} = 1 + \tan^2 \frac{v}{2} = 1 + z^2,$$

then
$$r_1 + r_2 = q\left(2 + z_1^2 + z_2^2\right).$$

On the other hand, summing both equations in (51), we obtain
$$r_1 + r_2 = \frac{A + B}{2}.$$

Therefore
$$q?(2 + z_1^2 + z_2^2) = \frac{A + B}{2}.$$

By adding and deducting $2\,z_1\,z_2$ in the parentheses from the left, we obtain
$$2\left(1 + z_1\,z_2\right) + \left(z_2 - z_1\right)^2 = \frac{A + B}{2q},$$

or, in view of (53), we obtain finally for the second combination
$$z_2 - z_1 = \frac{\sqrt{A} \pm \sqrt{B}}{(2q)^{1/2}}. \tag{54}$$

Now it remains to substitute both combinations, (53) and (54), into our initial equation (49). Then we get
$$\frac{3k(t_2 - t_1)}{q^{3/2}} = \frac{\sqrt{A} \pm \sqrt{B}}{(2q)^{1/2}}\left[\mp 3\,(AB)^{1/2} + (A + B) \pm 2\,(AB)^{1/2}\right]\frac{1}{2q}$$

or, after some manipulations,
$$6k(t_2 - t_1) = \left(\sqrt{A} \pm \sqrt{B}\right)\left[A + B \mp (AB)^{1/2}\right],$$

and by multiplication
$$6k(t_2 - t_1) = A^{3/2} \pm B^{3/2}.$$

Substituting here the expressions for A and B from (51), we obtain finally
$$6k(t_2 - t_1) = (r_1 + r_2 + s)^{3/2} - (r_1 + r_2 - s)^{3/2}. \tag{55}$$

This equation was first discovered by Euler in 1743, and therefore carries his name. It is also known under the name of **Euler's theorem**. It is remarkable first of all because of the absence of one of the elements of a parabolic orbit, namely, q, and at the same time represents a symmetrical interdependence between the radius-vectors r_1 and r_2 for the moments of time t_1 and t_2 and the chord s. Euler's equation will be used for the determination of elements of the parabolic orbit from geocentric observations (Chapter VII).

Later, Lambert derived an analog of Eq. (55) – a relationship for an elliptic orbit. By expanding the first part of the equation derived by him, a coincidence of the first member of this expansion with the right-hand side of Euler's equation occurs. Therefore sometimes Eq. (55) is also called **the Euler–Lambert equation.**

8   Motion of a Celestial Body in a Hyperbolic Orbit

From the mathematical point of view, the treatment of the problem of determination of the position of a celestial body in a hyperbolic orbit is analogous with that of the elliptic orbit. In particular, in this case the integral containing a trigonometric function with an argument v (the true anomaly) is also transcendent and cannot be calculated. As a consequence of this fact it is necessary to introduce a new variable of integration, F, and some analogy with the eccentric anomaly E in the case of an elliptic orbit arises. However, the form of this variable, F, should be completely different.

We have for a hyperbola $e > 1$ and $p = a(e^2 - 1)$. Only comets and some meteorite clouds move in hyperbolic orbits, therefore again we can take $M = 1$ for the total mass of the system. Then the initial equations – the integral of areas and the equation of the conical section in polar coordinates, will be written in the form

$$r^2 \frac{dv}{dt} = k \left[a \left(e^2 - 1 \right) \right]^{1/2}, \tag{56}$$

$$r = \frac{a \left(e^2 - 1 \right)}{1 + e \, \cos v}, \tag{57}$$

where v is the true anomaly, as before.

In the case of a hyperbolic orbit, the motion takes place only through one branch of its half hyperbola, just in the branches occupied by the Sun at the focus. In Fig. 4.6, this corresponds to the left branch, where S is the Sun.

The auxiliary variable F mentioned above can be obtained from the following geometric construction. We draw a circle centered at O with a radius equal to the semimajor axis of hyperbola a. We continue the perpendicular from the position of the comet M up to the intersection with the semimajor axis at the point G. Then we take the tangent GK to the circle and connect the point of contact K with the center O. The angle KOG is the sought variable F.

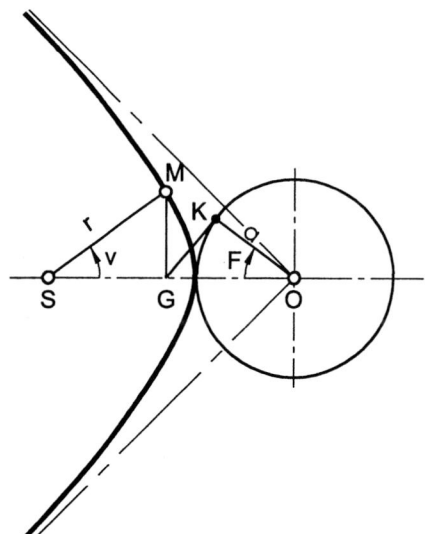

Fig. 4.6. *A geometrical construction explaining the transition from the true anomaly v to the angle F for motion in a hyperbolic orbit.*

Now, we shall derive the formulas for the radius-vector r and the true anomaly v dependent on the variable F.

We have, from Fig. 4.5, that $GS = SO - GO$ or

$$r \, \cos v = ae - \frac{a}{\cos F}, \tag{58}$$

where $SO = ae$ is half of the focal distance. On substituting the value of $\cos v$ from Eq. (58) into Eq. (57) we obtain

$$r = a \left(\frac{e}{\cos F} - 1 \right). \tag{59}$$

Now a formula should be derived connecting v with F. Equating Eq. (59) with Eq. (57) we have

$$\cos v = \frac{e \, \cos F - 1}{e - \cos F}. \tag{60}$$

Having the combinations $1 + \cos v$ and $1 - \cos v$, we arrive at

$$\cos \frac{v}{2} = \left(\frac{e - 1}{e - \cos F} \right)^{1/2} \cos \frac{F}{2}, \tag{61}$$

$$\sin \frac{v}{2} = \left(\frac{e + 1}{e - \cos F} \right)^{1/2} \sin \frac{F}{2}, \tag{62}$$

and hence

$$\tan \frac{v}{2} = \left(\frac{e+1}{e-1}\right)^{1/2} \tan \frac{F}{2}. \tag{63}$$

This formula is in complete analogy with Eq. (15) for an ellipse with the only difference being that here we have $e \pm 1$ instead of $1 \pm e$ for the ellipse.

So, the value of F for some t is known, and, hence, we can find without any difficulty r and v with the help of Eqs. (59) and (63), i.e. the position of the comet on the hyperbolic orbit.

We must now deduce an analogy of Kepler's equation for the hyperbolic orbit, i.e. the equation connecting the variable F with time t. For a short cut and in order to avoid confusion, we call this equation the **hyperbola equation**.

Differentiating Eq. (63) we obtain

$$dv = \left(\frac{e+1}{e-1}\right)^{1/2} \frac{\cos^2 \frac{v}{2}}{\cos^2 \frac{F}{2}} \, dF. \tag{64}$$

On the other hand, we have from Eq. (61):

$$\frac{\cos^2 \frac{v}{2}}{\cos^2 \frac{F}{2}} = \frac{e-1}{e - \cos F}. \tag{65}$$

Therefore

$$\frac{dv}{dt} = \frac{(e^2 - 1)^{1/2}}{e - \cos F} \frac{dF}{dt}. \tag{66}$$

On substituting this expression for dv/dt, as well as the value of r from Eq. (59) into Eq. (56), we obtain

$$\left(\frac{e}{\cos^2 F} - \frac{1}{\cos F}\right) dF = \frac{k}{a^{3/2}} \, dt. \tag{67}$$

The first term in parentheses on the left-hand side can be integrated easily. For the second term, a further substitution should be made:

$$\cos F = \sin \left(\frac{\pi}{2} + F\right) = 2 \sin \left(\frac{\pi}{4} + \frac{F}{2}\right) \cos \left(\frac{\pi}{4} + \frac{F}{2}\right)$$

$$= 2 \tan \left(\frac{\pi}{4} + \frac{F}{2}\right) \cos^2 \left(\frac{\pi}{4} + \frac{F}{2}\right).$$

Substituting this into Eq. (67) and integrating, we finally arrive at

$$e \tan F - \ln \left[\tan \left(\frac{\pi}{4} + \frac{F}{2} \right) \right] = M , \qquad (68)$$

where the following notation is used:

$$M = \frac{k}{a^{3/2}} \left(t - T \right) . \qquad (69)$$

The sequence of determination of the position of a celestial body in a hyperbolic orbit is as follows. First of all we determine the numerical value of M with the help of Eq. (69) for the given moment of time t. Then, with the help of the **hyperbola equation** (68) we obtain the value of the auxiliary variable F, after which we determine the polar coordinates $r(t)$ and $v(t)$ from Eqs. (59) and (63).

As we see, the situation with the hyperbola is analogous with that of the elliptic orbit: the hyperbola equation itself is also transcendent with respect to the sought magnitude F, hence, in this case also the calculations should be performed either by computer or by the method of sequential approximation.

Transition from polar coordinates r and v to the Cartesian heliocentric system x, y, z in the case of a hyperbolic orbit is performed in the same manner as in the case of an elliptic orbit, namely, by means of Eq. (36). The only difference is that we use Eqs. (59) and (63) for the determination of r and v, which are directly included in Eq. (36) in the case of a hyperbolic orbit.

The elements determining the motion of a celestial body through a hyperbolic orbit are as follows.

i – the inclination of the orbital plane to that of the ecliptic. Numerically i is within $\pm 90°$.

Ω – the longitude of the line of nodes or longitude of the ascending node of the orbital plane. It has the limits $0°–360°$.

ω – the longitude of the perigee from the line of nodes. It is within $0°–360°$.

a – the semimajor axis of the hyperbolic orbit.

e – the eccentricity of the hyperbolic orbit: $e > 1$.

T – the time of transit of the celestial body through its perigee.

Thus, the motion of a celestial body in a hyperbolic orbit is determined by six elements, and these elements are the same as those we had in the case of the elliptic orbit.

9 Rectilinear Orbit

Consider a situation when the semimajor axis remains constant, $a = const$, when the eccentricity tends to unity ($e \to 1$). In this case the ellipse will become more and more elongated, and the minimum distance of the perigee from the focus (Earth), equal to $r_{min} = a(1 - e)$, will tend to zero. In this limit the ellipse will be degenerated into a line, i.e. one will have a rectilinear ellipse or rectilinear orbit. It is not difficult to see that in this case, i.e. when $e \to 1$, both other orbits, parabolic and hyperbolic, will be transformed into rectilinear orbits as well.

Such orbits at first glance may seem abstractions. However, it is not so. There are comets with eccentricities of orbits so close to unity that their motion seems aligned. In most cases the orbits of those celestial bodies have transformed to rectilinear from initial parabola or hyperbola.

At $e \to 1$ the working formulae are as follows.

For a rectilinear ellipse, i.e. at $e \to 1$ we have from Eq. (12): $\cos v = -1$. Then we find from Eq. (10) for the radius-vector

$$r = a(1 - \cos e).\tag{70}$$

From (23) we shall have again at $e \to 1$

$$M = E - \sin E.\tag{71}$$

From the expression for velocity V, Eq. (71) of Chapter III, we find, having in view (70), for r

$$V = k(M/a)^{1/2} \cot(E/2).\tag{72}$$

Also in this case $e \to 1$ and $v = \pi$.

Both for rectilinear and normal parabolas the main conditions are the same: i.e. $e = 1$, so all formulas (40), (43) and (44) as well as (53) from Chapter III remain unchanged.

By the same manner, as in the case of elliptic orbits, we find for rectilinear hyperbolas, analogous with (70), (71) and (72)

$$r = a\left(\frac{1}{\cos F} - 1\right),\tag{73}$$

$$M = \tan F - \ln \tan\left(\frac{\pi}{4} + \frac{F}{2}\right),\tag{74}$$

$$V = k(M/a)^{1/2} \cot(F/2)\tag{75}$$

with $\cos v = -1$ and $v = \pi$.

As to the determination of the numerical value of eccentric anomaly E in the case of an elliptic orbit and function F for a hyperbolic one for given moment of the time t, the solution of this problem should be realized by the same manner as before for arbitrary values of eccentricity e. For detailed and advanced technique see the monograph by V.Brumberg and E.Brumberg (1999).

Chapter V

Ephemerides of Celestial Bodies

1 Statement of the Problem

In previous chapters the problem of the determination of the position of a celestial body in its orbit has been examined, and for all three types of orbits – elliptical, parabolic and hyperbolic – rigid solutions were obtained. In all cases the polar coordinates used enabled the determination of the position of the celestial body with respect to the central body, namely, the Sun. The transition from polar coordinates to Cartesian ones cannot change in essence the situation insofar as the center remains the same – the Sun. Therefore the discussion up to now on the motion of celestial bodies, strictly speaking, is true only for an imaginary observer located on the Sun.

Meanwhile, the observer is located on the Earth, where the system of measuring positions of celestial bodies is completely different. Hence, one has to perform the transition from the coordinate system associated with the Sun to that of the Earth. The study of the motion of a celestial body and the ability to determine its position for any moment of time pursue quite definite, often purely practical aims, so that one has to know beforehand the position of the celestial body for any definite moment of time. In other words, one should create tables of the positions of the celestial body indicating, say, the right ascension (α) and declination (δ) for any given moment. A number of reasons determine the need for having tables for equal time intervals: the convenience for interpolations, creation of navigation maps, the study of characters of motion, etc. The sequence of positions of a celestial body compiled for equidistant time intervals is called the *ephemerides*.

The ephemerides should be composed in the Earth's system of coordinates, or more definitely, in the geocentric equatorial system, though one can determine the position of a celestial body with respect to the heliocentric ecliptic system as well. Evidently the problem is the representation of the coordinates of the celestial body in a system convenient for the observer on the Earth, namely, in the form of right ascension α, declination δ and the distance ρ from the Earth. The present chapter is dedicated to this problem, so the main considerations and final results are equally true for all three types of orbits – elliptic, parabolic and hyperbolic.

2   Coordinate Systems

The position of any celestial body within the Solar system can be determined in two coordinate systems – **ecliptic** and **equatorial**. The basic plane in the ecliptic system is that of the ecliptic – the plane of the Earth's orbit. In the equatorial system the basic plane is the plane of the Earth's equator. The declination angle of the equatorial plane is constant and is $\varepsilon = 23°27'$. The zero-point in both systems is the same – the point of the vernal equinox, i.e. the point of intersection, from south to north, of the orbit's plane with the ecliptic: it is denoted by γ.

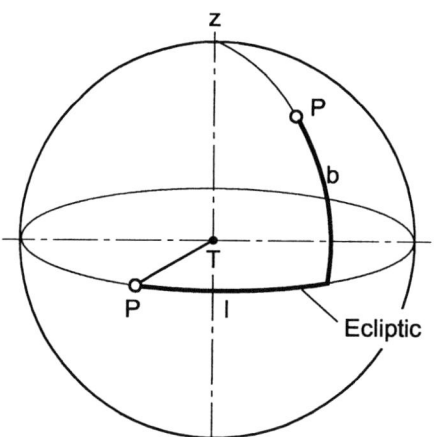

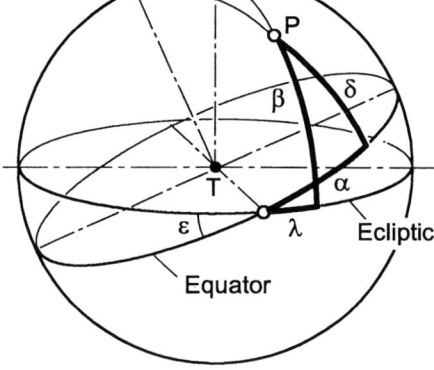

Fig. 5.1. *The Ecliptic heliocentric system of coordinates (l, b).*

Fig. 5.2. *Ecliptic (λ, β) and equatorial (α, δ) systems of coordinates.*

In the case in which the center of the coordinate system is the Sun, we talk about the heliocentric system; in the case in which the center is the Earth, we have the geocentric system. The polar coordinates in the ecliptic system are called **longitude** and **latitude**, in the equatorial system they are **right ascension** and **declination** with corresponding special notations. If the center of the coordinate system is the Sun's center, then Latin letters are used to denote polar and Cartesian coordinates, while in the case of the Earth's center, Greek letters are used. More definitely, the following notations (see also Fig. 5.1 and Fig. 5.2), are accepted to determine the location of the celestial body:

THE HELIOCENTRIC ECLIPTIC SYSTEM

Center of the coordinate system		*Sun*
Basic plane		*ecliptic*
Spherical coordinates:	longitude	l
	latitude	b
	distance	r
Cartesian coordinates		x, y, z

THE HELIOCENTRIC EQUATORIAL SYSTEM

Center of coordinate system		*Sun*
Basic plane		*equator*
Spherical coordinates:	right ascension	l_1
	latitude	b_1
	distance	r
Cartesian coordinates		x', y', z'

THE GEOCENTRIC EQUATORIAL SYSTEM

Center of coordinate system		*Earth*
Basic plane		*equator*
Spherical coordinates:	right ascension	α
	declination	δ
	distance	ρ
Cartesian coordinates		ξ, η, ζ

THE GEOCENTRIC ECLIPTIC SYSTEM

Center of coordinate system		*Earth*
Basic plane		*ecliptic*
Spherical coordinates:	right ascension	λ
	latitude	β
	distance	ρ
Cartesian coordinates		ξ', η', ζ'

3 The Heliocentric Ecliptic Coordinates of a Celestial Body

We describe the position of a celestial body in its orbital plane by means of radius-vector r and true anomaly v, with the Sun at the center of the coordinate system. If the coordinate plane $X_0 Y_0$ is combined with the orbital plane and the axis $O X_0$ is directed along the line of nodes, then the Cartesian coordinates of the position of the celestial body x_0, y_0, z_0 will be as follows:

$$x_0 = r \cos(v + \omega),$$
$$y_0 = r \sin(v + \omega), \tag{1}$$
$$z_0 = 0.$$

Now rotate now our system around the axis OX_0 with the angle i, which is the declination of the orbital plane with respect to the ecliptic one, in such a manner that, in the new system, the plane X_1OY_1 will be combined with that of the ecliptic one. The axis OX_1 is directed, as before, in the direction of the ascending node. Performing with the help of formulas of analytical geometry the transition from the system $X_0\,Y_0\,Z_0$ to that of $X_1\,Y_1\,Z_1$, we will have

$$x_1 = x_0,$$
$$y_1 = y_0 \cos i - z_0 \sin i, \tag{2}$$
$$z_1 = y_0 \sin i + z_0 \cos i.$$

On substituting the values of x_0, y_0 and z_0 from (1), we obtain

$$x_1 = r \cos(v + \omega),$$
$$y_1 = r \sin(v + \omega) \cos i, \tag{3}$$
$$z_1 = r \sin(v + \omega) \sin i.$$

Rotate the system $X_1\,Y_1\,Z_1$ around OZ_1 with the angle Ω in such a way that the new axis OX of the system XYZ will be directed towards the point of vernal equinox. Then the formulas of the transition from the coordinates x_1, y_1, z_1 to those of x, y, z will be

$$x = x_1 \cos \Omega - y_1 \sin \Omega,$$
$$y = x_1 \sin \Omega + y_1 \cos \Omega, \tag{4}$$
$$z = z_1.$$

On substituting the values from Eq. (4) instead of x_1, y_1, z_1 we will have

$$x = r\left[\cos(v + \omega) \cos \Omega - \sin(v + \omega) \sin \Omega \cos i\right],$$
$$y = r\left[\cos(v + \omega) \sin \Omega + \sin(v + \omega) \cos \Omega \cos i\right], \tag{5}$$
$$z = r \sin(v + \omega) \sin i.$$

These formulas determine the **Cartesian heliocentric ecliptic coordinates** of a celestial body. They coincide with those of deduced by methods of spherical trigonometry – Eq. (36) of Chapter IV.

Thus, we have the Cartesian coordinates of a celestial body with respect to the ecliptic. Now the transition from the Sun to the Earth on the one hand, and from the ecliptic to the equator on the other hand, can be realized. First, let us deal with the second problem.

4 Transition from Ecliptic to Equatorial

Observations from the Earth usually give the coordinates of a celestial body with respect to the plane of the equator. Therefore, before moving from the Sun to the Earth, one first has to perform a rotation around the coordinate axis OX by an angle $\varepsilon = 23°27'$ in such a way that the plane XOY, i.e. the plane of the ecliptic, coincides with the plane of the equator, with the Sun at the center as before. Denote the new system, again Cartesian and heliocentric, but equatorial, as x', y', z'. The axis OX' is directed towards the point of vernal equinox, and the axis OY' towards the point of the right ascension, $90°$. Then the formulas of transition from coordinates x, y, z to x', y', z' will be

$$x' = x \,,$$
$$y' = y \cos \varepsilon - z \sin \varepsilon \,, \qquad (6)$$
$$z' = y \sin \varepsilon + z \cos \varepsilon \,.$$

On substituting the expressions for x, y, z from Eq. (5), we obtain

$$x' = r \left[\cos \left(v + \omega \right) \cos \Omega - \sin \left(v + \omega \right) \sin \Omega \cos i \right],$$
$$y' = r \left[\cos \left(v + \omega \right) \sin \Omega \cos \varepsilon + \sin \left(v + \omega \right) \left(\cos \Omega \cos i \right.\right.$$
$$\left.\left. - \sin i \sin \varepsilon \right) \right], \qquad (7)$$
$$z' = r \left[\cos \left(v + \omega \right) \sin \Omega \sin \varepsilon + \sin \left(v + \omega \right) \left(\cos \Omega \right.\right.$$
$$\left.\left. \times \cos i \sin \varepsilon + \sin i \cos \varepsilon \right) \right].$$

These are the formulas for computation of the heliocentric **Cartesian equatorial coordinates** x', y', z' of a celestial body.

5 The Transition of the Center of Coordinate Frame from the Sun to the Earth

We have to perform one last operation – the parallel transition of the center of the coordinate system x', y', z' from the center of the Sun to the center of the Earth. In Fig. 5.3 the axis x' as well the axis ξ' are directed perpendicular to the picture plane, M denotes the celestial body, and r and ρ are its heliocentric and geocentric distances, correspondingly.

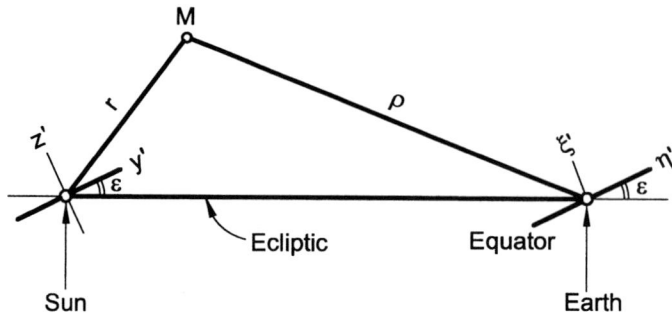

Fig. 5.3. *The transition from the heliocentric system to the geocentric system of coordinates.*

Denote by X', Y', Z' the geocentric equatorial coordinates of the Sun; these coordinates are available in the form of tables in all astronomical handbooks for certain epochs of time. For the parallel transition of the coordinate system from the Sun to the Earth, we can write for the Cartesian geocentric equatorial coordinates of a celestial body ξ, η, ζ

$$\xi = x' + X',$$
$$\eta = y' + Y',$$
$$\zeta = z' + Z'.$$
(8)

On passing from the Cartesian coordinates ξ, η, ζ to polar ones α, δ, ρ, we can write

$$\xi = \rho \cos \alpha \cos \delta,$$
$$\eta = \rho \sin \alpha \cos \delta,$$
$$\zeta = \rho \sin \delta,$$
(9)

where ρ is the distance of the celestial body from the center of the Earth, α is the right ascension, and δ is the declination.

Taking into account Eq. (9), we will have from Eq. (8)

$$\rho \cos \alpha \cos \delta = x' + X',$$
$$\rho \sin \alpha \cos \delta = y' + Y',$$
$$\rho \sin \delta = z' + Z'.$$
(10)

From the first two formulas we obtain

$$\tan \alpha = \frac{y' + Y'}{z' + Z'}.$$
(11)

With the help of this formula we find the first coordinate – the right ascension of the celestial body α. Having α, it is not difficult to obtain the second coordinate – the declination δ, by means of one of the following formulas:

$$\tan \delta = \frac{z' + Z'}{y' + Y'} \sin \alpha,$$

$$\tan \delta = \frac{z' + Z'}{x' + X'} \cos \alpha. \tag{12}$$

Now the determination of the geocentric distance ρ by means of one of the formulas in Eq. (10), for example,

$$\rho = \frac{z' + Z'}{\sin \delta}. \tag{13}$$

is not difficult. This completes the derivation of the formulas that are necessary for the computation of ephemerides of celestial bodies.

Chapter VI

Determination of the Orbit of a Celestial Body from Observations

1 General Outline

For determination of the position of a celestial body, it is necessary to know, as was shown in previous chapters, six elements in the case of elliptic and hyperbolic orbits and five elements for the parabolic one. These elements are simultaneously the constants of integration appearing during the solution of the differential equations in the two-body problem. Now we will find the numerical values of these constants, i.e. consider the new problem stated in the title of the present chapter, because the determination of the orbit of a celestial body implies the estimation of the values of these six elements from observations. This question obviously also includes the manner in which the corresponding observations are to be carried out.

For any celestial body these six elements are determined separately and can be used in principle for the computation of its position for any epoch. Hence, it is necessary to develop a method of determination of these elements directly from observations. The results of the observations can be presented, e.g. in the form of Cartesian heliocentric coordinates x, y, z related to the elements of the orbit, by the following relationships:

$$x = r[\cos(v + \omega)\cos\Omega - \sin(v + \omega)\sin\Omega\cos i],$$
$$y = r[\cos(v + \omega)\sin\Omega - \sin(v + \omega)\cos\Omega\cos i], \qquad (1)$$
$$z = r\sin(v + \omega)\sin i.$$

On the right-hand side all six unknown elements are present, three of which – Ω, i, and ω explicitly, the other three – a, e, and T indirectly, are represented through r and v. For the left-hand side of Eq. (1), the coordinates x, y, z, should be obtained from observations. Hence, the problem of determination of the elements of the orbit is reduced to the solution of an inverse problem, namely: from known coordinates of the celestial body find the elements of its orbit. Newton was the first to formulate the problem in this formulation and to solve it using a graphical method. Later Laplace, Lagrange and especially Gauss developed highly precise theoretical and practical methods of determination of orbital elements from observations.

Thus, if we carry out a single observation, for Eq. (1) we will have three equations with six unknown parameters. Actually, the number of unknown parameters is larger.

The point is that observations from the Earth give the values of only two coordinates of the body out of three, namely, α and δ, both in angles. They characterize only the direction of the line connecting the observer with the given object. The third quantity – the distance of the celestial body from the Earth, ρ, the geocentric distance, remains unknown; it enters into Eq. (1) at least through ρ while moving from the system related to the Sun to that of the Earth.

The transition of coordinates from the Sun to the Earth is performed by means of already known equations:

$$\begin{aligned} \xi' &= x + X \,, \\ \eta' &= y + Y \,, \\ \zeta' &= z + Z \,, \end{aligned} \tag{2}$$

where X, Y, Z are the coordinates of the Sun, i.e. known parameters, and ξ', η', ζ' are the geocentric ecliptic coordinates of the celestial body

$$\begin{aligned} \xi' &= \rho \cos \lambda \cos \beta \,, \\ \eta' &= \rho \sin \lambda \cos \beta \,, \\ \zeta' &= \rho \sin \beta \,, \end{aligned} \tag{3}$$

where λ and β are the longitude and latitude of the object, i.e. the angular coordinates with respect to the ecliptic. However, the observations give α and δ, i.e. the angular equatorial coordinates with respect to the observer on the Earth. Hence, it is necessary to move from the equatorial system of coordinates to the ecliptic one: such a transition may be carried out with the help of well-known formulas of spherical trigonometry as shown via the spherical triangle in Fig. 5.3 (Chapter V). As a result, we obtain

$$\cos \beta \cos \lambda = \cos \delta \cos \alpha \,, \tag{4}$$

$$\cos \beta \sin \lambda = \sin \varepsilon \sin \delta + \cos \varepsilon \cos \delta \sin \alpha \,, \tag{5}$$

$$\sin \beta = \cos \varepsilon \sin \delta - \sin \varepsilon \cos \delta \sin \alpha \,, \tag{6}$$

where ε is the angle between the equatorial and the ecliptic planes.

Eqs. (4)–(6) solve our problem uniquely, i.e. λ and β are determined via known α and β. From Eqs. (5) and (4) we have

$$\tan \lambda = \frac{\sin \varepsilon \sin \delta + \cos \varepsilon \cos \delta \sin \alpha}{\cos \delta \, \cos \alpha} \,. \tag{7}$$

From that we find the longitude of the object λ and its latitude β from Eq. (6). Substituting Eqs. (1) and (3) into Eq. (2), we obtain

$$\rho \cos \lambda \cos \beta = r[\cos(v + \omega) \cos \Omega - \sin(v + \omega) \sin \Omega \cos i] - X,$$
$$\rho \sin \lambda \cos \beta = r[\cos(v + \omega) \sin \Omega - \sin(v + \omega) \cos \Omega \cos i] - Y, \tag{8}$$
$$\rho \sin \beta = r \sin(v + \omega) \sin i \,.$$

One can represent the heliocentric distance r, as shown below, through ρ, so that in Eq. (8) one will have seven unknown parameters – six elements and ρ; the other ones, λ, β, X, Y, and Z, are known. Therefore in functional form Eq. (8) may be written as follows:

$$f_1(\Omega, \ i, \ \omega, \ a, \ e, \ T, \ \rho, \ t) = 0 \,,$$
$$f_2(\Omega, \ i, \ \omega, \ a, \ e, \ T, \ \rho, \ t) = 0 \,, \tag{9}$$
$$f_3(\Omega, \ i, \ \omega, \ a, \ e, \ T, \ \rho, \ t) = 0 \,,$$

where t is the time of observation (it enters through v and r).

Eqs. (8) and (9) represent the basic system for the determination of elements of the orbit. From a single observation one has three equations with seven unknown parameters. So the problem is not yet solved. Thus, the point is to find the minimum number of observations enabling the solution of the main problem – the determination of the orbital elements.

2　The Minimum Number of Observations

One observation carried out at the moment of time t_1 enables one to rewrite the three equations in Eq. (9) with seven unknown parameters:

$$\Omega, \ i, \ \omega, \ a, \ e, \ T, \ \rho_1 \,.$$

Two observations carried out at t_1 and t_2 lead to six equations of the form of Eq. (9) – three equations for each observation; however, now with eight unknown parameters:

$$\Omega, \ i, \ \omega, \ a, \ e, \ T, \ \rho_1, \ \rho_2 \,.$$

Again, even in this case the number of equations is less than the number of unknown parameters.

Now assume that three observations are realized at t_1, t_2 and t_3. As a result one has nine equations of the type of Eq. (9) with nine (!) unknown parameters:

$$\Omega, \ i, \ \omega, \ a, \ e, \ T, \ \rho_1, \ \rho_2, \ \rho_3 \, .$$

Thus, to obtain the elements of the orbits of celestial bodies one has to carry out at least three observations. The corresponding branch of theoretical astronomy was known as the "Determination of orbital elements by three observations", which was associated mainly with the name of Gauss.

Nowadays, in the epoch of radio ranging of celestial bodies, the situation is obviously changed. For example, if simultaneously with optical observations, i.e. while determining α and β, radio ranging measurements can be carried out as well, then one will have two quantities, namely, the geocentric distance ρ and the velocity of the object v (combined with the angular motion of the body). From the known position of the Sun $(L_\odot, R_\odot)$ one can easily move from ρ to the heliocentric distance r by means of the equations

$$r \cos l \cos b = \rho \cos \lambda \cos \beta - R_\odot \cos L_\odot \, ,$$

$$r \sin l \cos b = \rho \sin \lambda \cos \beta - R_\odot \sin L_\odot \, , \qquad (10)$$

$$r \sin = \rho \sin \beta \, ,$$

where l and b are the heliocentric ecliptic polar coordinates of the celestial body. Here the only unknown parameters are r, l and b, which are determined uniquely by known values of ρ, λ and β.

Using the known relationship

$$V^2 = k^2 M \left(\frac{2}{r} - \frac{1}{a} \right)$$

one can obtain a, so that one parameter, the major half-axis a, is already known. In other words, from seven unknown quantities in Eq. (9) two are already known – a and ρ, and therefore only five elements remain unknown. It is easy to realize that only two observations suffice for their determination.

One more example. Assume that an object is discovered about which the only available information is that the declination of its orbital plane is zero, i.e. $i = 0$, and, hence, $z = 0$. Then, the third equation in (8) vanishes, and hence so does the element ω, the angular distance of the perihelion from the line of nodes, and instead of Ω and ω we shall have their sum, $\varphi = \Omega + \omega$, determining the angular distance of the perihelion from the point of vernal equinox. Only two equations from Eq. (9) remain:

$$f_1(\varphi, \ a, \ e, \ T, \ \rho, \ t) = 0 \, ,$$

$$f_2(\varphi, \ a, \ e, \ T, \ \rho, \ t) = 0 \, , \qquad (11)$$

i.e. one has two equations with five unknown parameters – four elements and one geocentric distance ρ. In this case one observation gives two equations with five unknowns, two observations give four equations with six unknown parameters, and three observations give six equations with seven unknown parameters. Only four observations carried out at t_1, t_2, t_3, and t_4 enable us to write eight equations with eight unknowns – φ, a, e, T, ρ_1, ρ_2, ρ_3, ρ_4.

This example is also interesting for another reason: it may restrain one from drawing quick conclusions concerning the number of the observations at a specific orbital orientation. In particular, not in all cases can the preliminary knowledge concerning a given element lead to a reduction in the number of optical observations required for the determination of the remaining elements.

3 Derivation of the Main Equation

Three observations being required is a condition common for the determination of elements for all types of orbits, elliptic, parabolic and hyperbolic. The question, however, arises, of how the observation data should be used. Such a statement of the problem may seem strange and even in contradiction to the above-mentioned conclusions, that the observed coordinates of the object λ and β are related to the sought orbital elements and the unknown geocentric distance ρ by means of Eq. (8). However, the point is that Eq. (8) is a transcendent one and all six elements, Ω, i, ω, a, e, and T, appear within the trigonometrical functions in rather inconvenient combinations. Eq. (8) is linear, as we shall see later, only with respect to ρ, and it is just this circumstance that one has to exploit.

It is clear that, besides Eq. (8) one also needs some additional starting equations. Our problem is the search for and derivation of those equations. The peculiarity of the situation is that even the latter equations are common for all types of orbits and the specific properties of orbits reveal themselves only in concrete applications.

Thus, three observations of the celestial body have been carried out at t_1, t_2 and t_3. Obviously all three series of observed coordinates x, y, z must satisfy Eq. (12) of the orbital plane passing through the Sun (Chapter III):

$$a_1 x + a_2 y + a_3 z = 0 \,. \tag{12}$$

Then we can write

$$a_1 x_1 + a_2 y_1 + a_3 z_1 = 0 \,,$$

$$a_1 x_2 + a_2 y_2 + a_3 z_2 = 0 \,, \tag{13}$$

$$a_1 x_3 + a_2 y_3 + a_3 z_3 = 0 \,,$$

where x_i, y_i, and z_i denote the coordinates of the body at t_1, t_2, and t_3, respectively. Eqs. (13) are homogeneous and linear with respect to unknown coefficients a_1, a_2, and a_3 which are the constants of integration of differential equations (1) and (2) (Chapter III). However, the unknown parameters a_1, a_2, and a_3 cannot be determined readily, except in the trivial case, i.e. $a_1 = 0$, $a_2 = 0$, $a_3 = 0$, which is obviously not of interest. As is well known, the general condition is

$$\begin{vmatrix} x_1, & y_1, & z_1 \\ x_2, & y_2, & z_2 \\ x_3, & y_3, & z_3 \end{vmatrix} = 0. \tag{14}$$

By such a maneuver one can exclude quantities a_1, a_2 and a_3 from Eq. (13). Now rewrite Eq. (14) in the form

$$x_1(y_2 z_3 - z_2 y_3) - x_2(y_1 z_3 - z_1 y_3) + x_3(y_1 z_2 - z_1 y_2) = 0,$$

$$y_1(x_2 z_3 - z_2 x_3) - y_2(x_1 z_3 - z_1 x_3) + y_3(x_1 z_2 - z_1 x_2) = 0, \tag{15}$$

$$z_1(x_2 y_3 - y_2 x_3) - z_2(x_1 y_3 - y_1 x_3) + z_3(x_1 y_2 - y_1 x_2) = 0.$$

In fact, these three equations are various representations of one and the same equation. However, if we can by means of some additional considerations express the relations within the parentheses $(x_i y_j - y_i x_j)$ through known quantities then we shall have three independent equations. To do this, let us analyze the geometrical content of the coefficients within the parentheses. Consider e.g. the coefficient $x_1 y_3 - y_1 x_3$. It is not difficult to show that this combination is the doubled area of a triangle projected on the XOY plane and with the Sun and two positions of the celestial body, M_1 and M_3, situated at its vertices as shown in Fig. 6.1.

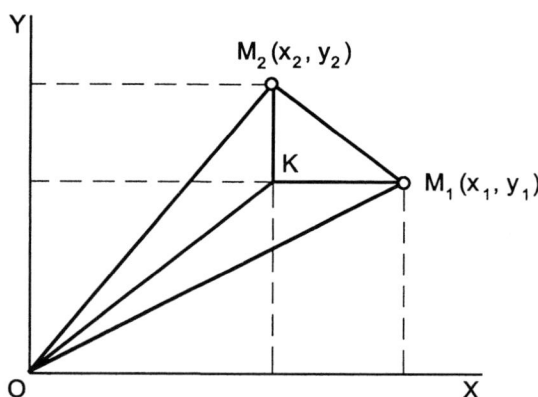

Fig. 6.1. *The determination of the ratio of areas of triangles.*

The area of the triangle $OM_1 M_3$ can be represented as the sum of three triangles, namely,

$$\Delta OM_1 M_3 = \Delta OKM_1 + \Delta OKM_3 + \Delta KM_1 M_3$$

or

$$\Delta OM_1M_3 = \frac{1}{2}(x_1 - x_3)y_1 + \frac{1}{2}(y_3 - y_1)x_3 + \frac{1}{2}(x_1 - x_3)(y_3 - y_1)$$

or

$$\Delta OM_1M_3 = \frac{1}{2}(x_1y_3 - y_1x_3). \tag{16}$$

Denote the area of the triangle formed by radius-vectors r_i and r_j by $[r_ir_j]$; obviously we will have three triangles $[r_1, r_2]$, $[r_1, r_3]$, $[r_2, r_3]$ corresponding to various combinations of the three radius-vectors. Let i_{xy}, i_{yz}, and i_{xz} be the inclination angles of the orbit's plane to the planes of coordinates XOY, YOZ and XOZ, respectively. Then we will have for the coefficients of Eq. (15) the following expressions, taking into account also Eq. (16):

$$x_1y_2 - y_1x_2 = 2[r_1r_2]\cos i_{xy},$$
$$x_1y_3 - y_1x_3 = 2[r_1r_3]\cos i_{yz}, \tag{17}$$
$$x_2y_3 - y_2x_3 = 2[r_2r_3]\cos i_{zx},$$

$$y_1z_2 - z_1y_2 = 2[r_1r_2]\cos i_{xy},$$
$$y_1z_3 - z_1y_3 = 2[r_1r_3]\cos i_{yz}, \tag{18}$$
$$y_2z_3 - z_2y_3 = 2[r_2r_3]\cos i_{zx},$$

$$x_1z_2 - z_1x_2 = 2[r_1r_2]\cos i_{xy},$$
$$x_1z_3 - z_1x_3 = 2[r_1r_3]\cos i_{yz}, \tag{19}$$
$$x_2z_3 - z_2x_3 = 2[r_2r_3]\cos i_{zx}.$$

Note that, in this representation, first, the values of the cosines in each of these groups (17), (18) and (19) are one and the same, and, second, the triangle area $[r_ir_j]$ appears in three combinations in all nine equations in (17)–(19).

On substituting these expressions into Eq. (15) we obtain

$$[r_2r_3]x_1 - [r_1r_3]x_2 + [r_1r_2]x_3 = 0,$$
$$[r_2r_3]y_1 - [r_1r_3]y_2 + [r_1r_2]y_3 = 0, \tag{20}$$
$$[r_2r_3]z_1 - [r_1r_3]z_2 + [r_1r_2]z_3 = 0.$$

Here the coefficients at x, y and z with a common subscript in contrast to Eq. (16) coincide within all three equations, i.e. instead of nine unknown coefficients $(x_i y_j - y_i x_j)$ one has only three, $[r_i r_j]$. Note that all coefficients have the same dimension, that of an area.

In the next step let us rewrite Eq. (20) in the following form:

$$\frac{[r_2 r_3]}{[r_1 r_3]} x_1 + \frac{[r_1 r_3]}{[r_1 r_3]} x_3 = x_2 \,,$$

$$\frac{[r_2 r_3]}{[r_1 r_3]} y_1 + \frac{[r_1 r_2]}{[r_1 r_3]} y_3 = y_2 \,, \tag{21}$$

$$\frac{[r_2 r_3]}{[r_1 r_3]} z_1 + \frac{[r_1 r_3]}{[r_1 r_3]} z_3 = z_2 \,,$$

This system differs advantageously from Eq. (20); first, the coefficients of x_i, y_i, and z_i, expressed through the ratio of the triangle areas, are now dimensionless quantities, and second, their number has turned out to be two instead of three.

Below, a number of formulas will be derived enabling one to obtain the numerical values of the coefficients in Eq. (21), i.e. the ratios of areas of the triangles. On the other hand, the coordinates x_i, y_i and z_i in Eq. (21) depend on orbital elements, six in total, so that three equations are not enough for the determination of all of them. However, replacing x_i, y_i and z_i by their expressions

$$x_i = \xi_i - X_i \,,$$

$$y_i = \eta_i - Y_i \,, \tag{22}$$

$$z_i = \zeta_i - Z_i$$

one can represent all three expressions in Eq. (21) by unknown geocentric distances ρ_1, ρ_2, and ρ_3 through coordinates ξ_i, η_i, and ζ_i. As a result the system (21) with the six unknown parameters turns out to be a system of three equations with only three unknown parameter, ρ_1, ρ_2 and ρ_3. Thus, the first three unknown parameters from nine are obtained, the other six unknown ones, i.e. the orbital elements, will be determined from other considerations.

Thus, the problem of determination of an orbit from observations is split into two steps. First, all three geometric distances of the celestial body should be obtained from Eq. (21); this problem is solved relatively easily due to the linear dependence among all three distances ρ_1, ρ_2 and ρ_3. Second, from the already known geocentric distances of the celestial body all six orbital elements should be determined. In contrast to geocentric distances, the dependencies among the orbital elements are nonlinear.

4   Calculation of the Ratio of Triangle Areas

One can simplify the solution of the problem of determination of the ratio of the triangle areas via superposition of the orbital plane upon the plane XOY. Then the triangle areas will be represented by the following formulas:

$$[r_1 r_2] = \frac{1}{2}(x_1 y_2 - y_1 x_2),$$

$$[r_1 r_3] = \frac{1}{2}(x_1 y_3 - y_1 x_3), \tag{23}$$

$$[r_2 r_3] = \frac{1}{2}(x_2 y_3 - y_2 x_3).$$

Because these areas depend on the coordinates x_i and y_i, one can formulate the problem in a slightly different manner: to obtain the coordinates x_i and y_i for each of these three moments t_1, t_2, and t_3 dependent on x_0, y_0, dx_0/dt and dy_0/dt for some initial moment of time t_0. To make the equations more compact we perform the following transformations of time t:

$$\tau = k\sqrt{M}\,(t - t_0), \tag{24}$$

assuming that $t = t_0$ at $\tau = 0$, and $d\tau = k\sqrt{M}\,dt$. Then the usual differential equations of relative motion

$$\frac{d^2 x}{dt^2} = -k^2 M \frac{x}{r^3},$$

$$\frac{d^2 y}{dt^2} = -k^2 M \frac{y}{r^3}$$

will be written in the form

$$\frac{d^2 x}{d\tau^2} = -\frac{x}{r^3},$$

$$\frac{d^2 y}{d\tau^2} = -\frac{y}{r^3}. \tag{25}$$

These equations will be the basic ones while obtaining x_i and y_i as function of x_0 and y_0. The solution of this problem is reached via Maclaurin series:

$$x = x_0 + \tau \left(\frac{dx}{d\tau}\right)_0 + \frac{\tau^2}{1 \times 2}\left(\frac{d^2x}{d\tau^2}\right)_0 + \frac{\tau^3}{1 \times 2 \times 3}\left(\frac{d^3x}{d\tau^3}\right)_0 + \cdots$$
$$\frac{\tau^n}{n!}\left(\frac{d^nx}{d\tau^n}\right)_0 + \cdots,$$

$$y = y_0 + \tau \left(\frac{dy}{d\tau}\right)_0 + \frac{\tau^2}{1 \times 2}\left(\frac{d^2y}{d\tau^2}\right)_0 + \frac{\tau^3}{1 \times 2 \times 3}\left(\frac{d^3y}{d\tau^3}\right)_0 + \cdots$$
$$\frac{\tau^n}{n!}\left(\frac{d^ny}{d\tau^n}\right)_0 + \cdots,$$

$$(26)$$

where x_0, y_0, $dx_0/d\tau$, $dy_0/d\tau$, etc. are the values of coordinates and their derivatives with respect to τ at $\tau = 0$. The higher order members in Eq. (26) can be ignored since in practice the observations of newly discovered celestial bodies are realized at moments t_1, t_2, and t_3 within small time intervals, i.e. with small τ.

Let us express the derivatives in Eq. (26) higher than the second order through x_0, y_0, $dx_0/d\tau$ and $dy_0/d\tau$. For the second order derivatives we have from Eq. (25)

$$\frac{d^2x_0}{d\tau^2} = -\frac{x_0}{r_0^3}, \quad \frac{d^2y_0}{d\tau^2} = -\frac{y_0}{r_0^3}. \tag{27}$$

Differentiating Eq. (25) once more, we obtain at $\tau = 0$

$$\frac{d^3x_0}{d\tau^3} = \frac{3x_0}{r_0^4}\frac{dr_0}{d\tau} - \frac{1}{r_0^3}\frac{dx_0}{d\tau},$$
$$\frac{d^3y_0}{d\tau^3} = \frac{3y_0}{r_0^4}\frac{dr_0}{d\tau} - \frac{1}{r_0^3}\frac{dy_0}{d\tau}. \tag{28}$$

On substituting Eqs. (27) and (28) into Eq. (26) and combining the terms with x_0 and y_0 on one side, and terms with $dx/d\tau$ and $dy/d\tau$ on the other, we obtain

$$x = \left(1 - \frac{1}{2}\frac{\tau^2}{r_0^3} + \frac{\tau^3}{2r_0^4}\frac{dr_0}{d\tau} + \cdots\right)x_0 + \left(\tau - \frac{\tau^3}{6r_0^3} + \cdots\right)\frac{dx_0}{d\tau},$$
$$y = \left(1 - \frac{1}{2}\frac{\tau^2}{r_0^3} + \frac{\tau^3}{2r_0^4}\frac{dr_0}{d\tau} + \cdots\right)y_0 + \left(\tau - \frac{\tau^3}{6r_0^3} + \cdots\right)\frac{dy_0}{d\tau}. \tag{29}$$

Note the following two circumstances. First, the coefficients within the parentheses in x_0 and y_0 on the one hand, and coefficients in $dx_0/d\tau$ and

$dy_0/d\tau$ on the other, are the same. Second, both expressions within parentheses in Eq. (29) contain r_0 and $dr_0/d\tau$ – both unknown quantities.

Now one has to choose the initial moment of time. The second observation seems to be the most convenient, i.e. $t_0 = t_2$. We adopt also

$$x_0 = x_2, \quad y_0 = y_2, \quad r_0 = r_2,$$

$$\frac{dx_0}{d\tau} = \frac{dx_2}{d\tau}, \quad \frac{dy_0}{d\tau} = \frac{dy_2}{d\tau}, \quad \frac{dr_0}{d\tau} = \frac{dr_2}{d\tau} \tag{30}$$

and introduce the following notations

$$\tau_1 = k\sqrt{M}(t_3 - t_2),$$
$$\tau_2 = k\sqrt{M}(t_3 - t_1), \tag{31}$$
$$\tau_3 = k\sqrt{M}(t_2 - t_1)$$

together with the condition $\tau_2 = \tau_1 + \tau_3$.

Now we can derive the expressions for x_1, x_3, y_1, and y_3 using Eq. (29), and proceed to the computation of areas $[r_i r_j]$ with the help of Eq. (23). As a result we obtain, for example, for $[r_1 r_3]$, preserving only the third order terms with respect to τ

$$[r_1 r_3] = \frac{\tau_2}{2} \left(1 - \frac{\tau_2^2}{6r_2^3} + \cdots \right) \left(x_2 \frac{dy_2}{d\tau} - y_2 \frac{dx_2}{d\tau} \right). \tag{32}$$

Proceeding from the theorem of areas, i.e. Kepler's first law (see Eq. (18), Chapter III) we have

$$x \frac{dy}{dt} - y \frac{dx}{dt} = c_1 \cos i. \tag{33}$$

Adopting $i = 0$ since in our case the plane XOY coincides with the orbital plane, and $c_1 = k \, (pM)^{1/2}$, and passing from t to τ, Eq. (24), we will have instead of Eq. (33)

$$x \frac{dy}{d\tau} - y \frac{dx}{d\tau} = \sqrt{p}. \tag{34}$$

Then the expression for $[r_1 r_3]$ takes the following final form:

$$[r_1 r_3] = \frac{\tau_2}{2} \left(1 - \frac{\tau_2^2}{6r_2^3} + \cdots \right) \sqrt{p}. \tag{35}$$

Note that the parameter p depends on the orbital elements and therefore is an unknown quantity.

Analogously one can obtain

$$[r_1 r_2] = \frac{\tau_3}{2}\left(1 - \frac{\tau_3^2}{6r_2^3}\right)\sqrt{p},$$

$$[r_2 r_3] = \frac{\tau_1}{2}\left(1 - \frac{\tau_1^2}{6r_2^3}\right)\sqrt{p}. \tag{36}$$

However, according to the statement of the problem, we need only the ratios of the triangle areas appearing in Eq. (21). Therefore finally we obtain

$$\frac{[r_2 r_3]}{[r_1 r_3]} = \frac{\tau_1}{\tau_2}\left(1 - \frac{\tau_1^2 - \tau_2^2}{6r_2^3} + \cdots\right),$$

$$\frac{[r_1 r_2]}{[r_1 r_3]} = \frac{\tau_3}{\tau_2}\left(1 - \frac{\tau_3^2 - \tau_2^2}{6r_2^3} + \cdots\right). \tag{37}$$

In these expressions the unknown quantity, p, no longer exists, which is an essential advantage. However, Eq. (37) contains another unknown parameter, r_2, the heliocentric distance of the position of the celestial body at the moment $t = t_2$. Hence, one has either to add one more equation containing r_2 to Eq. (21) or to find a way of replacing r_2. As we shall see in forthcoming chapters, both ways can be useful depending on the type of the orbit.

In the present chapter, which is completely devoted to the preliminary operations of the determination of the orbit from three observations, the basic concept and the general strategy to solve this problem have been formulated. It is split into two steps: first, the determination of three geocentric distances ρ_1, ρ_2 and ρ_3, and second, the estimation of orbital elements. Both steps will be realized in a different manner depending on the type of the orbit: the approach is common for elliptic and hyperbolic orbits and different for parabolic ones. Therefore these types of orbits will be examined separately.

Chapter VII

Determination of Elements of the Orbits by Three Observations

1 Elliptic Orbit. Basic Equations

As we saw in the previous chapter, for the determination of elements of the elliptic orbit it is necessary to have three complete observations of the celestial body. The problem of the manner in which observational data should be used, and how they will be involved in the general theory in order to obtain the final quantitative results, is solved by a special method developed for elliptical orbits. Such a method of exhaustive completeness was created by Gauss in 1801. The present chapter is devoted to the description of this method, which has survived up to now almost without change.

The idea of splitting the problem of determination of the elliptical orbit into two stages is preserved: first, determination of the geocentric distances of the celestial body ρ_1, ρ_2, and ρ_3; second, determination of orbital elements themselves.

The system of three basic equations for the determination of geocentric distances, as was shown in the previous chapter, has the form

$$\frac{[r_2 r_3]}{[r_1 r_3]} x_1 + \frac{[r_1 r_2]}{[r_1 r_3]} x_3 = x_2 \,,$$

$$\frac{[r_2 r_3]}{[r_1 r_3]} y_1 + \frac{[r_1 r_2]}{[r_1 r_3]} y_3 = y_2 \,, \qquad (1)$$

$$\frac{[r_2 r_3]}{[r_1 r_3]} z_1 + \frac{[r_1 r_2]}{[r_1 r_3]} z_3 = z_2 \,.$$

The geocentric distances ρ_1, ρ_2, and ρ_3 are related to the geocentric coordinates ξ_i, η_i, and ζ_i in the following manner:

$$\xi_i = \rho_i \cos \lambda_i \cos \beta_i\,,$$
$$\eta_i = \rho_i \sin \lambda_i \cos \beta_i\,, \tag{2}$$
$$\zeta_i = \rho_i \sin \beta_i\,,$$

where $i = 1$, 2 and 3, and λ and β are the longitude and latitude of the celestial body with respect to the ecliptic, respectively, and are the observable quantities. On the other hand, the heliocentric coordinates of the celestial body x, y, and z are expressed through its geocentric coordinates ξ_i, η_i, and ζ_i and the geocentric coordinates of the Sun X, Y, and Z in the following manner:

$$x_i = \xi_i - X_i\,,$$
$$y_i = \eta_i - Y_i\,, \tag{3}$$
$$z_i = \zeta_i - Z_i\,.$$

By rewriting Eqs. (2) and (3) for $i = 1$, 2, and 3, and substituting the values of x_i, y_i and z_i into Eq. (1), we obtain

$$\frac{[r_2 r_3]}{[r_1 r_3]}(\rho_1 \cos \lambda_1 \cos \beta_1 - R_1 \cos L_1)$$

$$+\frac{[r_1 r_2]}{[r_1 r_3]}(\rho_3 \cos \lambda_3 \cos \beta_3 - R_3 \cos L_3) = \rho_2 \cos \lambda_2 \cos \beta_2 - R_2 \cos L_2\,,$$

$$\frac{[r_2 r_3]}{[r_1 r_3]}(\rho_1 \sin \lambda_1 \cos \beta_1 - R_1 \sin L_1) \tag{4}$$

$$+\frac{[r_1 r_2]}{[r_1 r_3]}(\rho_3 \sin \lambda_3 \cos \beta_3 - R_3 \sin L_3) = \rho_2 \sin \lambda_2 \cos \beta_2 - R_2 \sin L_2\,,$$

$$\frac{[r_2 r_3]}{[r_1 r_3]}\rho_1 \sin \beta_1 - R_1 + \frac{[r_1 r_2]}{[r_1 r_3]}\rho_3 \sin \beta_3 = \rho_2 \sin \beta_2\,,$$

where R and L are the heliocentric distance and ecliptic longitude of the Sun, respectively, at the moment of observation t and are known quantities; they are represented in astronomical annual handbooks for certain periods of time.

The ratios of triangle areas (the expressions in square brackets) are known and are given by Eq. (37), Chapter VI; they depend on the moments of observations t_1, t_2, and t_3 as well as on the heliocentric distance at the moment of the second observation of the celestial body r_2, which is also an unknown parameter.

Thus, we have obtained the system (4) of three equations with three unknown parameters – ρ_1, ρ_2, and ρ_3, so that the system, in principle, is quite solvable. One might even think that this system could be solved rather easily insofar as it is linear with respect to ρ_1, ρ_2, and ρ_3. However, it is not difficult to see that each of these quantities, ρ_1, ρ_2, and ρ_3, should be expressed either via r_2 or via r_2^3 (see Eq. (37), Chapter VI).

Turn now to the solution of system (4), assuming for a moment that the ratios of the triangle areas are known and are represented through τ_1, τ_2, and τ_3 as well as through r_2^3. One can realize this procedure by expressing first the unknown quantities ρ_1 and ρ_3 via ρ_2 and ρ_2^3 and then, excluding both ρ_1 and ρ_3 from Eq. (4), presenting the unknown ρ via ρ_2^3 and all τ.

Acting in such a manner and also excluding both ρ_1 and ρ_3 from Eq. (4), we obtain the following expression for ρ_2 in terms of its dependence on r_2:

$$\rho_2 = -m + \frac{l}{r_2^3}\,, \tag{5}$$

where m and l are known quantities that can be determined via rather simple formulae and where τ_i, R_i, L_i, λ_i, and β_i with subscripts $i = 1$, 2, and 3 appear in various combinations. However, Eq. (5) manifests itself as an equation with two unknown parameters, ρ_2 and r_2. It is clear, therefore, that for its solution we need one more equation:

$$f(\rho_2,\, r_2) = 0\,. \tag{6}$$

Let us find that equation.

2 Laplace Equation

Consider the triangle shown in Fig. 7.1, where the Sun is indicated by S_2, the Earth by T_2 and the celestial body by M_2; their mutual distances are also shown for the second moment of observations t_2. From this triangle we have

$$r_2^2 = \rho_2^2 + R_2^2 - 2R_2\rho_2 \cos\psi_2\,, \tag{7}$$

where ψ_2 is the angle between the directions of the Sun and of the celestial body with respect to the observer on the Earth. That angle can be determined from the spherical triangle $M_2 S_2 K$ as shown in Fig. 7.2, where S_2 is the position of the Sun on the ecliptic $S_2 K$, T_2 is that of the Earth, and M_2 that of the celestial body. Two sides of the spherical triangle, β_2 and $\lambda_2 - L_2$, are known quantities, and ψ_2 in Eq. (7) is the angle of visibility both of the Sun and of the celestial body from the Earth. The plane $S_2 M_2$ passing through the second positions of the Sun and the celestial body has no relation to the orbit's plane.

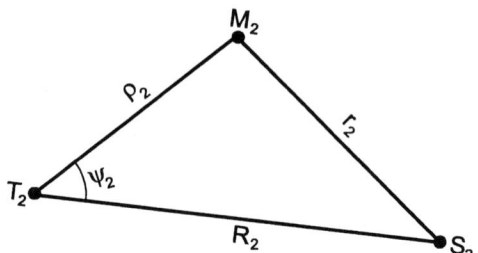

Fig. 7.1. *To the solution of Laplace equation.*

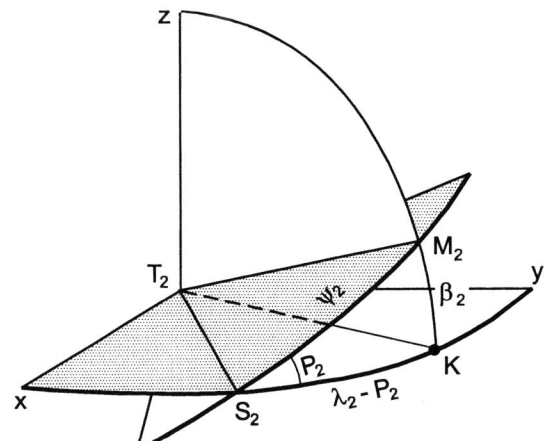

Fig. 7.2. *To the determination of auxiliary angle ψ_2.*

By means of basic equations of spherical trigonometry we obtain

$$\cos \psi_2 = \cos(\lambda_2 - L_2) \cos \beta_2,$$

$$\sin \psi_2 \cos P_2 = \sin(\lambda_2 - L_2) \cos \beta_2, \tag{8}$$

$$\sin \psi_2 \sin P_2 = \sin \beta_2.$$

From them one can uniquely determine both angles, ψ_2 and P_2, though the angle P_2 itself will not be used below.

Thus, the angle ψ_2 in Eq. (7) is known and the unknown parameters are ρ_2 and r_2. Hence, the relationship (7) can be used as the sought second equation of Eq. (6) type. Obviously, by solving Eqs. (5) and (7) jointly, we will find r_2 and ρ_2.

By excluding ρ_2 and substituting r_2 from Eq. (5) into (7), we finally obtain:

$$r_2^8 - r_2^6 (m^2 - R_2^2) + r_2^3 [2ml + 2R_2 \cos \psi_2 (mr_2^3 - l)^{1/2}]. \tag{9}$$

This equation is named after Laplace, who was the first to propose the determination of the heliocentric distance r_2, and afterwards of the geocentric distance ρ_2 by means of Eq. (5), angle ψ_2 and Eq. (8).

Thus, the problem of determination of the heliocentric distance r_2 is reduced to the solution of an algebraic equation of the eighth (!) degree with respect to r_2. The solution of this transcendental equation, though possible in principle, say, by the successive approximation method, should be associated with the choice of the only true solution among all eight roots. Formally one has to add eight more conditions, for example, excluding obviously the imaginary solutions or those that become unphysically large or very close to zero. However, to obtain all the necessary additional conditions, as a rule, appears to be impossible. Therefore this method of determination of r_2 and ρ_2 has not found wide application.

3   The Gauss Equation

Quite a different idea for the determination of r_2 and ρ_2 was proposed by Gauss. Gauss introduced a new independent quantity z, the angle between the directions of the Sun S_2 and the Earth E_2 for the observer located on the celestial body M_2, as shown in Fig. 7.3, which is apparently quite similar to Fig. 7.1. For this triangle one can write

$$\frac{\rho_2}{\sin(\Psi_2 + z)} = \frac{r_2}{\sin \Psi_2} = \frac{R_2}{\sin z}. \tag{10}$$

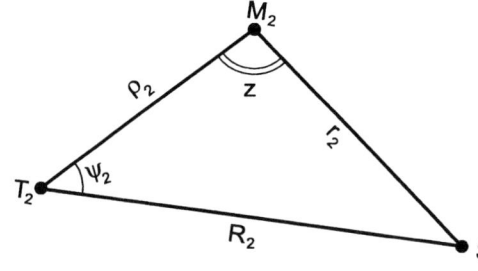

Fig. 7.3. *The transition from the Laplace equation to the Gauss equation.*

From that we obtain

$$\rho_2 = R_2 \frac{\sin(\Psi_2 + z)}{\sin z}, \tag{11}$$

$$r_2 = R_2 \frac{\sin \Psi_2}{\sin z}, \tag{12}$$

where ψ_2 is determined from Eq. (8) as before. On substituting Eqs. (11) and (12) into Eq. (5), we have

$$R_2 \frac{\sin(\Psi_2 + z)}{\sin z} = -m + l \frac{\sin^3 z}{R_2^3 \sin^3 \Psi_2}. \tag{13}$$

In such a manner Gauss managed to replace Eq. (5) containing two unknown variables, ρ_2 and r_2, by Eq. (13) with only one unknown parameter, z. Obviously, by determining z from this, one can find $\rho_2 r$ and r_2 from Eqs. (11) and (12).

Rewrite Eq. (13) in the following form:

$$R_2 \sin \Psi_2 \cos z + (R_2 \cos \Psi_2 + m) \sin z = \frac{l}{R_2^3 \sin \Psi_2} \sin^4 z \qquad (14)$$

and introduce new quantities, Ω and ω:

$$\begin{aligned} \Omega \sin \omega &= -R_2 \sin \psi_2, \\ \Omega \cos \omega &= R_2 \cos \psi_2 + m \,. \end{aligned} \qquad (15)$$

Then Eq. (14) takes the form:

$$\Omega \sin(z - \omega) = \frac{l}{R_2^3 \sin^3 \Psi_2} \sin^4 z \qquad (16)$$

or

$$M \sin^4 z = \sin(z - \omega) \,, \qquad (17)$$

where

$$M = \frac{l}{\Omega R_2^3 \sin^3 \Psi_2} \,. \qquad (18)$$

Eq. (17) is known as the Gauss equation.

The Gauss equation is of fourth degree with respect to z, and though it is also transcendental like the Laplace equation, nevertheless the number of roots is less by a factor of two – only four. Moreover, owing to the fact that z is an angular and not a distance variable like r_2 in Eq. (9), the formulation of conditions for the exclusion of useless roots becomes much simpler.

In Fig. 7.4 one of the two curves, y_2, is a normal sinusoid, though it does not pass through the center of the coordinate system, $z = 0$, and can be considered to be the geometrical representation of the right-hand side of the Gauss equation, i.e. of the function $y_2 = \sin(z - \omega)$. The second curve, y_1, represents its left-hand side, $y_1 = M \sin^4 z$. The two curves intersect at four points, corresponding to the values z_1, z_2, z_3 and z_4, these being the four roots of Eq. (17); the condition $y_1 = y_2$ is satisfied at those points.

Which of these four values of z should be chosen? To answer this question we can proceed from the following three conditions.

a. The value of z, as one of the angles of a triangle, cannot exceed $180°$. As can be seen from Fig. 7.4, $z_4 > 180°$, and therefore it cannot be accepted.

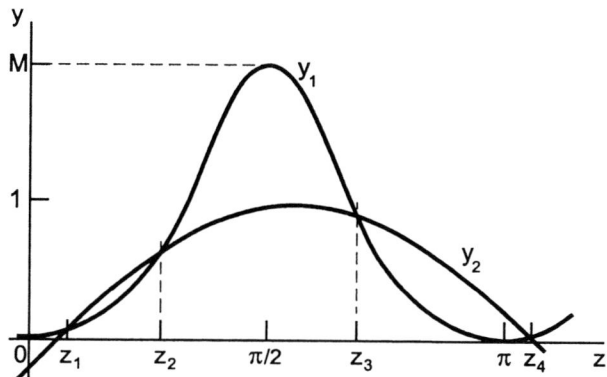

Fig. 7.4. *The solution of the Gauss equation.*

b. The solution that turns $\sin(z + \psi_2)$ into a negative quantity cannot be accepted, since both ρ_2 and r_2 are positive.

c. The value of z when $\sin(z + \psi_2)$ turns out to be zero or close to zero should also be excluded, because in this case ρ_2 becomes equal to zero which is again impossible.

In practice, these conditions appear to be enough for the choice of the required solution. In rare cases when two values of z both seem to be suitable, the final choice is realized with the help of additional observations.

4   Determination of Geocentric Distances

When ρ_2 is already obtained, one can turn to the determination of ρ_1 and ρ_3, expressing them via ρ_2. To do this, it is necessary to return to the basic Eq. (4) and, excluding ρ_3 from the first two equations, we can represent ρ_1 via ρ_2. Then we obtain

$$\rho_1 = \frac{\rho_2 \sin(\lambda_3 - \lambda_2)\cos\beta_2 - (N_1 - n_1)h_1 - (N_3 - n_3)h_3}{n_1 \sin(\lambda_3 - \lambda_1)\cos\beta_1}, \qquad (19)$$

where

$$h_1 = R_1 \sin(\lambda_3 - L_1) \,,$$

$$h_3 = R_3 \sin(\lambda_3 - L_3) \,,$$

$$n_1 = \frac{\tau_1}{\tau_3} \left(1 - \frac{1}{6} \frac{\tau_1^2 - \tau_2^2}{r_2^3} + \cdots \right) \,,$$

$$n_3 = \frac{\tau_3}{\tau_2} \left(1 - \frac{1}{6} \frac{\tau_3^2 - \tau_2^2}{r_2^3} + \cdots \right) \,.$$

Analogously, excluding ρ_1 from the first two equations of Eq. (4), we find for ρ_3

$$\rho_3 = \frac{\rho_2 \sin(\lambda_2 - \lambda_1) \cos \beta_2 + (N_1 - n_1)g_1 + (N_3 - n_3)g_3}{n_2 \sin(\lambda_3 - \lambda_1) \cos \beta_3} \,, \tag{20}$$

where

$$g_1 = R_1 \sin(\lambda_1 - L_1) \,,$$

$$g_3 = R_3 \sin(\lambda_1 - L_3) \,.$$

Thus, the first step – the determination of the geocentric distances ρ_1, ρ_2, and ρ_3, is finished. We turn now to the second stage of the problem – the determination of orbital elements themselves.

5   Determination of the Elements of an Elliptic Orbit

We start from the determination of heliocentric ecliptic coordinates r, l, and b of a celestial object by means of the known geocentric ecliptic coordinates ρ, λ, and β as well as R and L for the Sun using the following relationships:

$$r_i \cos l_i \cos b_i = \rho_i \cos \lambda_i \cos \beta_i - R_i \cos L_i \,,$$

$$r_i \sin l_i \cos b_i = \rho_i \sin \lambda_i \cos \beta_i - R_i \sin L_i \,, \tag{21}$$

$$r_i \sin b_i = \rho_i \sin \beta_i \,.$$

Rewriting this system for $i = 1$, 2 and 3, we obtain all nine quantities: r_1, l_1, b_1, r_2, l_2, b_2, and r_3, l_3, b_3, corresponding to the three moments of time t_1, t_2 and t_3. Note that r_2 must coincide with its value obtained above with the help of Eq. (12).

For the spherical triangle NMK in Fig. 7.5, where S is the Sun, NK is the ecliptic, NM is the orbital plane, M is the position of celestial body, and l and b are its ecliptic coordinates at a given moment of time, we can write

$$\cos u = \cos(l - \Omega) \cos b ,$$

$$\sin u \cos i = \sin(l - \Omega) \cos b , \qquad (22)$$

$$\sin u \sin i = \sin b .$$

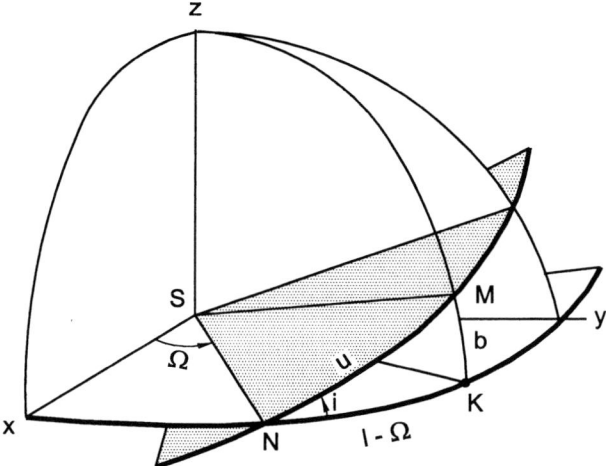

Fig. 7.5. *The discovery of elements Ω and i determining the orbit's position in space.*

From that we find:

$$\tan i = \frac{\tan b}{\sin(l - \Omega)} . \qquad (23)$$

and, rewriting it for the moments t_1 and t_3,

$$\tan i \, \sin(l_1 - \Omega) = \tan b_1 ,$$
$$\tan i \, \sin(l_3 - \Omega) = \tan b_3 . \qquad (24)$$

we obtain a system with only two unknown parameters, i and Ω. In order to simplify the calculations, we substitute $l_3 - \Omega = (l_3 - l_1) + (l_1 - \Omega)$, so that Eq. (24) takes the form

$$\tan i \, \sin(l_1 - \Omega) = \tan b_1 , \qquad (25)$$

$$\tan i \, \cos(l_1 - \Omega) \sin(l_3 - l_1) = \tan b_3 - \tan b_1 \, \cos(l_3 - l_1) . \qquad (26)$$

From that we obtain

$$\tan(l_1 - \Omega) = \frac{\tan b_1 \, \sin(l_3 - l_1)}{\tan b_3 - \tan b_1 \, \cos(l_3 - l_1)} . \qquad (27)$$

With the help of this formula we easily obtain the first element Ω, and, thereafter, the second one, i from Eq. (25).

Thus, the first two elements of the elliptic orbit, Ω and i, are obtained. Simultaneously the numerical values of the so-called latitude argument u for the moments t_1 and t_3, i.e. the values of u_1 and u_3, have been obtained as well. Note that

$$u_1 = \omega + v_1 \,,$$
$$u_3 = \omega + v_3 \,,$$
(28)

and in general

$$v_3 - v_1 = u_3 - u_1 \,.$$
(29)

We will return to these relationships later.

6   Determination of the Semimajor Axis: The Gauss Method

The first method for determination of the semimajor axis a of an elliptic orbit was proposed by Gauss in 1801. The essence of this method is as follows.

Let S be the area of a sector between radius-vectors r_1 and r_3 and the corresponding arc of the orbit. According to Kepler's first law we can write

$$S = \frac{1}{2}c_1(t_3 - t_1) = \frac{1}{2}k\sqrt{p}(t_3 - t_1) \,,$$
(30)

where $c_1 = k(pM)^{1/2} = k\sqrt{p}$ has been used for the constant of the ellipse.

The $[r_1 r_3]$ triangle area equals

$$[r_1 r_3] = \frac{1}{2}r_1 r_3 \sin(u_3 - u_1) \,.$$
(31)

The ratio η of these two areas is

$$\eta = \frac{S}{[r_1 r_3]} = \frac{k\sqrt{p}\,(t_3 - t_1)}{r_1 r_3 \sin(u_3 - u_1)} \,.$$
(32)

The ellipse equation in polar coordinates gives

$$\frac{p}{r_1} = 1 + e\cos v_1 \,, \qquad \frac{p}{r_3} = 1 + e\cos v_3 \,.$$

On taking the sum of the latter two expressions, we obtain

$$p\frac{r_1 + r_3}{r_1 r_3} = 2 + e(\cos v_1 + \cos v_3) = 2 + 2e\cos\frac{v_3 + v_1}{2}\cos\frac{v_3 - v_1}{2} \,.$$
(33)

Because $v_3 - v_1 = u_3 - u_1$ is a known quantity, the only unknown one on the right-hand side of this equation that remains is $e \cos \left[(v_3 + v_1)/2 \right]$, which should be excluded.

Combining Eqs. (11), (13) and (14) from Chapter IV, we obtain

$$\sqrt{r} \cos \frac{v}{2} = \left[a \left(1 - e \right) \right]^{1/2} \cos \frac{E}{2} ,$$

$$\sqrt{r} \sin \frac{v}{2} = \left[a \left(1 + e \right) \right]^{1/2} \sin \frac{E}{2} .$$

Rewriting these equations for the moments t_1 and t_3 we find

$$(r_1 r_3)^{1/2} \cos \frac{v_3 - v_1}{2} = a \cos \frac{E_3 - E_1}{2} - a e \cos \frac{E_3 + E_1}{2} , \qquad (34)$$

$$(r_1 r_3)^{1/2} \cos \frac{v_3 + v_1}{2} = a \cos \frac{E_3 + E_1}{2} - a e \cos \frac{E_3 - E_1}{2} . \qquad (35)$$

Excluding $e \cos \left[(E_3 + E_1)/2 \right]$ and solving this system with respect to $e \cos \left[(v_3 + v_1)/2 \right]$, we obtain

$$e \cos \frac{v_3 + v_1}{2} = \frac{p}{(r_1 r_3)^{1/2}} \cos \frac{E_3 - E_1}{2} - \cos \frac{u_3 - u_1}{2} ,$$

where $v_3 - v_1 = u_3 - u_1$. Substituting this equality into Eq. (33), we have

$$p = \frac{2 r_1 r_3 \sin^2 \dfrac{u_3 - u_1}{2}}{r_1 + r_3 - 2 (r_1 r_3)^{1/2} \cos \dfrac{u_3 - u_1}{2} \cos \dfrac{E_3 - E_1}{2}} . \qquad (36)$$

Solving Eqs. (36) and (32) jointly and excluding ρ, we find for η

$$\eta^2 = \frac{m^2}{l + \sin^2 \dfrac{g}{2}} , \qquad (37)$$

where

$$u_3 - u_1 = 2f , $$
$$E_3 - E_1 = 2g , \qquad (38)$$

$$m = \frac{k(t_3 - t_1)}{\left[2 (r_1 r_3)^{1/2} \cos f \right]^{3/2}} , \qquad (39)$$

$$l = \frac{r_1 + r_3}{4\,(r_1 r_3)^{1/2} \cos f} - \frac{1}{2}\,. \tag{40}$$

Recall that, in these expressions, r_1, r_3 and f are known quantities. The only unknown ones are η and g. Hence, it is necessary to have, besides Eq. (37), one more equation for their determination. The latter can be derived in the following manner.

We have from Kepler's equation

$$E_1 - e \sin E_1 = \frac{k(t_1 - T)}{a^{3/2}}\,,$$

$$E_3 - e \sin E_3 = \frac{k(t_3 - T)}{a^{3/2}}\,.$$

From that we find, taking into account Eq. (38), that

$$\frac{k(t_3 - t_1)}{a^{3/2}} = 2g - 2e \sin g \, \cos \frac{E_3 + E_1}{2}\,. \tag{41}$$

Both a and $e \cos\left[(E_3 + E_1)/2\right]$ should be excluded from this equation. On substituting $e \cos\left[(E_3 + E_1)/2\right]$ from Eq. (34) into Eq. (41), we obtain

$$\frac{k(t_3 - t_1)}{a^{3/2}} = 2g - \sin 2g + 2\frac{(r_1 r_3)^{1/2}}{a} \sin g \, \cos f\,. \tag{42}$$

The exclusion of a still remains to be achieved. From the formula for the radius-vector r we have

$$\frac{r_3}{a} = 1 - e \cos E_1\,, \qquad \frac{r_3}{a} = 1 - e \cos E_3$$

or

$$\frac{r_1 + r_3}{a} = 2 - 2e \cos g \, \cos \frac{E_3 + E_1}{2}\,. \tag{43}$$

By solving Eqs. (42) and (43) jointly we obtain for the second sought equation connecting η with g

$$\eta^3 - \eta^2 = \frac{m^2}{\sin^3 g}(2g - \sin 2g)\,. \tag{44}$$

Thus, the values of η and g can be obtained by solving Eqs. (37) and (44) jointly.

We denote by x and $X(x)$:

$$x = \sin^2 \frac{g}{2}\,, \qquad X(x) = \frac{2g - \sin(2g)}{\sin^3 g}\,. \tag{45}$$

Then Eqs. (44) and (37) can be rewritten in the following form:

$$\eta^3 - \eta^2 = m^2 X(x),$$
$$x = m^2 \eta^{-2} - l. \tag{46}$$

These are the equations obtained by Gauss for determination of the ratio η, and the half-difference of eccentric anomalies, $g = (E_3 - E_1)/2$. Having g, one can easily find the element a from Eq. (41).

The solution of Eq. (46) is reduced after some modifications (in particular, expanding the function $X(x)$ initially in series of g, and thereafter of x), to the solution of the following equation

$$\eta^3 - \eta^2 - h\eta - \frac{h}{g} = 0, \tag{47}$$

where

$$h = \frac{m}{5/6 + l + \xi}, \tag{48}$$

m and l are given by Eqs. (39) and (40), and ξ is given by

$$\xi = \frac{2}{35}x^2 + \frac{52}{1575}x^3 + \dots . \tag{49}$$

Eq. (47) can be solved via iteration. Assuming initially that $x = 0$ and, hence, $\xi = 0$, for h in Eq. (48) we obtain to a first approximation

$$h = \frac{m}{5/6 + l}. $$

Then finding η from the cubic equation (47), we obtain x from the second equation in (46), and thereafter $\xi(x)$ from Eq. (49) and, hence, a more precise value of h from Eq. (48). For this new value of h the process of determination of η is repeated up to the final results. In practice this process converges rather quickly. To simplify the calculations special tables exist for the function $\xi(x)$, which is the original one created by Gauss himself.

Thus, the ratio η is obtained, after which one can find the value of x from the second equation in (46). Finally, from the first equation of (45), i.e.

$$g = 2 \arcsin \sqrt{x} \tag{50}$$

we obtain g, our final aim.

Now, without any difficulty, one can obtain the major half-axis a from Eq. (43)

$$a = \frac{r_1 + r_3}{2 - 2e \cos g \cos \dfrac{E_3 + E_1}{2}}. \tag{51}$$

This is the final relationship of the Gauss method of determination of the major half-axis of an elliptic orbit.

7　The Moulton Method

The Gauss method is perfect from all points of view; however, it is somewhat cumbersome. A more practical and more convenient method for the determination of the semimajor axis was proposed by Moulton in 1901, exactly a hundred years after Gauss. The essence of the Moulton method is as follows.

First of all one determines the parameter p, which in the case of an ellipse is related to a and e by the relationship $p = a(1 - e^2)$. Adopting $M = 1$ from Kepler's first law, we have

$$k\sqrt{p}\, dt = r^2\, du\,.$$

Integrating within the limits u_1 and u_3, we obtain

$$k\sqrt{p}\,(t_3 - t_1) = \int_{u_1}^{u_3} r^2\, du\,. \tag{52}$$

The integrand on the right-hand side represents the doubled area demarcated by the radius-vectors r_1 and r_3 and by an arc of the elliptic orbit, namely that covered by the celestial body during a time interval $t_3 - t_1$. To integrate Eq. (52) it is necessary to express r via u. We can expand r^2 via series in $u - u_2$, where u_2 is its value at the second moment of time t_2: the latter is convenient to take as an initial value. Then

$$r^2 = r_2^2 + c_1(u - u_2) + c_2(u - u_2)^2 + c_3(u - u_2)^3 + \dots\,. \tag{53}$$

Here the interval $t_2 - t_1$ is assumed to be not too large and correspondingly the difference $u_2 - u_1$ is small.

The coefficients c_1, c_2, and c_3 are unknown; however, they can be obtained using the fact that the quantities v_1 and v_3 are known for given u_1 and u_3. Therefore, rewriting Eq. (53) for two particular cases r_1 and r_3, respectively, and neglecting third and higher orders of $u - u_2$, we obtain

$$\begin{aligned}
r_1^2 &= r_2^2 + c_1\sigma_3 + c_2\sigma_3^2\,, \\
r_3^2 &= r_2^2 + c_1\sigma_1 + c_2\sigma_1^2\,,
\end{aligned} \tag{54}$$

where the following notations are used:

$$\sigma_1 = u_3 - u_2\,, \qquad \sigma_3 = u_2 - u_1\,.$$

Eq. (54) uniquely determines c_1 and c_2.

By substituting the value of r^2 from Eq. (53) into (52) and integrating, we obtain

$$k\sqrt{p}\,(t_3 - t_1) = r_2^2\sigma_2 - \frac{c_1}{2}\sigma_2(\sigma_3 - \sigma_1) + \frac{c_2}{3}(\sigma_1^3 + \sigma_3^3)\,, \tag{55}$$

where $\sigma_2 = u_3 - u_1$. From that we determine the numerical value of p with an accuracy that is quite adequate in the majority of cases.

However, for determination of the major half-axis a it is necessary also to know e. Therefore our problem should now be the determination of e and ω.

Using the equation of an ellipse in polar coordinates:

$$e \cos v = \frac{p - r}{r}$$

for the first and third moments, we obtain

$$e \cos v_1 = \frac{p - r_1}{r_1},$$
$$e \cos v_3 = \frac{p - r_3}{r_3}. \tag{56}$$

From Eq. (28) we have

$$v_3 = v_1 + (u_3 - u_1),$$

and, substituting it into the second equation in (56), we obtain

$$e \cos v_1 \cos(u_3 - u_1) - e \sin v_1 \sin(u_3 - u_1) = \frac{p - r_3}{r_3}.$$

Replacing $e \cos v_1$ by its expression from Eq. (56), we have finally

$$e \cos v_1 = \frac{p - r_1}{r_1},$$
$$e \sin v_1 = \frac{1}{\sin(u_3 - u_1)} \left(\frac{p - r_1}{r_1} \cos(u_3 - u_1) - \frac{p - r_3}{r_3} \right). \tag{57}$$

The only unknown quantities in this system are e and v_1; the others r_1, r_3, u_1, and u_3 as well as p, have been determined earlier. Then we obtain v_1 from the equation

$$\tan v_1 = \frac{1}{\sin(u_3 - u_1)} \left(\cos(u_3 - u_1) - \frac{r_1}{r_3} \frac{p - r_3}{p - r_1} \right), \tag{58}$$

and thereafter e from one of the equations in (57). For known v_1 and v_2 we obtain ω from one of the expressions in (28).

Thus, four quantities from six are obtained: Ω, i, e and p. From the known e and p we obtain a from the formula

$$a = \frac{p}{1 - e^2}. \tag{59}$$

The sixth and last element T still remains. Obtaining the eccentric anomaly E from the expression

$$\tan \frac{E}{2} = \left(\frac{1-e}{1+e}\right)^{1/2} \tan \frac{v}{2}, \qquad (60)$$

and using it for one of the moments t_1, t_2 or t_3, we find the mean anomaly M from Kepler's equation

$$M = E - e \sin E.$$

The quantity T is related to the mean anomaly M by the following relationship:

$$M = \frac{k}{a^{3/2}}(t - T). \qquad (61)$$

which gives for sixth element, T, the moment of the passage of the celestial body through the perihelion. Finally

$$T = t - \frac{a^{3/2}}{k}M. \qquad (62)$$

The number of elements in the case of a hyperbolic orbit is again six, and therefore the described method of determination of elements of the elliptic orbit can be used without any changes in that case as well. The only difference concerns the application of concrete formulas, especially in the final stage of calculations when instead of the eccentric anomaly E in Kepler equation the function F, together with its corresponding equation, should be used.

8   Parabolic Orbit. The Basic Equations

The parabolic orbit is determined by five elements, one element less than the elliptic and hyperbolic orbits. However, the number of observations necessary for determination of the parabolic orbit is the same – three complete observations. These observations enable one to write the following equations:

$$\rho \cos \lambda \cos \beta = x + X,$$
$$\rho \sin \lambda \cos \beta = y + Y, \qquad (63)$$
$$\rho \sin \beta = z + Z.$$

Note that the number of unknown variables is eight, five of which are the orbital elements Ω, i, ω, q, and T; the other three are the heliocentric distances of the celestial body ρ_1, ρ_2, and ρ_3. In other words, in the case of a parabolic orbit one equation might seem not to be necessary. However, it is not so.

The point is that the real number of unknown parameters in the case of elliptic and hyperbolic orbits is not nine but ten, taking into account the indirect presence of an extra unknown parameter, namely, the heliocentric distance of the celestial body r_2 for the second moment of observation t_2. The tenth equation in the case of an elliptic orbit turns out to be the Gauss equation.

The heliocentric distance r_2 also exists in the case of a parabolic orbit, again within the basic equations. Hence, the real number of unknown variables in that case is not eight but nine, i.e. five elements, three geocentric distances and one heliocentric distance. There are nine equations, exactly as many as the number of unknown parameters. Everything seems all right, and no additional equation like the Gauss equation has to be built up for a parabolic orbit. However, the problem in fact includes the heliocentric distance r_2 in the nonlinear form of r_2^3, as distinct from the geocentric distances ρ_1, ρ_2, and ρ_3. This difficulty should be overcome.

As it turns out, one of the observations in the case of a parabolic orbit can be used in such a manner – owing to the clever method proposed by Olbers, whereby r_2 vanishes completely. However, this is reached using one of the three equations of our basic system relating the geocentric distances ρ_1, ρ_2, and ρ_3 to the ratios of areas of triangles, which are required for the determination of the geometric distances. The other two equations with three unknown parameters ρ_1, ρ_2 and ρ_3 enable one to find in the best case the relation between the two unknown ones, e.g., $\rho_3 = f(\rho_1)$. Hence, it is necessary to have one more equation of the same structure. Such an equation already exists – the Euler equation derived earlier, in Chapter IV. By solving jointly these two equations, we obtain ρ_1, ρ_2 and then ρ_3. Thus, the first stage of computation is completed and one can move to the main problem – the determination of the orbit's elements.

These are the general features of the formulation of the problem of the determination of elements of the parabolic orbit. As for the basic equations, they again involve Eq. (63), as in the case of an elliptic orbit. With the help of these equations, the first stage of the problem – the determination of the geocentric distances of the celestial body, ρ_1, ρ_2, and ρ_3, is realized.

Turn to Fig. 7.5, in which γNK is the ecliptic, and M_2 is the position of celestial body at the moment of the second observation t_2. The center of the coordinate system coincides with that of the Earth T.

Imagine an arbitrary plane passing through the observed position of the celestial body M_2 in such a manner that it intersects with the ecliptic. The resulting curve TN obviously has no relation to the line of nodes because the plane NM_2 has no relation to the orbital plane. Therefore we denote

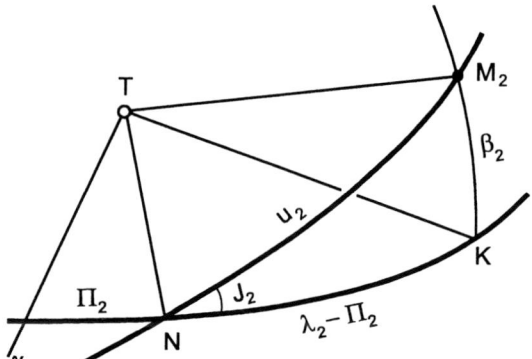

Fig. 7.6. *The discovery of the elements of a parabolic orbit.*

by Π_2 the angular distance of the point N from the point γ, and by J_2 the inclination of the plane NM_2 with respect to the ecliptic; of course, both Π_2 and J_2 are still unknown quantities and should be determined later.

The ecliptic geocentric coordinates of the celestial body M_2 are represented by latitude β_2 and longitude λ_2 counted from the point γ. For the spherical triangle NKM_2 we have $NK = \lambda_2 - \Pi_2$, and $NM_2 = u_2$, where u_2 is also unknown. From this triangle we can write

$$\cos u_2 = \cos(\lambda_2 - \Pi_2) \cos \beta_2 \, ,$$
$$\cos J_2 \sin u_2 = \sin(\lambda_2 - \Pi_2) \cos \beta_2 \, , \tag{64}$$
$$\sin J_2 \sin u_2 = \sin \beta_2 \, .$$

Dividing the third equation by the second one, we obtain

$$\tan J_2 = \frac{\tan \beta_2}{\sin(\lambda_2 - \Pi_2)} \, . \tag{65}$$

For any arbitrary value of Π_2 one can determine J_2 from Eq. (65). The peculiarity of the solution of the problem of determination of geocentric distances in the case of a parabolic orbit is that the quantities λ_2 and β_2 should later be replaced by Π_2 and J_2.

Below, we can decrease all angles in the plane of the ecliptic by Π_2. This is equivalent to rotation of the coordinate system around the vertical axis z by Π_2; as a result the axis OX will be directed not towards the point γ, but towards N. Then Eq. (63) will take the form

$$x = \rho \cos(\lambda - \Pi_2) \cos \beta - R \cos(L - \Pi_2) \, ,$$
$$y = \rho \sin(\lambda - \Pi_2) \cos \beta - R \sin(L - \Pi_2) \, , \tag{66}$$
$$z = \rho \sin \beta \, ,$$

where R and L are the geocentric distance and longitude of the Sun.

From the left-hand sides of Eq. (66), x, y, and z are obviously functions of the orbital elements Ω, i, ω, q, T as well as of time t. However, if we substitute the expressions for x, y, and z from Eq. (66) into the basic system (1) with corresponding subscripts, then the orbital elements will be excluded completely, and the only unknown variables that remain will be the geocentric distances ρ_1, ρ_2, and ρ_3.

Thus, rewriting Eq. (66) with the corresponding subscripts for the moments of observations t_1, t_2, and t_3, and substituting the expressions for the coordinates into the basic equation (1), we have

$$\frac{[r_2 r_3]}{[r_1 r_3]}[\rho_1 \cos(\lambda_1 - \Pi_2)\cos\beta_1 - R_1\cos(L_1 - \Pi_2)]$$

$$+\frac{[r_1 r_2]}{[r_1 r_3]}[\rho_3 \cos(\lambda_3 - \Pi_2)\cos\beta_3 - R_3\cos(L_3 - \Pi_2)]$$

$$= \rho_2 \cos(\lambda_2 - \Pi_2)\cos\beta_2 - R_2\cos(L_2 - \Pi_2)\,,$$

$$\frac{[r_2 r_3]}{[r_1 r_3]}[\rho_1 \sin(\lambda_1 - \Pi_2)\cos\beta_1 - R_1\sin(L_1 - \Pi_2)] \qquad (67)$$

$$+\frac{[r_1 r_2]}{[r_1 r_3]}[\rho_3 \sin(\lambda_3 - \Pi_2)\cos\beta_3 - R_3\sin(L_3 - \Pi_2)]$$

$$= \rho_2 \sin(\lambda_2 - \Pi_2)\cos\beta_2 - R_2\sin(L_2 - \Pi_2)\,,$$

$$\frac{[r_2 r_3]}{[r_1 r_3]}\rho_1 \sin\beta_1 + \frac{[r_1 r_2]}{[r_1 r_3]}\rho_3 \sin\beta_3 = \rho_2 \sin\beta_2\,.$$

The unknown parameters in this system are ρ_1, ρ_2, and ρ_3 together with Π_2 and r_2 appearing in indirect form, through the ratios of the triangles' areas.

For the determinations of the elements of a parabolic orbit it is enough to know, as we shall see later, two geocentric distances, ρ_1 and ρ_3. Formally, it is necessary first of all to exclude ρ_2 from Eq. (67). However, that is not enough since even by excluding ρ_2 we still do not get rid of r_2 in the other two equations of the system. Meanwhile the point is to apply a method enabling one to achieve the final exclusion of r_2. For that, one first of all has to replace λ_2 by J_2 within Eq. (67). By substituting, therefore, the value of $\sin(\lambda_2 - \Pi_2)$ from Eq. (65) into the second equation in (67), we can write

$$\frac{[r_2 r_3]}{[r_1 r_3]}[\rho_1 \sin(\lambda_1 - \Pi_2) \cos \beta_1 - R_1 \sin(L_1 - \Pi_2)]$$

$$+ \frac{[r_1 r_2]}{[r_1 r_3]}[\rho_3 \sin(\lambda_3 - \Pi_2) \cos \beta_3 - R_3 \sin(L_3 - \Pi_2)]$$

$$= \rho_2 \frac{\sin \beta_2}{\tan J_2} - R_2 \sin(L_2 - \Pi_2), \tag{68}$$

$$\frac{[r_2 r_3]}{[r_1 r_3]} \rho_1 \sin \beta_1 + \frac{[r_1 r_2]}{[r_1 r_3]} \rho_3 \sin \beta_3 = \rho_2 \sin \beta_2.$$

Here λ_2 is excluded though J_2 has appeared instead.

By multiplying the first equation by $\sin J_2$, and the second one by $\cos J_2$, and taking the difference, we exclude not only ρ_2 but β_2 as well. Substituting also the values of the ratios of triangle areas from Eq. (37), Chapter VI, we obtain

$$\rho_3 = \psi(r_2) + M\rho_1. \tag{69}$$

where $\psi(r_2)$ depends on r_2

$$\psi(r_2) = \sin(L_2 - \Pi_2) \left(\frac{1}{r_2^3 - 1/R_2^3} \right) Q, \tag{70}$$

where Q is a rather complicated combination of trigonometric functions containing λ_3, β_2, and J_2 as well as τ_1 and τ_3. M is given by

$$M = \frac{\tau_1}{\tau_3} \frac{\sin \beta_1 \cot J_2 - \sin(\lambda_1 - L_2) \cos \beta_1}{\sin(\lambda_3 - L_2) \cos \beta_3 - \sin \beta_3 \cot J_2}, \tag{71}$$

i.e. M is independent of r_2. Eq. (69) contains three unknown variables, ρ_1, ρ_3, and r_2.

9 The Olbers Equation

Up to now the position of the plane passing through the second position of the celestial body M_2 (Fig. 7.6) and determined by parameters J_2 and Π_2 has been chosen arbitrarily. Olbers proposed the idea of using the plane not only passing through the second position of the celestial body M_2 but also through the second position of the Sun. In other words, the point N in Fig. 7.6 coincides with the second position of the Sun on the ecliptic.

It is easy to see that, in this case, the parameter Π_2 coincides with L_2, i.e. $\Pi_2 = L_2$. Then the first term on the right-hand side of Eq. (69), i.e. the function $\psi(r_2)$, becomes zero in accord with Eq. (70), and hence, the term containing r_2 in Eq. (69) vanishes completely(!). As a result, Eq. (69) takes the form

$$\rho_3 = M\rho_1 \,. \tag{72}$$

This equation is known as the **Olbers equation**.

The coefficient M, as follows from Eq. (71) depends besides on τ_1 and τ_3, also on β_1, β_3, λ_1, and λ_3 as well as on J_2, which is as yet unknown. Because $\Pi_2 = L_2$, the value of J_2 can be obtained directly from Eq. (65):

$$\tan J_2 = \frac{\tan \beta_2}{\sin(\lambda_2 - L_2)} \,. \tag{73}$$

Thus, the second observation of the celestial body, i.e. the quantities λ_2 and β_2, enters into the Olbers equation through J_2.

10   Joint Solution of the Olbers and Euler Equations

The Olbers equation (72) is a relationship between two unknown variables, ρ_1 and ρ_3, whereas r_2 is absent in Eq. (72). To find ρ_1 and ρ_3, it is necessary to have one more relationship between these two geocentric distances. For that we will use the Euler equation (see Chapter IV) which in our case has the form

$$6k(t_3 - t_1) = (r_1 + r_3 + s)^{3/2} - (r_1 + r_3 - s)^{3/2} \,. \tag{74}$$

The left-hand side of this equation is known, and on the right-hand side we have two heliocentric distances, r_1 and r_3, corresponding to the first and third positions of the celestial body, as well as the chord s connecting the first and the third positions of the celestial body. All these three quantities are unknown, and one has to interrelate them in terms of ρ_1 and ρ_3, i.e. it is necessary to move from r_1, r_3 and s in Eq. (74) to ρ_1 and ρ_3. As a result from Eq. (74) we obtain

$$\rho_3 = \varphi(\rho_1) \,, \tag{75}$$

so that Eq. (75) together with the Olbers equation (72) forms a system of two algebraic equations with two unknown variables, ρ_1 and ρ_3, which is solvable in principle.

However, in contrast to the Olbers equation, in which ρ_1 and ρ_3 are related linearly, the dependence between ρ_1 and ρ_3 in Eq. (75) is nonlinear. These equations can indeed be solved, however; due to the radicals in the Euler equation, while determining ρ_1 or ρ_3 one arrives at an algebraic equation of rather high order. The situation is quite similar to that of the Laplace equation while determining the geocentric distances for an elliptical orbit, so that this method loses its practical interest.

At present the method of successive approximations of determination of r and ρ while performing the joint solution of the Olbers and Euler equations is considered to be the most reliable and convenient. The essence of this method is as follows.

First, r_1, r_3 and s are expressed through the Cartesian heliocentric ecliptic coordinates, namely,

$$r_1^2 = x_1^2 + y_1^2 + z_1^2 , \tag{76}$$

$$r_3^2 = x_3^2 + y_3^2 + z_3^2 , \tag{77}$$

$$s^2 = (x_3 - x_1)^2 + (y_3 - y_1)^2 + (z_3 - z_1)^2 , \tag{78}$$

where the Cartesian coordinates x, y, and z are themselves expressed through ρ_1 and ρ_3 with the help of the following relationships:

$$x_1 = \rho_1 \cos(\lambda_1 - L_2) \cos \beta_1 - R_1 \cos(L_1 - L_2) ,$$
$$y_1 = \rho_1 \sin(\lambda_1 - L_2) \cos \beta_1 - R_1 \sin(L_1 - L_2) , \tag{79}$$
$$z_1 = \rho_1 \sin \beta_1 .$$

$$x_3 = \rho_3 \cos(\lambda_3 - L_2) \cos \beta_3 - R_3 \cos(L_3 - L_2) ,$$
$$y_3 = \rho_3 \sin(\lambda_3 - L_2) \cos \beta_3 - R_3 \sin(L_3 - L_2) , \tag{80}$$
$$z_3 = \rho_3 \sin \beta_3 .$$

It follows from these relationships that the right-hand side of the Euler equation can be represented as a function of ρ_1 and ρ_3; i.e. one has to substitute the expressions for x, y, and z from Eqs. (79) and (80) into Eqs. (76), (77) and (78), and then, as a result, r_1, r_3 and s should appear as functions of ρ_1 and ρ_3. On substituting these values of r_1, r_3 and s into Eq. (74), one will have a relationship of Eq. (75) type. Before turning to the main steps of the application of this method for determination of ρ_1 and ρ_3, let us write the necessary working formulae.

The Dependence of s on $r_1 + r_3$. The aim is the solution of Euler's equation, i.e. the determination of s in terms of its dependence on $r_1 + r_3$ in Eq. (74). The solution is obtained by means of an auxiliary quantity γ (an angle) in the following form:

$$s = (r_1 + r_3) \sin (2\gamma) , \tag{81}$$

where

$$\sin \gamma = \sqrt{2} \sin \theta , \tag{82}$$

$$\sin(3\theta) = \frac{6k}{2^{3/2}} \frac{t_3 - t_1}{(r_1 + r_3)^{3/2}} , \tag{83}$$

i.e. for the given sum $r_1 + r_3$ we obtain, first, θ from Eq. (83), then γ from Eq. (82), and finally s from Eq. (81).

The Dependence of ρ_1 on s. Substituting Eqs. (79) and (80) into Eq. (78) and replacing ρ_3 by $M\rho_1$ according to the Olbers equation (72), we find for the dependence of ρ_1 on s

$$\rho_1 = \frac{1}{h} \left[g \cos \varphi + (s^2 - g^2 \sin^2 \varphi)^{1/2} \right],\tag{84}$$

where

$$\cos \varphi = \cos \xi \cos(G - H).\tag{85}$$

The auxiliary quantities g, G, h, H and ξ are determined from the following equations

$$g \cos G = R_3 \cos(L_3 - L_2) - R_1 \cos(L_1 - L_2),$$
$$g \sin G = R_3 \sin(L_3 - L_2) - R_1 \sin(L_1 - L_2),\tag{86}$$

$$h \cos \xi \cos H = M \cos(\lambda_3 - L_2) \cos \beta_3 - \cos(\lambda_1 - L_2) \cos \beta_1,$$
$$h \cos \xi \sin H = M \sin(\lambda_3 - L_2) \cos \beta_3 - \sin(\lambda_1 - L_2) \cos \beta_2,\tag{87}$$
$$h \sin \xi = M \sin \beta_3 - \sin \beta_1,$$

where M is the same coefficient as in the Olbers equation.

The Dependence of $r_1 + r_3$ on ρ_1. Substituting Eqs. (79) and (80) into Eqs. (76) and (77) and replacing ρ_3 by $M\rho_1$ again, we obtain

$$r_1 = R_1 \sin \psi_1 \sec \theta_1,$$
$$r_3 = R_3 \sin \psi_3 \sec \theta_3,\tag{88}$$

where the auxiliary quantities θ_1, θ_3, ψ_1 and ψ_3 are determined from the following relationships:

$$\tan \theta_1 = \frac{\rho_1}{R_1} \csc \psi_1 - \cot \psi_1,$$
$$\tan \theta_3 = \frac{\rho_3}{R_3} \csc \psi_3 - \cot \psi_3,\tag{89}$$

$$\cos \psi_1 = \cos(\lambda_1 - L_1) \cos \beta_1,$$
$$\sin \psi_1 \cos P_1 = \sin(\lambda_1 - L_1) \cos \beta_1,$$
$$\sin \psi_1 \sin P_1 = \sin \beta_1,$$
$$\cos \psi_3 = \cos(\lambda_3 - L_3) \cos \beta_3,\tag{90}$$
$$\sin \psi_3 \cos P_3 = \sin(\lambda_3 - L_3) \cos \beta_3,$$
$$\sin \psi_3 \sin P_3 = \sin \beta_3,$$

where the angles P_1 and P_3 will not be used further.

We turn now to the main problem, namely, the application of the formulas presented above and the sequence of realization of the successive approximation method for the determination of ρ_1 and ρ_3. The successive approximation method includes the following four steps.

Step I. One starts by adopting some values for r_1 and r_3. Usually it is assumed that $r_1 = r_3 = 1$, for comets moving in parabolic orbits; as a rule they are discovered at the same distance from the Sun as from the Earth. From known values of r_1 and r_3, the chord s is determined with the help of Eq. (81).

Step II. From the known magnitude of s from Eq. (84), first ρ_1, and thereafter, ρ_3, are obtained from the Olbers equation.

Step III. From the known magnitudes of ρ_1 and ρ_3 one finds the heliocentric distances r_1 and r_3 by means of Eq. (88).

Step IV. Obviously, the obtained r_1 and r_3 differ from their initial values, i.e. from unity. Therefore, with these new values for r_1 and r_3 one has to repeat all the calculations in the same sequence, unless the values of r_1 and r_3, obtained from Eq. (88) after n-cycle calculations, coincide with their values before that cycle. Obviously, the obtained values of ρ_1 and ρ_3 should be the desirable solution both of the Olbers and of the Euler equations. Simultaneously, the heliocentric distances r_1 and r_3 are determined as well.

11   Determination of the Elements of a Parabolic Orbit

With the determination of the geocentric distances of the celestial body ρ_1 and ρ_3 the solution of the first part of our problem is completed. The second part concerns the determination of the elements of the parabolic orbit.

First, one has to perform the transition from the geocentric ecliptic coordinates ρ, λ, β to the heliocentric ecliptic coordinates r, l, b. This can be done in the same manner as in the case of an elliptic orbit, i.e. with the help of the system of three equations (10) of Chapter VI. Using this system for the moments t_1 and t_3, we can obtain all the necessary quantities r, l and b with subscripts 1 and 3.

Afterwards, using Eq. (22) for the moments t_1 and t_3, we obtain uniquely the numerical values of the latitude arguments u_1 and u_3 (see Fig. 7.6) as well as the first two elements, Ω and i, determining the spatial orientation of the orbital plane by means of the formulas

$$\tan(l_1 - \Omega) = \frac{\tan b_1 \sin(l_3 - l_1)}{\tan b_3 - \tan b_1 \cos(l_3 - l_1)}, \qquad (91)$$

$$\tan i = \frac{\tan b_1}{\sin(l_1 - \Omega)} . \tag{92}$$

The latitude argument u can be obtained from the expression

$$\tan u_1 = \frac{\tan(l_1 - \Omega)}{\cos i} ,$$

$$\tan u_3 = \frac{\tan(l_3 - \Omega)}{\cos i} . \tag{93}$$

Up to now everything was the same as in the case of an elliptic orbit. The other three elements of a parabolic orbit, ω, q and T, should be determined in the following way.

Consider the equation of the parabola in polar coordinates and apply it for the first and third moments

$$r_1 = \frac{q}{\cos^2 \dfrac{v_1}{2}} , \qquad r_3 = \frac{q}{\cos^2 \dfrac{v_3}{2}} . \tag{94}$$

From that we obtain

$$\frac{1}{\sqrt{q}} \cos \frac{v_1}{2} = \frac{1}{\sqrt{r_1}} ,$$

$$\frac{1}{\sqrt{q}} \cos \frac{v_3}{2} = \frac{1}{\sqrt{r_3}} , \tag{95}$$

where r_1 and r_3 are known quantities, and the only unknown ones are v_1 and v_3. However, in view of

$$u_1 = v_1 + \omega ,$$

$$u_3 = v_3 + \omega , \tag{96}$$

one can replace v_3 in the second equation in (95) by $v_3 = v_1 + (u_3 - u_1)$, thus having instead of Eq. (95)

$$\frac{1}{\sqrt{q}} \sin \frac{v_1}{2} = \frac{1}{\sqrt{r_1}} \cot \frac{u_3 - u_1}{2} - \frac{1}{\sqrt{r_3}} \operatorname{cosec} \frac{u_3 - u_1}{2} ,$$

$$\frac{1}{\sqrt{q}} \cos \frac{v_1}{2} = \frac{1}{\sqrt{r_1}} . \tag{97}$$

From that we obtain

$$\tan \frac{v_1}{2} = \cot \frac{u_3 - u_1}{2} - \left(\frac{r_1}{r_3}\right) \operatorname{cosec} \frac{u_3 - u_1}{2} . \tag{98}$$

From which we find v_1, and then, in view of Eq. (96), we obtain the third element of the orbit – ω:

$$\omega = u_1 - v_1 . \tag{99}$$

Having the values of v_1 and r_1 or v_3 and r_3, we find the fourth element, q, using the expression

$$q = r_1 \cos^2 \frac{v_1}{2} = r_3 \cos^2 \frac{v_3}{2} . \tag{100}$$

The fifth and last element of the parabolic orbit, T, the time of transition of the comet through the perihelion, is obtained from the equation connecting the true anomaly v with the moment of time t:

$$\frac{k(t-T)}{q^{3/2}\sqrt{2}} = \tan \frac{v}{2} + \frac{1}{3} \tan^3 \frac{v}{2} . \tag{101}$$

Applying this equation, say, for the first moment, we obtain

$$T = t_1 - \frac{q^{3/2}}{k}\sqrt{2} \left(\tan \frac{v_1}{2} + \frac{1}{3} \tan^3 \frac{v_1}{2} \right) . \tag{102}$$

Thus the determination of all five elements of a parabolic orbit, Ω, i, ω, q, and T by three observations is completed.

Chapter VIII

The N-Body Problem

1 The Differential Equations of Motion

Consider a system consisting of N bodies with the masses m_1, m_2, ..., m_N interacting according to Newton's law of gravity. Assume that all of the bodies are spherical and centrosymmetric, i.e. they consist of homogeneous spherical layers. Then the mutual attractions among them occur as if their masses were concentrated within their centers; in this case the bodies are considered as mathematical points.

Let us fix the coordinate system ξ, η, ζ of rest with an arbitrarily situated center. Let ξ_i, η_i, ζ_i be the coordinates of the body m_i and let r_{ij} be the distance between the bodies m_i and m_j

$$r_{ij} = [(\xi_i - \xi_j)^2 + (\eta_i - \eta_j)^2 + (\zeta_i - \zeta_j)^2]^{1/2} . \qquad (1)$$

The gravitational attraction exerted on the point mass m_i by another one m_j equals

$$k^2 \frac{m_i m_j}{r_{ij}^2} ,$$

with the direction coinciding with the line connecting the points m_i and m_j. The components of the force along the axes ξ, η, ζ, are equal, respectively, to

$$k^2 m_i m_j \frac{\xi_i - \xi_j}{r_{ij}^3} , \quad k^2 m_i m_j \frac{\eta_i - \eta_j}{r_{ij}^3} , \quad k^2 m_i m_j \frac{\zeta_i - \zeta_j}{r_{ij}^3} .$$

Therefore one can write the differential equations of motion for a body m_i

in the field of the other $n - 1$ bodies in the following form:

$$m_i \frac{d^2\xi_i}{dt^2} = -k^2 m_i \sum_{j=1}^{n} m_j \frac{\xi_i - \xi_j}{r_{ij}^3},$$

$$m_i \frac{d^2\eta_i}{dt^2} = -k^2 m_i \sum_{j=1}^{n} m_j \frac{\eta_i - \eta_j}{r_{ij}^3}, \qquad (2)$$

$$m_i \frac{d^2\zeta_i}{dt^2} = -k^2 m_i \sum_{j=1}^{n} m_j \frac{\zeta_i - \zeta_j}{r_{ij}^3},$$

where $i = 1, 2, \ldots, n$ and $i \neq j$.

Each of these equations contains directly and indirectly, through r_{ij}, all $3n$ variables ξ_i, η_i, and ζ_i, so that the system must be solved jointly. Eq. (2) contains $3n$ equations, each of them of second order, so that the order of the problem is $6n$, that being also the total number of integrals or constants of integration. The theoretical aim of the solution of this system is the determination of the motion of each of the N bodies, namely, the dependence of its coordinates on time t:

$$\xi_i = f_i(t), \quad \eta_i = \varphi_i(t), \quad \zeta_i = \psi_i(t). \qquad (3)$$

Having these functions, one can find all the parameters of the motion – the velocity, acceleration, trajectory etc. However, as we shall see later, this aim cannot be attained, and the functions in Eq. (3) cannot be obtained in explicit form. Nevertheless, the system leads to a number of interesting results.

Our equations can be rewritten in a more elegant and convenient form via the potential U:

$$U = \frac{1}{2} k^2 \sum_{i=1}^{n} \sum_{j=1}^{n} \frac{m_i m_j}{r_{ij}}, \qquad (4)$$

where the $1/2$ on the right-hand side appears in order to avoid consideration twice of the symmetrical combinations ij and ji. The partial derivative of U with respect to ξ_i is:

$$\frac{\partial U}{\partial \xi_i} = k^2 m_i \frac{\partial}{\partial \xi_i} \sum_{j=1}^{n} \frac{m_j}{r_{ij}},$$

or, in view of Eq. (1),

$$\frac{\partial}{\partial \xi_i} \left(\frac{1}{r_{ij}} \right) = -\frac{\xi_i - \xi_j}{r_{ij}^3}$$

we obtain

$$\frac{\partial U}{\partial \xi_i} = -k^2 m_i \sum_{j=1}^{n} m_j \frac{\xi_i - \xi_j}{r_{ij}^3} \, .$$

One obtains an analogous result after differentiation of Eq. (4) with respect to η_i and ζ_i. Therefore, Eq. (2) can be written in the much more compact form

$$m_i \frac{d^2 \xi_i}{dt^2} = \frac{\partial U}{\partial \xi_i},$$

$$m_i \frac{d^2 \eta_i}{dt^2} = \frac{\partial U}{\partial \eta_i}, \tag{5}$$

$$m_i \frac{d^2 \zeta_i}{dt^2} = \frac{\partial U}{\partial \zeta_i},$$

where $i = 1, 2, \ldots, n$. The system of Eq. (5) in this form was discovered by Lagrange in 1773.

2 The Motion of the Center of Mass of a System

The potential function depends only on the mutual distances between the bodies and, hence, it does not depend on the choice of coordinate system. This translational invariance implies that the shift of the center of the coordinate system parallel to the axis ξ onto α does not change the potential function:

$$\frac{\partial U}{\partial \alpha} = 0 \, . \tag{6}$$

Because U is a function of all coordinates ξ, we can write

$$U = U(\xi_1, \, \xi_2, \, \ldots, \, \xi_n) \, .$$

with the new coordinates ξ_i':

$$\xi_i' = \xi_i + \alpha \, . \tag{7}$$

The partial derivative of U with respect to α has the form

$$\frac{\partial U}{\partial \alpha} = \frac{\partial U}{\partial \xi_1'} \frac{d\xi_1'}{d\alpha} + \frac{\partial U}{\partial \xi_2'} \frac{d\xi_2'}{d\alpha} + \cdots + \frac{\partial U}{\partial \xi_n'} \frac{d\xi_n'}{d\alpha} = \sum_{i=1}^{n} \frac{\partial U}{\partial \xi_i'} \frac{d\xi_i'}{d\alpha} \, .$$

However, according to Eq. (7)

$$\frac{\partial \xi_i'}{\partial \alpha} = 1 \, .$$

Therefore, in view of Eq. (6), we have analogous equations also along the axes η_i and ζ_i for the shifts in β and γ respectively:

$$\frac{\partial U}{\partial \alpha} = \sum_{i=1}^{n} \frac{\partial U}{\partial \xi_i} = 0 \,,$$

$$\frac{\partial U}{\partial \beta} = \sum_{i=1}^{n} \frac{\partial U}{\partial \eta_i} = 0 \,,$$

$$\frac{\partial U}{\partial \gamma} = \sum_{i=1}^{n} \frac{\partial U}{\partial \zeta_i} = 0 \,.$$

On comparing with Eq. (5), we find

$$\sum_{i=1}^{n} m_i \frac{d^2 \xi_i}{dt^2} = 0 \,,$$

$$\sum_{i=1}^{n} m_i \frac{d^2 \eta_i}{dt^2} = 0 \,, \tag{8}$$

$$\sum_{i=1}^{n} m_i \frac{d^2 \zeta_i}{dt^2} = 0 \,.$$

These equations are easily integrated:

$$\sum_{i=1}^{n} m_i \frac{d\xi_i}{dt} = \alpha_1 \,,$$

$$\sum_{i=1}^{n} m_i \frac{d\eta_i}{dt} = \beta_1 \,,$$

$$\sum_{i=1}^{n} m_i \frac{d\zeta_i}{dt} = \gamma_1 \,,$$

where α_1, β_1, and γ_1 are the constants of integration. Integrating once more, we obtain

$$\sum_{1}^{n} m_i \xi_i = \alpha_1 t + \alpha_2 \,,$$

$$\sum_{1}^{n} m_i \eta_i = \beta_1 t + \beta_2 \,, \tag{9}$$

$$\sum_{1}^{n} m_i \zeta_i = \gamma_1 t + \gamma_2 \,,$$

where α_2, β_2, and γ_2 are the new constants of integration.

Denote by ξ, η, and ζ the coordinates of the center of mass of the system, and by $M = \sum_1^n m_i$ its total mass. Then, from the definition of the center of mass we obtain

$$\sum_1^n m_i \xi_i = M\xi,$$

$$\sum_1^n m_i \eta_i = M\eta,$$

$$\sum_1^n m_i \zeta_i = M\zeta$$

or, by comparison with Eq. (9), we obtain

$$M\xi = \alpha_1 t + \alpha_2,$$
$$M\eta = \beta_1 t + \beta_2, \qquad (10)$$
$$M\zeta = \gamma_1 t + \gamma_2.$$

This is the equation of the center of mass, i.e. the dependence of its coordinates ξ, η, ζ on time t. The coordinates of the center of mass change proportionally to time, and its velocity V is

$$V = \left[\left(\frac{d\xi}{dt} \right)^2 + \left(\frac{d\eta}{dt} \right)^2 + \left(\frac{d\zeta}{dt} \right)^2 \right]^{1/2} = \frac{1}{M} (\alpha_1^2 + \beta_1^2 + \gamma_1^2)^{1/2}, \qquad (11)$$

i.e. the velocity of the center of mass of the system is constant, i.e. independent of the positions of certain bodies of the system. The motion of the center of mass occurs without acceleration.

Finally, by excluding the time t from Eq. (10), we obtain the equation of the trajectory of the center of mass:

$$\frac{M\xi - \alpha_2}{\alpha_1} = \frac{M\eta - \beta_2}{\beta_1} = \frac{M\zeta - \gamma_2}{\gamma_1}. \qquad (12)$$

This is the equation of a straight line.

Thus, the center of mass of an isolated system of N bodies interacting only via Newton's law, moves in space with constant velocity and in a straight line. This result coincides with that obtained in the case of the two-body problem. The motion in both cases, i.e. for the two-body system and that of, say, a giant cluster containing hundreds or thousands of stars, occurs in one and the same manner – in a straight line and with constant velocity.

3   The Three Integrals of Sectorial Velocity

The potential function U is not changed by rotation of the coordinate axes because the distances between the bodies remain unchanged. Turn now to the coordinate system around the axis OZ at angle φ. Denoting the new coordinates via ξ_i', η_i', and ζ_i', one has their relation with the old coordinates ξ_i, η_i and ζ_i:

$$\xi_i' = \xi_i \cos\varphi - \eta_i \sin\varphi \,,$$
$$\eta_i' = \xi_i \sin\varphi + \eta_i \cos\varphi \,, \qquad (13)$$
$$\zeta_i' = \zeta_i \,.$$

The function U is not changed by rotation of the coordinate system:

$$\frac{\partial U}{\partial \varphi} = 0 \,. \qquad (14)$$

Therefore we can write

$$\frac{\partial U}{\partial \varphi} = \sum_1^n \frac{\partial U}{\partial \xi_i'}\frac{d\xi_i'}{d\varphi} + \sum_1^n \frac{\partial U}{\partial \eta_i'}\frac{d\eta_i'}{d\varphi} + \sum_1^n \frac{\partial U}{\partial \zeta_i'}\frac{d\zeta_i'}{d\varphi} = 0 \,. \qquad (15)$$

From Eq. (13) we find

$$\frac{d\xi_i'}{d\varphi} = -\xi_i \sin\varphi - \eta_i \cos\varphi = -\eta_i' \,,$$
$$\frac{d\eta_i'}{d\varphi} = \xi_i \cos\varphi - \eta_i \sin\varphi = \xi_i' \,, \qquad (16)$$
$$\frac{d\zeta_i'}{d\varphi} = 0 \,,$$

so that Eq. (15) takes the form

$$\sum_{i=1}^n \left(\xi_i'\frac{\partial U}{\partial \eta_i'} - \eta_i'\frac{\partial U}{\partial \xi_i'} \right) = 0 \,.$$

One can obtain analogous relationships by rotating the coordinate system around the axes OX and Oy respectively, i.e.

$$\sum_1^n m_i \left(\xi_i \frac{d^2\eta_i}{dt^2} - \eta_i \frac{d^2\xi_i}{dt^2} \right) = 0 \,,$$
$$\sum_1^n m_i \left(\eta_i \frac{d^2\zeta_i}{dt^2} - \zeta_i \frac{d^2\eta_i}{dt^2} \right) = 0 \,, \qquad (17)$$
$$\sum_1^n m_i \left(\zeta_i \frac{d^2\xi_i}{dt^2} - \xi_i \frac{d^2\zeta_i}{dt^2} \right) = 0 \,.$$

Each of these equations can be integrated once:

$$\sum_{1}^{n} m_i \left(\xi_i \frac{d\eta_i}{dt} - \eta_i \frac{d\xi_i}{dt} \right) = c_1 ,$$

$$\sum_{1}^{n} m_i \left(\eta_i \frac{d\zeta_i}{dt} - \zeta_i \frac{d\eta_i}{dt} \right) = c_2 , \qquad (18)$$

$$\sum_{1}^{n} m_i \left(\zeta_i \frac{d\xi_i}{dt} - \xi_i \frac{d\zeta_i}{dt} \right) = c_3 ,$$

where c_1, c_2 and c_3 are integration constants. The expressions within the parentheses are already familiar from the two-body problem: they have the dimensions of area covered in unit time and are the projections of sectorial velocities of different bodies ($i = 1, \ldots, n$) within corresponding coordinate planes.

Let A_i, B_i and C_i be the projections of areas covered by the radius-vector r_i of a body m_i during a time dt in three coordinate planes, respectively. Then we can write for the components of the sectorial velocity of the body m_i

$$\frac{dA_i}{dt} = \xi_i \frac{d\eta_i}{dt} - \eta_i \frac{d\xi_i}{dt},$$

$$\frac{dB_i}{dt} = \eta_i \frac{d\zeta_i}{dt} - \zeta_i \frac{d\eta_i}{dt}, \qquad (19)$$

$$\frac{dC_i}{dt} = \zeta_i \frac{d\xi_i}{dt} - \xi_i \frac{d\zeta_i}{dt}.$$

Then Eq. (18) takes the form

$$\sum_{1}^{n} m_i \frac{dA_i}{dt} = c_1,$$

$$\sum_{1}^{n} m_i \frac{dB_i}{dt} = c_2, \qquad (20)$$

$$\sum_{1}^{n} m_i \frac{dC_i}{dt} = c_3,$$

i.e. the sum of the products of the mass on the projection of corresponding sectorial velocities is a constant for an n-body system.

Moving from the components of sectorial velocities to the true sectorial velocity dW_i/dt of the motion of the body within the orbital plane, we have

$$
\sum_1^n m_i \frac{dW_i}{dt} = \left[\left(\sum m_i \frac{dA_i}{dt} \right)^2 + \left(\sum m_i \frac{dB_i}{dt} \right)^2 \right.
$$
$$
\left. + \left(\sum m_i \frac{dC_i}{dt} \right)^2 + \right]^{1/2} = (c_1^2 + c_2^2 + c_3^2)^{1/2} \tag{21}
$$

or

$$
\sum_1^n m_i \frac{dW_i}{dt} = c. \tag{22}
$$

This relationship can be considered as the analogy of Kepler's first law for an n-body system.

Kepler's first law in the two-body problem is reduced, as was shown in Chapter III, to the constancy of the sectorial velocity. In the case of an n-body system the situation is somewhat different, namely, here we have the constancy of the sum of products of mass and corresponding sectorial velocity for all n bodies.

One more difference exists between Kepler's law in the case of the two-body problem and the "Kepler law" for an n-body system: in the former case Kepler's law is derived for a relative motion, whereas in the second case Eq. (22) is derived for an arbitrary coordinate system. However, we can move from Eq. (22), derived for the n-body system to the particular case $n = 2$:

$$
m_1 \frac{dW_1}{dt} + m_2 \frac{dW_2}{dt} = c. \tag{23}
$$

For the relative motion, by adopting the body m_1 as central, and substracting dW_1/dt from both parts on the left-hand side of Eq. (23), we obtain

$$
m_2 \left(\frac{dW_2}{dt} - \frac{m_1}{m_2} \frac{dW_1}{dt} \right) = c'. \tag{24}
$$

The term in the parentheses is obviously the relative sectorial velocity dW_0/dt of the body m_2 executing an orbital motion around the central body m_1. Therefore we can write

$$
m_2 \frac{dW_0}{dt} = c'. \tag{25}
$$

Because according to the definition of the sectorial velocity,

$$
\frac{dW_0}{dt} = \frac{1}{2} r^2 \frac{d\theta}{dt},
$$

we will finally have instead of Eq. (25)

$$r^2 \frac{d\theta}{dt} = \frac{2c'}{m_2} = constant. \tag{26}$$

This is nothing other than Kepler's first law of relative motion for the two-body problem. By integrating Eq. (22) we obtain

$$\sum_{i=1}^{n} m_i W_i = ct + c''. \tag{27}$$

Here W_i is the area covered by the radius-vector of body m during a time t. Therefore, the content of Eq. (27) can be formulated in the following manner: the sum of products of mass and area covered by the corresponding radius-vectors is proportional to time. This formulation remains valid also in the case of the projections of the areas. Indeed, on integrating Eq. (20), we find

$$\sum m_i A_i = c_1 t + c_1',$$

$$\sum m_i B_i = c_2 t + c_2', \tag{28}$$

$$\sum m_i C_i = c_3 t + c_3'.$$

Laplace demonstrated the existence of a plane in which the sum of products of mass and area projection will be maximal so that both constants, c_1 and c_2, within Eq. (18) become zero, and the third one becomes equal to c_3. This plane is known as the *Laplace invariant plane* for the given system of bodies. In the case of e.g. the Solar system the Laplace invariant plane is inclined with respect to the ecliptic plane by approximately $2°$.

4 The Integral of Energy

On multiplying the basic equation (5) by $d\xi_i/dt$, $d\eta_i/dt$, $d\zeta_i/dt$, respectively, after summation we obtain

$$\sum_{i=1}^{n} m_i \left(\frac{d^2\xi_i}{dt^2} \frac{d\xi_i}{dt} + \frac{d^2\eta_i}{dt^2} \frac{d\eta_i}{dt} + \frac{d^2\zeta_i}{dt^2} \frac{d\zeta_i}{dt} \right)$$
$$= \sum_{i=1}^{n} \left(\frac{\partial U}{\partial \xi_i} \frac{d\xi_i}{dt} + \frac{\partial U}{\partial \eta_i} \frac{d\eta_i}{dt} + \frac{\partial U}{\partial \zeta_i} \frac{d\zeta_i}{dt} \right). \tag{29}$$

The right-hand side of this relationship represents the derivative of U with respect to t. Therefore we can write

$$\frac{1}{2} \frac{d}{dt} \sum_{i=1}^{n} m_i \left[\left(\frac{d\xi_i}{dt} \right)^2 + \left(\frac{d\eta_i}{dt} \right)^2 + \left(\frac{d\zeta_i}{dt} \right)^2 \right] = \frac{dU}{dt}. \tag{30}$$

Its left-hand side is the derivative of the kinetic energy T of the system. Therefore, after integration, we obtain

$$T = U + C, \tag{31}$$

where C is the constant of integration, the so-called **energy constant**.

The potential function has quite definite physical content, i.e. the characteristics of the transition from one energetic state to another. For example, in the two-body problem the amount of work necessary to move the distance from r_1 to r_2 is

$$A_{ij} = k^2 m_i m_j \int_{r_1}^{r_2} \frac{dr}{r^2} = k^2 m_i m_j \left(\frac{1}{r_1} - \frac{1}{r_2} \right). \tag{32}$$

If both bodies are initially situated at an infinite distance with respect to each other, i.e. $r_1 = \infty$, then Eq. (32) takes the well-known form $(r_2 = r_{ij})$

$$A_{ij} = -k^2 \frac{m_i m_j}{r_{ij}}.$$

On comparing it with the expression for the force, we find

$$U = -\frac{1}{2} k^2 \sum_{i=1}^{n} \sum_{j=1}^{n} A_{ij}. \tag{33}$$

Thus the function U is nothing other than the potential energy of the system taken with the opposite sign and for the initial positions at infinity. Now Eq. (31) can be written in the familiar form

$$T - U = C, \tag{34}$$

i.e. for an isolated n-body system the sum of kinetic and potential energies remains constant.

What consequences can Eq. (34) have, say, for stellar systems? Consider a star cluster containing N stars of some average mass. Usually N is of the order of hundreds or thousands in the case of open star clusters and tens or hundreds of thousands of stars for globular clusters. Eq. (34) is fully applicable for such systems insofar as all external influences can be neglected.

However, the stars within the cluster perform smooth motion under the action of the gravitational forces from other members of the cluster. At first sight, any increase in velocity and, hence, in kinetic energy, of a given star should be compensated by a decrease in potential energy of the system as a result of internally self-consistent redistribution of stars.

However, as a result of stellar mutual encounters, certain stars can gain rather high velocities, even exceeding the parabolic velocity of the system, i.e. enough to escape from the given cluster. The escaped star obviously carries away a certain amount of kinetic energy $T_\star = mV_p^2/2$, where m is the mean stellar mass and V_p is the parabolic velocity of the cluster. As a result the total kinetic energy of the cluster decreases by the same amount $T_\star$. Obviously this is a secular effect, though it leads to a certain instability in the system. Indeed, as follows from Eq. (34), the loss of kinetic energy T has to be compensated via the potential energy U of the system, which is possible only by means of a decrease in mean distances between the stars. In other words, the escape of the stars from the cluster will lead to its inevitable compression. The described effect of "evaporation" of stars from the cluster (Ambartzumian, 1938) and the simultaneous "core collapse" are the basic dynamical effects characterizing the evolution and the cosmogonical essence of star clusters.

5   The Problem of Additional Integrals

Thus, from all $6n$ integrals required for the complete solution of the n-body problem, we obtained only ten – six integrals of motion of the center of mass, three integrals, of areas and the integral of energy. Besides these ten integrals no other integrals are known within celestial mechanics. In this connection a question arises concerning the possibility of discovering the missing $6n - 10$ integrals.

Already in 1887 Bruns had shown that, even in the case of the three-body problem, every first integral is the consequence of the ten integrals mentioned above, so that even in the case of $n = 3$ no new algebraic integrals exist. Poincaré (1896) generalized this conclusion, proving the absence of a wider class of first integrals and obtaining the fundamental result of the nonintegrability of the n-body problem when $n > 2$. It became clear that further progress in the n-body problem would have to be associated with perturbation theory, which we will discuss in the following chapter.

6   Differential Equations of Relative Motion

Up to now we have been dealing with the behavior of a body m_i within the n-body system, i.e. under the influence of the gravitational forces of the other $n - 1$ bodies in an arbitrary coordinate system. The main results obtained were the ten mentioned integrals, describing, however, some global properties of n-body **system** itself. Regarding the behavior of an individual body within the system, in particular its motion, no information has been obtained.

The situation was the same as in the case of the two-body problem. However, as soon as we moved to the **relative** motion, the situation changed sharply: together with the information concerning the system itself (though only of two bodies), we were able to derive the **equation of motion** for the second body in the gravitational field of the central body.

Thus, that consideration of the two-body problem was due to the fact that we moved from the examination of **absolute** motion to that of relative motion. Moreover, the great advantage in consideration of the relative motion has been shown from the methodical point of view.

It would be logical, therefore, to act in the same manner in the case of the n-body system, i.e. via localizing the center of the coordinate system on one of the bodies of the system. Obviously, even in this case one can never hope for a comprehensive solution of the problem as in the case of the two-body problem; nevertheless, it is interesting to investigate the possibilities of that trick.

The Solar system itself represents a typical example of an n-body system. Therefore, in further analysis let us choose the corresponding number of objects, m_n, and identify the center of the coordinate system x, y, z with the center of the Sun. Denoting by x_i, y_i, z_i the coordinates of objects m_i with respect to this new system, we have their relation to its old coordinates ξ_i, η_i, ζ_i:

$$\xi_i = x_i + \xi_n,$$
$$\eta_i = y_i + \eta_n, \tag{35}$$
$$\zeta_i = z_i + \zeta_n,$$

where ξ_n, η_n, ζ_n are the coordinates of the Sun in the old system. The coordinates of the Sun in the new system x_n, y_n, z_n will be obtained via the substitution $i = n$ in Eq. (35),

$$x_n = y_n = z_n = 0. \tag{36}$$

During such a transition of the coordinate system, of course, the distances between the bodies do not change, and the differences between the old variables (coordinates) equal those of the new variables. Therefore we can write

$$\frac{\partial U}{\partial \xi_i} = \frac{\partial U}{\partial x_i},$$
$$\frac{\partial U}{\partial \eta_i} = \frac{\partial U}{\partial y_i}, \tag{37}$$
$$\frac{\partial U}{\partial \zeta_i} = \frac{\partial U}{\partial z_i},$$

Substituting Eq. (35) into Eq. (5) and taking into account Eq. (37), we will have

$$\frac{d^2 x_i}{dt^2} + \frac{d^2 \xi_n}{dt^2} = \frac{1}{m_i} \frac{\partial U}{\partial x_i},$$

$$\frac{d^2 y_i}{dt^2} + \frac{d^2 \eta_n}{dt^2} = \frac{1}{m_i} \frac{\partial U}{\partial y_i}, \tag{38}$$

$$\frac{d^2 z_i}{dt^2} + \frac{d^2 \zeta_n}{dt^2} = \frac{1}{m_i} \frac{\partial U}{\partial z_i},$$

where $i = 1, 2, \ldots, n - 1$.

From the first equation of the system (2) we can obtain, assuming that $i = n$ and taking into account Eqs. (35) and (36),

$$\frac{d\xi_n}{dt^2} = k^2 \sum_{j=1}^{n-1} m_j \frac{x_j}{r_{jn}^3}. \tag{39}$$

This equation, together with the corresponding equations for η and ζ from Eq. (38), in fact completes the transition to the new system x, y, z. However, it is more convenient to combine the terms in a slightly different form, i.e. we separate the terms determined by the Solar gravity from the others. Moreover, for simplicity we rewrite the differential equations of relative motion for the body with the number $n = 1$, so that we will have from Eq. (39)

$$\frac{d^2 \xi_n}{dt^2} = k^2 m_1 \frac{x_1}{r_{1n}^3} + k^2 \sum_{j=2}^{n-1} m_j \frac{x_j}{r_{jn}^3}. \tag{40}$$

In Eq. (4) one can separate out the term describing the interactions of all $n - 1$ bodies with the Sun:

$$U = k^2 m_n \sum_{j=1}^{n-1} \frac{m_j}{r_{jn}} + \frac{k^2}{2} \sum_{j=1}^{n-1} \sum_{i=1}^{n-1} \frac{m_i m_j}{r_{ij}}.$$

Moreover, we separate out the first term on the right-hand side associated with the object $j = 1$:

$$U = k^2 \frac{m_n m_1}{r_{1n}} + k^2 m_n \sum_{j=2}^{n-1} \frac{m_j}{r_{jn}} + \frac{k^2}{2} \sum_{i=1}^{n-1} \sum_{j=1}^{n-1} \frac{m_i m_j}{r_{ij}} \tag{41}$$

or

$$U = k^2 \frac{m_n m_1}{r_{1n}} + k^2 m_n \sum_{j=2}^{n-1} \frac{m_j}{r_{jn}} + U'.$$

From that we obtain for the derivative with respect to x_1

$$\frac{\partial U}{\partial x_1} = -k^2 m_n m_1 \frac{x_1 - x_n}{r_{1n}^3} + \frac{\partial U'}{\partial x_1}, \tag{42}$$

while the derivative in the second term on the right-hand side in Eq. (41) which does not contain x_1 is zero.

In view of Eq. (36), we have

$$\frac{1}{m_1} \frac{\partial U}{\partial x_1} = -k^2 m_n \frac{x_1}{r_{1n}^3} + \frac{1}{m_1} \frac{\partial U'}{\partial x_1}. \tag{43}$$

By substituting Eqs. (40) and (43) into Eq. (38), we obtain

$$\frac{d^2 x_1}{dt^2} + k^2(m_1 + m_n)\frac{x_1}{r_{1n}^3} = \frac{1}{m_1}\frac{\partial U'}{\partial x_1} - k^2 \sum_{j=2}^{n-1} m_j \frac{x_j}{r_{jn}^3},$$

$$\frac{d^2 y_1}{dt^2} + k^2(m_1 + m_n)\frac{y_1}{r_{1n}^3} = \frac{1}{m_1}\frac{\partial U'}{\partial y_1} - k^2 \sum_{j=2}^{n-1} m_j \frac{y_j}{r_{jn}^3}, \tag{44}$$

$$\frac{d^2 z_1}{dt^2} + k^2(m_1 + m_n)\frac{z_1}{r_{1n}^3} = \frac{1}{m_1}\frac{\partial U'}{\partial z_1} - k^2 \sum_{j=2}^{n-1} m_j \frac{z_j}{r_{jn}^3}.$$

This is the system of differential equations of relative motion of the object m_1 under Solar gravity and that of the other $n - 2$ members of the system.

We can now rewrite Eq. (44) for arbitrary bodies:

$$\frac{d^2 x_i}{dt^2} + k^2(m_i + m_n)\frac{x_i}{r_{in}^3} = \frac{1}{m_i}\frac{\partial U'}{\partial x_i} - k^2 \sum_{j=1}^{n-1} m_j \frac{x_j}{r_{jn}^3},$$

$$\frac{d^2 y_i}{dt^2} + k^2(m_i + m_n)\frac{y_i}{r_{in}^3} = \frac{1}{m_i}\frac{\partial U'}{\partial y_i} - k^2 \sum_{j=1}^{n-1} m_j \frac{y_j}{r_{jn}^3}, \tag{45}$$

$$\frac{d^2 z_i}{dt^2} + k^2(m_i + m_n)\frac{z_i}{r_{in}^3} = \frac{1}{m_i}\frac{\partial U'}{\partial z_i} - k^2 \sum_{j=1}^{n-1} m_j \frac{z_j}{r_{jn}^3},$$

where the condition $i \neq j$ should be satisfied. Eq. (45) can be rewritten in more compact form by means of the function R_{ij} in the following manner:

$$R_{ij} = k^2 \left(\frac{1}{r_{ij}} - \frac{x_i x_j + y_i y_j + z_i z_j}{r_{jn}^3} \right). \tag{46}$$

Then the equations of relative motion of the body m_i take the following final form:

$$\frac{d^2x_i}{dt^2} + k^2(m_i + m_n)\frac{x_i}{r_{in}^3} = \sum_{j=1}^{n-1} m_j \frac{\partial R_{ij}}{\partial x_i},$$

$$\frac{d^2y_i}{dt^2} + k^2(m_i + m_n)\frac{y_i}{r_{in}^3} = \sum_{j=1}^{n-1} m_j \frac{\partial R_{ij}}{\partial y_i}, \qquad (47)$$

$$\frac{d^2z_i}{dt^2} + k^2(m_i + m_n)\frac{z_i}{r_{in}^3} = \sum_{j=1}^{n-1} m_j \frac{\partial R_{ij}}{\partial z_i}.$$

where $i \neq j$, and $i = 1, \ldots, n-1$. The system of equations (47) has wide application in astronomy.

7 The Transition to Two Bodies

We now rewrite Eq. (47) for $n = 2$, i.e. as for the two-body problem, when, for example, one of the bodies is a planet of mass m_1 and the other is the Sun of mass $m_n = m_0$. Then, on the left-hand side of Eq. (47) we have $i = 1$, $m_i = m_1$, and $m_n = m_0$, while on the right-hand side the number $j = 2$ should be absent in the sum. Thus the right-hand side of Eq. (47) becomes zero:

$$\frac{d^2x_1}{dt^2} + k^2(m_1 + m_0)\frac{x_1}{r_{10}^3} = 0,$$

$$\frac{d^2y_1}{dt^2} + k^2(m_1 + m_0)\frac{y_1}{r_{10}^3} = 0, \qquad (48)$$

$$\frac{d^2z_1}{dt^2} + k^2(m_1 + m_0)\frac{z_1}{r_{10}^3} = 0,$$

or, denoting $M = m_1 + m_0$, we can simply write

$$\frac{d^2x}{dt^2} = -k^2 M \frac{x}{r^3},$$

$$\frac{d^2y}{dt^2} = -k^2 M \frac{y}{r^3}, \qquad (49)$$

$$\frac{d^2z}{dt^2} = -k^2 M \frac{z}{r^3}.$$

These are nothing other than the well-known differential equations of Keplerian motion, i.e. those for the two-body problem derived in Chapter II.

Keplerian motion is called **unperturbed**; every deviation from that motion is called **perturbed**. Therefore, the function R_{ij} from Eq. (46) is known as a **perturbation function**.

In celestial mechanics a problem is considered integrable if one can find a general integral from the system of differential equations containing constants of integration with the number exactly equal to that of the order of the system. As mentioned above, the two-body problem turns out to be the **only** integrable one: for $n = 2$, and the number of the constants of integration equals $6n = 12$. However only ten integrals are discovered.

The transition to the relative coordinate system leads to the reduction of the number of integrals from $6n$ to $6n - 6$ by means of elimination from the system of six quantities related to the Sun, namely, the three coordinates ξ_n, η_n, and ζ_n, and three components of velocity $\dot{\xi}_n$, $\dot{\eta}_n$, and $\dot{\zeta}_n$. Therefore the solution of the n-body problem in the relative motion version comes down to finding the $6n - 6$ integration constants. In the case of $n = 2$, the number of constants is six and that many integrals have been discovered in the two-body problem. Even for $n = 3$ the number of integrals becomes 12, but one cannot obtain more than four integrals. Hence, the fatal conclusion that not only the n-body problem, but even the three-body problem, cannot be solved in the sense mentioned above, i.e. by obtaining $6n - 6$ integrals.

However, there is another approach within celestial mechanics concerning the study of the system of differential equations of the N-body problem: the analysis of that system by means of **series**, if the latter are known to be convergent for all initial conditions (the masses and velocities of the bodies, etc.). From the point of view of the final aim of obtaining the motion of a given body at any given moment of time under the gravitational field of the remaining $n - 1$ bodies by this method, every problem of celestial mechanics can be attributed to the category of integrable ones. From a more formal point of view, one can conclude that celestial mechanics has no unsolvable problems.

8   The Dynamical Essence of the Perturbation Function

The perturbation function R_{ij} has definite physical content, namely, its derivatives with respect to coordinates, i.e. the quantities $\partial R_{ij}/\partial x_i$, $\partial R_{ij}/\partial y_i$, $\partial R_{ij}/\partial z_i$ describe the components of the acceleration of the body m_i with respect to the other bodies of the system.

Assume for simplicity that $n = 3$, i.e. one has two planets of masses m_1 and m_2 and the Sun of mass m_0, with given distances between the Sun and the planets being r_1 and r_2, respectively. Then Eq. (47) can be rewritten, taking into account also that, from two members ($n - 1 = 2$), one is excluded due to the condition $i \neq j$, in the form

$$\frac{d^2x_1}{dt^2} + k^2(m_1 + m_0)\frac{x_1}{r_1^3} = k^2 m_2 \frac{\partial}{\partial x_1}\left(\frac{1}{r_{12}} - \frac{x_1 x_2 + y_1 y_2 + z_1 z_2}{r_2^3}\right),$$

$$\frac{d^2y_1}{dt^2} + k^2(m_1 + m_0)\frac{y_1}{r_1^3} = k^2 m_2 \frac{\partial}{\partial y_1}\left(\frac{1}{r_{12}} - \frac{x_1 x_2 + y_1 y_2 + z_1 z_2}{r_2^3}\right), \quad (50)$$

$$\frac{d^2z_1}{dt^2} + k^2(m_1 + m_0)\frac{z_1}{r_1^3} = k^2 m_2 \frac{\partial}{\partial z_1}\left(\frac{1}{r_{12}} - \frac{x_1 x_2 + y_1 y_2 + z_1 z_2}{r_2^3}\right),$$

and

$$\frac{d^2x_2}{dt^2} + k^2(m_1 + m_0)\frac{x_2}{r_2^3} = k^2 m_1 \frac{\partial}{\partial x_2}\left(\frac{1}{r_{21}} - \frac{x_1 x_2 + y_1 y_2 + z_1 z_2}{r_1^3}\right),$$

$$\frac{d^2y_2}{dt^2} + k^2(m_1 + m_0)\frac{y_2}{r_2^3} = k^2 m_1 \frac{\partial}{\partial y_2}\left(\frac{1}{r_{21}} - \frac{x_1 x_2 + y_1 y_2 + z_1 z_2}{r_1^3}\right), \quad (51)$$

$$\frac{d^2z_2}{dt^2} + k^2(m_1 + m_0)\frac{z_2}{r_2^3} = k^2 m_1 \frac{\partial}{\partial z_2}\left(\frac{1}{r_{21}} - \frac{x_1 x_2 + y_1 y_2 + z_1 z_2}{r_1^3}\right),$$

where $r_{12} = r_{21}$ is the distance between m_1 and m_2:

$$r_{12} = [(x_1 - x_2)^2 + (y_1 - y_2)^2 + (z_1 - z_2)^2]^{1/2},$$

$$r_1 = (x_1^2 + y_1^2 + z_1^2)^{1/2},$$

$$r_2 = (x_2^2 + y_2^2 + z_2^2)^{1/2}.$$

If the mass m_2 is zero then all three equations in (49) will become independent from Eq. (50) being the equations of the relative motion of m_1 with respect to m_0. Similarly, if $m_1 = 0$ in Eq. (50), its dependence on Eq. (49) should vanish. As a result, in both cases the problem should be reduced to the known two-body problem. All divergences from purely elliptic motion are caused by the presence of the right-hand sides in Eqs. (49) and (50). The latter in the first three equations in (49) are the partial derivatives of R_{12} with respect to the variables x_1, y_1 and z_1, respectively, while in Eq. (50) they are with respect to the variables x_2, y_2 and z_2.

Partial derivatives of the first terms within the parentheses in Eq. (50) with respect to x, y, and z, respectively are equal to

$$-k^2 m_2 \frac{x_1 - x_2}{r_{12}^3}, \quad -k^2 m_2 \frac{y_1 - y_2}{r_{12}^3}, \quad -k^2 m_2 \frac{z_1 - z_2}{r_{12}^3}$$

and are the components of the acceleration of m_1 under the influence of m_2. The partial derivatives of the second terms within the parentheses in Eq. (50) are equal to

$$-k^2 m_2 \frac{x_2}{r_2^3}, \quad -k^2 m_2 \frac{y_2}{r_2^3}, \quad -k^2 m_2 \frac{z_2}{r_2^3},$$

and are the components of the acceleration of the Sun with the opposite sign, influenced by m_2. Therefore, the right-hand sides of the three equations in (49) represent the differences in the acceleration components of m_1 and of the Sun. Similarly, the right-hand sides of the three equations in (50) are the differences between the components of acceleration of m_2 and the Sun.

Thus, expressions of the type

$$m_j \frac{\partial R_{ij}}{\partial x_i}$$

applied to the body m_i represent the difference between two accelerations – the acceleration of m_i caused by m_j, and the acceleration of the Sun caused by the same m_j.

9   The Lagrange–Jacobi Formula

In some problems of celestial mechanics one has to deal with the so-called **moment of inertia** of a celestial body J_i, i.e. the product of the mass m_i and the squared distance from the center of the coordinate system r_i: $J_i = m_i r_i^2$. The Lagrange–Jacobi formula relates the moment of inertia J of the n-body system to its potential energy U. Given by Eq. (4), U is a homogeneous function of the coordinates. Such functions satisfy the well-known Euler theorem:

$$\sum_0^{n-1} \left(\xi_i \frac{\partial U}{\partial \xi_i} + \eta_i \frac{\partial U}{\partial \eta_i} + \zeta_i \frac{\partial U}{\partial \zeta_i} \right) = -U \ . \tag{52}$$

Multiplying Eqs. (5) by ξ_i, η_i, and ζ_i, respectively, and taking their sum, we obtain

$$\sum_0^{n-1} m_i \left(\xi_i \frac{d^2\xi_i}{dt^2} + \eta_i \frac{d^2\eta_i}{dt^2} + \zeta_i \frac{d^2\zeta_i}{dt^2} \right)$$
$$= \sum_0^{n-1} \left(\xi_i \frac{\partial U}{\partial \xi_i} + \eta_i \frac{\partial U}{\partial \eta_i} + \zeta_i \frac{\partial U}{\partial \zeta_i} \right) \tag{53}$$

or, comparing with Eq. (52), we have

$$\sum_{0}^{n-1} m_i \left(\xi_i \frac{d^2\xi_i}{dt^2} + \eta_i \frac{d^2\eta_i}{dt^2} + \zeta_i \frac{d^2\zeta_i}{dt^2} \right) = -U. \tag{54}$$

Summing this equality with the doubled integral of energy in Eq. (31), we obtain

$$\sum_{0}^{n-1} m_i \left[\left(\xi_i \frac{d^2\xi_i}{dt^2} + \eta_i \frac{d^2\eta_i}{dt^2} + \zeta_i \frac{d^2\zeta_i}{dt^2} \right) \right.$$
$$\left. + \left(\frac{d^2\xi_i}{dt} \right)^2 + \left(\frac{d^2\eta_i}{dt} \right)^2 + \left(\frac{d^2\zeta_i}{dt} \right)^2 \right] = U + 2C$$

or

$$\frac{d}{dt} \sum_{0}^{n-1} m_i \left[\xi_i \frac{d\xi_i}{dt} + \eta_i \frac{d\eta_i}{dt} + \zeta_i \frac{d\zeta_i}{dt} \right] = U + 2C ,$$

or, substituting, for example,

$$\xi_i \frac{d\xi_i}{dt} = \frac{1}{2} \frac{d\xi_i^2}{dt},$$

we obtain

$$\frac{d^2}{dt^2} \sum_{0}^{n-1} m_i(\xi_i^2 + \eta_i^2 + \zeta_i^2) = 2U + 4C. \tag{55}$$

The expression

$$J = \sum_{0}^{n-1} m_i(\xi_i^2 + \eta_i^2 + \zeta_i^2) = \sum_{0}^{n-1} m_i r_i^2 \tag{56}$$

is the moment of inertia of an n-body system. Then Eq. (55) takes the form

$$\frac{d^2 J}{dt^2} = 2U + 4C. \tag{57}$$

This is the Lagrange–Jacobi formula, discovered originally by Lagrange in 1772, for the case of two and three bodies. Later, in 1842, it was generalized by Jacobi to an arbitrary number of bodies. Jacobi also showed that one can obtain general conclusions on the character of motion from this formula. The above derivation of the formula (57) is that of Jacobi. The formula derived by Lagrange for two bodies,

$$\frac{d^2 r^2}{dt^2} = 2k^2(m_0 + m_1) \left(\frac{1}{r} - \frac{1}{a} \right). \tag{58}$$

represents a particular case of Eq. (57) .

10   The Virial Theorem

The so-called **virial theorem** has a direct relation to n-body systems. This theorem plays an essential role in statistical mechanics and in stellar dynamics, enabling one to obtain the dynamical characteristics of the systems.

Consider an n-body system of masses m_1, m_2, ..., m_n. Let the system be in a stationary state, i.e. it is closed, isolated from external influences, its members do not leave the system, and all types of internal secular effects proceed rather slowly.

Our aim is to rewrite the expression for the moment of inertia J with respect to the center of mass. Denoting by r_i the radius-vector of the body or of the star with mass m_i, we have

$$J = \sum_1^n m_i r_i^2 = \sum_1^n m_i(x_i^2 + y_i^2 + z_i^2). \tag{59}$$

In general, J can: a) preserve its constant value corresponding to the stationary state of the system, or b) change with a constant rate, i.e.

$$\frac{dJ}{dt} = constant \tag{60}$$

or

$$J = a + bt, \tag{61}$$

i.e. the changes in the moment of inertia J, and hence, all kinds of internal changes, occur linearly with respect to time.

It follows from Eq. (60) that

$$\frac{d^2 J}{dt^2} = 0. \tag{62}$$

Differentiating Eq. (59) twice with respect to t and taking into account Eq. (62), we obtain

$$\sum_1^n m_i(\dot{x}_i^2 + \dot{y}_i^2 + \dot{z}_i^2) + \sum_1^n m_i(x_i\ddot{x}_i + y_i\ddot{y}_i + z_i\ddot{z}_i) = 0, \tag{63}$$

where the points over the letters imply derivatives with respect to time.

The first term in Eq. (63) is obviously the doubled kinetic energy of the system. Therefore we can write

$$\sum_1^n m_i v_i^2 = 2T. \tag{64}$$

The second term in Eq. (63),

$$-\sum_{1}^{n} m_i(x_i\ddot{x}_i + y_i\ddot{y}_i + z_i\ddot{z}_i) \,, \tag{65}$$

is known as **virial**. It turned out that the virial term has a direct relation to the potential function U. Indeed, by writing the expression for the potential energy of the n-body system in the form

$$U = \frac{1}{2}G\sum_{i=1}^{n}\sum_{j=1}^{n} \frac{m_i m_j}{r_{ij}}, \tag{66}$$

as well as the equation of motion of the i-th body,

$$m_i\ddot{x}_i = \frac{\partial U}{\partial x_1} \,,$$
$$m_i\ddot{y}_i = \frac{\partial U}{\partial y_1} \,, \tag{67}$$
$$m_i\ddot{z}_i = \frac{\partial U}{\partial z_1}$$

and substituting the values of $\ddot{x}_i$, $\ddot{y}_i$, and $\ddot{z}_i$ from Eq. (67) into Eq. (65), we obtain for the virial

$$-\sum_{1}^{n}\left(x_i\frac{\partial U}{\partial x_i} + y_i\frac{\partial U}{\partial y_i} + z_i\frac{\partial U}{\partial z_i} \right). \tag{68}$$

Owing to the fact that U is a homogeneous function of the coordinates, we can write in accord with Euler's theorem

$$\sum_{1}^{n}\left(x_i\frac{\partial U}{\partial x_i} + y_i\frac{\partial U}{\partial y_i} + z_i\frac{\partial U}{\partial z_i} \right) = -U \,, \tag{69}$$

i.e. the virial is nothing other than the potential energy of the system with a negative sign. On substituting the value of the virial from Eq. (69) into Eq. (63), in view of Eq. (64), we obtain

$$2T - U = 0. \tag{70}$$

This is the virial theorem, which reads: in a stationary stellar system the sum of the doubled kinetic energy and the potential energy equals zero, or, equivalently, the absolute value of potential energy for a closed stationary system must be equal to the doubled kinetic energy. This theorem was proved by Poincaré, in 1913.

However, for the closed stationary n-body system we have one more theorem on the conservation of kinetic energy:

$$T + U = constant = H. \tag{71}$$

Then Eq. (70) can be rewritten in the form

$$T = -H, \tag{72}$$

or

$$U = 2H. \tag{73}$$

The virial theorem has wide astrophysical applications. It defines the so-called **dynamical mass** of stellar systems, e.g. of open (scattered), globular star clusters, stellar associations, individual galaxies, clusters of galaxies, etc. The dynamical masses are therefore obtained via the measured line-of-sight velocities v_i of stars, galaxies etc. However, astrophysicists also have other methods for determination of the masses of the systems, e.g. by empirical relations between the mass and the luminosity discovered for a given category of objects such as binary stars, star clusters, galaxies, etc.

In the majority of cases remarkable discrepancies were discovered between the mass data obtained by different methods for the same object or class of objects, and, what is striking, as a rule the *luminous* mass appears to be smaller than the dynamical mass. In some cases the discrepancy is one or two orders of magnitude. This fact has led to the hypothesis of the existence of an essential amount of **missing or hidden mass** in the Universe.

The concept of this *dark matter*, based on the above determination of dynamical mass, however, encounters essential difficulties. The point is that the virial theorem Eq. (70), as mentioned above, is applicable only to stationary systems, i.e. to systems far from expansion, merging or decaying states, otherwise the doubled kinetic energy cannot fulfill Eq. (70), and the entire system can possess a **positive** total energy. The difficulty is obviously in the reliable observational confirmation of the stationary state of the systems.

Nevertheless, one can derive a working formula for determination of the dynamical mass of a system from the magnitude of its mean square velocity $\bar{v}$, the so-called velocity dispersion. Assuming for simplicity a system in the form of a homogeneous sphere of radius R and total mass $\mathcal{M}$, we can derive the following formula for the determination of the potential energy of such a sphere:

$$U = -\frac{3}{5}G\frac{\mathcal{M}^2}{R}. \tag{74}$$

From Eqs. (64), (70) and (74) we obtain for the dependence between $\mathcal{M}$ and $\bar{v}^2$

$$\bar{v}^2 = \frac{3}{5} G \frac{\mathcal{M}}{R},\tag{75}$$

where G is the gravitational constant in the CGS system. With the help of Eq. (75) one can obtain the total mass of the system $\mathcal{M}$ from the observed magnitude of the velocity dispersion $\bar{v}$ as well as the radius of the system R. Note that in Eq. (75) there are no limitations on the quantitative values of the parameters.

For the globular clusters, which are peripheral members of our Galaxy, we have $\bar{v} = 4\ km\ s^{-1}$, and this gives for the mass of a globular cluster $\mathcal{M}$ of diameter about 70 pc a value $\mathcal{M} = 1.6 \times 10^5 \mathcal{M}_\odot$, i.e. of the order of a hundred thousand Solar masses.

Consider now a rotating stellar system. Obviously, owing to the rotation every body in the system m_i gains an additional kinetic energy $m_i(\omega_i r_i)^2$, where ω is the angular rotational velocity, and the total kinetic energy turns out to be larger than that which is predicted by Eq. (64), i.e.

$$2T = m\bar{v}^2 + J\omega^2.\tag{76}$$

By substituting Eq. (76) into Eq. (70) and neglecting the positive quantity $m\bar{v}^2$, we obtain for the upper limit of the angular velocity

$$\omega^2 < \frac{U}{J}.\tag{77}$$

This inequality can be used as a condition for the stationarity of a rotating n-body system. Eq. (77) is known as the **Poincaré inequality**.

Consider the simplified case in which the masses of all N bodies of the system are equal, i.e. $m_i = m_j = m$ and, hence, $\mathbf{M} = nm$, and the bodies are distributed within the system's volume homogeneously with a mean distance between them of ρ. For the potential energy of such a system we have

$$U = \frac{1}{2} G \sum_{i=1}^{n} \sum_{j=1}^{n} \frac{m_i m_j}{r_{ij}} = Gm^2 \sum_{i=1}^{n} \sum_{j=1}^{n} \frac{1}{r_{ij}} = Gm^2 \frac{n(n-1)}{2\rho}$$

or, taking $n - 1 \approx n$ and $m = \mathcal{M}/n$ we obtain

$$U = G \frac{\mathcal{M}^2}{2\rho},\tag{78}$$

i.e. the potential energy of the system does not depend on the number of its members, and depends only on their total mass and the mean distance between them, ρ.

Finally, we have for the moment of inertia of the system

$$J = \mathcal{M}s^2, \tag{79}$$

where s is the radius.

By substituting Eq. (78) into Eq. (77), we can rewrite the Poincaré inequality in the following form

$$\omega^2 < G\,\frac{\mathcal{M}}{2\rho s^2}. \tag{80}$$

Expressions analogous to inequality (80) can be derived also for other, non-spherical configurations of N bodies, e.g. for ellipsoids with various elongations, like the majority of elliptical galaxies.

The kinetic energy of a system $\mathcal{M}\bar{v}^2$ is determined by the proper motions of its members (stars); in this case one speaks of **irregular** motions. In contrast to this case, the kinetic energy $J\omega^2$ is determined by rotation of the whole system, independent of the law or character of rotation; in this case one deals with the **regular** (rotational) motions. Thus, the Poincaré inequality (80) is satisfied more strongly when the energy of irregular motions $\mathcal{M}v^2$ is larger than the energy of regular (rotational) motions.

In stellar systems of the type of spiral galaxies, including our Galaxy, the regular component of kinetic energy ($J\omega^2$) is much larger than the irregular component ($\mathcal{M}\bar{v}^2$). In other words, the kinetic energy of the Galaxy as a whole is determined by its rotation and not by the peculiar motions of the stars. Therefore, one can think that the angular velocity of spiral galaxies cannot differ greatly from the Poincaré limit. In the limiting case, we can rewrite from (80) approximately

$$\omega^2 \approx G\,\frac{\mathcal{M}}{2\rho s^2}. \tag{81}$$

To obtain the magnitude of the angular velocities ω of galaxies by spectroscopic observations is relatively easy. Then, the expression (81) can be used for determination of the masses of the galaxies.

Chapter IX

The Restricted Three-Body Problem

1   Statement of the Problem

As mentioned in the previous chapter, celestial mechanics in principle can successfully deal with the n-body problem, especially, the three body problem, i.e. with systems of $n = 3$ comparable masses. The system of corresponding differential equations can be solved by means of series convergent for arbitrary parameters of the system. For a three body problem with arbitrary masses such a solution had already been found by Sundman in 1912.

However, in general, obtaining the integral of differential equations containing the constants of integration of the three-body problem with comparable masses of the components is impossible. Owing to this fact, even in the early days of celestial mechanics as a mathematical discipline, the direction of searching for particular solutions of the three-body problem was developed.

One particular case from the very beginning turned out to be of particular interest, i.e. when the masses of two bodies are much smaller than that of the third one. This problem obviously reduces to the solution of two separate two-body problems. Our century of space flights offers new applications for this case: the artificial satellite, an object of infinitely small mass, moves around the Earth according to the law of the two-body problem quite independently of similar motions of other satellites. Another example is the motion of three bodies of arbitrary masses situated along a line – such motions are called **collinear**. The latter problem was originally examined by Euler in 1767. Euler showed that collinear motion is possible for arbitrary masses; each body executes a Keplerian (i.e. elliptical) orbit with respect to the other. For three arbitrary masses with three similar motions this problem is known as the **Euler case of the three-body problem**.

Among Euler cases are considered motions in which the configuration of three bodies, i.e. their mutual location in a plane, remains similar to itself. Lagrange analyzing this problem in 1772, showed that, if three bodies form an equilateral triangle, then their motion both with respect to their common center of inertia and with respect to each other, occurs in Keplerian orbits. Therefore the motions are called **Lagrangian**, if the moving masses always stay within the vertices of an equilateral triangle. In the triangle two masses can be situated with respect to the third one in two different ways, so one has two possible versions of the Lagrange motion.

These five cases, three Eulerian and two Lagrangian, are in fact the only ones in which the distances of the bodies preserve their constant ratios. The interest of Euler and Lagrange motions is due to the possibility of their quantitative treatment within the framework of the classical two-body problem, though they do not lead to any new methodical gain for the three-body problem. The practical importance of both types of motions is also not clear.

Quite different is the case in which the two bodies, m_1 and m_2, have finite mass, but the mass of the third one, Δm, is infinitely small. The behavior of the two finite mass objects in this case is obviously independent of the existence of the third one, whereas the motion of the latter is completely determined by the first two objects. In other words, the problem is only the determination of the motion of an infinitely small body in the gravitational field of two finite masses. In contrast with Eulerian and Lagrangian motions, the practical value of this case is incomparably larger. One can recall a number of examples of such motion: a comet in the gravitational fields both of the Sun and of Jupiter, a spacecraft in the gravitational fields of the Earth and the Moon, etc.

The three-body problem in this formulation is known as the **restricted three-body problem**. Important results on this problem, as we will see below, were obtained by Jacobi in 1836, while the date of birth of the restricted three-body problem goes back to Euler, regarding his second theory of the motion of the Moon, which he developed in 1772. The term "restricted problem" was introduced by Poincaré in 1892.

Usually the problem of determination of the motion of any celestial body implies the traditional sequence starting from the statement of the problem up to achievement of the final aim, including the derivation of the system of differential equations, their integration for the given boundary or initial conditions and the estimation of the constants of integration. From the point of view of the general aim, i.e. obtaining rigorous solutions of the equations of motion, even the restricted version of the three-body problem appears to be unsolvable. The solution of this problem can be followed only up to some stage revealing certain properties of the infinitely small mass in the gravitational fields of the two finite bodies. However, even those qualitative results played a key role in the development of celestial mechanics, so that acquaintance with these results seems quite useful.

2 Differential Equations of Motion of an Infinitely Small Body

Thus, we have a system of two bodies of finite mass, and a body Δm of infinitely small mass. Insofar as the third body has no influence on the dynamics of the first two bodies, the latter perform Keplerian motion around their common center of mass. Let the sum of the two finite masses be $M = 1$, so that each of them will be $1 - \mu$ and μ. The constant distance a between the finite bodies we adopt as unity; the time unit we choose in a such way as to have k^2 also equal to unity.

Consider the coordinate system ξ, η, ζ of rest with center O coinciding with the center of mass of the finite bodies (Fig. 9.1) and the coordinate plane $\xi O \eta$ coinciding with the plane of their motion. Denote the coordinates of the finite bodies $1 - \mu$ and μ by ξ_1, η_1, 0 and ξ_2, η_2, 0, respectively, and that of the infinitely small mass as ξ, η, ζ. Denote by r_1 and r_2 the distances between the mass Δm and masses $1 - \mu$ and μ, respectively, where

$$r_1 = [(\xi - \xi_1)^2 + (\eta - \eta_1)^2 + (\zeta - \zeta_1)^2]^{1/2},$$

$$r_2 = [(\xi - \xi_2)^2 + (\eta - \eta_2)^2 + (\zeta - \zeta_2)^2]^{1/2} .$$

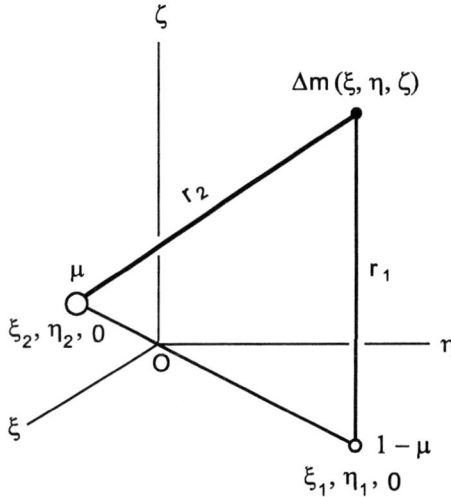

Fig. 9.1. *The restricted three-body problem.*

Then the differential equations of motion of the infinitely small mass Δm can be rewritten in the form

$$\frac{d^2\xi}{dt^2} = -(1-\mu)\frac{\xi-\xi_1}{r_1^3} - \mu\frac{\xi-\xi_2}{r_2^3} \ ,$$

$$\frac{d^2\eta}{dt^2} = -(1-\mu)\frac{\eta-\eta_1}{r_1^3} - \mu\frac{\eta-\eta_2}{r_2^3} \ , \tag{1}$$

$$\frac{d^2\zeta}{dt^2} = -(1-\mu)\frac{\zeta}{r_1} - \mu\frac{\zeta}{r_2^3} \ ,$$

where, as one might expect, the mass Δm is absent. For the adopted units the mean angular velocity n of the finite mass also turns out to be unity:

$$n = \frac{k\sqrt{M}}{a^{3/2}} = 1 \ .$$

Now turn from the coordinates ξ, η, ζ to the rotating system x, y, z with the same center at the point O, so that the plane XOY coincides with the plane $\xi O\eta$, and rotation takes place around the axis OZ with constant angular velocity n. During a time interval t the angular shift of the axes ξ and η will be $\varphi = nt = t$. Correspondingly, the coordinates of Δm in the new system XYZ will be determined by the formulae

$$\xi = x \cos t - y \sin t \ ,$$

$$\eta = x \sin t + y \cos t \ , \tag{2}$$

$$\zeta = z \ ,$$

Analogous expressions may be written also for the finite bodies $1-\mu$ and μ.

By substituting the second derivatives from Eq. (2) into Eq. (1), we obtain

$$\left(\frac{d^2x}{dt^2} - 2\frac{dy}{dt} - x\right)\cos t - \left(\frac{d^2y}{dt^2} + 2\frac{dx}{dt} - y\right)\sin t$$

$$= -\left((1-\mu)\frac{x-x_1}{r_1^3} + \mu\frac{x-x_2}{r_2^3}\right)\sin t$$

$$+ \left((1-\mu)\frac{y-y_1}{r_1^3} + \mu\frac{y-y_2}{r_2^3}\right)\sin t \ ,$$

$$\left(\frac{d^2x}{dt^2} - 2\frac{dy}{dt} - x\right)\sin t + \left(\frac{d^2y}{dt^2} - 2\frac{dx}{dt} - y\right)\cos t \tag{3}$$

$$= -\left((1-\mu)\frac{x-x_1}{r_1^3} + \mu\frac{x-x_2}{r_2^3}\right)\sin t$$

$$-\left((1-\mu)\frac{y-y_1}{r_1^3}+\mu\frac{y-y_2}{r_2^3}\right)\cos t\;,$$

$$\frac{d^2z}{dt^2}=-(1-\mu)\frac{z}{r_1^3}-\mu\frac{z}{r_2^3}\;.$$

Multiplying the first two equations correspondingly by $\cos t$ and $\sin t$ and then by $-\sin t$ and $\cos t$ and summing, we obtain

$$\frac{d^2x}{dt^2}-2\frac{dy}{dt}=x-(1-\mu)\frac{x-x_1}{r_1^3}-\mu\frac{x-x_2}{r_2^3}\;,$$

$$\frac{d^2y}{dt^2}+2\frac{dx}{dt}=y-(1-\mu)\frac{y-y_1}{r_1^3}-\mu\frac{y-y_2}{r_2^3}\;,\tag{4}$$

$$\frac{d^2z}{dt^2}=-(1-\mu)\frac{z}{r_1^3}-\mu\frac{z}{r_2^3}\;.$$

One can rewrite Eq. (4) in a more convenient form by choosing the co-ordinate system in such a way that the axis OX is directed along the line passing through the center of the finite bodies. Then $y_1=0$, $y_2=0$, and Eq. (4) takes the following form

$$\frac{d^2x}{dt^2}-2\frac{dy}{dt}=x-(1-\mu)\frac{x-x_1}{r_1^3}-\mu\frac{x-x_2}{r_2^3}\;,$$

$$\frac{d^2y}{dt^2}+2\frac{dx}{dt}=y-(1-\mu)\frac{y}{r_1^3}-\mu\frac{y}{r_2^3}\;,\tag{5}$$

$$\frac{d^2z}{dt^2}=-(1-\mu)\frac{z}{r_1^3}-\mu\frac{z}{r_2^3}\;.$$

This is the final form of the differential equations of motion for the infinitely small body Δm in the gravitational fields of two finite bodies $1-\mu$ and μ in the rotating coordinate system that is occasionally known as the **synodic system**. The essential advantage of Eq. (5) compared with Eq. (1) is that in the latter case the quantities ξ_1, ξ_2, η_1, and η_2 are functions of time t, whereas in Eq. (5), due to the rotation of the coordinate axes together with all three bodies, the quantities ξ_1, η_1, ξ_2 and η_2 are constants.

This simplification of Eq. (5) is still not enough to provide integrable equations, insofar as the radius-vectors r_1 and r_2 themselves depend on the coordinates of the body Δm. The derivation of the equations of motion in the form $x=f_1(t)$, $y=f_2(t)$, $z=f_3(t)$ is impossible even in the case of the restricted three-body problem. In general the problem of the determination of the motion of an infinitely small body is of sixth order. In the case in which its motion takes place in the same plane as the motion of the finite bodies, the problem is of fourth order.

3   The Jacobi Integral

Eqs. (5) contain an integral derived first by Jacobi in 1836, which was later the subject of a number of studies revealing the interesting properties of the motion of an infinitely small body. For

$$U = \frac{x^2 + y^2}{2} + \frac{1 - \mu}{r_1} + \frac{\mu}{r_2} \,, \tag{6}$$

Eqs. (5) have the form

$$\frac{d^2x}{dt^2} - 2\frac{dy}{dt} = \frac{\partial U}{\partial x} \,,$$

$$\frac{d^2y}{dt^2} + 2\frac{dx}{dt} = \frac{\partial U}{\partial y} \,, \tag{7}$$

$$\frac{d^2z}{dt^2} = \frac{\partial U}{\partial z} \,.$$

By multiplying these equations by $2dx/dt$, $2dy/dt$ and $2dz/dt$, respectively, and summing, we obtain a new equation, which can be integrated

$$\left(\frac{dx}{dt}\right)^2 + \left(\frac{dy}{dt}\right)^2 + \left(\frac{dz}{dt}\right)^2 = 2U - C \,. \tag{8}$$

The left-hand side of this equation is the squared velocity of the small body. Substituting U from Eq. (6), we have

$$V^2 = x^2 + y^2 + \frac{2(1 - \mu)}{r_1} + \frac{2\mu}{r_2} - C \,, \tag{9}$$

where C is the constant of integration. This relationship is known as the **Jacobi integral**; the constant C is called the **Jacobi constant**.

4   Null Velocity Surfaces

We still know nothing about the law of motion of the infinitely small body Δm, i.e we cannot determine its coordinates – its position with respect to the finite bodies – at a given moment in time. However, we have a relationship, i.e. Eq. (9), which can be used to obtain the velocity of this body at the arbitrary position x, y, z. This velocity depends also on the numerical value of the Jacobi constant C, which itself is determined by the initial conditions. Therefore, for various values of C one can have different values of V for one and the same position (x, y, z) of Δm. As a matter of fact, Eq.

(9) is nothing other than the equation of a surface on which the infinitely small body has a fixed velocity.

However, Eq. (9) is valid also in the case when $V = 0$:

$$x^2 + y^2 + \frac{2(1 - \mu)}{r_1} + \frac{2\mu}{r_2} = C , \qquad (10)$$

which combined with the expressions

$$\begin{aligned} r_1 &= [(x - x_1)^2 + y^2 + z^2]^{1/2} , \\ r_2 &= [(x - x_2)^2 + y^2 + z^2]^{1/2} , \end{aligned} \qquad (11)$$

gives the geometrical location of points at which the velocity of the infinitely small body is zero. Obviously, on one side of this surface the velocity will be real, and though we cannot say anything about the orbit of the body Δm, at least we can be sure that, in that region, regular motions are possible.

Thus, within the restricted three-body problem, a new and rather interesting problem arises, namely, the study of surfaces with null relative velocity of the infinitely small body. The Jacobi integral (9) in this problem was considered in detail by Hill (1878) as a part of his theory of the Moon's motion (Hill, 1907).

Insofar as only the squares of the coordinates x and y appear in Eq. (10), the surfaces defined by this equation will be symmetrical with respect to the plane YOZ. From the existence of the coordinate z in Eq. (10) it follows that any line parallel to the axis z will intersect the null velocity surface at either two or zero points.

As one can see, the form of the family of null velocity surfaces defined by Eq. (10) and corresponding to various values of the Jacobi constant C and various ratios μ of the masses of the finite bodies should be rather complicated. To reveal the geometry of these surfaces, we will use the well-known method of analysis of surfaces, namely, the study of their intersections with three coordinate planes. The forms of these intersection curves should reveal the structure of null velocity surfaces. These are three such intersections. We start from the curves of intersection with the plane XOY.

5   Forms of Null Velocity Surfaces (Plane XOY)

To obtain the curves born at the intersection of the null velocity surface of the mass Δm with the plane XOY, i.e. with the plane of motion of the finite bodies $m_1(1-\mu)$ and $m_2(\mu)$, it is necessary to substitute $z = 0$ into the basic equation (10), as well into Eq. (11). Then we obtain

$$(x^2 + y^2) + \frac{2(1 - \mu)}{[(x - x_1)^2 + y^2]^{1/2}} + \frac{2\mu}{[(x - x_2)^2 + y^2]^{1/2}} = C . \qquad (12)$$

Consider two extreme cases.

a) Coordinates x and y are large. This corresponds to large remoteness of the infinitely small body Δm from the center of the coordinate system. In this case the second and third terms in Eq. (12) are small so that the equation can be written in the form

$$x^2 + y^2 = C - \varepsilon(x,\, y)\,, \tag{13}$$

where $\varepsilon(x,\, y)$ is a small quantity that is weakly dependent on x and y.

Eq. (13) defines the curve rather close to a circle of radius $[C - \varepsilon(x,\, y)]^{1/2}$. At large values of the Jacobi constant C, the curves (13) asymptotically tend to the circle of the radius $\sqrt{C}$. With the decrease in C, the curves of intersection of the null velocity surface with the plane XOY will be ovals within the mentioned asymmetrical circle; we will call them **great ovals**. Two such ovals corresponding to $C = C_1$ and $C = C_2$, where $C_2 < C_1$, are shown in Fig. 9.2: both ovals tend to the most external circle.

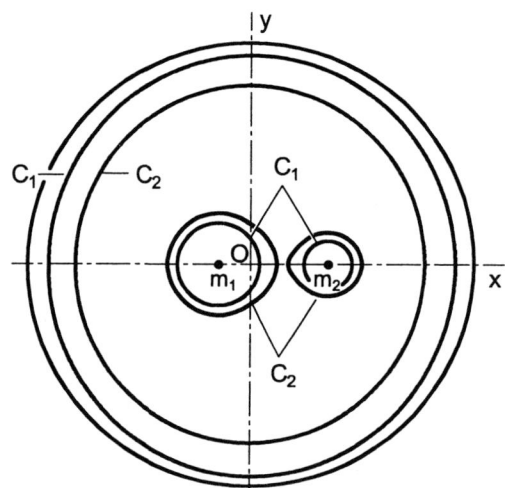

Fig. 9.2. *Null velocity curves in the coordinate plane XOY.*

In the considered case, in which x and y are large, both of the finite bodies are situated within the great ovals. On decreasing the Jacobi constant C, the great ovals contract.

b) Coordinates x and y are small. The infinitely small mass Δm lies close to finite bodies m_1 and m_2. In this case the first term in Eq. (12) is small compared with the second and third terms, so that we have:

$$\frac{1 - \mu}{[(x - x_1)^2 + y^2]^{1/2}} + \frac{\mu}{[(x - x_2)^2 + y^2]^{1/2}} = \frac{C}{2} - \varepsilon(x,\, y)\,. \tag{14}$$

This is the equation of the **equipotential curves** for two gravitating centers. For large values of the Jacobi constant C the null velocity curves are the two relatively lesser circles, which are slightly deformed, i.e. ovals, **small ovals**, containing m_1 and m_2. On decreasing C the small ovals increase, simultaneously deforming more and more. As a result, two oppositely directed pear-shaped figures are formed; see the ovals in Fig. 9.2 for the same C_1 and C_2 as above.

Both the great and the small ovals in Fig. 9.2 correspond to the value $\mu = 0.3$; note that the ovals and pear-shaped figures are not symmetrical. However, at $\mu = 0.5$, the great ovals are absolutely symmetrical with respect to the axis y. The small ovals are symmetrical with respect to both axes, X and Y, and are the same for all values of μ.

Thus, the two limiting cases of null velocity curves have been examined. The diversity of their forms depends on three factors: the "mass parameter" μ, the Jacobi constant C, and the coordinates x and y of the infinitely small mass.

The role of the mass parameter μ in the null velocity curves can be seen rather easily. The possible values of μ begin from 0.5 and decrease to 0.1, 0.01 etc. Correspondingly, the values of $1 - \mu$ start from 0.5 and increase to 0.90, 0.99, etc. The influence of μ on the null velocity curves is as follows.

1. Both types of oval, great and small, are closer to the circle the smaller the value of μ. The largest divergence from the circle is at $\mu = 0.5$, while e.g. at $\mu = 0.01$ all of the ovals practically do not differ from the circle.

2. The sizes of small ovals are the same only when $\mu = 0.5$. With decreasing μ the ratio of their sizes increases approximately as $\mu^{1/2}$; e.g. at $\mu = 0.1$ the oval surrounding m_1 will be almost three times larger than that around m_2. At $\mu = 0.01$ the ratio between the two ovals is about ten.

Even stronger is the influence of the Jacobi constant on the form of the null velocity curves. Therefore it seems reasonable to consider this point together with the role of the coordinates of the infinitely small body.

6   The Jacobi Constant and the Forms of the Null Velocity Curves

The variation of the Jacobi constant leads to essential qualitative changes in the null velocity curves, i.e. even a relatively small change in C reveals new and rather interesting properties of these curves. At rather high values of C, as we saw above, the null velocity curves are ovals of different sizes. Consider now the evolution of these curves at intermediate values of C and intermediate positions of the infinitely small body with respect to the finite bodies. In contrast to the above qualitative analysis, the real forms of these curves can be obtained only by concrete computations. It is necessary to follow the evolution of the ovals, both great and small, with decreasing C.

First consider the great ovals. With decreasing C one arrives at a moment corresponding to some $C = C_1$ called "critical", at which the highly deformed oval turns into a complicated curve of a rather strange form (Figs. 9.3 and 9.4). Moreover, this curve can at some points be self-contacting, i.e. quite a new property appears as shown in Fig. 9.5 in which the contact of inner ovals takes place at the libration point L_2 (see section 9 in this chapter). Another example is shown in Fig. 9.6 for $\mu = 0.4$ and with contact at the libration point L_3. In Fig. 9.4 the contacting point is the libration point L_1.

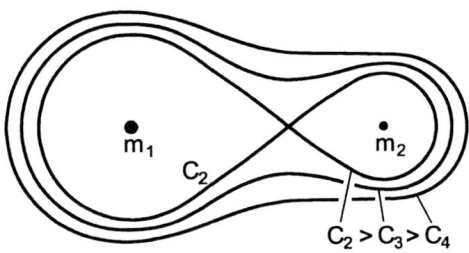

Fig. 9.3. *Null velocity curves surrounding both bodies m_1 and m_2 at intermediate magnitudes of the Jacobi constant C.*

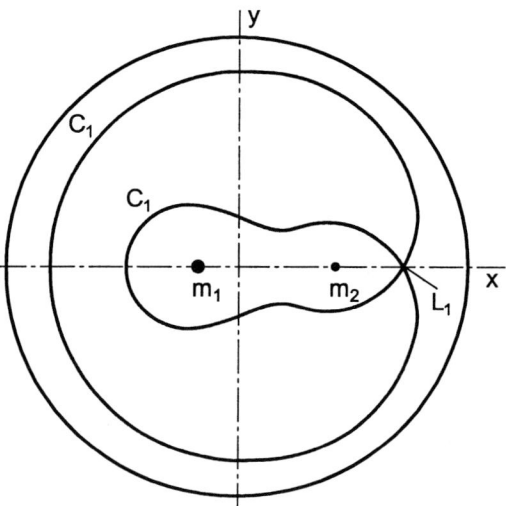

Fig. 9.4. *Null velocity curves, showing the transition through libration point L_1, i.e., at the Jacobi constant value $C_1 = 3.6$.*

In fact, however, at all three libration points, L_1, L_2 and L_3, contact occurs between the two halves of the null velocity curves corresponding to one and the same value of the Jacobi constant C. However, these contacts take place at rather acute angles for one to speak of their intersection. Incidentally, the libration points are also known as null velocity points.

Consider now the evolution of small ovals with further decreasing of the Jacobi constant C. These ovals are intersecting, in fact contacting, for strongly definite combinations of μ and C. More precisely, for a given μ there is a definite $C = C_2$, the second critical value of the Jacobi constant at which contact occurs. Then the null velocity curve is **figure eight**-shaped. The intersection occurs again on the line m_1 and m_2, thus the second singular null velocity point L_2 appears. In contrast to the point L_1, the point L_2 is always situated between the bodies m_1 and m_2, and is always close to the smaller mass m_2. Only in one case, when $\mu = 0.5$, is point L_2 situated exactly in the center between m_1 and m_2, and the figure-eight-shape is symmetrical for both axes. In Fig. 9.5 three curves, the figure eight-shaped zero velocity curves are shown for $\mu = 0.5$, 0.3, and 0.1 corresponding to the following values of $C = C_2(\mu)$: $C_2(0.5) = 4.25$, $C_2(0.3) = 4.13$ and $C_2(0.1) = 3.69$. It follows from these curves, in particular, that the point L_2 moves quickly towards the small mass m_2 with decreasing μ. In the case of the Moon, for example, we have $\mu = 0.012$ and the point L_2 lies at nearly a sixth of the Earth–Moon distance, i.e. nearly $66\,000\ km$ from the Moon and on the Earth–Moon line. Obviously the orbit of the critical point is a circle surrounding the Earth.

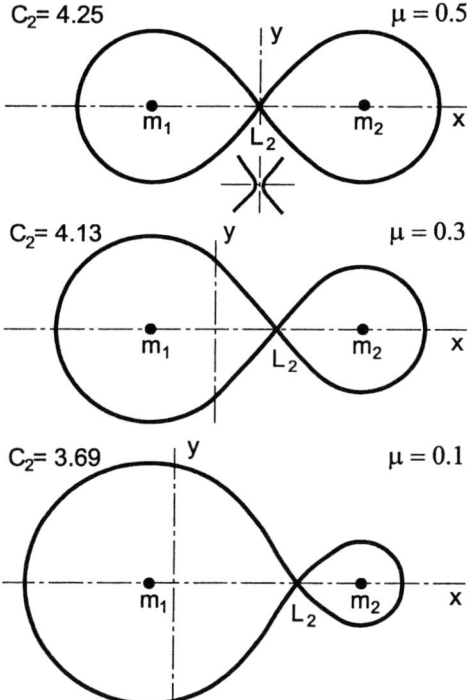

Fig. 9.5. *Null velocity curves, showing the transition through libration point L_2 at three values of the mass parameter μ: 0.1, 0.3, and 0.5.*

At values of C smaller than C_2, the ovals can never intersect (come into contact); in this case the null velocity ovals form opened dumbbell or cocoon-shaped figures, as shown in Fig. 9.3 for $\mu = 0.3$.

Thus, unless $C > C_2$, the final bodies m_1 and m_2 have their own separate ovals isolated from each other. In the case in which $C < C_2$, the ovals are merged into single closed dumbbell-shaped curve, containing both bodies, m_1 and m_2.

No less interesting is the evolution of the great ovals. With further decrease of the Jacobi constant, below C_2, one more critical point $C = C_3$ appears, implying the existence of the third libration point L_3 to the right of m_1. An example of such a horseshoe-shaped curve is shown in Fig. 9.5 for $\mu = 0.3$ with the critical point $C_3 = 3.50$.

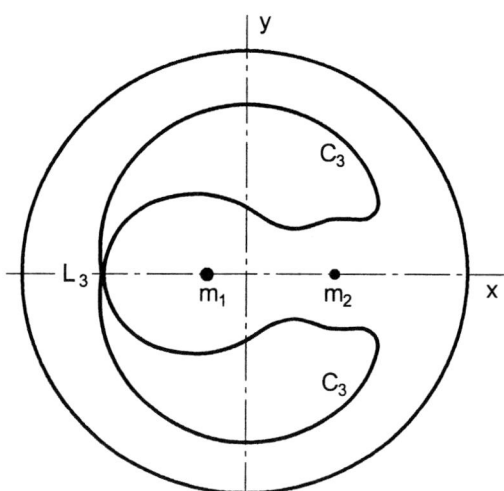

Fig. 9.6. *Null velocity curves, showing the transition through libration point L_3 at the value $C_3 = 3.5$ of the Jacobi constant.*

The last value, $C = C_3 = 3.50$, is not the final point. With further increase of the Jacobi constant, to values larger than C_3, the singularity at L_3 disappears since the two halves of the null velocity curves diverge. A pair of closed curves appears on both sides of the axis X, with a slightly elongated tadpole-like form. With decreasing C these tadpole shapes diminished very quickly, and, at least, at $C = C_4 = C_5 = 3.00$, they turn into a point, or more precisely two points, L_4 and L_5, symmetrically situated near the axis X. In such a way the fourth and fifth libration points, L_4 and L_5, appear.

In Fig. 9.7 a family of closed curves, tadpole shapes related to the described cases $C_4 < C < C_5$, are shown. The curve denoted by C_3 represents the starting phase of a series of null velocity curves. This series ends with the points L_4 and L_5.

These are the general properties of the intersection of the null velocity surfaces of an infinitely small mass Δm with the XOY plane. Fig. 9.8

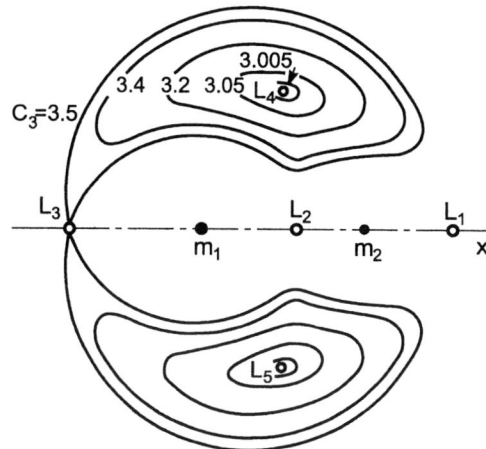

Fig. 9.7. *Null velocity curves at values of the Jacobi constant within the limits from $C_3 = 3.5$ (libration point L_3) to $C_4 = C_5 = 3.05$ (libration points L_4 and L_5).*

summarizes the types of curves for $\mu = 0.3$ and the libration points L_1, L_2, L_3, L_4 and L_5 are shown as well.

7 Regions of Real and Imaginary Velocities

On all of the curves shown in Figs. 9.2–9.8 the velocity of relative motion of the infinitely small mass is zero. This means that the first term on the right-hand side in the relationship

$$V^2 = \left((x^2 + y^2) + \frac{2(1 - \mu)}{r_1} + \frac{2\mu}{r_2} \right) - C = 0 \qquad (15)$$

exactly compensates the second term, i.e. $-C$.

Obviously, a real regular motion with a velocity different from zero is possible only within the region where $V^2(x, y) > 0$, i.e. when

$$\left((x^2 + y^2) + \frac{2(1 - \mu)}{r_1} + \frac{2\mu}{r_2} \right) - C > 0 . \qquad (16)$$

It is easy to see that there exists a definite region within which expression in square brackets in Eq. (16) will become smaller than C, and then $V^2 < 0$, i.e. we will have imaginary velocities. In this connection there arises the question of the determination of the regions where the motion is real and where it is imaginary.

Let us start from the great ovals, i.e. when x and y are large, and large values of the Jacobi constant. The equation of the ovals satisfies the condition

$$f(x_0, y_0) = C , \qquad (17)$$

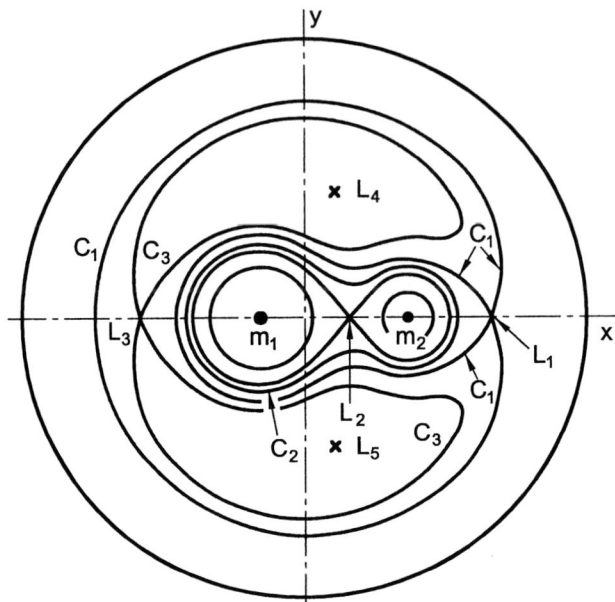

Fig. 9.8. *Null velocity curves in limiting cases including those transiting through the libration points L_1, L_2, and L_3.*

where

$$f\,(x,\,y) = (x^2 + y^2) + \frac{2(1-\mu)}{r_1} + \frac{2\mu}{r_2}\,, \tag{18}$$

and where x_0 and y_0 are related to the oval. For great ovals

$$f\,(x_0,\,y_0) = x_0^2 + y_0^2\,. \tag{19}$$

Real Keplerian motion with a velocity different from zero is possible within a region satisfying the condition

$$f\,(x,\,y) > C\,. \tag{20}$$

By comparing this with Eq. (19) we see that the motion is possible ($V > 0$) within the region where $x > x_0$ and $y > y_0$. Hence, the real motion is possible in the regions **external** to the great ovals.

Turn now to the small ovals:

$$f\,(x_0,\,y_0) = \frac{2(1-\mu)}{[(x_0 - x_1)^2 + y_0^2]^{1/2}} + \frac{2\mu}{[(x_0 - x_2)^2 + y_0^2]^{1/2}}\,, \tag{21}$$

where x_0 and y_0 are the coordinates of the small oval. The condition (20) can be satisfied only if x and y in the equation

$$f\,(x,\,y) = \frac{2(1 - \mu)}{[(x - x_1)^2 + y^2]^{1/2}} + \frac{2\mu}{[(x - x_2)^2 + y^2]^{1/2}}\,, \qquad (22)$$

are $x < x_0$ and $y < y_0$. This corresponds to the inner regions of the small ovals.

These regions of permitted motion – outside the great ovals and within the small ones, are shown in Fig. 9.9. When an infinitely small body Δm is located far from the center of the system, i.e. within the shaded area outside the large ovals, its motion will be Keplerian and performed around the center of mass of the finite mass bodies m_1 and m_2. Then the motion of Δm is determined by some hypothetical body of equivalent mass $M = m_1 + m_2$ situated at the center of mass of the system. However, if Δm lies close either to m_1 or to m_2, then Keplerian motion is possible if Δm lies within the shaded zone of either m_1 or m_2. In the first case the motion of Δm is controlled by m_1, as if m_2 were absent and vice versa in the second case.

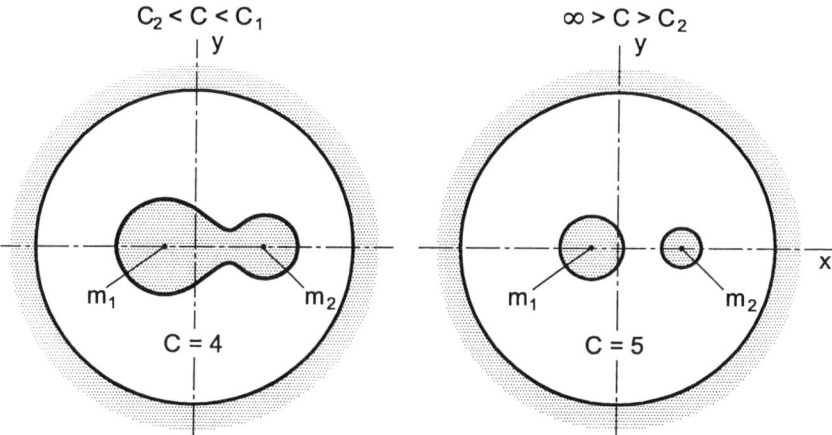

Fig. 9.9. *Regions of real motions (shaded) at values of Jacobi constant $C = 4$ (right) and $C = 5$ (left) at $\mu = 0.3$.*

The left-hand figure in Fig. 9.9 corresponds to a Jacobi constant larger than C_2, i.e. when $\infty > C > C_2$. At $C = C_2$, as was shown above, a libration point L_2 appears, and the null velocity curve has a figure eight-shaped form. When $C < C_2$, that form is replaced by a dumbbell figure. This concerns the small ovals. In the case of the great ovals, though the ovals themselves are preserved, their sizes decrease. The regions of permitted motions in this case $(C < C_2)$ are shown in the right-hand figure of Fig. 9.9.

The Jacobi constant has a dimension of squared velocity. Within the dumbbell figure (Fig. 9.9, right) the motion is permitted, i.e. here $V > 0$. Hence, there are definite values of C, i.e. definite values of the initial velocity V_0, for which the motion of the infinitely small body Δm is performed either around the body m_1 or around m_2. This corresponds, for example, to the motion of an artificial satellite either around the Earth only or around the Moon (Fig. 9.9, left). For such values of C, motion of Δm within the intermediate unshaded area between m_1 and m_2 is forbidden.

There also exist values of C' and hence of V', for which motion is possible not only around m_1 and m_2, but also within the area between these bodies, as shown in the right-hand figure in Fig. 9.9. In this case trajectories for artificial satellites surrounding both the Earth and the Moon do exist.

For lower values of C the region of permitted motions can take rather complicated forms – two examples are shown in Fig. 9.10, corresponding to $\mu = 0.3$. Note that the difference in the values of C is not large, $C = 3.6$ and 3.3, although the configurations differ greatly. This indicates the sensitivity of the forms of regions of permitted motions to the value of the Jacobi constant, if the latter lies within the interval 3–4.

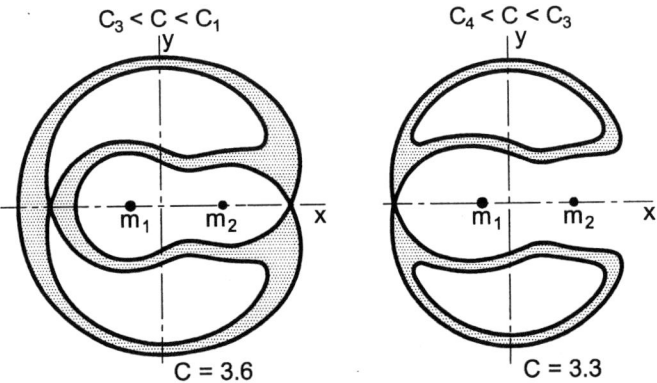

Fig. 9.10. *Regions of real motions (shaded) at values of the Jacobi constant $C = 3.3$ (right) and 3.6 (left) at $\mu = 0.3$.*

8 Forms of Null Velocity Surfaces (Planes ZOX and ZOY)

Now we will examine the curves of intersection of the null velocity surfaces of an infinitely small mass with the other coordinate planes. The equation of the intersection curves of the null velocity surfaces with the coordinate plane XOZ can be derived by substituting $y = 0$ into Eqs. (10) and (11):

$$x^2 + \frac{2(1-\mu)}{[(x-x_1)^2 + z^2]^{1/2}} + \frac{2\mu}{[(x-x_2)^2 + z^2]^{1/2}} = C . \tag{23}$$

At high values of x and z the second and third terms are small, therefore Eq. (23) can be rewritten in the form

$$x^2 = C - \varepsilon(x, z) \tag{24}$$

whence

$$x = \pm[C - \varepsilon(x, z)]^{1/2} . \tag{25}$$

If x and z are large, then $\varepsilon(x, z) \approx 0$ and Eq. (25) turns into a pair of straight lines parallel to the axis z and at equal distances from the center of the coordinate system: $x = \pm\sqrt{C}$. The larger C is, the larger will x be to satisfy a given value of z in Eq. (24), and therefore the smaller will $\varepsilon(x, z)$ be. In other words, at high values of C the intersection lines will be close to the asymptotic cylinder $x = \pm\sqrt{C}$, as is shown in Fig. 9.11.

For lower values of x and z satisfying Eq. (23), the first term is small, and therefore we have

$$\frac{2(1-\mu)}{[(x-x_1)^2 + z^2]^{1/2}} + \frac{2\mu}{[(x-x_2)^2 + z^2]^{1/2}} = \frac{C}{2} - \varepsilon(x) . \tag{26}$$

This is the equation of the same equipotential curves and has the same properties as before. Therefore, the shapes of the curves of intersection with the plane XOZ (Fig. 9.11) at small values of x and z will be qualitatively analogous to those derived for low values of x and y (Fig. 9.8).

Finally, for the equation of intersection of null velocity surfaces with the coordinate plane YOZ we have, substituting $x = 0$ into Eqs. (10) and (11),

$$y^2 + \frac{2(1-\mu)}{[x_1^2 + y^2 + z^2]^{1/2}} + \frac{2\mu}{[x_2^2 + y^2 + z^2]^{1/2}} = C . \tag{27}$$

For high values of y and z the second and third terms are small, and Eq. (27) turns into the equation of a pair of parallel lines near to the asymptotic cylinder at the distance

$$y = \pm[C - \varepsilon(y, z)]^{1/2} \tag{28}$$

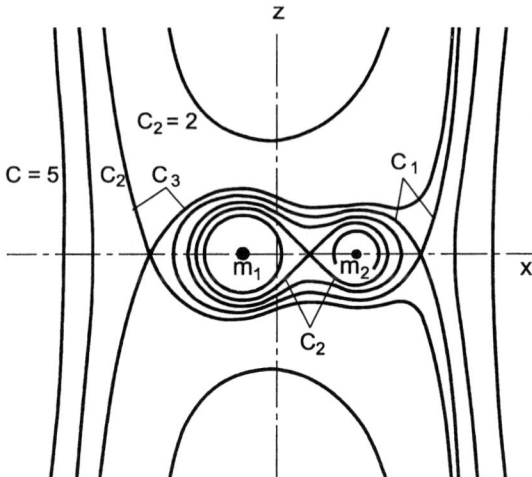

Fig. 9.11. *Null velocity curves in the coordinate plane ZOX.*

from the axis z. For lower (y, z) values the equation (27) takes the form

$$\frac{2(1-\mu)}{[x_1^2 + y^2 + z^2]^{1/2}} = C - \varepsilon\,(y)\;, \tag{29}$$

insofar as usually $x_2 > x_1$ when $1 - \mu > \mu$. Eq. (29) is the equation of a circle of increasing radius with decreasing C. These curves are shown in Fig. 9.12.

The curves passing close to the asymptotic cylinders, i.e. those defined by Eqs. (25) or (29) and continuing along the axis z from $-\infty$ to $+\infty$, are occasionally called **curtains**.

Thus, for high values of C the family of curves of the intersections of null velocity surfaces with the coordinate planes YOZ and XOZ form two closed nearly spherical surfaces surrounding each of the finite bodies m_1 and m_2, and curtains symmetrically located within the asymptotic cylinder along the axis z. For low values of C, the central hemispheric surfaces around m_1 and m_2 are expanding, forming dumbbell figures.

9   Libration Points

The analysis of the curves of intersection of null velocity surfaces with the coordinate plane XOY leads, as we have seen above, to the discovery of five

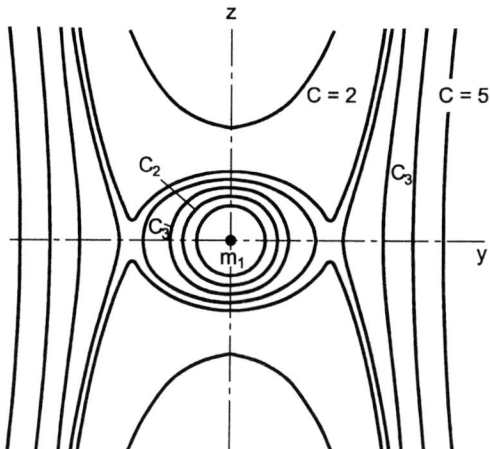

Fig. 9.12. *Null velocity curves in the coordinate plane ZOY.*

singular points where the velocity equals zero. All these points denoted by L_1, L_2, L_3, L_4 and L_5 are situated on the plane XOY' three of them, L_1, L_2 and L_3 – along the axis OX, or more precisely, along the line m_1m_2, and the last two, L_4 and L_5 – far from both of the axes OX and OY.

These five singular points have been given various names. Most often they are called **stationary points** or the **points of relative equilibrium** as well as Laplace points, Lagrange points, and Euler points. Terms such as **centers of libration** and especially **libration points** are also widely used.

The equations of the curves of intersection of null velocity surfaces with the coordinate planes have the following form:

$$F\left(x,\ y,\ z\right) = (x^2 + y^2) + \frac{2(1-\mu)}{r_1} + \frac{2\mu}{r_2} - C = 0 \ . \qquad (30)$$

The coordinates of the singularity points are defined by the equations

$$\frac{\partial F}{\partial x} = x - (1-\mu)\frac{x-x_1}{r_1^3} - \mu\frac{x-x_2}{r_2^3}\ ,$$

$$\frac{\partial F}{\partial y} = y - (1-\mu)\frac{y}{r_1^3} - \mu\frac{y}{r_2^3}\ , \qquad (31)$$

$$\frac{\partial F}{\partial z} = -(1-\mu)\frac{z}{r_1^3} - \mu\frac{z}{r_2^3}\ .$$

Meanwhile, Eq. (30) enables one to obtain the corresponding values of the Jacobi constant C. The coordinates of points L_1, L_2 and L_3 differ from each other, so that for them we will have three different values of C. The coordinates of points L_4 and L_5 are numerically the same, the only difference being in the sign of y. Therefore we have $C_4 = C_5$.

From these relationships one can easily reveal the dynamical content of the singularity points. The expressions $\partial F/\partial x$, $\partial F/\partial y$ and $\partial F/\partial z$ are proportional to the direction cosines of the normal at all points on the curve (i.e. to expressions of the type $(x - x_1)/r_1$, y/r_1 etc.) However, insofar as

$$\frac{dx}{dt} = 0, \quad \frac{dy}{dt} = 0, \quad \frac{dz}{dt} = 0$$

on the null velocity surface, then from Eq. (4) we readily obtain

$$\frac{d^2x}{dt^2} = 0, \quad \frac{d^2y}{dt^2} = 0, \quad \frac{d^2z}{dt^2} = 0$$

i.e. at the singularity point, the body of infinitely small mass has a velocity and acceleration equal to zero, and, hence, a body that once reaches this point will remain there forever. Thus, the singularity points of null velocity surfaces are the equilibrium points of the infinitely small masses in our coordinate system.

The fact that the body of an infinitely small mass finding itself at the libration point does not influence any acceleration is illustrated by a diagram in Fig. 9.13, in which the positions of m_1 and m_2 as well as of the libration points L_1, L_2, L_3 and L_4 are shown. Consider for example, the case of the point L_4. Here the gravitational forces of m_1 and m_2 are equal to F_1 and F_2 and lead to a resulting force R_g directed along the line OL_4 ($O\Delta m$). Recall that the center of the coordinate system O coincides with the center of mass of the system $m_1 + m_2$, and at the same time is the center of rotation of the coordinate system around the axis OZ (perpendicular to the picture's plane) with constant velocity V_0 for m_1 and m_2. The centrifugal force R_c that appeared as a result of rotation will also be directed along the line OL_4 but in the opposite direction. Thus the essence of the libration point is in the fact that $F_c = F_g$.

The situation is analogous for the other libration points situated on the line m_1m_2. For example, the gravitation forces F_1 and F_2 at the point L_3 from both m_1 and m_2 are directed only the line m_1m_2 and give $R_g = F_1 + F_2$. However, the centrifugal force R_c, also directed along the line m_1m_2 though in the opposite direction, is exactly balanced by R_g, i.e. $R_c = R_g$.

Now we turn to the determination of the coordinates of the libration points. From the last equation in (31) it follows that $z = 0$. Hence, the first and most important conclusion: all five centers of the libration must be situated on the plane XOY.

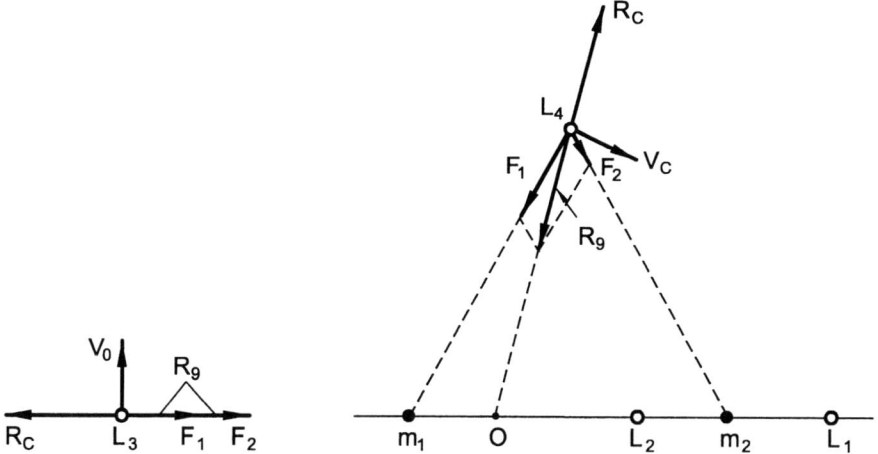

Fig. 9.13. *A vectorial diagram illustrating the origin of the null velocity at the libration points L_4 (left) and L_3 (right).*

Let us start by searching for the libration points not situated on the axis OX, i.e. those for which y is non zero. Then the second equation in (31) gives

$$1 - \frac{1-\mu}{r_1^3} - \frac{\mu}{r_2^3} = 0 \; . \tag{32}$$

Multiplying this equation by x and subtracting the result from the first equation (31), we obtain

$$-(1-\mu)\frac{x_1}{r_1^3} + \mu\frac{x_2}{r_2^3} = 0 \; . \tag{33}$$

However, according to the determination

$$-(1-\mu)\,x_1 + \mu x_2 = 0 \; , \tag{34}$$

we will have from Eq. (33)

$$-\frac{1}{r_1^3} + \frac{1}{r_2^3} = 0 \; . \tag{35}$$

This equation has only one solution, namely, $r_1 = r_2 = 1$. Recall that r_1 and r_2 are the distances of the infinitely small body from the finite bodies m_1 and m_2, respectively.

Thus, the libration point L_4 (and L_5) lies, first, at the same distance from m_1 and from m_2, and, second, the latter is equal to the distance between the bodies m_1 and m_2. In other words, the three points, L_4, m_1 and m_2 form an equilateral triangle.

This result, which is amazing in its simplicity, is remarkable for its complete independence from the masses of the finite bodies m_1 and m_2, as well as from their ratio m_1/m_2, and therefore it is universal. The coordinates of the points L_4 and L_5 are given (in the adopted system of units of mass and distance) by

$$x = \frac{1}{2} - \mu \,, \qquad y = \pm \frac{\sqrt{3}}{2} \qquad z = 0 \,. \tag{36}$$

As for the determination of the positions of the other three libration points L_1, L_2 and L_3, situated on the axis X, i.e. with $y = 0$, we obtain their abscissa from the first equation in (31):

$$x - (1 - \mu)\frac{x - x_1}{r_1^3} - \mu\frac{x - x_2}{r_2^3} = 0 \,. \tag{37}$$

This equation has three roots – one for each interval $(-\infty,\ x_2)$, $(x_1,\ x_2)$ and $(x_1,\ +\infty)$. The singularity points corresponding to these roots are just the libration points L_1, L_2 and L_3. We must obtain the distances r_1 and r_2 for these three points.

Because r_1 and r_2 depend only on the mass ratio $\mu/(1 - \mu)$, we will have $r_2 = 1 - r_1$. Substituting this into Eq. (37), we will have for r_2, after rewriting Eq. (37) in another form;

$$r_2^5 - (3 - \mu)r_2^4 + (3 - 2\mu)r_2^3 - \mu r_2^2 + 2\mu r_2 - \mu = 0 \,. \tag{38}$$

This equation reveals the first essential difference of the first three libration points L_1, L_2 and L_3 from the last two, L_4 and L_5; in the first case their coordinates (abscissa) depend on the masses of the finite bodies, i.e. on μ, in the second case they do not. Eq. (38) has five roots though only one of them is positive and is related to μ in the following manner:

$$r_2^3 = \mu\,\frac{1 - 2r_2 + r_2^2 + 2r_2^3 - r_2^4}{3 - 3r_2 + r_2^2} \,. \tag{39}$$

On expanding Eq. (39) by series, we obtain finally the position (abscissa) of the first point $L_1\,(-\infty,\ m_2)$

$$r_2(L_1) = \nu - \frac{1}{3}\nu^2 - \frac{1}{9}\nu^3 \ldots \,, \tag{40}$$

where

$$\nu = \left(\frac{\mu}{3}\right)^{1/3} . \tag{41}$$

Analogously we find for the positions of the other two libration points $L_2\,(m_1,\ m_2)$ and $L_3\,(m_1,\ +\infty)$

$$r_2(L_2) = \nu + \frac{1}{3}\nu^2 - \frac{1}{9}\nu^3 + \ldots \tag{42}$$

or in slightly different form,

$$r_2(L_3) = \frac{7}{12}\mu + \frac{23 \times 7^2}{12^2}\mu^3 + \ldots . \tag{43}$$

Having the positions of L_1, L_2 and L_3 in terms of their dependence on μ, we can obtain the numerical values of the Jacobi constant C corresponding to these points:

$$
\begin{aligned}
C_1 &= (1 - \mu)(3 + 9\nu^2 - \nu^3 + \ldots) \,, \\[4pt]
C_2 &= (1 - \mu)(3 + 9\nu^2 - 5\nu^3 + \ldots) \,, \\[4pt]
C_3 &= (1 - \mu)\left(3 + 4\mu^2 + \frac{191}{48}\mu^3 + \ldots\right) .
\end{aligned}
\tag{44}
$$

For example, in the case of $\mu = 0.3$ we will have for the points L_1, L_2 and L_3 respectively,

$$C_1 = 3.92, \quad C_2 = 3.56, \quad C_3 = 3.30 \,,$$

and for the points L_4 and L_5, respectively,

$$C_4 = C_5 = 3.00.$$

With the help of these formulas, special tables are composed for determination of values of the Jacobi constant in terms of its dependence on μ. The curves as functions of r and $C/(1-\mu)$ are represented in Figs. 9.14 and 9.15 for the points L_1, L_2 and L_3.

Strictly speaking, the libration points are not mathematical points. The singularity points, e.g. L_4 or L_5, as we will see below, cover some area within the limits where the zero velocity of an infinitely small mass can be preserved.

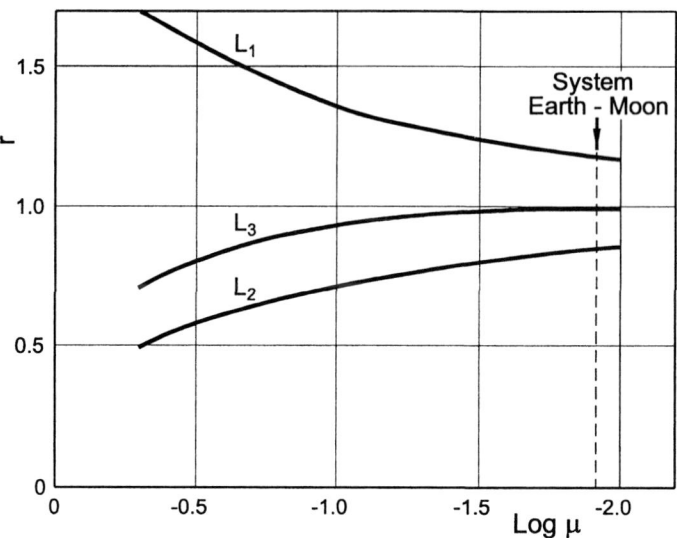

Fig. 9.14. *The dependence of the location* (r_1) *of the libration points* L_1, L_2 *and* L_3 *on the mass parameter* μ. *The vertical broken line on the right is related to the Earth–Moon system.*

10   Stability of Motion

The stability of motion is one of the central problems of celestial mechanics. The essence of the problem is as follows. As is well known, differential equations of motion with constant coefficients also have partial solutions as well as the general solution. Two cases are possible: a) the partial solutions are expressed via periodic functions – in this case the motion will be stable; b) the solutions involve nonperiodic functions, i.e. the motion will be unstable.

    These criteria of stability and instability of motion are rather general and at the same time are too far from being complete. The problem of classification of planetary motions in terms of their degree of stability (instability), including the definitions of the concept of instability itself, will be considered in more detail in section 14 of Chapter X.

    The fundamental results of the problem of stability of motion of celestial bodies are associated mainly with the names of Poincaré and Lyapunov. Briefly the mathematical treatment of the problem is as follows.

    Consider the differential equations of motion (5) in a rotating system of

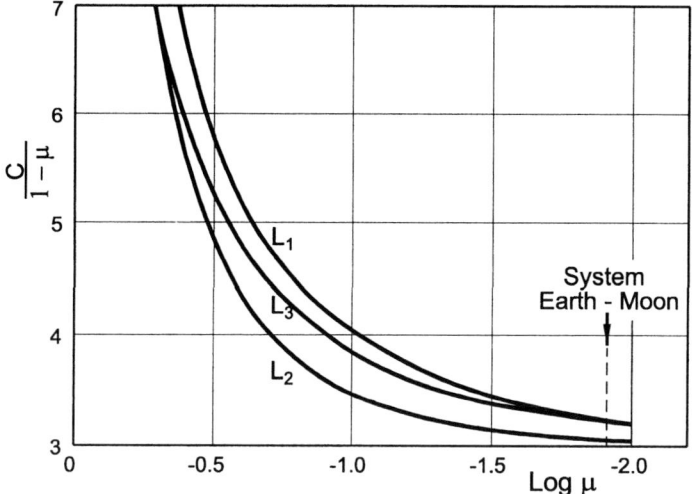

Fig. 9.15. *The dependence of the Jacobi constant C on the mass parameter μ for the libration points L_1, L_2 and L_3. The vertical broken line is related to the Earth–Moon system.*

coordinates rewritten in the following form:

$$
\frac{d^2x}{dt^2} - 2\frac{dy}{dt} = f(x, y) ,
$$

$$
\frac{d^2y}{dt^2} + 2\frac{dx}{dt} = g(x, y) .
$$

$$(45)$$

Let x_0 and y_0 be the partial solutions of this system satisfying the equations

$$
\frac{d^2x_0}{dt^2} - 2\frac{dy_0}{dt} = 0 ,
$$

$$
\frac{d^2y_0}{dt^2} + 2\frac{dx_0}{dt} = 0 .
$$

$$(46)$$

Obviously this is equivalent to the condition

$$
f(x_0, y_0) = 0 ,
$$

$$
g(x_0, y_0) = 0 .
$$

$$(47)$$

On shifting the body by x' and y' so that its coordinates and corresponding velocities dx'/dt and dy'/dt are

$$x = x_0 + x', \qquad \frac{dx}{dt} = \frac{dx'}{dt} \, ,$$

$$y = y_0 + y', \qquad \frac{dy}{dt} = \frac{dy'}{dt} \, . \tag{48}$$

and substituting this into Eq. (45), we obtain

$$\frac{d^2 x'}{dt^2} - 2\frac{dy'}{dt} = f\left(x_0 + x', \ y_0 + y'\right) ,$$

$$\frac{d^2 y'}{dt^2} + 2\frac{dx'}{dt} = g\left(x_0 + x', \ y_0 + y'\right) . \tag{49}$$

The expansion of the right-hand side of this system in Taylor series gives

$$f\left(x_0 + x', \ y_0 + y'\right) = f\left(x_0, \ y_0\right) + \frac{\partial f}{\partial x'}x' + \frac{\partial f}{\partial y'}y' + \ldots \, ,$$

$$g\left(x_0 + x', \ y_0 + y'\right) = g\left(x_0, \ y_0\right) + \frac{\partial g}{\partial x'}x' + \frac{\partial g}{\partial y'}y' + \ldots \, , \tag{50}$$

where $x = x_0$ and $y = y_0$ should be used for the expressions of partial derivatives.

According to Eq. (47) the first terms on the right-hand side of Eq. (50) are equal to zero. Therefore Eq. (49) takes the form

$$\frac{d^2 x'}{dt^2} - 2\frac{dy'}{dt} = \frac{\partial f}{\partial x'}x' + \frac{\partial f}{\partial y'}y' + \ldots \, ,$$

$$\frac{d^2 y'}{dt^2} + 2\frac{dx'}{dt} = \frac{\partial g}{\partial x'}x' + \frac{\partial g}{\partial y'}y' + \ldots \, . \tag{51}$$

If the initial values of shifts x' and y' are small, their influence will be negligible at least during some period. Confining ourselves to the first degrees of derivatives x' and y', we will have the system of linear differential equations

$$\frac{d^2 x'}{dt^2} - 2\frac{dy'}{dt} = \frac{\partial f}{\partial x'}x' + \frac{\partial f}{\partial y'}y' \, ,$$

$$\frac{d^2 y'}{dt^2} + 2\frac{dx'}{dt} = \frac{\partial g}{\partial x'}x' + \frac{\partial g}{\partial y'}y' \, . \tag{52}$$

The general solution of this system can be represented in the form

$$x' = a_1 e^{\lambda_1 t} + a_2 e^{\lambda_2 t} + a_3 e^{\lambda_3 t} + a_4 e^{\lambda_4 t} \ ,$$
$$y' = b_1 e^{\lambda_1 t} + b_2 e^{\lambda_2 t} + b_3 e^{\lambda_3 t} + b_4 e^{\lambda_4 t} \ , \tag{53}$$

where $a_1, \ldots, a_4, b_1, \ldots, b_4$ are the constants of integration and $\lambda_1, \ldots, \lambda_4$ are the roots of the characteristic equations; these roots depend on the constants of integration as well as on the parameters on the right-hand side of Eq. (45).

The character of the motion obviously depends on $\lambda_1, \ldots, \lambda_4$. If they all are imaginary numbers then the solutions x' and y' are expressed via periodic functions of t; in this case the motion is **stable**. If, however, even one of them is real or complex, x' and y' will never be periodic functions; such motion is considered **unstable**. In the first case the motion is Keplerian, i.e. is described by known laws. In the second case the body will leave the system.

Formally, there can be cases in which the exponential functions in the solution of the equations could be represented via constant numbers; such motion is also stable if all power indexes are imaginary. In the case in which the exponential functions contain factors of certain degrees of t, the motion is unstable, etc.

Thus, in every concrete case the problem of the stability or instability of motion is determined by the structure of the right-hand side of Eq. (45), i.e. by the structure of the functions $f(x, y)$ and $g(x, y)$.

The fact that, on the right-hand side of Eq. (52), derivatives of second and higher degrees were neglected, leaves open the question of the character of the motion in the sense of its stability or instability for large t, i.e. for the distant future or the distant past. However, taking into account the second derivative will lead to serious mathematical difficulties. Therefore the conclusions obtained from Eq. (52), strictly speaking, are valid for periods of time that are not too large.

Below, as an example, we will apply these considerations to the problem of the stability of motion of a body of an infinitely small mass within the libration points in the framework of the restricted three-body problem.

11   Motion Around Libration Points

A body situated at a libration point must remain in absolute rest in the rotating coordinate system. However, it can be localized not exactly at this point but slightly shifted away from it. In this connection the question of the fate of such a body arises: either it should oscillate with respect to the libration point with a definite frequency or it should escape relatively quickly.

The first case concerns the stability of the motion of the mass around the libration point; the second one concerns the unstable motion. In both cases the problem reduces to the study of a system of differential equations in the restricted three-body problem.

Consider first the motion of an infinitely small mass around a triangle libration point L_4 or L_5 in the coordinate plane XOY with coordinates determined by Eq. (36):

$$r_1 = 1 , \quad r_2 = 1 ,$$

$$x = \frac{1}{2} - \mu , \quad y = \pm \frac{\sqrt{3}}{2} , \quad z = 0 .$$

On substituting these coordinates into the right-hand side of Eq. (5) we have for the functions $f(x, y)$ and $g(x, y)$

$$f(x, y) = x - (1 - \mu)\frac{x - x_1}{r_1^3} - \mu\frac{x - x_2}{r_2^3} = \frac{3}{4}x + \frac{3\sqrt{3}}{4}(1 - 2\mu)y ,$$

$$g(x, y) = y - (1 - \mu)\frac{y}{r_1^3} - \mu\frac{y}{r_2^3} = \frac{3\sqrt{3}}{4}(1 - 2\mu)x + \frac{9}{4}y .$$

$$(54)$$

Then Eq. (52) will take the form

$$\frac{d^2 x'}{dt^2} - 2\frac{dy'}{dt} = \frac{3}{4}x' + \frac{3\sqrt{3}}{4}(1 - 2\mu)\, y' ,$$

$$\frac{d^2 y'}{dt^2} + 2\frac{dx'}{dt} = \frac{3\sqrt{3}}{4}(1 - 2\mu)\, x' + \frac{9}{4}y' ,$$

$$(55)$$

$$\frac{d^2 z'}{dt^2} = -z' .$$

The last equation, as before, does not depend on the first two equations, and its solution can be readily written:

$$z' = c_1 \cos t + c_2 \sin t ,$$

$$(56)$$

i.e. the motion of an infinitely small mass along the axis z and around the libration point L_4 (L_5) must be periodic. The only coefficient in t is nothing other than the angular velocity of rotation of the coordinate system or the rotation of the finite bodies themselves: $n = 1$. Hence, the periodic motion takes place with the same period as the rotation of the finite bodies m_1 and m_2 around their center of mass.

Let us search for the solution of the first two equations in Eq. (55) in the form

$$x' = Ke^{\lambda t} , \quad y' = Le^{\lambda t} .$$

$$(57)$$

On substituting these expressions into Eq. (55) we obtain

$$\left(\lambda^2 - \frac{3}{4}\right) K - \left(2\lambda + \frac{3\sqrt{3}}{4}(1 - 2\mu)\right) L = 0 \,,$$
$$\left(2\lambda - \frac{3\sqrt{3}}{4}(1 - 2\mu)\right) K + \left(\lambda^2 - \frac{9}{4}\right) L = 0 \,. \tag{58}$$

In spite of the linearity of these equations with respect to both coefficients K and L, their joint solution obtained by equating the corresponding determinant to zero, leads to the following characteristic equation for λ:

$$\lambda^4 + \lambda^2 + \frac{27}{4}\mu(1 - \mu) = 0 \,. \tag{59}$$

The roots of this biquadratic equation are

$$\lambda_1 = -\lambda_2 = \frac{1}{\sqrt{2}} \left\{-1 + [1 - 27\mu(1 - \mu)]^{1/2}\right\}^{1/2} \,,$$
$$\lambda_3 = -\lambda_4 = \frac{1}{\sqrt{2}} \left\{-1 - [1 - 27\mu(1 - \mu)]^{1/2}\right\}^{1/2} \,. \tag{60}$$

Our first aim is to find the condition under which the motion will be stable. However, it can be stable only in the case in which all the roots (60) are imaginary and conjugate pair by pair. This is possible if

$$1 - 27\mu(1 - \mu) \geq 0 \,. \tag{61}$$

Denoting by μ_0 the limiting value of μ corresponding to the case in which the equality sign in Eq. (61) is fulfilled, we have

$$27\mu_0^2 - 27\mu_0 + 1 = 0 \,. \tag{62}$$

Then we find

$$\mu_0 = \frac{1}{2} \pm \left(\frac{23}{108}\right)^{1/2} \,. \tag{63}$$

If the right-hand side in Eq. (61) is nonzero and is positive, ε, then we have, instead of Eq. (63)

$$\mu_0 = \frac{1}{2} \pm \left(\frac{23 + 4\varepsilon}{108}\right)^{1/2} \,. \tag{64}$$

The parameter μ can never be larger than 0.5, so that one can take a negative sign in Eqs. (63) and (64). Then for the limiting case ($\varepsilon = 0$) we obtain from Eq. (63)

$$\mu_0 = 0.0385 \,. \tag{65}$$

Thus, the motion of an infinitely small mass may be stable only when the parameter μ for the system with two finite bodies is smaller than μ_0. Therefore the condition

$$0 < \mu < \mu_0 = 0.0385 \tag{66}$$

can be adopted as a condition of stability of the motion of an infinitely small mass around the libration points L_4 or L_5. One can arrive at exactly the same conclusions from analysis of the motion around the libration points L_1, L_2 and L_3.

If $\mu = \mu_0$ or $\mu = 1 - \mu_0$, the corresponding partial solution (57) will have the form

$$x' = x'_0 \cos(\sigma t) + \frac{y'_0}{c} \sin(\sigma t) ,$$
$$y' = c x'_0 \sin(\sigma t) - y'_0 \cos(\sigma t) , \tag{67}$$

where σ is a real number, and c is a constant. Eqs. (67) are in fact equations of the motion of an infinitely small mass in the vicinity of a libration point. By excluding from them the time t we obtain the equation of the orbit of such a motion in the following form:

$$\frac{x^2}{x_0^2 + y_0^2/c^2} + \frac{y^2}{c^2 x_0^2 + y_0^2} = 1 . \tag{68}$$

This is the equation of an ellipse with its axes coinciding with the coordinate axes, as do their centers.

Thus, when the condition (66) is fulfilled, then the infinitely small mass cannot escape fromm the libration point under the influence of small perturbations but realizes a periodic motion in an elliptic orbit with a period equal to that of the rotation of the finite bodies around their mass center. The approximate values both of the major and of the minor axes may be obtained from the following formulas:

$$a = \left(\frac{4\Delta}{9\mu}\right)^{1/2} , \quad b = \left(\frac{4\Delta}{12 - 9\mu}\right)^{1/2} , \tag{69}$$

where Δ is the deviation from the given value of the Jacobi constant C at the libration point, i.e. $C = C_i \pm \Delta$.

Returning to the condition (66), one should stress that for all planets of the Solar system as well as for the Moon–Earth system this condition is well satisfied: the maximal value of μ among the planets is that of the Jupiter–Sun system ($\mu \approx 0.001$); the Moon–Earth system has $\mu \approx 0.01$.

12 Libration Points of the Earth–Moon System

The Moon's mass is 81.4 times smaller than that of the Earth, therefore we have for the Earth–Moon system: $\mu = 0.01214$ or $\log \mu = -1.916$. Correspondingly, we have from Figs. 9.14 and 9.15 for the positions of the libration points, i.e. for r_1 (or r_2) and the corresponding Jacobi constants (in units of the Earth–Moon distance, the results in Table 9.1.

T a b l e   9.1

Parameters of all five libration points for
the Earth–Moon system

Libration point	r_1	r_2	C
L_1	1.175	0.175	3.184
L_2	0.850	0.150	3.200
L_3	1.00	2.00	3.024
L_4	1	1	3.002
L_5	1	1	3.002

Thus, the nearest libration point L_2 lies at a distance $r_1 = 0.850 \times 386\,000 \approx 328\,000\ km$ from the Earth or $r_2 = 0.150 \times 386\,000 \approx 58\,000\ km$ from the Moon. The next libration points in remoteness, L_4 and L_5, are situated at equal distances $r_1 = 1 = 380\,000\ km$ both from the Earth and from the Moon, and at an angle of 60° on both sides of the Earth–Moon line. The most distant point is L_1; it lies at a distance $r_1 = 1.175 \times 380\,000 = 446\,500\ km$ from the Earth or at a distance $r_2 = 0.175 \times 380\,000 = 66\,500\ km$ from the Moon. All of the libration points of the Earth–Moon system are shown in Fig. 9.16 with an indication of their mutual distances. In view of the smallness of μ, the center of the coordinate system is located rather close to m_1, i.e. to the Earth.

Numerous attempts have been made to discover some bodies or diffuse matter at the libration points of the Earth–Moon system regarding the collinear libration points, but these attempts had no success. In all cases the luminosity of the diffuse matter that was supposed to be at the libration points L_1, L_2 and L_3 was much lower than the brightness of the night sky. Various expeditions were organized to different regions of the Earth to find the places with the lowest brightness of the night sky; however, again with negative results: no local excess of brightness at the libration points has been discovered. The final hope of solving this problem should obviously concern observations from space. If the results are again negative, then one can conclude the absence of diffuse matter at the libration points.

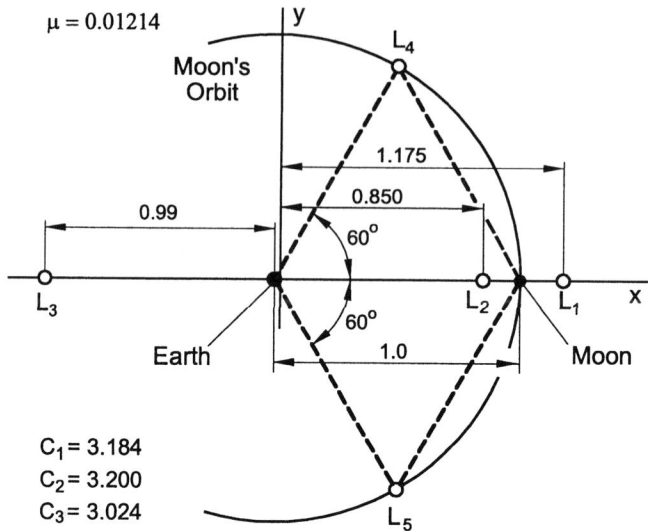

Fig. 9.16. *Reconnaissance of the collinear libration points L_1, L_2, and L_3 and the triangle libration points L_4 and L_5 in the Earth–Moon system.*

However, interplanetary matter does really exist, and it must appear in some way at the libration points. In all cases, if this matter is not accumulated at these points, the cause should be other effects acting in interplanetary space. The latter could be e.g. the radiation pressure of Solar radiation, the electromagnetic forces induced in the interplanetary space again under the influence of the Sun, and especially, the Solar wind, the flow of charged particles, mainly protons, accelerated up to rather high velocities, of the order of a few thousand $km\ s^{-1}$; such a storm could blow away all the matter that would otherwise appear at the libration points.

The considerations just described, apparently, are not purely speculative: there were indications of the existence of clouds of interplanetary dust in the vicinity of the triangle libration points L_4 and L_5 of the Earth–Moon system (Kordylevski, 1961). However, in view of the inevitable perturbations mentioned above, these clumps must be unstable – being temporally renewed by the flows of the interplanetary matter.

The libration points L_1 and especially L_2 could have another much more interesting role as the locations of future, permanent space stations – scientific research centers and space colonies. At the libration points one can construct giant platforms, stabilized facilities, space factories and laboratories for special technologies. The insignificant shifts inevitable for those platforms could be easily compensated by reactive engines. Corresponding projects for the creation of such bases, particularly at the libration point L_2, are being considered by NASA. We will return to the libration points of the Earth–Moon system in Chapter XIII.

13   The Libration Points of Planets

By means of the method described above one can determine the positions of the libration points for every planet combined with its natural satellites or the Sun. Of special interest among them are the libration points of the Jupiter–Sun system, which were predicted by Lagrange as early as 1772. The parameter μ for the Jupiter–Sun system is $\mu = 0.0009539$, and the semimajor axis for Jupiter's orbit is $\alpha = 5.2027$ a.u. From these data the positions of the triangle libration points L_4 and L_5 were obtained; the point L_4 is located outside Jupiter's orbit, L_5 is within it. In 1906, the first Trojan group asteroid, 588 Achilles, was discovered near the libration point L_5, at a distance of 1.28×10^6 km from Jupiter. Later, other asteroids were discovered not only at the point L_5 but also at L_4. At present more than a dozen asteroids are known in the vicinity of Jupiter's libration points L_4 and L_5.

Thus, the first observational indication of the existence of libration points and the possibility of the presence of matter even in the form of large objects of asteroid size has been obtained in the case of the Jupiter–Sun system.

Numerically, the parameter μ for other planets is much smaller than that for Jupiter. Even though the stability criterion for triangle libration points, namely, $\mu < 0.0385$, for all other planets is satisfied if one considers the Sun–planet system in the framework of the restricted three-body problem, so far asteroids have been observed at the points L_4 and L_5 only in Jupiter's case. The possible explanation of this fact, i.e. the absence of asteroids at the libration points of other planets, could be related to the conditions of the restricted problem, namely, the complete absence of external perturbations. In the case of other planets, this condition is satisfied much less strongly. Probably, the source of such perturbations for these planets is Jupiter itself, especially in view of the violation of equilibrium at libration points even at small perturbations.

14   Libration Points of the Earth–Sun System

In the previous section, the question was about the libration points of the Earth–Moon system. Replacing the Earth by the Sun and the Moon by the Earth, we have a new system of libration points L_1, L_2 and L_3. All three points are situated again on the line connecting the Earth with the Sun, as shown in Fig. 9.17 (without keeping the scale). Particularly, the points L_1

and L_2 are situated now on both sides from the Earth; however, both points are on the left side from the Sun, and the point L_3 on the opposite side, on the right. The point O, the origin of the coordinate system, lies on the mass center of the system.

In Fig. 9.17 the point L_1 lies at a distance $EL_1 = 1.51\,million\,km$ to the left from the Earth, and the point L^2 at a distance $EL_2 = 1.50\,million\,km$ to the right from the Earth, i.e. both these points, L_1 and L_2 are located out of the sphere of the Earth's action, almost exactly on the boundary of Hill's sphere for the Earth relative to the Sun, the radius of which is equal to $1.5\,million\,km$. Spacecraft at these points, L_1 and L_2, may be counted as the Earth's satellites with a one-year period of orbiting around the Earth. However, these spacecraft may be the Sun's satellites as well with the same duration of orbiting around the Sun. The point L_3 lies at a distance $EL_3 = 299\,million\,km$ from the Earth. The mass parameter for the Sun–Earth system is $\mu = 3.0359.10^{-6}$ taking into account the Moon's mass as well. The numerical value of the Jacobi constant for an equipotential curve passing through the point L_2 is $C_2 = 3.00900$. Equipotential curves in Fig. 9.17 are related to the Moon with Jacobi constant $C_0 = 3.0012$ which is larger as compared with the Jacobi constant for the system Earth–Moon–Sun.

As to flights to the libration points of the Earth–Sun system L_1, L_2 and L_3, they cannot be realized by the same manner as in the case of libration points of the Earth–Moon system, first of all since the collinear libration points for the Earth–Sun system are located near the boundary of the planet's action and, therefore, in the calculations of flight trajectories to these points one should take into account the gravitation actions not only of the Sun but of the planets as well, and, in the some cases, of their more massive satellites.

The peculiarity of the libration points of the Earth–Sun system is also due to the necessity to realize the launch not with a hyperbolic velocity relative to the Earth but with an elliptic starting velocity.

Any object located in the libration points L_1 or L_2 in the case of Earth–Sun system may be considered as a satellite both of the Sun and of the Earth with the same orbiting period, one year. The heliocentric velocity, for example, of the point L_1 is equal to $29.5\,km\,s^{-1}$ but the geocentric velocity is only $0.3\,km\,s^{-1}$. The situation is almost the same with the point L_2. In the reference system connected with the Earth–Sun line the velocities of both points, L_1 and L_2, are zero. All these points obviously should be taken into consideration during the calculations of the flight trajectory from the Earth to the libration point L_1 or L_2 of the Earth–Sun system.

One of the versions of the calculated trajectories of the flight of a spacecraft from the Earth to the libration point L_1 of the Earth–Sun system is shown in Fig. 9.18 (Farquar, Richardson, 1976). The calculations have been

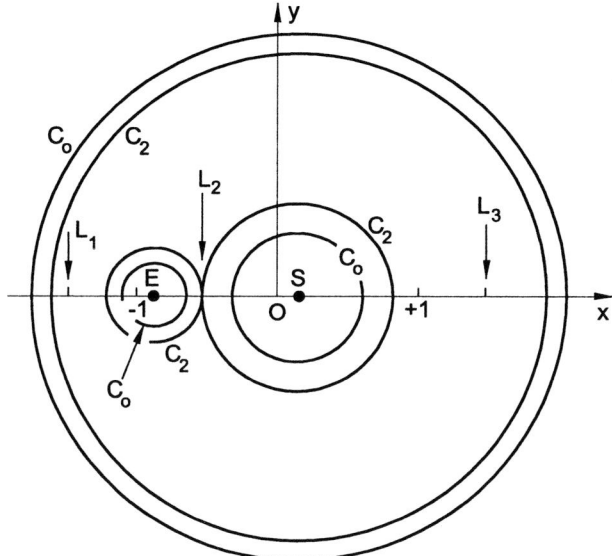

Fig. 9.17. *A schematic view of zero velocity curves for the Earth–Sun system with* $\mu = 3.03591.10^{-6}$ *and the indication of locations of the Lagrangian points* L_1, L_2 *and* L_3. *Not to scale.*

realized in order to be more specific for the appointed time of the launch of the spacecraft (24 July, 1978), with the scheduled 106-day duration of the flight from the Earth to the point L_1 of the Earth–Sun system with an initial launch velocity of $10.989 \, km \, s^{-1}$, with corrections on the flight path etc. The spacecraft is allocated to the point L_1 and held on a closed halo-orbit with the point L_1 in its center. In order to keep the spacecraft on that orbit certain maneuvers would be required.

15   The Tisserand Criterion

Comets moving in elliptic orbits should obviously appear periodically in the Solar vicinity. For these comets both the eccentricities and semimajor axes are, as a rule, rather large so that they can move far away from the Sun, up to the distance of invisibility. Under such conditions no comet can ever be followed visually in the same way as a planet, i.e. be permanently observable at all points of its orbit. In the case of the appearance of a new comet, one cannot easily identify it with any previously observed ones. The problem of

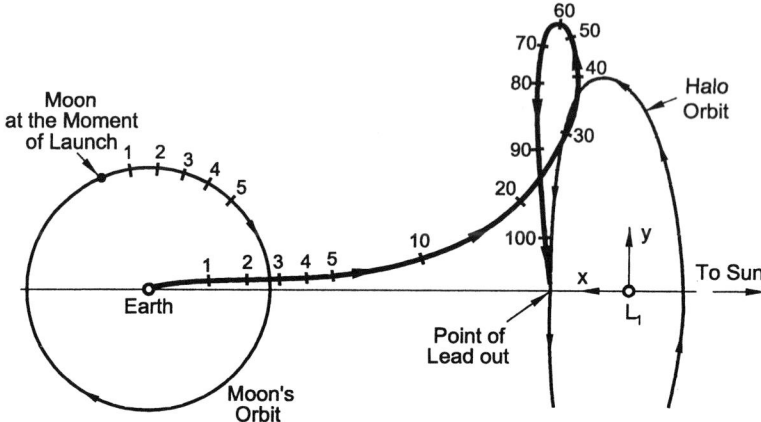

Fig. 9.18. *The launch of a spacecraft into halo-orbit near the libration point L_1 of the Earth–Sun system. The numbers denote the the days after the launch.*

identification of newly discovered periodic comets with earlier known ones, therefore, arises.

The simplest method for such an identification is comparison of the orbital elements of the previously discovered comet with those of the orbit of the newly discovered comet. However, this method is not always acceptable since the comet, during its motion in an elliptic orbit around the Sun, is inevitably passing near the giant planets, and especially, near Jupiter, the perturbation of which can disturb the elements of its orbit drastically: there are known cases of the transition of the orbit of a comet under such a perturbation from elliptic to parabolic, and vice versa.

The comets do not possess characteristic features that are constantly preserved on the cosmogonical time scale such as we have for the planets. Every passage of the comet near the Sun can be fatal for it. Under the powerful action of Solar radiation, both of photon and especially of other elementary particles, essential changes in the physical nature of the comet can take place. As a result, many features of comets can change at their next appearance, such as their brightness, and the form and the size of the tail and the core. Therefore, identification based on those characteristics is not suitable.

In principle, a simple and reliable method for such an identification does exist, namely, the determination by direct computation of the elements of the orbit for an earlier comet, which is a possible candidate for identification, taking into account all perturbations caused by its passage near the giant planets, and the comparison of the obtained elements with those of the newly discovered comet. However, in view of the complicated nature and time-consuming character of such computations, this method is not considered to be the most efficient.

Thus, the search for another, principally new criterion could be a way out of the situation. For example, the search for some combination or relationship between the elements of the orbit which remain unchanged after any perturbations. It appears that such a combination does exist. It is the combination between the elements a, e and i – the function $\Phi(a, e, i)$ composed from these three elements preserving their constant values, $\Phi(a, e, i) = constant$ irrespective of any perturbations of the orbit. In other words, for this combination the condition $\Phi(a_1, e_1, i_1) = \Phi(a_2, e_2, i_2)$ is satisfied for all appearances of a given comet. Violation of this condition implies that the identification is wrong.

The mentioned function was discovered by Tisserand in 1889. Since that time this function has been successfully used for the identification of newly discovered comets.

The strongest changes in the elements of a comet's orbit usually occur at transits close to Jupiter, when the comet is in fact captured within Jupiter's sphere of influence. Even a short transit of the comet within this sphere is enough to create perturbations greatly exceeding the perturbations caused by all of the other planets taken together. The motion of the comet can obviously be described in terms of motion of infinitely small mass in the restricted three-body problem where the finite bodies are the Sun and Jupiter.

The eccentricity of Jupiter's orbit is very small, so that we can take $\varepsilon = 0$, which is adequate for our purposes. Then we can write the Jacobi integral

$$\left(\frac{dx}{dt}\right)^2 + \left(\frac{dy}{dt}\right)^2 + \left(\frac{dz}{dt}\right)^2 = (x^2 + y^2) + \frac{2(1 - \mu)}{r_1} + \frac{2\mu}{r_2} - C \ . \qquad (70)$$

Obviously, under the influence of Jupiter's perturbation the elements of the comet's orbit will be changed. If these elements are known, then the velocity and the coordinates may be computed for any moment of time. Then we can obtain from Eq. (70) the value of the Jacobi constant C, since this constant must remain unchanged for any moment of time. Therefore, if the identification of two comets is correct, we will have one and the same value for the constant C. In other words, if the two values of C are one and the same or differ only slightly, then there is a high probability that we have the same comet in both cases. This is the scheme for practical identification of two comets as being the same.

The problem reduces to the representation of Eq. (70) through the elements of the orbit and, hence, the presentation of C not via the velocity and the coordinates of a comet but via the elements of its orbit. We start the solution of this problem from the presentation of Eq. (70) in the rest system of coordinates ξ, η, ζ, insofar as it is written with respect to the rotating system of coordinates x, y, z. One can use the reverse equations to Eq. (2), i.e.

$$x = \xi \cos nt + \eta \sin nt \ ,$$
$$y = -\xi \sin nt + \eta \cos nt \ , \qquad (71)$$
$$z = \zeta \ ,$$

where n is the mean daily motion of Jupiter; we take $n = 1$. Then we obtain from these two equations

$$x^2 + y^2 = \xi^2 + \eta^2 \ ,$$

$$\left(\frac{dx}{dt}\right)^2 + \left(\frac{dy}{dt}\right)^2 + \left(\frac{dz}{dt}\right)^2 = \left(\frac{d\xi}{dt}\right)^2 + \left(\frac{d\eta}{dt}\right)^2 + \left(\frac{d\zeta}{dt}\right)^2$$
$$+ (\xi^2 + \eta^2) - 2\left(\xi\frac{d\eta}{dt} - \eta\frac{d\xi}{dt}\right) \ .$$

In new coordinates ξ, η, ζ, the Jacobi integral takes the following form:

$$\left(\frac{d\xi}{dt}\right)^2 + \left(\frac{d\eta}{dt}\right)^2 + \left(\frac{d\zeta}{dt}\right)^2 - 2\left(\xi\frac{d\eta}{dt} - \eta\frac{d\xi}{dt}\right) = \frac{2(1-\mu)}{r_1} + \frac{2\mu}{r_2} - C \ , \quad (72)$$

where $1 - \mu \approx 1$ for the Sun and $\mu \approx 0.001$ for Jupiter are taken.

We have for V^2 the well-known expression

$$V^2 = k^2 M \left(\frac{2}{r} - \frac{1}{a}\right)$$

or in our notation ($k = 1$, $M = 1$)

$$\left(\frac{d\xi}{dt}\right)^2 + \left(\frac{d\eta}{dt}\right)^2 + \left(\frac{d\zeta}{dt}\right)^2 = \frac{2}{r} - \frac{1}{a} \ , \qquad (73)$$

and also from the areas theorem (see Eq. (18), Chapter III)

$$\xi\frac{d\eta}{dt} - \eta\frac{d\xi}{dt} = k\,[a(1 - e^2)M]^{1/2} \cos i = [a(1 - e^2)]^{1/2} \cos i \ . \qquad (74)$$

Then Eq. (72), in view of Eq. (73), takes the form

$$\frac{2}{r} - \frac{1}{a} - 2\,[a(1 - e^2)]^{1/2} \cos i = \frac{2(1-\mu)}{r_1} + \frac{2\mu}{r_2} - C \ .$$

For Jupiter $\mu \sim 0.001$ and, hence, the center of the coordinates lies very close to the center of the Sun, therefore $r \approx r_1$. Usually the elements are to be determined when the comet is far both from the Sun and from Jupiter. Then the expression

$$\frac{2}{r_1} + \frac{2\mu}{r_2}$$

will be small and can be neglected. Taking also on the left-hand side $r \approx a$, we will have

$$\Phi\,(a,\ e,\ i) = C\ , \tag{75}$$

where

$$\Phi\,(a,\ e,\ i) = \frac{1}{a} + 2[a(1 - e^2)]^{1/2} \cos i\ . \tag{76}$$

In fact the function $\Phi(a,\ e,\ i)$ is equivalent to the Jacobi constant. The final Eq. (75) should be understood in the following manner: if two orbits with different elements $a_1,\ e_1, i_1$ and $a_2,\ e_2, i_2$ belong to one and the same comet, then the condition

$$\Phi\,(a_1,\ e_1,\ i_1) = \Phi\,(a_2,\ e_2,\ i_2) \tag{77}$$

should be satisfied despite to its apparent paradoxical nature.

Eq. (77) is the Tisserand criterion. Formally, it is nothing but a modified Jacobi integral, written for the rest system of coordinates. The Tisserand criterion also represents an interesting example of an application of the Jacobi integral for the solution of problems of practical interest.

%corrected 16.01.2001, 18.10.2001

PLATE I

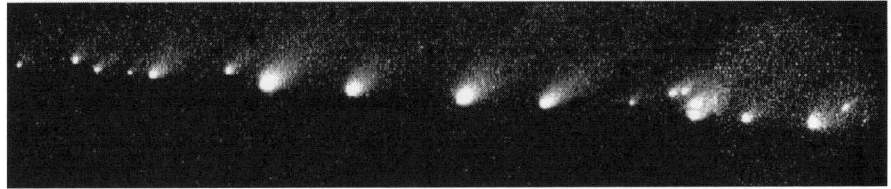

SHOEMAKER-LEVY 9 COMET BEFORE ITS CRASH
WITH JUPITER

Top: The first image of Shoemaker-Levy 9 obtained on March 23, 1993.

Middle: The multiple nuclei of Comet Shoemaker-Levy 9 three days later, on March 27, 1993. The length of the "train" is about 50 arc sec.

Below: Hubble Space Telescope image of the 21 fragments of Comet Shoemaker-Levy 9, destroyed within Jupiter's Roche lobe before their impact at velocity 60 km/sec with the atmosphere of Jupiter. Each nuclei is 1-2 km across.

Plates I, IX are credited to STScI, II-VIII, X-XII to NASA.

PLATE II

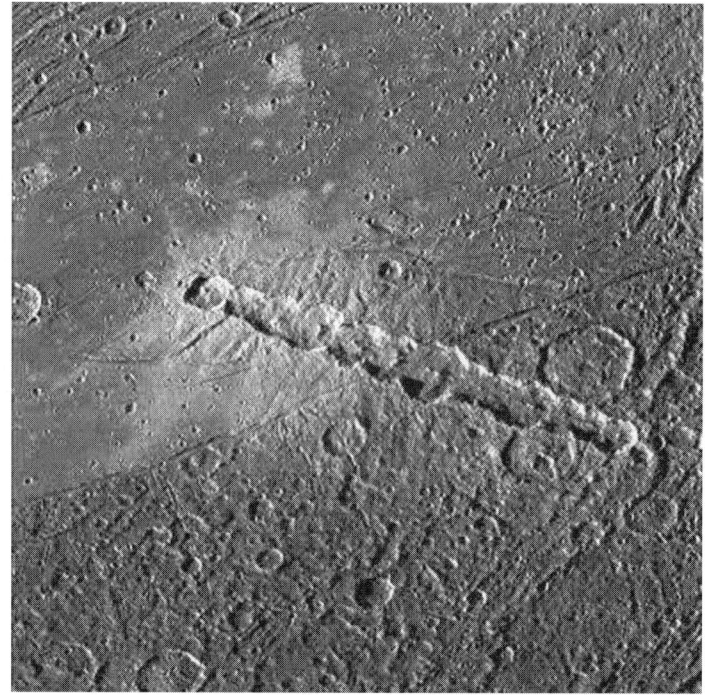

A CRATER CHAIN ON GANYMEDE

Jupiter's largest moon, Ganymede, taken by the Galileo Spacecraft. Enki Catena, a 150-kilometer-long chain of craters is seen. Three such features are known to exist on Ganymede, even more are on Callisto. Crater chains are believed to arise due to tidal forces of Jupiter.

PLATE III

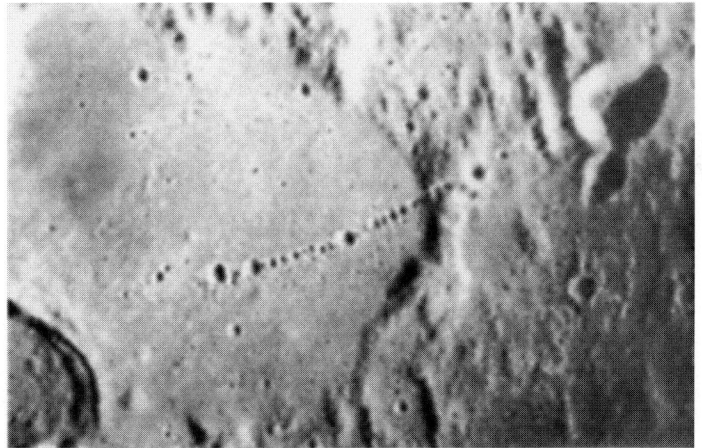

A CHAIN OF SMALL CRATERS ON THE MOON

A chain of craters, each of 1-3 km, near the lunar crater Davy in Mare Nubium. This can be a result an impact of an object disrupted in the Roche lobe, analogous to that of the 'train' of Shoemaker-Levy-9 with Jupiter (PLATE I).

PLATE IV

PHOBOS, MARS'S NEAREST AND LARGEST MOON

Phobos, the satellite of Mars, of 26 x 18 km size, as taken by Mars Global Surveyer on August 19, 1998, with the 10-km wide crater. The powerful impact causing such a crater, had to change the initial orbit of Phobos.

PLATE V

MIRANDA, SATELLITE OF URANUS

Miranda, 480 km across, as imaged by Voyager 2. Features as small as 5.6 km are visible. "The chevron" is in the centre and the several km high cliff is at lower right.

PLATE   VI

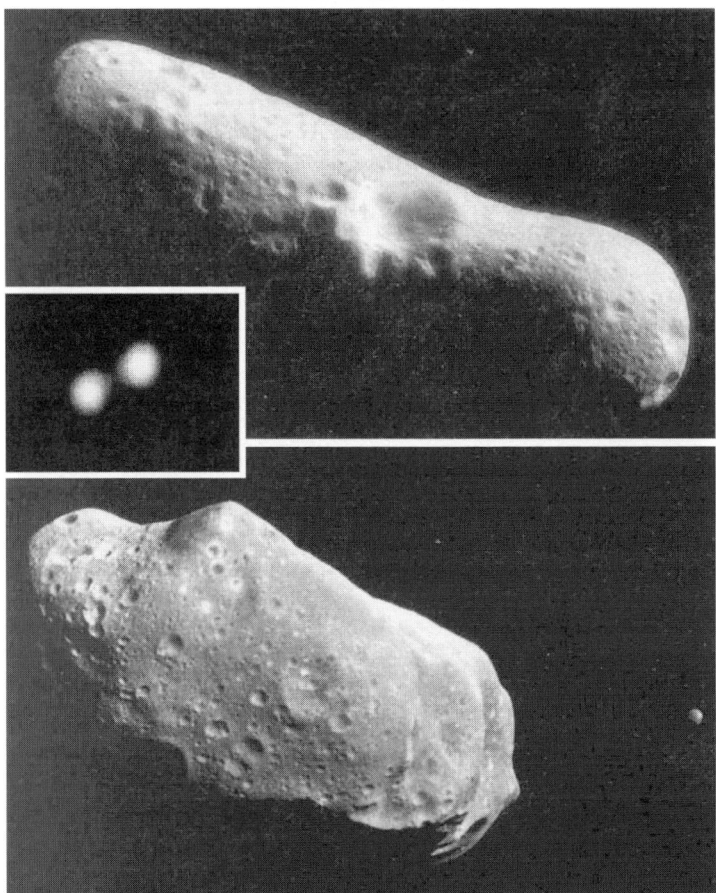

ASTEROIDS EROS, ANTIOPE and IDA.

Top: **Eros**, 33 km and 13 km in size, as imaged by NEAR-Shoemaker probe from 330 km distance. The 5.5 km wide crater is a result of an impact with a body of 5 km across at least.

Middle: **Antiope**, a binary asteroid with equal mass components of about 80 km across. The components are in 170 km or in 0.12 arcsecond as seen from Earth.

Below: **Ida**, 56 km across with a small, 1.5 km across, satellite (to the right). It is discovered by Galileo spacecraft in August, 1993. Both Ida and the satellite are heavily cratered.

PLATE VII

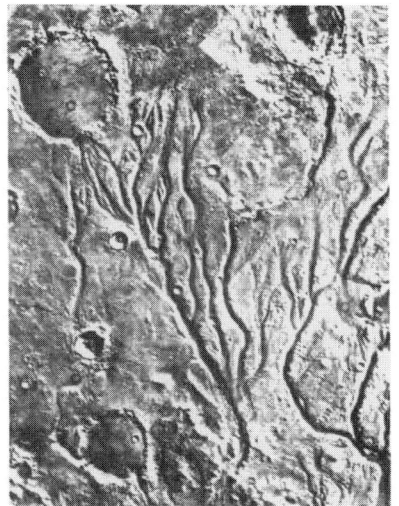

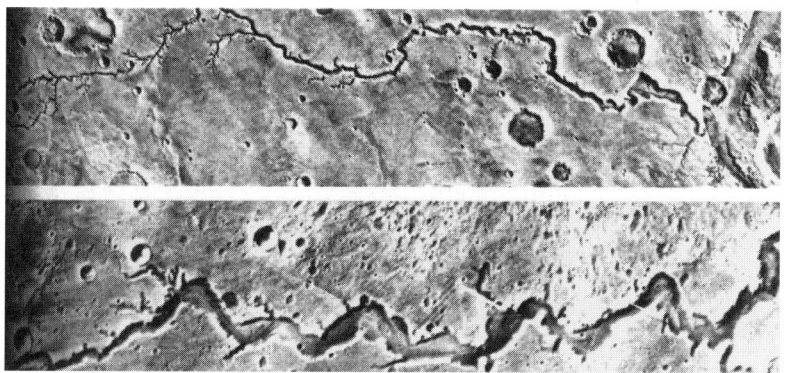

CHANNELS OF MARS

Top: The structures Mars strongly resembling Earth's river beds, imaged by Viking 1 Orbiter in August, 1976.

Below: A 170-km segment of a pair of channels Nanedi Valles imaged the Viking 2 Orbiter in June, 1980.
The width of channels is from 200 m up to 1.5 km (top) and from 1.5 km up to 6 km.

P L A T E   VIII

MARS: WIND SHAPED PATTERNS

A 2.3-km wide area, with ripped dunes and wind shaped hills as imaged by Mars Global Surveyor. Rocks as small as 20 m across are visible.

PLATE IX

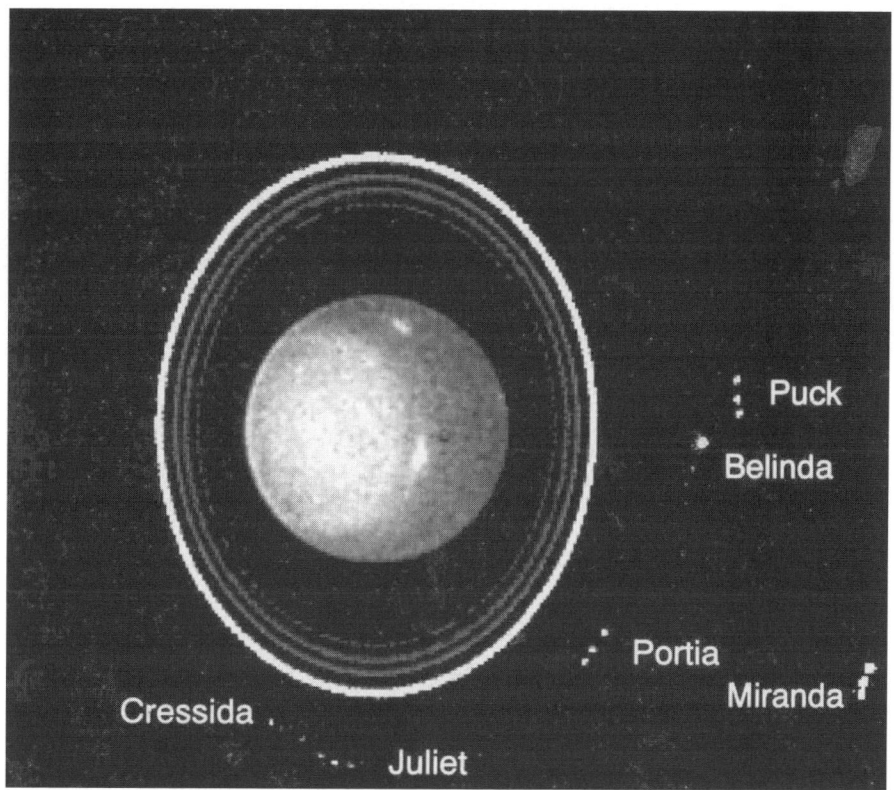

URANUS: RINGS AND SATELLITES

Planet Uranus, its four rings, six satellites and the pair of bright clouds in the planet's atmosphere, imaged by Hubble Space Telescope on August 14, 1994. In total Uranus has 11 concentric rings of dark dust and 20 satellites of various diameters, from more than 1500 km (Titania, Oberon) up to 20 km (Prospero, Setebos). The rotation axis of Uranus lies in its orbital plane. The inner moons, Cressida, Juliet, Portia, Belinda, Puck and Miranda, are shown as a string of three dotes since the picture is composed of three images, taken six minutes apart (Credit STScI).

PLATE X

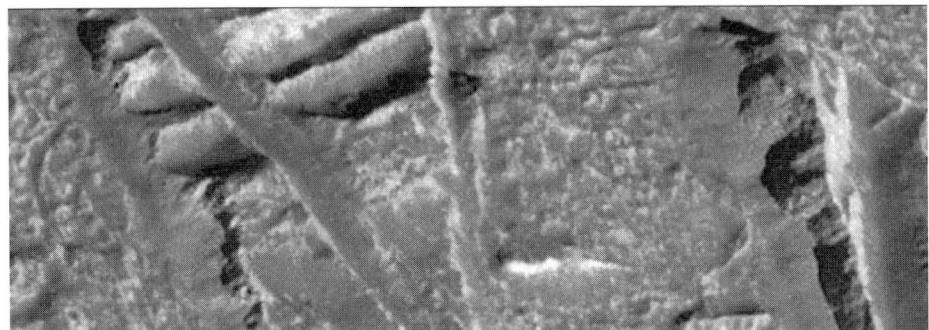

EUROPA, JUPITER'S SATELLITE

A fragment from the surface of Europa, one of the largest Jupiter's moon, imaged by Galileo's spacecraft. Apparently we see a surface of an ice-covered ocean. The linear resolution on the image is about 20 m.

PLATE XI

Mimas • Herschel Crater   © Copyright 1999 by Calvin J. Hamilton

MIMAS, SATURN'S SATELLITE

The crater **Herschel** on Saturn's moon Mimas imaged by the Voyager 2 in August, 1981. The diameter of Mimas is 398 km, of the crater is 145 km. The circular shape of the crater indicates a radial collision, which had to modify the orbit of the moon, as in the case of Mars' satellite Phobos (PLATE IV).

PLATE XII

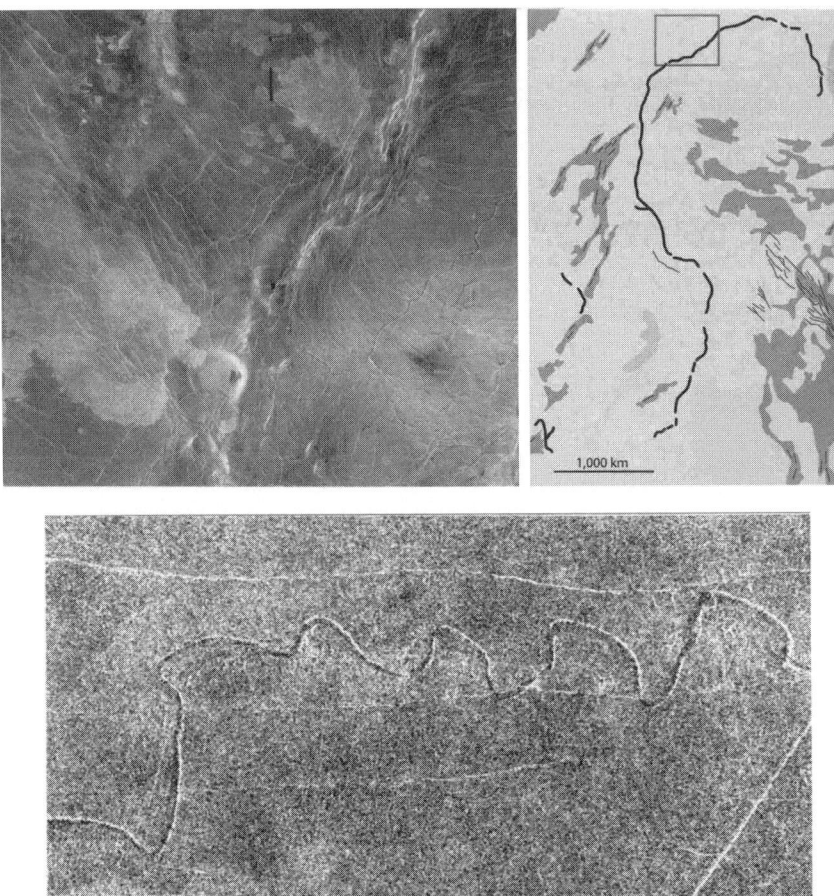

RIVERS OF VENUS

Radar images from the Magellan spacecraft.

Above: A fragment from the Venus's cannel Baltis Vallis (left) of 6800 km long, the longest channel known in the Solar system. The box (in the right) indicates the area of the Magellan radar image.. The channel fragment shown here is about 10 km wide.

Below: Lava river on Venus.

Chapter X

Equipotential Surfaces, Binary Stars and Star Clusters

1 Statement of the Problem

The equipotential surfaces examined in the previous chapter can have an interesting application in stellar systems, particularly for close binary star systems and even star clusters and galaxies. Indeed, it appears that the figure eight-like closed equipotential surfaces can lead to large enough emission volumes around the secondary components of the close binary stars, thus providing an additional and powerful source of the generation of chromospheric-type emission. This and associated phenomena will be the subject of this chapter.

Starting with consideration of binary star systems, first let us note that so-called W UMa type contact binary stellar systems have a common photosphere with an outer boundaries lying between the inner and outer Lagrangian zero-velocity surfaces for a particle which is on the null equipotential surface with respect to both components of the binary system. Such *common envelopes* blend both components and have the same photospheric structure as the photospheres of both stars. Such phenomena may take place only in contact binary systems, i.e. in very close binary systems or those with the smallest periods of orbital rotation, from $0^d.2$ up to $0^d.6$.

At the same time, the blending of **chromospheres** of both components of the binary system may take place as well, but without the blending or contact of the photospheres if the separation between the photospheres is bigger, i.e. for binary systems with orbital periods exceeding $0^d.5$ up to $5 - 10^d$ and even up to 100^d.

The characteristic property of this second category of objects, i.e. of so-called RS CVm type binary systems with blended chromospheres, is the extraordinarily powerful emission, as well as the strongly expressed non-stationarity connected mainly with the transfer of gaseous matter from one component, usually a giant or supergiant of the latter type.

Originally the idea of such common envelopes, *roundchroms* appeared while examining the problem of 2800 MgII emission in RS CVn type close binaries. The characteristics of the roundchroms is possible to obtain, via identification of the outer boundary of the roundchrom with an equipotential zero-velocity surface at some value of the Jacobi constant C_1. The corridor near the Lagrangian point L_1 enables the transition of the gaseous matter from the primary component to the secondary one (Gurzadyan, 1998).

Further, the contacting photosphere in the case of W UMa systems does bring a noticeable increase in the effective luminosity of the system: in this case the common envelope appears as a geometrical concept. Quite different is the situation in the case of roundchroms: although the roundchrom is a common chromosphere enveloping both chromospheres of the system, its emission capability can exceed by one, two or even three orders of magnitude the total power of chromosphere emissivity of both components as two individual stars. In other words, if the isthmus connecting both photospheres in contact binary systems does not influence the energy balance of the system, the roundchrom appears as an independent and extremely powerful source of chromospheric emission stipulated, as we shall see below, below, by a strong increase in the effective volume of the chromosphere mainly in the space surrounding the secondary component.

Coming to the equipotential surfaces, now we will use them to obtain the configuration of the roundchrom, i.e. the form of its outer boundary. As to the inner boundary, it is identified with the surfaces of the photospheres of the components of the system.

As shown in Fig. 10.1, the roundchrom has a figure-eight-shape whose inner boundary corresponds to the sizes of the photospheres of both components and whose outer boundary we shall identify with the zero-velocity equipotential surface. The sought curve lies between two equipotential curves corresponding, first, to a critical value of the Jacobi constant C_1 at which the contact between two isolated equipotential curves around the centers of both components of the system takes place, and second, to some value of C of an equipotential surface close to the Roche lobe of the primary component.

Fig. 10.1 corresponds to the mass ratio of components $\mu = M_B/(M_A + M_B) = 0.41$, where M_A and M_B are the masses of the components, and Jacobi constant $C_1 = 3.9846$ and $C_2 = 3.75$, with an intermediate Lagrangian point L_1 between the components (unlike planetary systems, in stellar systems the point L_1 lies between the components of the pair).

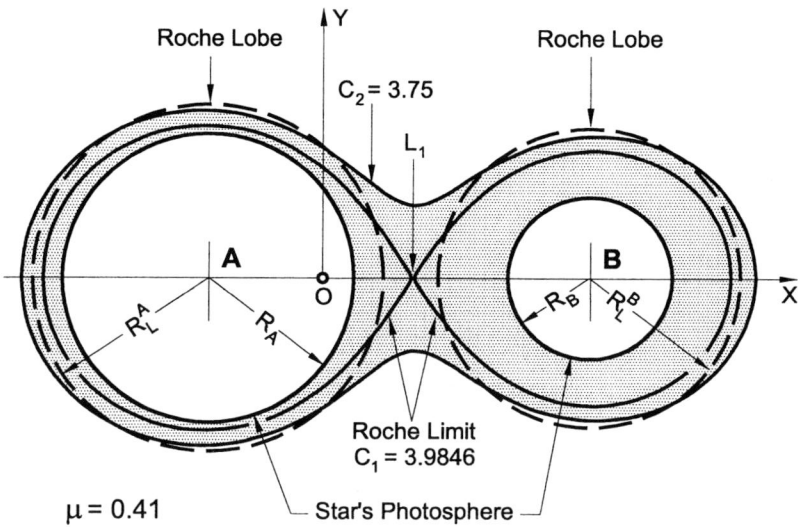

Fig. 10.1. *The common chromosphere, roundchrom, (shaded area) enveloping both components of a close binary stellar system at mass ratio $\mu = 0.41$. The equipotential figure-eight-like curve for Jacobi constant $C_1 = 3.9841$ corresponds to the Roche lobe, and the curve of $C_2 = 3.75$ corresponds to the roundchrom's outer boundary. R_A and R_B are the radii of components A and B of the binary system, and R_A^L and R_B^L are the radii of Roche limits (dashed circles). O is the mass center of the binary system.*

The outer boundary of the roundchrom is given by the equipotential curve at Jacobi constant $C < C_1$. The equation of this equipotential surface in Cartesian coordinates has the following form, in they XOY plane ($Z = 0$):

$$f(x, y) \equiv x^2 + y^2 + \frac{2(1 - \mu)}{\sqrt{(x - x_1)^2 + y^2}} + \frac{2\mu}{\sqrt{(x - x_2)^2 + y^2}} = C \qquad (1)$$

where the intercomponent distance between the components is assumed as unity, $1 - \mu = \mathcal{M}_1/\mathcal{M}$ and $\mu = \mathcal{M}_2/\mathcal{M}$ where $\mathcal{M}_1$ and $\mathcal{M}_2$ are the masses of the components and $\mathcal{M} = \mathcal{M}_1 + \mathcal{M}_2$. The origin of the coordinate system XOY is at the mass center O, and the axis OX is directed along the line connecting both centers.

At larger values of the Jacobi constant, C, both chromospheres should be isolated, while at some critical value of C_0 a contact òccurs between both chromospheres at Lagrangian point L_2.

We must find the value of C, smaller than C_0, at which the equipotential surface reaches a chromosphere, and finds itself within the Roche lobe of the radius

$$R_L = \frac{0.49\,q^{2/3}}{0.6\,q^{2/3} + \ln(1 + q^{1/3})}\,a \qquad (2)$$

where a is the intercomponent distance (in units of solar radius $R_\odot$)

$$a = 4.21 \cdot P^{2/3}\,\mathcal{M}^{1/3}\,, \qquad (3)$$

and P is the period of orbital rotation in days and $\mathcal{M} = \mathcal{M}_A + \mathcal{M}_B$ in solar mass, and $q = \mathcal{M}_B/\mathcal{M}_A$ is the ratio of the masses.

When the Lagrangian point L_1 is out of the Roche lobe (Fig. 10.1), the upper layers of the chromosphere should be near L_2, enabling the transfer process of the chromospheric matter. As a result the chromosphere, as a viscous liquid, will be stretched to the narrow corridor in L_2.

For further discussion it is necessary to know the roundchrom's volume V. We shall consider the roundchrom as a rotating body of a cross-section on the plane XOY around the axis OX. Then we have for the roundchrom's volume

$$V = V_\star a^3 - \frac{4\pi}{3}\,(R_A^3 + R_B^3)\,, \qquad (4)$$

where $V_\star$ is the dimensionless, at $a = 1$, volume of the roundchrom given by the relation

$$V_\star = \pi \int y^2\,dx\,. \qquad (5)$$

The limits of integration obviously are the boundaries of the roundchrom. $f(x, y)$ is the function of the equipotential surface, depending on q and given by equation (1). In Fig. 10.2 the dependence of $V_\star$ on the mass ratio q is shown. In most cases of close binary systems $V_\star \approx 0.50$.

In further discussions we shall operate by mass ratio $\mu = \mathcal{M}_B/(\mathcal{M}_A + \mathcal{M}_B))$ or ratio of mass $q = \mathcal{M}_B/\mathcal{M}_B$, i.e.

$$\mu = \frac{q}{1 + q} \qquad \text{or} \qquad q = \frac{\mu}{1 - \mu}\,. \qquad (6)$$

Fig. 10.3 represents the roundchroms for three RS CVn type close binary systems, AR Lac, V 841 Cen and AR Mon, based on the data given in the first seven columns of Table 10.1. The last four columns contain the derived parameters of the roundchroms. The radii of Roche lobes R_L^A and R_L^B are obtained with the help of Eq. (2).

From Fig. 10.3 one can note that in all three cases the roundchrom's radius for the secondary star is significantly larger than its radius R^B, while for the primary star the Roche's radius R_L^A is only sightly larger than its radius R_A. As a result, large roundchroms are formed around the secondary components of binary systems.

T a b l e 10.1

Initial parameters of three close binary systems, AR Lac, V 841 Cen and AR Mon; periods P, intercomponent distances a, spectral types of the components, masses, radii, mass ratio μ. The last four columns contain the derived parameters: the Roche radii R_L, Jacobi constant C_1, volumes $V_\star$ and V. The masses are in $M_\odot$, linear parameters in $R_\odot$.

N a m e	P	a	Spectral type A / B	M_A M_B	R_A R_B	μ	R_L^A R_L^B	C_1	$V_\star$	V $10^{35}cm^3$
AR Lac	$1.^d98$	9.03	K0 IV	1.35	2.82	0.50	3.4	4.00	0.44	0.59
			G2 IV	1.35	1.54		3.4			
V 841 Cen	6.0	19.5	K1 IV	2.0	8.0	0.33	8.5	3.94	0.50	5.6
			K1 V	1.0	1.0		6.2			
AR Mon	21.2	47.8	K3 III	0.8	12.0	0.77	13	3.84	0.58	307
			G8 III	2.7	14.2		22			

Fig. 10.3 also indicates the coincidence of the roundchrom's outer boundaries for the components A and B with their Roche lobes (dashed circles).

Thus, the chromospheric matter transferred through the corridor or mouth in the point L_1 fills the Roche lobe around the component B, the volume f, which is much larger than the volume of the chromosphere of B.

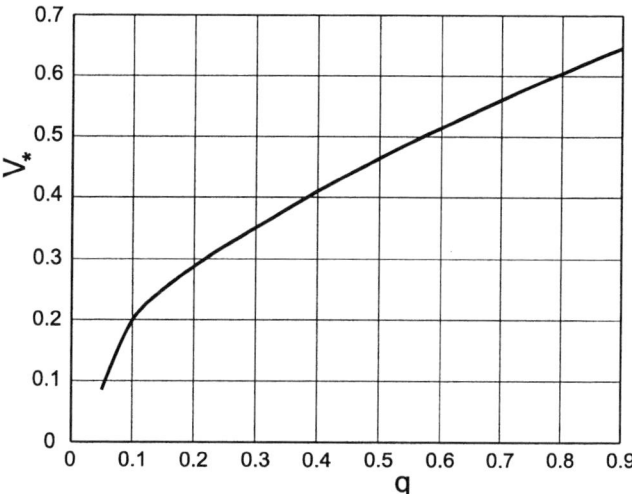

Fig. 10.2. *The dimensionless volume $V_\star$ depending on the mass ratio q.*

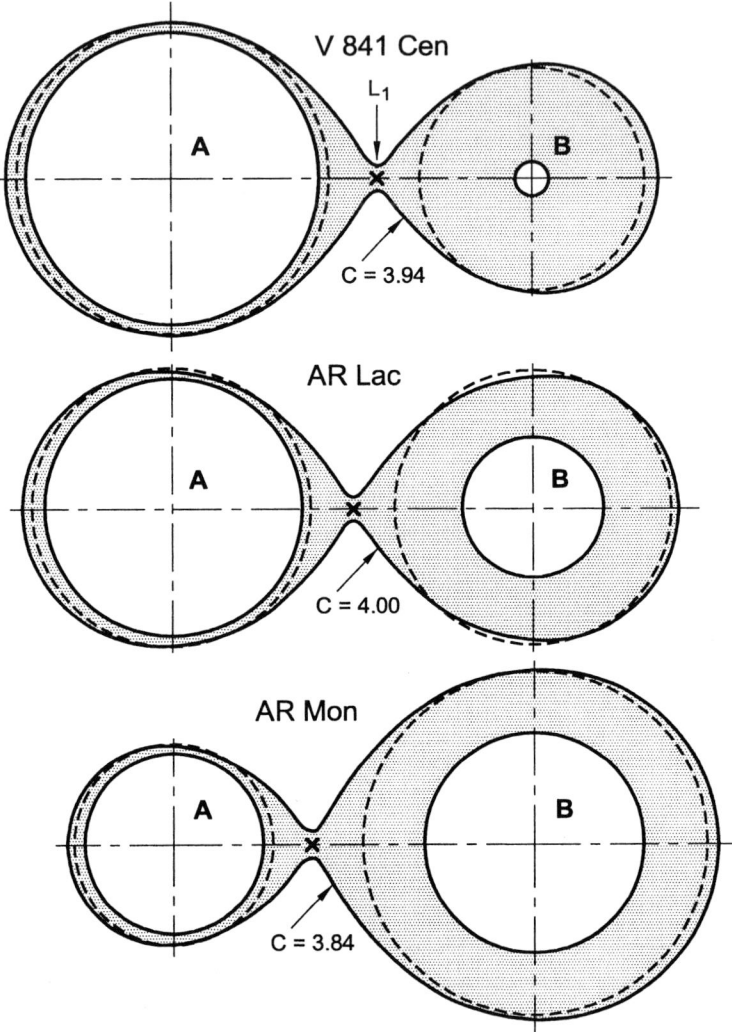

Fig. 10.3. *Roundchroms (shaded) for close binary systems V 841 Mon, AR Lac and AR Mon. L_1 is the intermediate Lagrangian point, the dashed circles denote the Roche lobes.*

The situation is changed strongly: now the main contribution in the observed chromospheric emission is stipulated not by the primary component of the system nor even by intercomponent space around the corridor in the point L_1 but by the second half of the roundchrom, around B.

2 Roundchroms of Close Binary Stars

The enhanced character of the power of the emission in ultraviolet doublet of ionized magnesium 2800 MgII is one of the peculiarities of the majority of RS CVn type close binary systems. Indeed, the ratio $L(\text{MgII})/L_{bol}$, i.e. of the luminosity in magnesium doublet emission $L(\text{MgII})$ to the bolometric luminosity L_{bol}, for the single G or K type giants, in which this magnesium doublet emission has a completely chromospheric origin, is usually $10^{-5} - 10^{-6}$.

Quite different is the situation for RS CVn type close binary systems usually having G or K type giants as a primary: for many representatives of this category of binary systems the relative power of magnesium emission is ten, in some cases even one hundred times higher, that is, of the order of 10^{-4} and 10^{-3}. Such strong emission might have not a chromospheric but a **roundchrom** origin.

The geometry of the roundchroms is determined by the Jacobi constant C, being slightly smaller than its critical value C_2 and satisfying the conditions:

a. Existence of a corridor at the Lagrangian point L_1, to ensure the transition of the matter from the main component to the secondary.

b. The equipotential curve must be located not far from the surface of the primary, to envelope the main volume of its chromosphere.

c. The equipotential curve must envelope the Roche lobe of the primary, or, at least, they should overlap.

d. The roundchrom's outer boundary around secondary component B must be far enough from the star's surface.

In Table 10.2 the parameters of 16 RS CVn type binary stars are given, for which the roundchroms were constructed in the way described in the previous section. In this table, the initial data for these binaries are presented (first seven columns) with derived parameters of their roundchroms in last four columns.

A graphical view of the roundchroms for eight binaries from this list is presented in Fig. 10.4. As we see, practically all the binaries satisfied the above four conditions, hence, the roundchroms around secondary components really may be powerful sources of enhanced chromospheric-type emission.

In Tables 10.1 and 10.2, the roundchroms' volumes V, as one of their important parameters, are given as well; this parameter should be used when obtaining the electron concentrations (see Section 6).

The structure of a roundchrom is mainly determined by its relative volume around the secondary component. From this point of view, one can establish three types or classes of roundchrom (Fig. 10.5).

Type A. The volume of the secondary star is much smaller than that of its roundchrom, $R_B \ll R_L^B$, such as for the binary star HR 4492.

Type B. The radius of the secondary is nearly twice as small as the radius of the roundchrom: $R_B \approx 1/2\, R_L^B$. A typical case is the binary LX Per.

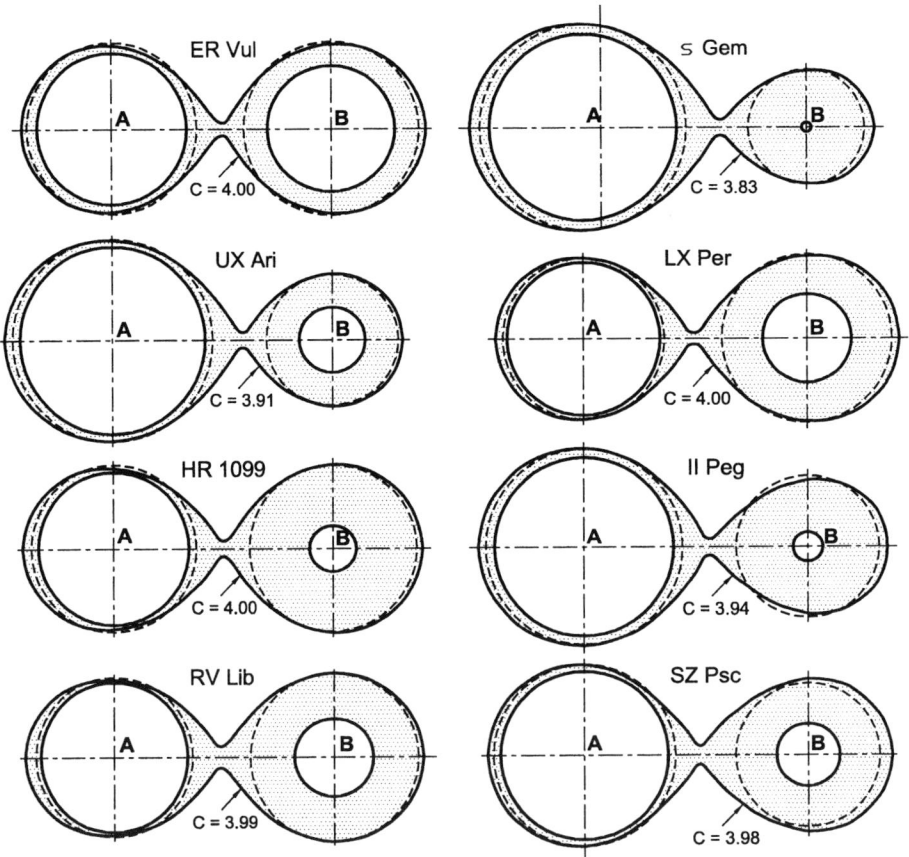

Fig. 10.4. *Roundchroms (shaded) for a sample of RS CVn type binary stars.*

T a b l e   10.2

Parameters of a sample of RS CVn type binary stars (first seven columns) and of their roundchroms (last four columns). Masses and linear sizes are in solar units, $M_\odot$ and $R_\odot$

N a m e	P	a	Spectral type A / B	M_A M_B	R_A R_B	μ	R_L^A R_L^B	C_1	V $10^{35} cm^3$	Type of Round
ER Vul	$0^d.70$	4.2	G0 V	1.10	1.4	0.46	1.6	4.00	0.06	B
			G5 V	1.05	1.1		1.6			
DH Leo	1.07	5.0	K0 V	0.9	2.1	0.44	2.0	3.99	0.016	C
			K5 V	0.7	1.7		1.8			
EI Eri	1.95	10.2	G5 IV	2.0	3.5	0.50	3.9	4.00	0.86	B
			G5 V	2.0	2.2		3.9			
HR 1099	2.84	13.0	K0 IV	2.0	4.8	0.50	5.0	4.00	1.71	B
			G5 IV	2.0	1.3		5.0			
SZ Psc	3.97	14.9	K1 IV	1.7	5.5	0.43	6.0	3.98	2.62	B
			F8 V	1.3	2.0		4.5			
RT Lac	5.10	16.2	G9 IV	0.8	4.5	0.67	5.2	3.94	5.30	B
			K1 IV	1.6	3.4		7.1			
UX Ari	6.44	20.7	K0 IV	2.2	9.0	0.27	9.2	3.91	4.95	B
			G5 V	0.9	3.0		6.2			
II Peg	6.70	21.0	K2 V	2.0	8	0.48	9.3	3.94	6.4	B
			–	1.0	1:		6.8			
LX Per	8.10	26.3	K0 IV	2.0	9	0.50	10	4.00	16.4	B
			G0 IV	2.0	9		10			
RV Lib	10.7	32.2	K3 IV	2.0	11	0.52	12	3.99	30.5	B
			G3 IV	2.2	10		12.6			
IL Hya	12.9	37.3	K1 III	3.5	17	0.22	18.4	3.83	36.1	A
			–	1:	1:		9.7			
V 350 Lac	13.8	39.0	K2 III	2.5	18	0.22	19.0	3.83	74	B
			G5 V	1:	1.0		10.9			
σ Gem	19.6	49.3	K1 III	3.5	17	0.22	17.6	3.83	94	A
			–	1:	1:		12.8			
TW Lep	28.3	57.8	G8 III	3.3	25	0.31	25.5	3.92	223	A
			F	1.5	1:		18			
ε UMi	39.5	76.2	G5 III	2.8	33	0.33	34	3.93	242	B
			F0 V	1.3	1.7		23			
HR 4492	61.4	105	K3 III	2.5	40	0.44	42	8.99	1336	A
			A0 V	2.0	2.3		38			

Type C. The outer boundary of the roundchrom practically coincides with the photosphere (chromosphere) of the secondary: $R_B \approx R_L^B$. A typical representative is DH Leo.

The type of the roundchrom characterizes the power of chromospheric emission, e.g. the ratio $L(MgII)/ L_{bol}$, the largest being for type A: $L(MgII)/ L_{bol} \sim 10^{-3}$, the smallest for type C: $L(MgII)/ L_{bol} \sim 10^{-5}$. For intermediate type B we have: $L(MgII)/ L_{bol} \sim 10^{-4}$.

As follows from Table 10.2, the majority of roundchroms are of type B,

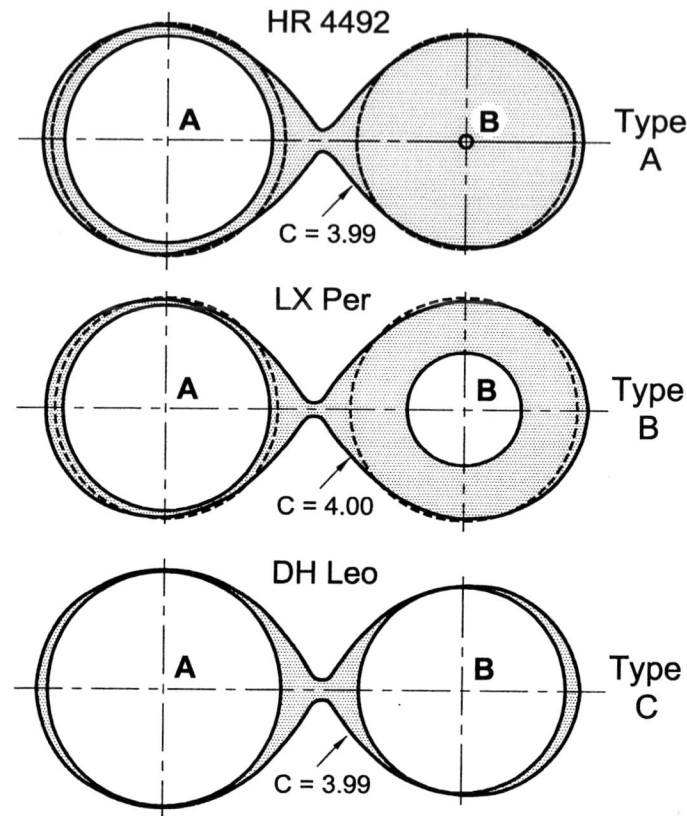

Fig. 10.5. *The three types of roundchroms (shaded).*

with type A in second place. Roundchroms of type C occur only in a few binaries. Nevertheless roundchroms of this type are of special interest. Note also that roundchroms of type B are found mainly in binaries with relatively small intercomponent distances (UV Psc, ER Vul), and those of type A in the largest binaries (12 Cam, HR 7428).

3   Contact Binary Stars

Another class of binary stars are W UMa type contact binary systems. Their equipotentials and roundchroms have certain peculiarities. Though the equation of the equipotential surfaces is the same, i.e. Eq. (1), the boundary conditions are different (Fig. 10.6).

Namely, in the case of contact binaries we have:

a. The inner boundary of the roundchrom should coincide with the Roche lobe with the physical contact in the Lagrangian point L_1. The Roche lobe represents the surface of the star.

b. The corridor in the Lagrangian point L_1 is essentially broader.

Thus the physical contact of the photospheres occurs in L_1; this configuration corresponds to a Jacobi constant C_1. At values of C smaller than the Jacobi constant C_1, figure-eight-shaped configurations envelop both components of the system. However, we also have dynamically stable roundchroms when $C = C_2$, i.e. when both halves of the equipotential curve contact in the second Lagrangian point L_2, with outflow of the matter.

The originality of the situation is in the fact that at a definite value of C slightly smaller than C_2 a narrow corridor may be formed in the point L_2 as well, which, in contrast to the point L_1 in the case of close binaries, must act as a door for the outflow of gaseous matter, i.e. for an active loss of roundchrom matter.

So, for contact binaries one has both types of roundchrom, closed with $C_1 < C < C_2$, and open with $C \approx C_2$, with the largest emission volume.

Below, we shall concentrate our analysis on W UMa objects with known observed fluxes in ultraviolet doublet 2800 MgII, which is necessary for further analysis. The data of such objects are represented in Table 10.3 (first four columns).

The first parameter characterizing the roundchrom is its linear size, determined by intercomponent distance a given by Eq. (2); the real sizes of roundchrom is proportional to a.

Then, the equipotential curves are constructed for these objects with the help of Eq. (1) with the aim, particularly, of finding the values of both Jacobi constants, C_1 and C_2.

The obtained numerical values of a, C_1 and C_2 are given in Table 10.3, and in Fig. 10.7, as an illustration, examples of roundchroms are shown for four contact binaries with increasing values of mass ratio q: ε CrA, TZ Boo, TW Cet and OO Aql. All figures are drawn for $a = 1$.

As follows from the data of Table 10.3, in the case of contact binaries, no roundchroms with large emission volumes can be formed around the secondary component, and no high MgII doublet emission can be expected for them, as confirmed by observations. Comparing the data of Table 10.3 with the data in Tables 10.1 and 10.2, we see the negligible volume of roundchroms for contact binaries.

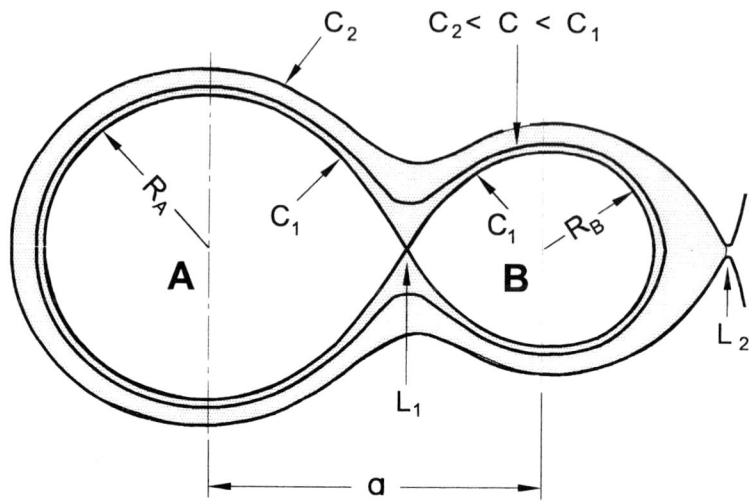

Fig. 10.6. *Outer boundaries for a roundchrom around the contact binaries.   Closed-type roundchroms may be formed at values of Jacobi constant C intermediate between C_1 and C_2, i.e. when $C_1 < C < C_2$. At $C = C_2$ an open-type roundchrom will be formed with a maximum emission volume and with a narrow corridor in the outer Lagrangian point L_2.*

T a b l e   10.3

Initial data for a sample of ten W UMa type contact binaries with the derived parameters of their roundchroms (last five columns)

Name		P	Spectral types of components	q	$V_\star$	a $R_\odot$	C_1	C_2	V $10^{33}cm^3$
44i	Boo	$0^d.268$	G2 + G2	0.50	0.465	1.94	3.945	3.547	1.17
VW	Cep	0.278	G8 + K0	0.41	0.415	1.82	3.911	3.558	0.86
TZ	Boo	0.297	G2 +	0.22	0.300	2.07	3.773	3.544	0.91
TW	Cet	0.317	G5 + G5	0.53	0.480	2.23	3.954	3.543	1.84
W	UMa	0.334	F8 + F8	0.49	0.460	2.31	3.943	3.549	1.96
AH	Vir	0.408	K0 + K0	0.42	0.420	2.64	3.917	3.557	2.67
V566	Oph	0.410	F0 + F4	0.24	0.320	2.74	3.794	3.550	2.26
ER	Ori	0.423	G1 +	0.76	0.585	2.83	3.991	3.501	4.58
OO	Aql	0.507	G5 +	0.83	0.620	3.40	3.995	3.490	8.34
ε	CrA	0.591	F2 +	0.11	0.220	3.62	3.599	3.468	3.40

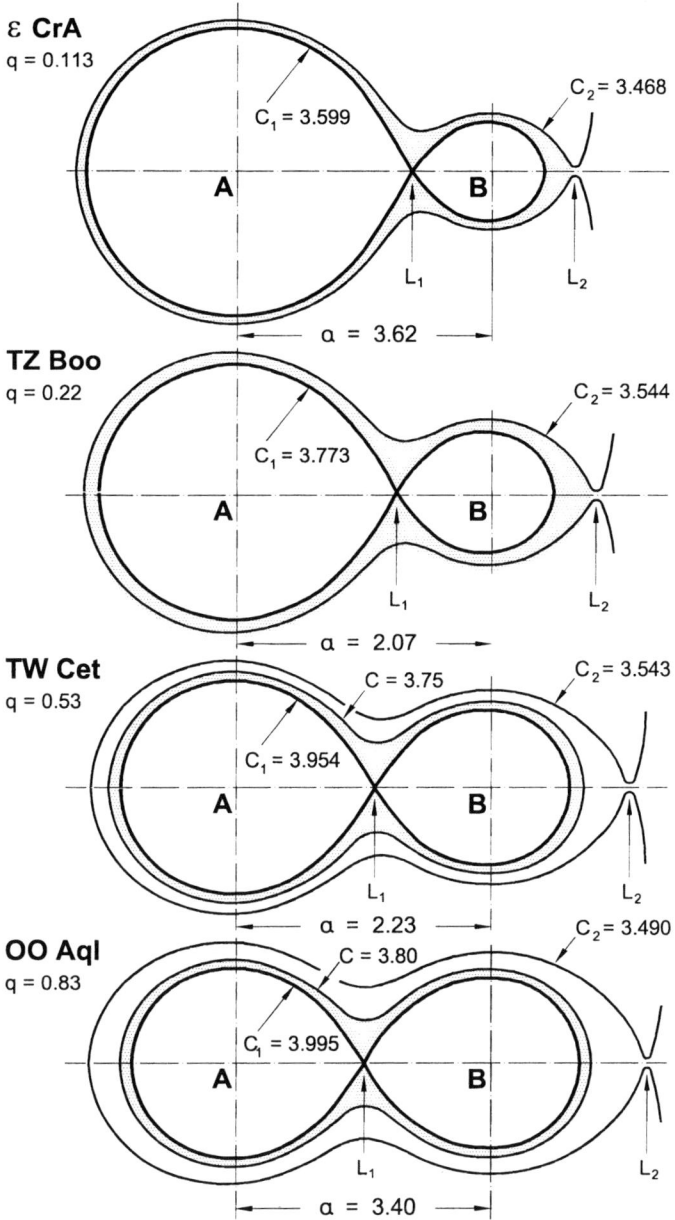

Fig. 10.7. *The roundchroms (shaded) with the largest possible emission volumes for four contact binaries with increasing mass ratio q and various intercomponent distances a.*

The roundchrom's next parameter of interest is the dimensionless volume $V_\star$, which determines the actual volume of roundchroms, V. The numerical values of all these parameters, a, C_1, C_2 and V, are given in the last four columns of Table 10.3.

Let us examine the behavior of the **relative** volume of roundchroms with respect of the summary volume of both components.

For the dimensionless ratio

$$\kappa = \frac{3}{4\pi} \frac{V_r}{(R_A^3 + R_B^3)} , \qquad (7)$$

where V_r is the effective volume of the roundchrom, R_A and R_B are the radii of components, we find the numerical values of κ for both groups of binaries, close and contact, which are given in Table 10.4.

T a b l e   10.4

Numerical values of δ, of the relative volume of the roundchrom, for objects of two groups, close binaries and contact binaries. The sequence of the objects is by increasing intercomponent distance a. R_A, R_B and a are in units of $R_\odot$.

Name	Close Binaries		Name	Contact Binaries	
	a	κ		a	κ
SV Cam	3.4	0.60	VW Cep	1.82	4.78
RT And	4.1	1.00	44i Boo	1.94	7.80
V824 Cam	8.8	0.25	TZ Boo	2.07	4.25
AR Lac	9.0	1.03	TW Cet	2.23	11.4
AS Dra	17.6	0.15	W UMa	2.81	9.75
V841 Cen	19.5	0.73	AH Vir	2.64	10.4
AR Mon	47.8	4.5	V566 Oph	2.74	9.15
ξ And	57.8	0.62	ER Ori	2.83	28.8
IM Peg	64.3	0.92	OO Aql	3.40	44.8
RZ Eri	75.0	1.96	ε CrA	3.62	10.8

The peculiarity of the situation is in the fact that no correlation between κ and a can be observed for close binaries, while the following correlation is noticeable for contact binaries:

$$\kappa = 1.10 \times a^{2.4} , \qquad (8)$$

where a is in units of $R_\odot$. Then the emission volume of the roundchrom V_r for a contact binary with an intercomponent distance a and radii of components R_A and R_B can be derived with the help of the following relationship

$$V_r = 4.60 \times a^{2.4} (R_A^3 + R_B^3) , \qquad (9)$$

where a is in units $R_\odot$. For example, we have for W UMa: $a = 2.31\,R_\odot$, $R_A = 0.467\,R_\odot$ and $R_B = 0.337\,R_\odot$, and then we obtain from Eq. (9) for the volume of the roundchrom around this contact binary: $V_r = 1.65 \times 10^{33}$ cm^3 which is not far from the directly obtained value 1.96×10^{33} cm^3 (Table 10.4).

The discovered empirical relationship in the form of Eq. (8) is related to roundchroms with maximum possible volumes, i.e. with unclosed outer boundaries, when the open corridor in the point L_2 is formed. However, it is not difficult to ensure that the relation $\kappa \sim a^{2.4}$ is valid also for the closed binary type roundchroms with outer boundaries identified with an equipotential curve corresponding to an intermediate value of C, i.e. when $C_1 < C < C_2$.

However, the most important conclusion for contact binaries is related to the open corridor in Lagrangian point L_2: the formation of a narrow corridor is possible not only in the intermediate Lagrangian point L_1 but also in the outer Lagrangian point L_2, with the outflow by natter.

The probability of this event, i.e. of gas outflow from the point L_2, is larger for contact binaries with small q, namely, $q < 0.50$.

The further study of this problem seems to be interesting from the point of view of the evolution of contact binaries.

According to the 4th edition of the General Catalogue of Variable Stars, the number of registered binaries classified as W UMa type contact binaries is more than 500. Therefore, the above conclusions on the nature of the outer regions of their atmospheres can reveal one of the remarkable properties of contact binaries in general.

4 Roundchroms with Periodic Breaking of Contact

Among RS CVn type systems there are binaries with rather large eccentricities, i.e. with extended elliptical orbits. In this case, the intercomponent distance can vary significantly, and while the distance between the components is greatly increased, the surface layers of the main component will find themselves too far from the point L_1. This can lead to the breaking of the roundchrom, i.e. the stopping of the transfer of gaseous matter from one component of the system to another. In this case, one will have, as physical configurations, two isolated chromospheres or emission clouds around each component.

After some time, the intercomponent distance will again decrease, due to the components approaching such other and as a result, the physical contact between both chromospheres will be renewed, and the roundchrom will be restored to its former shape. This new roundchrom will preserve its state for some time – up to the moment of the next break. Thus, the problem of periodically breaking roundchroms of close binary systems arises.

A selection of nine RS CVn type close binaries with large eccentricities from $e = 0.27$ up to $e = 0.68$ is presented in Table 10.5, along with the parameters necessary for further discussion in the present section. In its last column the numerical values of the ratio $r_{max}/r_{min} = (1 + e)/(1 - e)$ are given, but they vary too much for careful conclusions to be drawn.

T a b l e　　10.5

Initial data for a sample of RS CVn type close binary systems with large eccentricities and with periodic breaking of their roundchroms. Intercomponent distance a, the radii R_A and R_B and masses $\mathcal{M}_A$ and $\mathcal{M}_B$ are in solar units, $M_\odot$ and $R_\odot$.

N a m e	e	P	Spectral type A / B	$\mathcal{M}_A$ $\mathcal{M}_B$	R_A R_B	μ	a	$\dfrac{r_{max}}{r_{min}}$
33 Psc	0.27	72.9	K0 III A5 V	3.5 2.0	37 2:	0.36	130	1.7
LU Hya	0.31	35.0	K1 IV ...	2.0 1:	14 1:	0.33	63	1.9
TY Pic	0.32	43.8	K1 III G8	3.5 1:	27 1:	0.22	84	1.9
RZ Eri	0.36	39.3	K0 IV Am	2.0 2:	17 1:	0.50	77	2.1
UV For	0.39	15.0	K0 IV ...	2.0 1:	9.3 1:	0.33	36	2.3
AZ Psc	0.50	17.1	K0 IV ...	3.5 1:	10 1:	0.22	45	3.1
LS TrA	0.516	49.4	K0 IV ...	2.0 1:	16 1:	0.33	80	3.1
BM Mic	0.521	3.97	K1 III ...	3.5 1:	3.8 1:	0.22	17	3.2
HR 6469	0.68	83.2	G4 III F2 V	2.0 1.5	27 1:	0.43	124	5.2

In Fig. 10.8 for the given value of the mass ratio $\mu = \mathcal{M}_B/(\mathcal{M}_A + \mathcal{M}_B)$ and known magnitude of major semi-axis a, the equipotential surfaces (curves) are shown for four positions, 1, 2, 3 and 4, of the secondary component B on the elliptical orbit.

The situations on these positions are as follows:

Position 1. The intercomponent distance is the smallest and is equal to $r_1 = r_{min} = a(1 - e)$, and the true anomaly is zero: $v_1 = 0$.

Position 2. The true anomaly is $v_2 = \pi/2$, and intercomponent distance is $r_2 = a(1 - e^2)$.

Position 3. The secondary component is at the point of intersection of the orbit with the small axis of the ellipse. In this case, we have for the

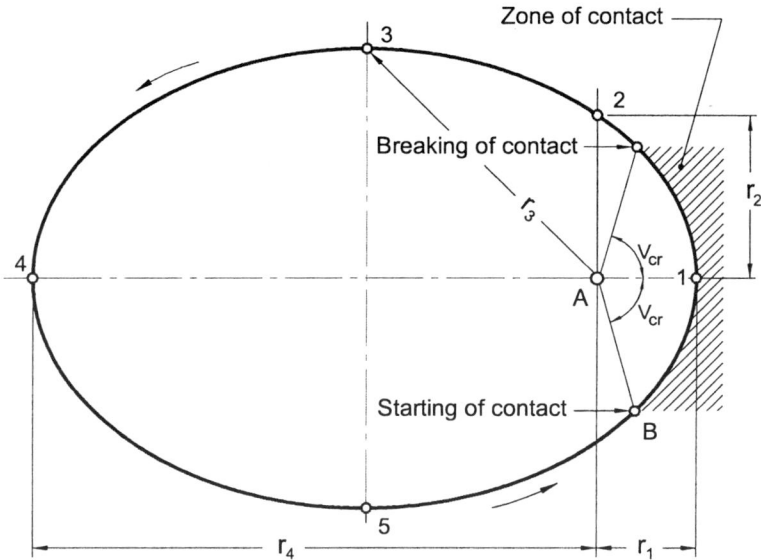

Fig. 10.8. *Basic parameters of motion of secondary component by elliptic orbit of close binary system with large eccentricity. The zone of active physical contact and the positions of breaking and restarting of contact are shown as well.*

intercomponent distance $r_3 = a$, and the true anomaly v_3 is determined from the relationship $\cos(180 - v_3) = e$.

Position 4. The secondary component is located at the edge of the major semi-axis, the intercomponent distance is the largest and is equal to $r_4 = a(1 + e)$ and $v_4 = 180°$.

As an illustration, four systems of equipotential surfaces, or more precisely, their cross-sections in the XOY plane, are shown in Fig. 10.9 at all four positions, for a binary system 33 Psc, with the smallest eccentricity in our sample ($e = 0.27$). These four positions correspond to the intercomponent distances (in units of solar radius): $r_1 = 92$, $r_2 = 117$, $r_3 = 127$ and $r_4 = 161$. The equipotential curves correspond to the same value of Jacobi constant, namely, $C = 3.96$, i.e. slightly smaller than its critical value $C_2 = 3.965$. The formation of the roundchrom is possible only in position 1 (shaded area above, in Fig. 10.9, Position 1).

Let Δ be the distance of Lagrangian point L_1, i.e. of the corridor from the surface of the star. At position 1 we have the smallest value of the ratio Δ/R_A at which the transfer of gaseous matter, and hence, the formation of a roundchrom around component B is still possible. This roundchrom will preserve its state as long as the ratio Δ/R_A remains smaller than some critical value $(\Delta/R_A))_{cr}$, when the breaking of the roundchrom in the point L_1 will become inevitable. Indeed, this can happen at some position of secondary B on the orbit, characterized by critical values either of the radius r_cr or of the true anomaly v_cr. Then, we can obtain the moment of breaking of the roundchrom and duration of the contact, t_0. We shall deal with the relative duration of the contact, i.e. by ratio $2t_0/P$ assuming that both processes, the breaking of the contact and its renewal, occur in the point of the orbit situated symmetrically relative to the position 1 (periastron). This relative duration can be determined from the relationship

$$\frac{2t_0}{P} = \frac{1}{\pi}\left(E_{cr} - e\sin E_{cr}\right) \tag{10}$$

with

$$\tan\frac{E_{cr}}{2} = \sqrt{\frac{1-e}{1+e}}\,\tan\frac{v_{cr}}{2}\,. \tag{11}$$

Below we shall assume that the breaking of the roundchrom takes place at critical values of the parameter $(\Delta/R_A)_{cr}$ equal to 0.6 and 0.8. From the dependence of Δ/R_A on v and the sequence of roundchrom positions in Fig. 10.9, we obtain the duration of contact $2t_0/P_0$ with the help of Eq. (10). The results are given in Table 10.6. As expected, the duration of active contact is assorter as larger is the eccentricity of the orbit of binary system.

T a b l e   10.6

Relative duration of active contact, $2t_0/P$, in percent, for two values of the parameter $(\Delta/R_A)_{cr}$, equal to 0.6 and 0.8, depending on eccentricity e

Binary	e	$2t_0/P$	
		$(\Delta/R_A)_{cr} = 0.6$	$(\Delta/R_A)_{cr} = 0.8$
33 Psc	0.27	26	41
UV For	0.39	15	25
HR 6469	0.68	7	10

The periodic character of the breaking and renewal of the contact of both halves of the roundchrom, i.e. the dynamical behavior of the figure eight-shaped, leads in its turn to a remarkable phenomenon of periodic pulsations of the roundchrom.

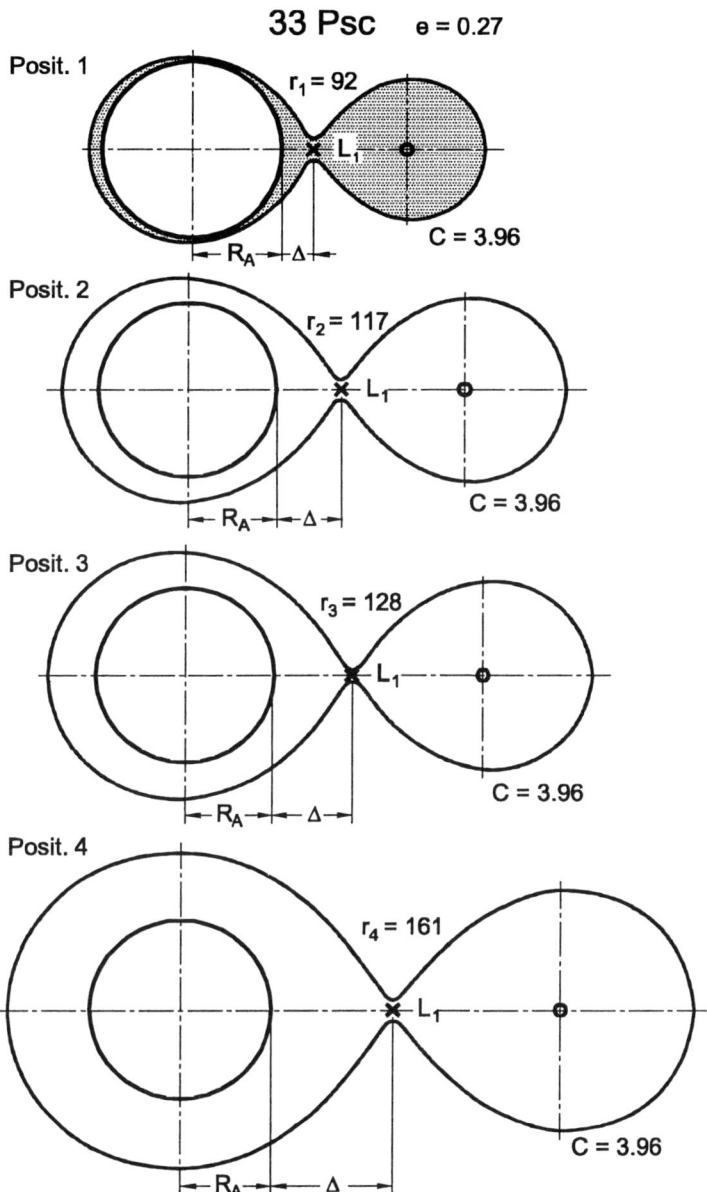

Fig. 10.9. *Close binary system 33 Psc with eccentricity of the elliptic orbit $e = 0.27$. Equipotential surfaces (curves) define the boundaries of roundchroms for four positions of the secondary. The Jacobi constant is the same in all cases, $C = 3.96$. We have a roundchrom in only one case, at position 1.*

What is the physical essence of the phenomenon of the periodic breaking of the roundchrom?

After the breaking of the contact in the Lagrangian point L_1, the outflow of gaseous matter from the main component towards the secondary should be stopped. The part of the roundchrom around the secondary, now isolated, in the form of an extended cloud, will turn into the state of reradiation and simultaneous expansion with a thermal velocity. This should lead to the decay of total luminosity in the emission lines. For example, for emission doublet of 2800 MgII, one will have for the total luminosity of the system: $L(\text{MgII}) \sim V\, n_e^2$, where V is the volume of the roundchrom, n_e is the electron concentration. At total number of electrons N_e, we have: $n_e \sim N_e/V$ and, hence, $L(\text{MgII}) \sim V^{-1}$.

At thermal expansion of the roundchrom with constant velocity, we shall have:

$$V \sim t^3 \text{and} L(\text{MgII}) \sim t^{-3},$$

i.e. we observe a weakening of emission. However, this cannot be continued too long, since with the renewing of the contact, the inflow of gaseous matter into the zone of the roundchrom around the secondary would lead to an increase in magnesium doublet emission and so on. This process takes place periodically, i.e. in correlation with the phase of orbital motion

Thus, we come to the effect when a correlation of emission lines with the phase of orbital motion of contact binaries occurs periodically.

5   Giant Roundchroms

Continuing discussion of the effects associated with equipotential surfaces and binary stars, we see that there are close binary systems with unusual observed properties. This concerns, particularly, their emission spectra, which are impossible to interpret within the usual schemes. Among such systems are the binaries SX Cas and 22 Vul.

The first, SX Cas, represents a close eclipsing system with components K3 III type cool giant and B 7 type hot normal star. The system reveals extraordinarily strong emission lines, e.g. the ratio $L(\text{MgII})/L_{bol}$ is $\sim 10^{-3}$, which is approximately two orders of magnitude larger than for normal close binaries.

The situation is analogous to 22 Vul, an interacting binary with a G8 Ib-II supergiant as primary and a B8 V normal hot star as secondary. At the orbital period $P = 249^d$ its intercomponent distance a is of the order of $300 R_\odot$, i.e. 4–5 times the radius of the primary. The high excitation emission lines, particularly 1240 N V, 1400 Si IV, 1550 CIV, 1640 He II, are rather strong and their variations are not correlated with the orbital rotation of the system.

The behavior of emission lines in the spectra of SX Cas and 22 Vul represents a challenging problem. The source of the ionizing energy cannot be connected with the primary component, but can be related to the mass transfer process from one component to another. The impression is that all these problems may be understood in the framework of the roundchrom concept.

In Table 10.7, the initial data for binary systems SX Cas and 22 Vul are given. Using these data, the roundchroms of these binary systems are constructed, as shown in Fig. 10.10. In last column of the table, the derived radii of the roundchroms of both binaries are given. As we can see, both the intercomponent distances and the sizes of the roundchroms are extraordinarily large. The volumes of the roundchroms, $V = 0.72\,10^{38}$ cm^3 for SX Cas and $V = 1.37\,10^{40}$ cm^3 for 22 Vul, on three orders of magnitude in the first case, and on five orders in the second, are larger as compared with the roundchroms of ordinary RS CVn type binary stars (Table 10.3).

T a b l e   10.7

Initial data for binary systems SX Cas and 22 Vul. The masses, the radii and intercomponent distance a are in solar units, $M_\odot$ and $R_\odot$

Name	P	D pc	Spectral type A / B	$\mathcal{M}_A$ $\mathcal{M}_B$	μ	a	R_A R_B	R_L^A R_L^B
SX Cas	$39^d.6$	500:	K3 III B7	3.5 4.0	0.53	91	32 3.4	35 33
22 Vul	249	500	GS Ib-II B8 V	4.3 3.0	0,41	323	100 3.3	112 133

Giant roundchroms have been discovered for at least five other binaries with sizes of intercomponent distances and roundchrom volumes as presented in Table 10.8.

T a b l e   10.8

Binary systems with giant roundchroms

Binary system	a $R_\odot$	V 10^{38} cm^3
λ And	90	0.6
SX Cas	91	0.7
HR 4492	105	1.3
12 Cam	126	1.0
HR 4665	126	2.0
22 Vul	323	140

The formation of common envelopes, roundchroms, in close binary systems is inevitable. They represent an essential physical formation in close binary systems, like the corona and the chromosphere in single stars.

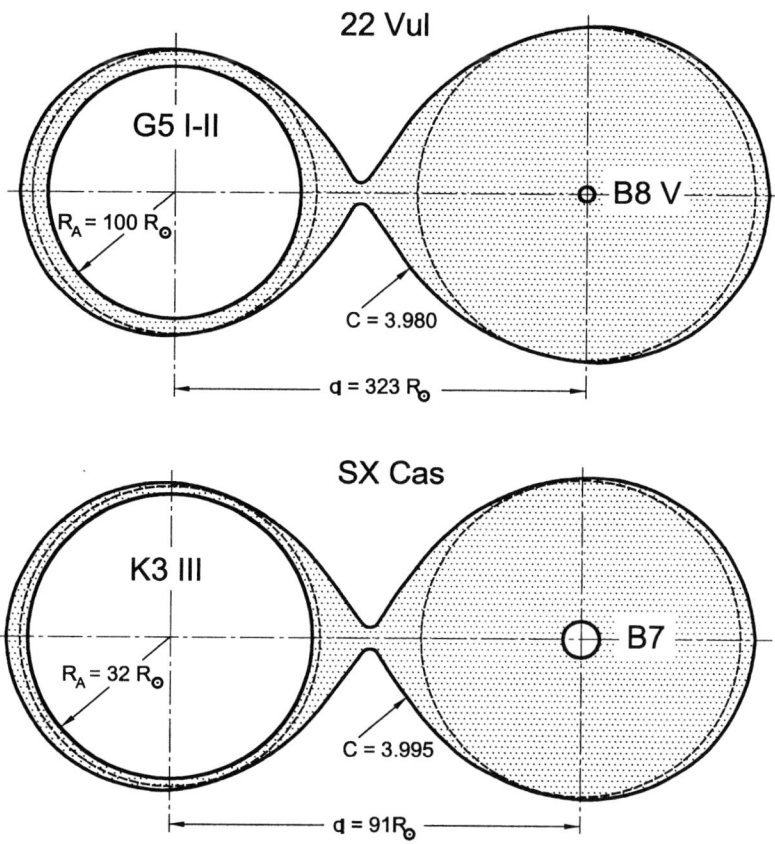

Fig. 10.10. *Giant roundchroms (shaded) for binary systems SX Cas and 22 Vul. Solid lines denote the outer and solid circles the inner boundaries of the roundchroms. Dashed circles are the outer limits of the Roche lobes.*

6   About Regularity of Roundchroms

Besides the geometrical characteristics of roundchroms discussed above, their physical parameters are of particular interest in understanding their role in observational features of close binary systems. The geometrical parameters of roundchroms, particularly their volumes V determined above for individual binaries, will now be used for the determination of one more parameter of roundchroms; we have in mind the electron concentration n_e in roundchroms.

The method of the determination of n_e is based on the use of one data constant well known for all close binary systems, namely, the observed flux $F(\mathrm{MgII})$ or luminosity $L(\mathrm{MgII})$ in emission of the doublet 2800 MgII. This doublet is excited by electron inelastic collisions whose volume emission coefficient in given by the relationship:

$$\varepsilon(\mathrm{MgII}) = 1.26 \times 10^{-20} n_e^2 T_e^{-1/2} \, \exp\left(-\frac{51\,400}{T_e}\right) \, , ergs\, cm^{-3} s^{-1} \qquad (12)$$

where T_e the electron temperature.

Assuming n_e constant throughout the roundchrom, we have for the determination of electron concentration in the roundchrom:

$$n_e = 1.16 \cdot 10^{13} \left[\frac{L(\mathrm{MgII})}{V}\right]^{1/2} \, cm^{-3} \qquad (13)$$

where V is the volume of the roundchrom in cm^3 and $L(\mathrm{MgII})$ is in $ergs\, s^{-1}$. The latter is determined by the known distance of the binary from the Sun D and the observed flux $F(\mathrm{MgII})$

$$L(\mathrm{Mg}) = 4\pi D^2 \, F(\mathrm{MgII}) \, ergs\, s^{-1} \, . \qquad (14)$$

The obtained values of n_e for a sample of 50 RS CVn type close binary systems are presented in Table 10.9. In Fig. 10.11, a relationship is drawn between n_e and a, according to the data of Table 10.9. The locations of SX Cas and 22 Vul are shown as well; as we see, 22 Vul is located in the low end of the $n_e \sim a$ relationship just on the correlation line $\log n_e \sim \log a$. The scatter of the points in Fig. 10.11 seems to be real and is caused, apparently, by various relative sizes of the radii of the secondary components as compared with the radius of their Roche lobes.

T a b l e   10.9

The electron concentrations n_e in roundchroms for 50 RS CVn type close binary systems. Intercomponent distances a are in units of solar radius, $R_\odot$

N a m e	a	n_e $10^{10} cm^{-3}$	N a m e	a	n_e $10^{10} cm^{-3}$	N a m e	a	n_e $10^{10} cm^{-3}$
CG Cyg	3.2	17.9	AS Dra	17.6	3.6	TW Leo	57.8	2.0
SV Cam	3.4	26.0	WW Dra	18.7	3.4	HR 7275	58.0	2.9
TR And	4.0	16.0	V841 Cen	19.5	4.7	TW Lep	63.0	2.0
ER Vul	4.2	15.0	UX Ari	20.7	4.7	IM Peg	64.3	2.1
UV Psc	4.7	18.6	II Peg	21.0	3.7	RZ Eri	75.0	0.61
DH Leo	5.0	14.0	LX Per	26.3	2.0	ε UMi	76.2	1.00
TZ GrB	5.8	17.1	AE Lyn	30.1	1.8	AY Cet	86.2	1.26
AR Lac	9.1	6.0	α Aur	31.8	3.7	SX Cas	91.0	3.2
V824 Ara	10.0	7.5	RV Lib	32.2	3.2	λ And	91.7	0.73
EI Eri	10.2	9.0	IL Hya	37.3	3.8	HR 4492	105	1.9
CF Tuc	11.9	8.6	V350 Lac	39.0	1.4	EQ Leo	121	1.00
HR 5110	12.0	7.0	TZ TriA	44.8	1.9	HR 4665	126	0.75
HR 1099	13.0	7.7	AR Mon	47.8	4.1	12 Cam	126	0.57
TY Pyx	14.2	7.0	σ Gem	49.3	2.9	o Dra	181	0.80
SZ Pic	14.9	7.0	ξ And	52.0	1.5	HR 7428	186	0.68
RT Lac	16.2	7.1	V792 Her	53.4	1.2	22 Vul	323	0.23
RS CVn	16.3	5.0	HK Lac	57.1	2.2			

Although the mean value of electron concentration seems to be the same and of the order of 10^{10} cm^{-3}, at least in the limits of one order of magnitude with extremely wide limits of roundchrom volumes, the existence of a real relationship in the limits of an order of magnitude also seems to be real. Then for the dependence of n_e on intercomponent distance a may be derived from Fig. 10.11:

$$n_e = 5.20 \, 10^{11} a^{-0.80} \ cm^{-3} \ . \tag{15}$$

This correlation is shown in Fig. 10.11 by an inclined broken line.

One more interesting mission is associated with the Lagrangian point L_1, namely, the formation of the so-called "Lagrangian cone", a cone-like region near this Lagrangian point, as a powerful area generating high-excitation emission lines, particularly of 1240 N V, 1400 Si IV, 1550 CIV etc. The existence of the "Lagrangian cone" is obvious when we follow the behavior of high-excitation emission lines at various phases of the eclipse in the case of binary systems SX Cas and 22 Vul. So, an important new component appears in the structure of roundchroms – the Lagrangian Cone as a powerful source of high-excitation emission lines.

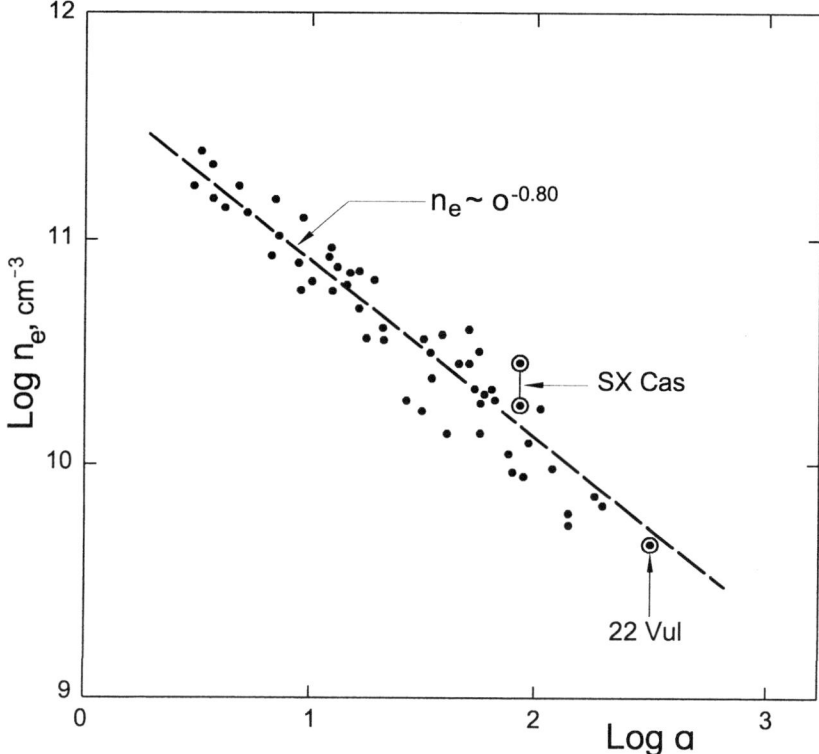

Fig. 10.11. *Empirical relation between the electron concentration n_e in the roundchrom and intercomponent distance a, in units of solar radius $R_\odot$, for 50 RS CVn type close binaries. The locations of SX Cas and 22 Vul are shown as well. The upper point of SX Cas corresponds to the distance of 500 pc, the lower to distance of 300 pc.*

7   On the Gas Transfer Problem

The problem of gas transfer between the components of close binary systems is one of the oldest in astrophysics, both theoretical and observational.

Historically, the first serious attempts concerning the mass transfer problem through the inner Lagrangian point L_1 date from the mid-1970s when detailed analysis was carried out concerning the flow of gaseous matter through the neck in the point L_1, the determination of the stream parameters, particularly the mass transfer rate, the movement of particles leaving through the corridor in the point L_1 and moving into the gravitation field of the binary system.

However, these analyses were carried out using a simplified ballistic approach to the examination of gas flow, without taking hydrodynamic effect into account. Very soon, the hydrodynamics of the flow of matter from one component to the other turns out to be a subject of extensive 3D numerical studies with the accounting of the morphology of gaseous flows (e.g. Armidage, Livio, 1996; Bisikalo *et al.*, 1998, 1999). The main conclusion is as follows: the transfer of gaseous matter from one component of a close binary system to another is possible without any principal restrictions. Such was the situation regarding the theory.

The final conclusion may be drawn by having confirmation of the possibility of a gas-transfer process by direct observation. Bearing in mind the extremely small angular separation between the components, to carrying out such observations by the usual methods of practical astrophysics is simply impossible. The situation may be changed by enlisting short-wavelength radio observations and the application of the so-called Doppler tomography technique.

The Doppler tomography technique is based on the reconstruction of three-dimensional (3D) images of any space formation from its two-dimensional (2D) pictures or "slices". In astronomy, this technique uses the one-dimensional information provided by the Doppler shifts of any emission line to generate 2D velocity images of the circumstellar material in the orbital plane of a binary, using the emission-line spectra of the binary seen at different positions in the orbit of the binary from one eclipse to the next. Since the Doppler shifts give information about the motions of the gas, these wavelength shifts may be converted into velocity shifts. The latter shifts are then used within the tomography program to produce the image of the circumstellar gas in velocity space. This image can be converted to the configuration space, if the velocity of the flow is known.

By mean of this technique spatial models and Doppler tomograms are produced for several Algol-type binary systems. On these pictures not only are the isthmuses connecting both components of the binary shown but the kinematical parameters on different points on these isthmuses are determined as well. As an example, in Fig. 10.12, the isthmuses or the paths of gas streams are shown for two Algol-type close binary systems, namely, β Per ($P = 2^d.87$) and U CrB ($P = 3^d.45$)(Richards *et al.*, 1995) in the real linear sizes of both components (solid line) and their Roche lobes (broken circles). The path, definitely fixed, of the gas stream is a curved line connecting both components of the system. Here the distinct flow of gas along the predicted gravitational free–fall path is clearly visible. Some parameters of the flow were determined along its path within small intervals.

In Fig. 10.13, the direct image is shown for a binary α PsA obtained with the help of the James Maxwell Telescope of Mauna Kea (Holland *et al.*, 1998). On this image the isthmus between the components, in fact in

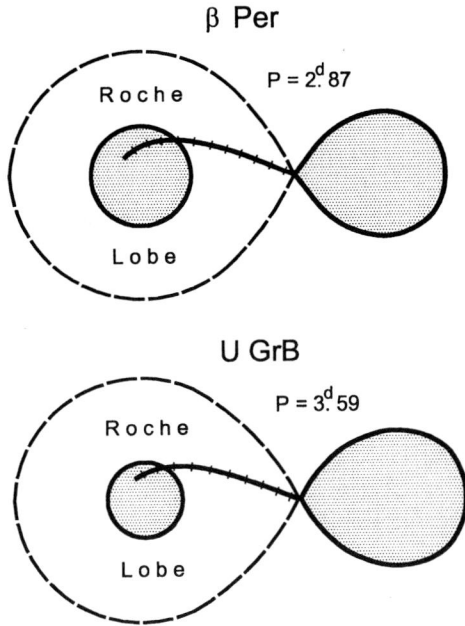

β Per

P = $2^{d}.87$

Roche

Lobe

U GrB

P = $3^{d}.59$

Roche

Lobe

Fig. 10.12. *Spatial models of close binaries β Per and U CrB elaborated by means of direct observations and of Doppler tomograms to reveal the gas streams between the components of binary system.*

the Lagrangian point L_1, is visible. Bearing in mind that the gaseous matter immediately on this isthmus or throat cannot be stable, the presence of too bright a background on this point should be accepted as direct confirmation that a gas stream process really does exist between the components.

Although both these binaries, β Per and α PsA, are not of RS CVn type, there cannot be any doubt that these results apply to all types of close binary systems, including the RS CVn type.

8   Roche Lobe

Roche lobe is a basic concept, forming a unique gravitational "surface" surrounding any two gravitating objects, in the present case two stars, that orbit one another. The originality of the Roche lobe is in the simultaneous accounting of two properties of celestial bodies – the gravity, i.e. the force, and the motion, i.e. the kinematics. The Roche surface takes into account the combined effect of gravity and orbital motion.

The orbit of two stars around their shared center of mass introduces a centrifugal force; this affects how we perceive the motions of nearby material points when we imagine the binary system as a stationary entity.

As we have seen in Chapter IX, examination of the problem of binary systems leads to the formation of surfaces of zero velocities whose form is

α PsA

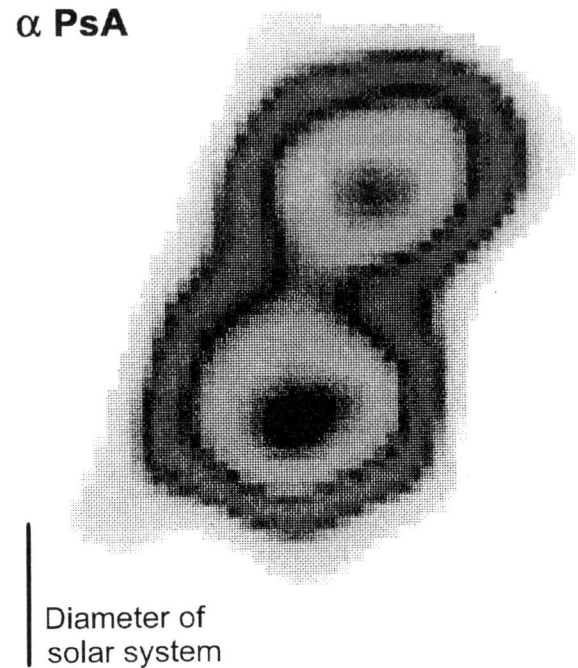

Fig. 10.13. *Direct image of the close binary system α PsA. The bright background between the components, at the narrow isthmus on the Lagrangian point L_1, is visible, which should be interpreted as a confirmation of the existence of a gas flow process between the components*

Diameter of
solar system

determined by the Jacobi constant C, in its turn depending on the position and velocity of the test mass – particles with negligible small mass. At various magnitudes of C the so-called zero-velocity surfaces consist of two separate parts – closed isolated curves, each of which like circles.

However, at certain values of the Jacobi constant C both closed isolated curves come into contact in the Lagrangian point L_2 which lies between them. Then, there is just one "figure eight-like" surrounding both components of the system. This "figure-eight-like" curve-surface of zero velocity, enveloping both components of the binary system, defines the Roche limit or Roche curve; examples of such curves are shown in Figs. 9.3, 9.5, 9.8, 9.9 and 9.11 of Chapter IX.

If we now take as massive bodies two components of binary stars, then this "figure-eight-like" characteristic zero-velocity curve-surface surrounding both components determines the upper limit of the sizes of the components. If a particle on the outer layers of the components of the binary system has an energy exceeding the limiting value C and crosses the zero-energy surface, then it can leave the system forever.

If the outer boundary of the star approaches close to this surface, then any particle having even negligible kinetic energy on this surface can leave the star forever.

In contact binary systems, as well in Algol-type systems, the satellite-component fills the Roche lobe completely. On the other hand, no binary stars are known where the radii of their components exceed the size of the Roche lobe.

The forming of Roche lobes around close binary systems is inevitable. This circumstance leads to an important consequence, namely, the outer layers of the star-components will have a tendency to stretch, especially when the gas particles are near to the zero-velocity surface. This tendency will be especially strong when the components of the binary system are non-stable stars with an outflow of gaseous matter. A particular case of binary systems is those with large eccentricities of their orbits, examined in Section 4 of this chapter, with the periodic rapprochement of the zero-velocity surfaces on the periastr, and resulting outflow of matter through the Lagrangian point L_2. If such a star is a component of a binary system it will fill its Roche lobe completely.

So, we arrive at an inevitable conclusion: close binary systems cannot be examined as isolated bodies – neither from the point of view of astrophysics, no of a celestial mechanics.

The Roche lobe is handsome and fruitful discovery in celestial mechanics, crucial for the dynamics of binary systems, both planetary and stellar, especially those with non-stationary or massive star components (neutron star, black hole). To this problem we shall return in Chapter XII.

9   Binary Globular Clusters

Equipotential surfaces can be a fruitful concept not only for binary stars, as discussed in previous sections, but also for binary globular clusters (G.Gurzadyan, 2000).

Contemporary views on the evolution of globular clusters include the following effects:

1. Evaporation of stars.
2. The radius (tidal) of a cluster is determined by the condition of compensation of the cluster's gravity and the field of the Galaxy.
3. Core collapse.
4. Core collapse may be halted by one or several hard binaries.
5. Tidal interaction or shocks to the Galaxy can accelerate the core collapse.

Consider now a binary system, both components of which are globular clusters, i.e. we have a *binary globular cluster*, particularly a *close binary globular cluster*. We will discuss heuristically the principal effects which arise when two ordinary globular clusters are components of a binary system.

The problem of binary globular clusters can have direct relation to galaxies with double nuclei, assuming that its both components are supermassive globular clusters. The problem of binary globular clusters may stand as a topic for special investigation, encompassing interactions between the components, tidal effects and the effect of axial rotation of the system, and the effect of the star-transfer process from one component to another, with the inevitable evolution of clusters.

Consider a binary system with globular clusters A and B as components, with masses M_A and M_B, mutual distance a and orbital period P by relation

$$a = 9.49 \cdot 10^{-8} \, P^{2/3} M^{1/3} \,, \tag{16}$$

where the summary mass $M = M_A + M_B$ is in units of solar mass, P is in days and a is in pc. For a binary globular cluster, for example, with $M = 10^6 \, M_\odot$, and intercomponent distance $a = 300$ pc, from (16) we have for the orbital period: $P = 5 \cdot 10^8$ years, i.e. comparable with the dynamical time of the galaxy. During that time scale such a binary system can perform two–three tens of revolutions around the mass center of the galaxy.

However, for reasons which we discuss later, we shall deal with masses of the order of 10^8–$10^9 \, M_\odot$, i.e. higher as compared with the mass of ordinary globular clusters in our Galaxy. Adopting, therefore, $M = 10^8 \, M_\odot$ and $a = 300$ pc, we find from (16): $P = 5 \cdot 10^7$ years, i.e. shorter as compared with the dynamical time scale of the galaxy.

The actual sizes of globular clusters in the Galaxy are essentially determined by the tidal field of the Galaxy. For an isolated globular cluster the tidal radius r_g is determined by the gravitation action of the Galaxy of a mass M_g and radius R_g; when the cluster is located in the environments of the Galaxy the tidal radius is given by the so-called Hoerner's formula

$$r_g = R_g \left(\frac{M_A}{2M_g} \right)^{1/3} . \tag{17}$$

As we see, the tidal radius r_g of a globular cluster depends on the masses both of cluster M_A and the Galaxy M_g. At $M_A = 10^7 \, M_\odot$, for example, for $M_g = 10^{11} \, M_\odot$ and $R_g = 15\,000$ pc one has $r_g = 550$ pc.

The situation is quite different when we deal with a close binary globular cluster. In this case, the tidal radius r_t for one of the components – the cluster A, is determined by the gravitation action of the second cluster B:

$$\frac{GM_B}{(a - r_t)^2} - \frac{GM_B}{a^2} = \frac{GM_A}{r_t^2} \tag{18}$$

or

$$\frac{1}{(1 - x)^2} - \frac{q}{x^2} = 1 \,, \tag{19}$$

where x and q denote

$$x = \frac{r_t}{a} \, , \qquad q = \frac{M_A}{M_B} \, . \qquad\qquad (20)$$

Eq. (19) is the correct relationship for the determination of x, i.e. of r_t. In the particular case, when $x \ll 1$, which corresponds to a "Galaxy–Globular cluster" combination, we have from (19): $x = (q/2)^{1/3}$ or $r_t = R_g(q/2)^{1/3}$, i.e. Hoerner's formula (17).

As follows from this relationship, the tidal radius r_t for the component A depends on the ratio of masses of both components, but not on the absolute magnitudes of their masses.

In Table 10.10 the numerical values of tidal radius r_t, obtained with the help of Eq. (19), are presented for three values $q = 1, 2, 3$ and the intercomponent distance $a = 500$ pc. As we see, in this case the tidal radii are 1.5–2 times smaller than those determined by the gravitational action of the Galaxy. Particularly for equal masses of both cluster-components, i.e. when $q = 1$, we have for the tidal radius $r_t = 265$ pc, essentially smaller than we have for the "Galaxy–Globular cluster" combination, i.e. $r_g = 550$ pc.

T a b l e   10.10

Tidal radius $r_t(A)$ for the component A of a bi-nary globular cluster determined by the gravita-tion action of the component B at three values of mass ratio q and intercomponent distance $a = 500$ pc

q	x	$r_t(A)$, pc
1	0.53	265
2	0.60	300
3	0.65	325

Thus, we arrive at the first remarkable, and important, conclusion: the tidal radius of a globular cluster that is a member of a binary cluster system is determined by the gravitational attraction of the second cluster and not of the Galaxy. Such clusters, therefore, are expected to be more compact than typical isolated globular clusters. In this case the real sizes of the globular clusters are determined not by tidal interaction of Galaxy mass but by mutual tidal interactions of globular clusters.

10   Equipotentials of a Binary Globular Cluster

In Fig. 10.14 the figure-eight-shaped zero-velocity equipotential curve enveloping both components of the binary cluster is shown for the masses of components $M_A = 4 \cdot 10^9 \, M_{\odot}$, $M_B = 2 \cdot 10^9 \, M_{\odot}$, i.e. for the mass ratio $\mu = 0.33$ or $q = 0.50$, intercomponent distance $a = 800$ pc and Jacobi constant $C = 3.94$, so that a narrow corridor can be formed in the intermediate Lagrangian point L_1 between components. The dashed circles denote the Roche lobes with radii R_{L} determined according to the above relationship (10.2), yielding $R_L(A) = 350$ pc and $R_L(B) = 175$ pc.

In the case of stars, as we have seen above, owing to the existence of strong outer boundaries – photospheres, one can compare the star's size with its Roche lobe. In the majority of cases the sizes of the main components in close binary systems coincide with their Roche lobe. Such a coincidence, again as a rule, is not the case for secondary components.

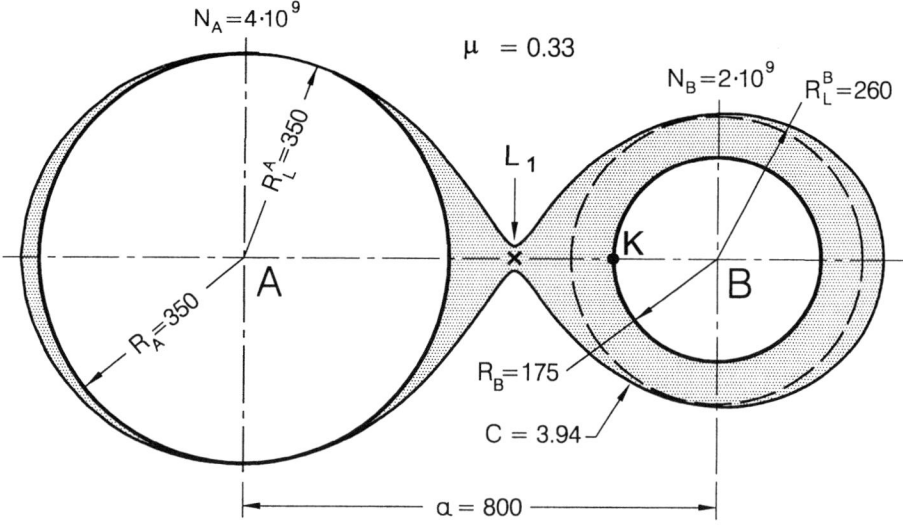

Fig. 10.14. *Roundchrom structure for close binary system with globular clusters as components with mass ratio $\mu = 0.33$ or $q = 0.50$, masses of components $M_A = 4 \, 10^9 \, M_{\odot}$, $M_B = 2 \, 10^9 \, M_{\odot}$ and intercomponent distance $a = 800$ pc, R_L^A and R_L^B are the radii of Roche lobes (dashed circles). All linear sizes are in parsecs.*

In the case of globular clusters the situation is obviously quite different since one is dealing with an N-body system and hence no clear boundary of the system exists. Nevertheless, the chaotic properties of the stellar orbits ensure that the basic effects can also arise for binary globular clusters.

Therefore, it appears possible to develop the principal idea of a roundchrom for the case of binary globular clusters, supposing for simplicity that the linear sizes of the component-clusters are proportional to N_A and N_B, and the outer boundary of the first component coincides with its Roche lobe, and is somewhat smaller in the case of the second component.

Then Fig. 10.14 should be interpreted in the following manner. If the outer boundary of the component A is not far from the point L_1, then one should expect a transfer or flow of stars from A to B. The surrounding volume, i.e. the "roundchrom" of the secondary, will be occupied mainly by these stars, resulting a monotonous decrease of the mass of the primary A with corresponding increase of the mass of the secondary B. As a result, the balance between the kinetic and potential energies cannot remain frozen for each of the clusters, along with the secular variation of their linear sizes.

11   Evolution of Clusters' Sizes

The problem now is to determine the rate of the expansion of the second component-cluster (B) of the binary system, and the rate of contraction of the first one (A). More definitely, we hope to deduce the law of the growth of the radius $R_B(t)$ of cluster B and the law of the decreasing of the radius $R_A(t)$ by time, i.e. during the star-transfer process from the first component to the second one.

It is convenient to start from the following basic dynamical principles. When the moment of inertia of the system can be considered as constant within some time interval, i.e. its variation is slow enough, one readily comes to the virial theorem

$$U(t) + 2T(t) = 0 \,. \tag{21}$$

Note that this relation is obtained not at the infinite time limit, but via averaging within a finite time interval.

Considering the slow stationary star-transfer process with conservation of the moment of inertia (if the clusters, for example, are far from the last phases of core collapse), we can assume that the energy conservation condition

$$T_0 + U_0 = T(t) + U(t) \,, \tag{22}$$

is fulfilled for each component cluster separately. Similarly, it is well known that for an isolated cluster, although the evaporated stars carry away some energy, the total energy of the cluster can still be considered constant within some long enough time interval while estimating the dynamical characteristics of the cluster.

Then, we have the well-known relationship between radius R, velocity dispersion v_o and mass $M = mN$ of the cluster

$$v_0^2 = \frac{G}{2} \frac{mN}{R},$$

(23)

where N is the number of stars, m is the mass of a star.

We can now obtain the variations of the sizes for each of these components of binary clusters A and B during of star-transfer process through the Lagrangian point L_1 from the component A to B.

We start from consideration of the behavior of the component B. For the initial values of kinetic T_0 and potential energy U_0 we have

$$T_0 = N_B \frac{m v_0^2}{2}$$

(24)

$$U_0 = -\frac{G}{2} \frac{m^2 N_B^2}{R_0},$$

(25)

where R_0 is the initial radius of the cluster B.

First, we will consider the star-transfer process with a constant rate. As a result of the star-transfer process the cluster B gains an additional kinetic energy

$$\Delta T = nt \frac{m v_*^2}{2},$$

where v_* is the velocity of stars in the transit point L_1, and n is the flux of stars with a constant rate n. Thus the total kinetic energy of cluster B at the moment t will be

$$T(t) = T_0 + \Delta T = N_B \frac{m v_0^2}{2} + nt \frac{m v_*^2}{2}.$$

(26)

Correspondingly, we have for the potential energy $U(t)$ of the cluster B at the moment t

$$U(t) = -\frac{G}{2} \frac{m^2 (N_B + nt)^2}{R_t},$$

(27)

where R_t is the modified radius of the cluster B.

The condition we will use is that the virial equilibrium remains valid during this process, i.e. for moment t. This condition has to be safely fulfilled during the stationary star-transfer process due to violent relaxation effects occurring within the dynamical (crossing) time scale. Then, substituting (26) and (27) into the virial expression (21) and bearing in mind (23), we obtain for the law of variation of the radius $R_t(B)$ of cluster B by time

$$\frac{R_t(B)}{R_0} = \frac{(1 + nt/N_B)^2}{1 + (v_*/v_0)^2 \, nt/N_B}.$$

(28)

This same equation can be obtained using the condition of conservation of total energy of each of these clusters during the star-transfer process.

Strictly speaking, v_* must be slightly larger than v_0. However, adopting as a first approximation $v_* = v_0$, we obtain

$$\frac{R_t(B)}{R_0} = 1 + \frac{n}{N_B} t \tag{29}$$

where the stellar flux rate n is considered constant according to the assumption on the stationarity of the star-transfer process.

In contrast with cluster B, cluster A contracts during the stellar transfer process, i.e. its radius should be decreased. A similar relationship can be derived for this case as well, i.e. for the law of decrease of the radius $R_t(A)$ of cluster A

$$\frac{R_t(A)}{R_0} = 1 - \frac{n}{N_A} t . \tag{30}$$

However, the star-transfer process from cluster A to cluster B cannot be stationary too long: with the decrease of the diameter of cluster A, its outer boundary will become more and more distant from the point L_1. Hence, the initially assumed constant star-transfer rate has to be substituted by a decreasing function

$$n_t = n_0 e^{-\gamma t} , \tag{31}$$

where γ in yr^{-1} characterizes the rate of variation of the star-transfer process from A to B.

In this case the total number of stars $n(t)$ transferred from A to B at the moment t, after starting the transfer process ($t = 0$), will be

$$n(t) = \frac{n_0}{\gamma} \left(1 - e^{-\gamma t}\right) . \tag{32}$$

As a result, we have for the evolution, i.e. for the growth of the radius of the cluster B at the moment t, the following relationship, again assuming $v_* = v_0$:

$$\frac{R_B(t)}{R_0} = 1 + \frac{n_0}{\gamma N_B} \left(1 - e^{-\gamma t}\right) \tag{33}$$

and for the decrease of the radius of A:

$$\frac{R_A(t)}{R_0} = 1 - \frac{n_0}{\gamma N_A} \left(1 - e^{-\gamma t}\right) . \tag{34}$$

Above, at the derivation of Eqs. (30) and (34), two extreme cases for the rate of the star-transfer process were considered, namely, with a constant rate and with an exponentially decreasing rate. The real situation has to be an intermediate one between these two limiting cases.

In Fig. 10.15 the curves of time variations are given for both radii, $R_A(t)$ and $R_B(t)$, for the second cases, i.e. at an exponentially decreasing star-transfer rate given by an expression (31), for two sets of initial parameters, namely, $N_A = 5 \cdot 10^9$, $N_B = 2 \cdot 10^9$, $\gamma = 2\,10^{-9}\ yr^{-1}$ and $n_0 = 1$, $n_0 = 2$ and $n_0 = 3\ str.yr^{-1}$ in the first case (at left), and $N_A = 5 \cdot 10^8$, $N_B = 2 \cdot 10^8$, $\gamma = 10^{-8}\ yr^{-1}$, and $n_0 = 0.5$, $n_0 = 1$ and $n_0 = 2\ str\,yr^{-1}$ in second (at right).

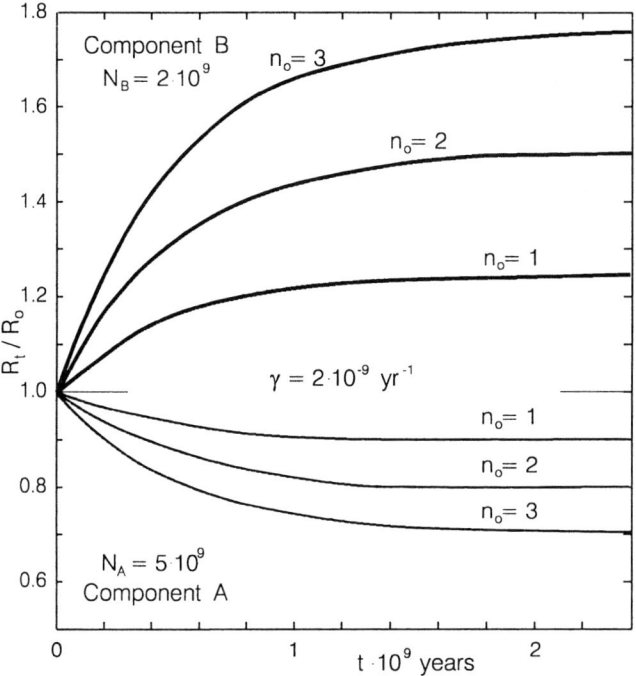

Fig. 10.15. *The time-dependent curves of the variations of linear radii R_t/R_0 of both components, A and B, of a binary system with globular clusters as components. For the initial moment, $t = 0$, it is assumed $R_t/R_0 = 1$. The curves are calculated for three values of star-transfer rate, n_0, 1, 2 and 3 str yr^{-1} and for $\gamma\ 2\,10^{-9}\ yr^{-1}$ and for the time scale 10^9 yr. During the star-transfer process from component A to component B, A undergoes a decrease of its radius, i.e. $R_t/R_0 < 1$ (lower half), and component B, on the contrary, an increase, $R_t/R_0 > 1$ (upper half).*

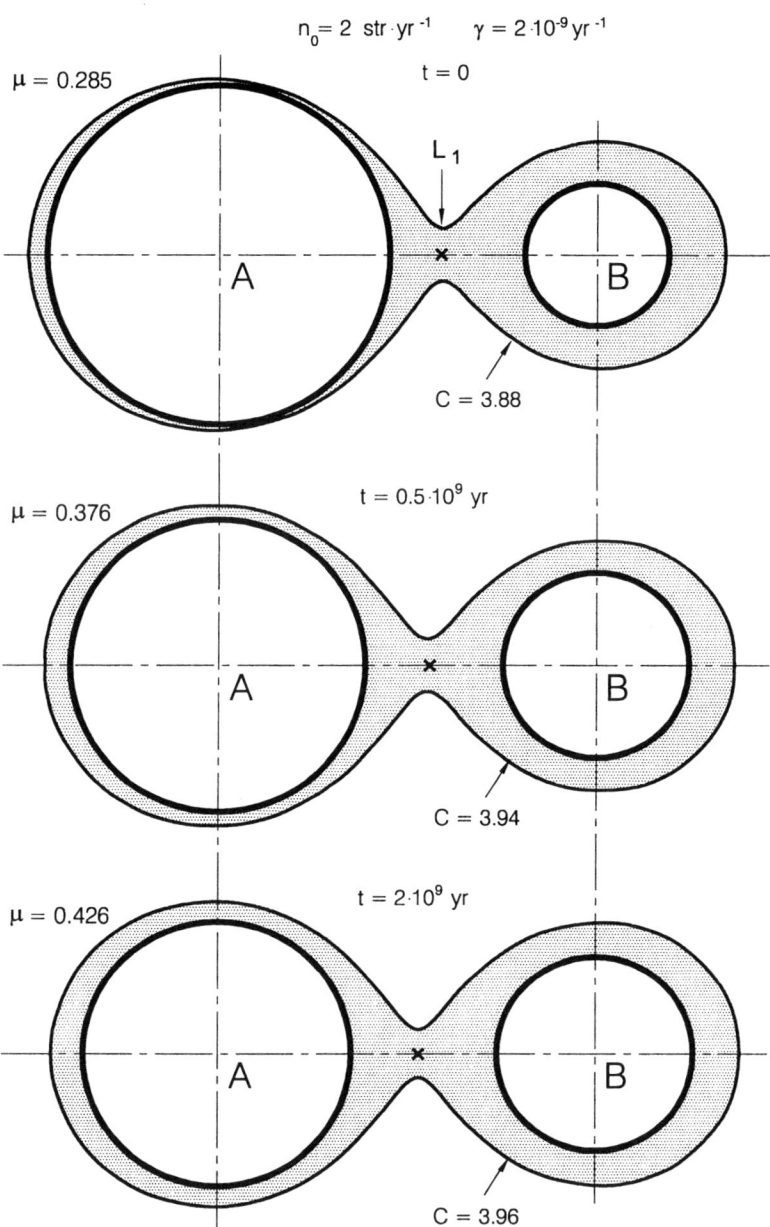

Fig. 10.16. *Three stages of the evolution of globular clusters' diameter as a function of time t, i.e. during the star-transfer process from component A to component B through the corridor in Lagrangian point L_1. All three configurations are calculated for one and the same star-flow rate $n_0 = 2$ str yr^{-1} and $\gamma = 2\,10^{-9}$ yr^{-1}. Initially, at $t = 0$, component B has a powerful roundchrom (pointed area) and component A is practically without a roundchrom.*

Note that the growth of the radius of cluster B takes place faster than the decrease of the radius of A. At the end of evolution, formally at $t \to \infty$, the radius of B is doubled while that of A is decreased only by one-fifth of its initial radius.

Also, the volume of the roundchrom around component B is decreased monotonously during the evolution of the system. The volume of the round-chrom around A increases, though not strongly.

Note that during the essential evolution of sizes of cluster A, i.e. over $2 \, 10^9 years$, the distance between its outer boundary and the point l_1 remains practically unchanged. This important property is explained also by the slow motion of the Lagrangian point L_1 in the direction of component A synchronously with the decreasing of its mass.

Despite the essential differences in the values of the initial numbers of stars in clusters as well in the rate of the star-transfer process, the main character of the curves in Fig. 10.15 is almost the same. This means that we can say nothing, for example, about the age of the binary system or the phase of evolution or star-transfer process of the pair, judging, say, by the direct image of a given interacting binary system in the center of a galaxy.

As an illustration, an evolution sequence of the sizes of a binary cluster is shown in Fig. 10.16. The initial data at $t = 0$ are as follows: $N_A = 5 \cdot 10^9$, $N_B = 2 \cdot 10^9$ with the same mass (solar) of stars in both cases, i.e. $m = M_\odot$, the ratio of initial radii of both components $R_A/R_B = 2.5$. The configurations of the roundchroms correspond to a star-transfer rate of $2 \, star \, yr^{-1}$ and $\gamma = 2 \cdot 10^{-9} \, yr^{-1}$. The second configuration from the top corresponds to $t = 0.5 \cdot 10^9 \, yr$, the third one – to $t = 2 \cdot 10^9 \, yr$.

In Fig. 10.16, the roundchrom-cluster concentric configurations allow us to predict the possible existence of binary globular clusters in the centers of galaxies, i.e. a two-component structure with an outer low-density ring around the central dense sphere.

Returning to the star-transfer problem, one should outline also the following aspects:

a. One can always find wandering stars in the vicinity of the corridor in L_1, ready for passage through this corridor.

b. The gravitational attraction of cluster B should support the accumulation of individual stars of the cluster A in the vicinity of the corridor in the point L_1.

12   The Tidal Effect

Independently of the star-transfer process from one component to another, each star, say, in cluster A, individually undergoes gravitational attraction by cluster B. As a result, the star may acquire an additional velocity. If so, the macrostructure of the energy balance of cluster A must be modified due to the variation of its linear size.

To evaluate these variations quantitatively, one has to estimate the amount of the additional kinetic energy gained due to the gravitational attraction of the total mass M_B of B located in its center on the distance a from the center of A.

Assuming N_A stars of equal mass m homogeneously distributed within a spherical cluster A with initial radius R_A and concentration

$$n_0 = \frac{3}{4\pi} \frac{N_A}{R_A^3} , \tag{35}$$

we can write the equation of the motion of a star at distance r from M_B

$$m\frac{d\Delta v}{dt} = -G\frac{mM_B}{r^2} , \tag{36}$$

which gives for the additional velocity Δv of the star

$$\Delta v = G\frac{M_B}{r^2} t . \tag{37}$$

In view of the gravitational action of cluster B on individual stars in cluster A, i.e. due to the **tidal** effect, we can write for the elementary kinetic energy dT_{tid} of a mass included into the element of a sphere of radius r and thickness dr

$$dT_{\text{tid}} = dm\frac{\Delta v^2}{2} = \pi\, n_0\, m\, G^2\, M_B^2 (1 - \cos\varphi)\frac{t^2}{r^2} dr , \tag{38}$$

where $M_B = m\, N_B$ and

$$\cos\varphi = \frac{a^2 - R_t^2 + r^2}{2ar} . \tag{39}$$

After integration of (38) over limits $a - R_t$ and $a + R_t$, we obtain for the additional kinetic energy acquired by cluster A under the action of cluster B

$$T_{\text{tid}}(t) = \frac{3}{4}\, G^2 m^3 \frac{N_A N_B^2}{R_t^4} \left(\frac{1}{u^2 - 1} - \frac{1}{2u} \ln\frac{u+1}{u-1} \right) t^2 , \tag{40}$$

where $u = a/R_t$.

Then the energy conservation law should be written in the form

$$T_0 + U_0 = T(t) + U(t) + T_{\text{tid}}(t) , \tag{41}$$

where T_0 and U_0 are the same as above. For $T(t)$ and $U(t)$ we have

$$T(t) = \frac{G}{4} \frac{m^2 N_A^2}{R_t}, \tag{42}$$

$$U(t) = \frac{G}{2} \frac{m^2 N_A^2}{R_t}. \tag{43}$$

Substituting (40), (42) and (43) into (41), and in view of (24) and (25), we obtain finally the radius R_t of cluster A at the moment of time t

$$\frac{N_A}{R_0}\left(1 - \frac{R_t}{R_0}\right) = Gm \frac{N_B^2}{R_t^4}\left(\frac{1}{u^2 - 1} - \frac{1}{2u}\ln\frac{u+1}{u-1}\right) t^2 \tag{44}$$

with $u = a/R_t$, and R_0 is the initial radius of cluster A.

The relationship (44) defines the law of variation of the cluster's radius R_t with time t. The tidal effect of cluster B leads to an increase of the radius R_t of cluster A.

As the analysis of the relationship (31) shows, the growth of the radius R_t/R_0 of cluster A due to the tidal effect of cluster B occurs rather rapidly – during a time scale of the order of 10^7 yrs. This result seems to be in disagreement with the dynamical nature of the globular clusters. However, as we will show below, the orbital rotation effect can essentially suppress the rapid growth of the radius of cluster A.

The peculiarity of the problem considered above, namely, the role of the tidal process in binary globular clusters, is that one is dealing not with atoms in the stellar atmospheres of interacting close binaries but with stars. The main difference is in the absence of collisions between stars, i.e. the absence of an energy exchange process between stars (particles).

Now consider the role of orbital rotation. Each star in both clusters will undergo the action of centrifugal force provoked by the orbital motion of the components. This force may compensate to a certain degree the gravitational attraction of the other cluster. It appears that this fact may radically change the above conclusions.

The centrifugal force depends on the tangential velocity V_t and the distance to the center of rotation of the binary system. Gravitational attraction provoked by cluster B can be compensated completely by centrifugal force only if the tangential velocity V_t of a star in the center of cluster A satisfies the condition

$$V_t = \left(G\frac{M_B}{a}\right)^{1/2}(1 + M_A/M_B)^{-1/2}, \tag{45}$$

where M_A and M_B are the masses of the clusters, as before.

On the other hand, for the orbital velocity V_{or} of a star we have

$$V_{\text{or}} = \frac{\pi a}{P} \tag{46}$$

where P is the orbital period of the binary system, or using the known expression for P,

$$V_{\text{or}} = \frac{1}{2} \left(G \frac{M_B}{a} \right)^{1/2} (1 + M_A/M_B)^{-1/2}. \tag{47}$$

From (47) and (45) we find for the ratio of both velocities

$$\frac{V_{\text{or}}}{V_t} = \frac{1}{2}(1 + M_A/M_B). \tag{48}$$

In the particular case when $M_A = M_B$, an important result will be achieved:

$$V_t = V_{\text{or}},$$

i.e. only at equal masses of both clusters do we have an equilibrium of both types of forces – centrifugal and gravitational; in this case the tidal acceleration of a star towards cluster A due to the gravitational attraction of cluster B will be compensated by the centrifugal force of an orbital rotation of the system. The further evolution of the sizes of clusters will be determined by the star-transfer process from one component to another.

So, orbital rotation plays an important regulating role in the dynamics of binary globular clusters.

Thus, the star-transfer process from cluster A to cluster B leads to the **decrease** of the size of cluster A. The *tidal effect* of cluster B on cluster A leads to the *growth* of the sizes of A. These are the consequences of the tidal effect in its "pure" form. However, the rotational effect due to the orbital motion of clusters A and B essentially *reduces* the rate of this process. As a result, the behavior of the size of component A has to be rather complex.

So, the orbital rotation of a binary system with globular clusters as its components plays an important regulating role in the dynamics of binary globular clusters.

The conclusion is clear: the star-transfer process from cluster A will definitely be accompanied by a decrease of the cluster's size, but at rate essentially dependent on various physical, dynamical and kinematical parameters of both clusters.

The latter two effects, the tidal effect and orbital rotation, lead to changes in the dynamical state of a globular cluster during a time scale of the order of tens of millions years, which is much shorter as compared with the time scale of a globular cluster's evolutionary processes. Thus the clusters in the binary globular systems must have different evolutionary time scales than the isolated globular clusters.

13   On the Binary Galactic Nuclei

Discoveries of galaxies with double nuclei are becoming more and more common. Due to their form and structure, in some cases the nuclei are interpreted as binary black holes. Nevertheless, the concept that nuclei are compact stellar systems cannot be excluded, i.e. supermassive globular clusters – of a mass of components of the order of $10^9\,M_\odot$ and more. An example of such a double nuclei is shown in Fig. 10.17; this Mrk 273 image was obtained by the Keck telescope (Knapen et al. 1997). It resembles a figure-eight-shaped binary configuration with different sizes of components. At the distance of this galaxy, 160 million pc, the intercomponent distance is estimated as 730 pc, with diameters of components 450 pc and 700 pc.

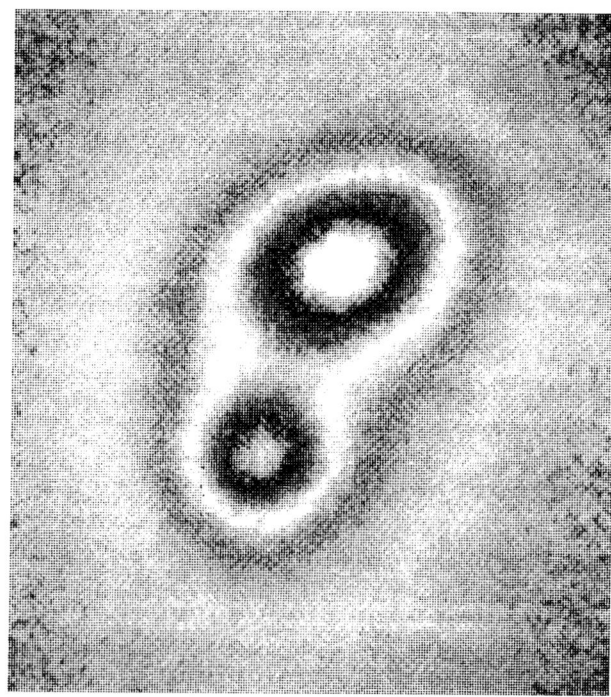

Fig. 10.17. *Double nuclei of Markarian 273. Components of nuclei are of different masses and different sizes. The separation between components is 730 pc. Note the background in the corridor between components.*

Note that in Fig. 10.17 the space between the components reveals a certain brightness. According to our concept, the point L_1 located there must provide a corridor for star flow from one component to another, so that the intercomponent space must acquire definite brightness, which can even be roughly estimated. At the flux 1 star per year with stellar velocity $100\ km\,s^{-1}$ we shall have 10^7 stars during $10^6\ yrs$ on 10 pc linear length on this area, thus forming a noticeable background brightness, as may be indicated in Mrk 273.

The number of known galaxies with binary nuclei is already over one hundred. For some of them, including Mrk 273 the absolute magnitudes of the components M_A and M_B have been obtained. Then, assuming complete stellar composition for these nuclei (by solar-type stars), we can evaluate the total number of stars N_A and N_B in the components; the results are presented in Table 10.11.

T a b l e   10.11

Total number of stars N_A and N_B in the components A and B of binary nuclei of six galaxies with known absolute magnitudes M_A and M_B of the components

Mrk, No	M_A	M_B	$N A$	N_B
266	-17.8	-17.5	$2.1 \cdot 10^9$	$1.6 \cdot 10^9$
273	-18.4	-17.7	3.6	2.0
463	-19.5	-19.3	10	8.3
673	-19.6	-19.6	10	10
739	-19.1	-18.3	6.9	3.3
789	-19.5	-17.5	10	1.6

In the case of Mrk 273, the total number of stars in the components is found to be $3.6 \cdot 10^9$ and $2.0 \cdot 10^9$, i.e. almost of the same order as those used above when drawing the figure-eight-shaped curve in Fig. 10.14. Moreover, the derived ratio $N_A/N_B \sim 2$ for this galaxy corresponds, at least qualitatively, to the photographic images of components in Fig. 10.17. Note the existence of nuclei with equal number of stars in the components (Mrk 673, ratio 1:1) as well with ratio up to 1:5 (Mrk 789). It is remarkable that the total number of stars in the binary nuclei, at least for these six galaxies is of the same order: 10^9–10^{10}.

We have another example in the case of IC 4553 (Arp 220) with a remarkable binary nuclei structure discovered at radio continuum 4.83 GHz, whose isophotic image is shown in Fig. 10.18 (Baan, Haschick, 1995). The projected separation between the components is 330 pc, the reconstructed separation 466 pc. At an orbital period of $7 \cdot 10^6$ yr for the two nuclei with equal masses, the estimated dynamical mass for both components is $10^{10} \, M_\odot$. Note that no clear hot spots are found in the centers of components within the line emission structure. The spectral properties confirm the starburst-dominated nature of the observed emission, and hence, the stellar content of both nuclei.

One more example of a galaxy with binary nuclei is shown in Fig. 10.19, a CCD image of the central part of the well-known extragalactic system

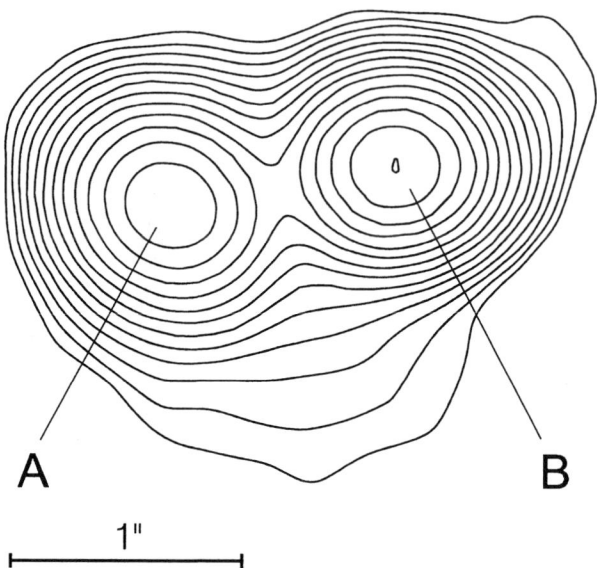

A   B

|—————— 1" ——————|

Fig. 10.18. *Double nuclei of galaxy IC 4553 imaged in radio continuum 4.83 GHz. The reconstructed separation between the centers of nuclei is 466 pc.*

Superantennae (Mirabel *et al.* 1991). At a distance of 250 Mpc, the inter-component separation of this binary is ~ 9000 *pc*; if so, then in this case the question may be about the existence of galaxies with supergiant binary nuclei.

The described effects can also be associated with the binary globular clusters in the Large Magellanic Cloud.

Thus at least four aspects of the dynamics of binary globular clusters can be outlined:

- The determining role of mutual tidal interaction of both components.

- The non-importance of the role of tidal interaction of the Galaxy mass on the dynamics of the binary globular cluster.

- The regulating role of the orbital rotation on the evolution of the binary globular system.

- The importance of the star-transfer process from one component of the system to the other.

The main conclusions can be summarized as follows:

• A globular cluster as a component of a binary system gains new dy-

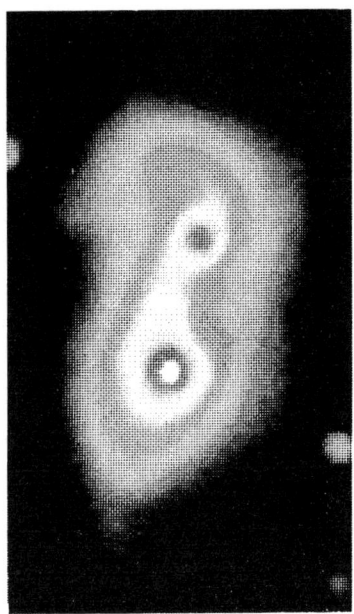

Fig. 10.19. *Double nuclei of extragalactic system Superantennae. The estimated separation between components is 9000 pc. Note the background, as in the case of Fig. 10.17, between the components.*

namical characteristics and properties.

- The dynamical evolution of a globular cluster as a member of a binary system occurs more rapidly as compared with an isolated cluster.
- The tidal radius of a globular cluster in a binary system is smaller as compared with the tidal radius determined by the Galactic field.
- The star-transfer process from one component to another in a binary cluster system leads to the reduction of linear size of the first component and the increase of the size of second one.
- In binary globular clusters the tidal interaction leads to the increase of their kinetic energy and, hence, to the expansion of clusters.
- The orbital rotation of a binary cluster system plays a regulating role on the variations of the sizes of the clusters.

This concludes the discussion aiming to reveal how inspiring the concept of equipotential surfaces is not only for planets, but also for binary stars, globular clusters and even galaxies.

Chapter XI

The Theory of Perturbations

1   Formulation of the Problem

The theory of motion of two bodies interacting according to Newtonian gravitation is based on the assumption that both bodies are of spherical form, possess axial symmetry and hence can be replaced by material points of masses equal to their masses. Moreover, both bodies are considered to be absolutely isolated from external interactions, i.e. the role of other planets, and of any other forces connected with the medium, of the resistance of interplanetary or interstellar matter, of the light pressure of the Sun (and other stars), hydrodynamical, electromagnetic and of all other effects, is neglected.

As for the motion under the action of Newtonian gravitation, here definite results are obtained only in the case in which the **relative** motion is examined. In this case the motion of a body is realized in a conical section – ellipse, parabola or hyperbola – at the focus of which the second, central body is located. It is important to stress that, from a mathematical point of view, the two-body problem is solved absolutely correctly, without any approximations, including the derivation of the explicit law of relative motion and the equations of the orbit, namely of the conical section. Owing to this, the position of the body can be determined definitively for any moment of time.

However, the problem of two bodies is an idealization that is occasionally rather far from the reality. In the Solar system, for example, no pair of absolutely isolated bodies exists. In a number of cases departures from sphericity cannot be ignored, i.e. the prolate shapes of the planets. A well-known example is the motion of an artificial satellite around the Earth, to which the higher layers of the atmosphere can resist, etc. The motion of a body influenced by Newtonian gravitation but with a variable mass is known for the binary stellar systems, in which the mass of one of the components (or even of both) changes with time: among the reasons for this could be the outflow of gaseous matter or its transfer from one component to another, accretion of diffuse matter from the surrounding medium, etc.

In the mentioned cases the orbits are never exact conical sections. Correspondingly, the coordinates and the components of the velocity of the body computed according to the two-body scheme will differ from the real parameters. Just this discrepancy between the coordinates and velocity components of the real orbit and those of the ideal conical sections is called the **perturbation**.

In order to describe the perturbations, i.e. to determine the true orbit taking into account the corresponding corrections to the unperturbed orbit – the pure conical section, one needs appropriate theory, including new mathematical concepts and technique. Thus, a principally new and broad branch of celestial mechanics was created, called **perturbation theory**.

The background of perturbation theory was created by Newton, and it was later developed in the brilliant works of Euler, Clairaut, Lagrange, and Laplace. As a matter of fact, perturbation theory in its present form, which is used for computing the orbits of modern artificial satellites, had already been formulated at the end of the XVIIIth century, i.e. two hundred years ago.

In the particular case in which the perturbations are caused by the gravitational interaction of the neighboring planets, the determination of the deviation of the true orbit from the pure conical sections is reduced to the n-body problem. In the other cases, the problem of the determination of perturbations remains open. Therefore, even were the n-body problem to be successfully overcome sometime, the perturbation theory would still retain its importance.

In this connection the following problem is certainly of interest: is it possible to reduce all the various types of perturbations caused, say, by the flattening of planets, the resistance of the medium, the outflow or accretion of matter etc., to the same n-body problem? The point is, obviously, whether it is possible to simulate those effects, i.e. the flattening of objects or the resistance of the medium by an infinite number of imaginary elementary bodies or sources with a certain spatial distribution. If the answer is negative for the general case, then its consideration even in particular cases could lead to an interesting direction in perturbation theory.

Perturbation theory is more general and more comprehensive then n-body theory; for perturbation theory the very nature of the perturbations or of their source is not important. Owing to its universal character, perturbation theory could include the most essential type of perturbations, i.e. the gravitational influence of neighboring planets. Though the perturbations are usually taken into account via successive approximations, the accuracy of the final results in principle can be made as high as necessary. The possibilities of perturbation theory are rather high and the main difficulties are of a technical character. Nowadays, due to the outstanding development of computer hardware and software, many difficulties are being successfully overcome (Valtonen, 1992; Morbidelli, 2002).

2 Two Trends of Perturbation Theory

There are two trends or two approaches to the computation of the perturbations, i.e. of the corrections to be added to the ideal, unperturbed trajectory – that on a conical section – in order to obtain the real position of a celestial body at a given moment in time.

In the first case, the aim is not the determination of the real, unperturbed trajectory of the celestial body but only of the corrections to the co-ordinates and components of the velocity corresponding to its unperturbed motion. This method enables one to predict the position of a celestial body with sufficient accuracy, though nothing reliable can be said concerning the trajectory of motion. This method is known as **variation of coordinates**, and is applied as a rule for objects that appear episodically, such as comets and small planets.

More general and more popular is the method called **variation of elements**; it is known also as variation of parameters, variation of constants of integration, etc. The essence of the method of variation of elements is as follows. The real perturbed motion of a celestial body can be considered as taking place in a classical conical section, i.e. in an ellipse, parabola or hyperbola, the elements of which, however, are not constants, as before in the case of the "pure" two-body problem. They are permanently changing in time, depending on the character and magnitude of the perturbations. In other words, all the results obtained for the ideal two-body problem, namely, the computation of the orbit, the determination of the position of the body at a given moment in time and the components of velocity etc., remain valid. Now, however, one needs also the numerical values of six orbital elements – Ω, i, ω, a, e, and T for the given moment of time. The essence of the method of variation of elements, and hence, of perturbation theory, lies therefore in the search for tools with which to determine the variation of these elements in time, i.e. the determination of the functions $\Omega(t)$, $i(t)$, $\omega(t)$, $a(t)$, $e(t)$, and $T(t)$.

For planets of the Solar system these elements change rather slowly, due to the weakness of the perturbations in the majority of cases. Therefore the method of variation of elements has a wide application for the bodies of the Solar system and is less efficient in the case of multiple stellar systems, where the perturbations are too large.

3   The Osculating Ellipse

Consider the motion of a celestial body m in an ellipse E_1 with elements Ω_1, i_1 ...T_1. In the absence of external perturbations such motion will be preserved unchanged, i.e. with unchanged elements, etc. The only force acting on the body m determining the elliptical orbit with these elements is obviously the gravity of the central body m_0.

Assume that an arbitrarily directed instantaneous force S_1 is acting on the body m at the point A_1, as shown in Fig. 11.1. As a result the body will move from the ellipse E_1 to another ellipse E_2 determined by quite different elements, Ω_2, i_2...T_2. With cessation of the action of the force, the further motion of the body m must take place in an ellipse with these new elements. Moreover, one can assume that a new action of similar force will move the body to a new ellipse, E_3 with elements Ω_3, i_3 ...T_3. This process can be considered to be repeated an infinite number of times. Then, within the intervals between the actions of the forces, the sequence of trajectories A_1A_2, A_2A_3, .., should be a sequence of arcs of ellipses determined by certain parameters. On the other hand, the path $A_1A_2A_3$, ..., is nothing but the real trajectory of motion of the body. Note, however that the focus containing the central body m_0 is common to all those ellipses.

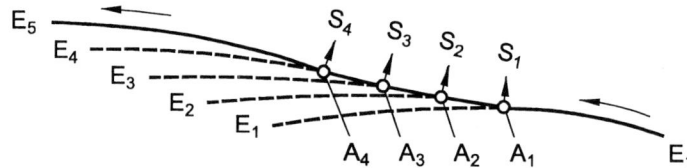

Fig. 11.1. *The formation of osculating ellipses A_1, A_2, A_3, A_4 ....*

Let the intervals between the instantaneous actions of the forces become ever smaller. Then the character of the motion just described, i.e. the transition from one ellipse to another, is such that the arcs of the sequence of ellipses become ever shorter. Obviously, in the limit one should have a continuously acting perturbation, and the real trajectory of motion turns to a conditional ellipse with continuously changing elements. The contribution of each of these pure ellipses should become ever smaller, reducing to a single point coming into contact with the real orbit. Thus, we arrive at a new and rather important formulation: the pure ellipses (conical sections) are coming into contact with the real orbit – osculating – at the contact points. These instantaneous ellipses smoothly coming into contact with the real orbit are called **osculating ellipses** (conical sections). Correspondingly, the elements of such an ellipse are called **osculating elements**.

Thus, we arrive at a new definition of the term **perturbation**, namely, the perturbation is the difference between the elements of the orbit in the initial stage of the motion and the elements of an osculating ellipse (conical section) at arbitrary moments of time.

At the point of contact – of osculation – of the real orbit and the osculating ellipse, two important conditions are satisfied:

i. The coordinates x, y, z of a body moving via a real orbit must coincide with the coordinates ξ, η, ζ of an imaginary point moving through the osculating ellipse at the moment of contact (osculation), i.e.

$$z = \xi, \quad y = \eta, \quad z = \zeta . \tag{A}$$

ii. At the point of contact the velocity vectors of the motion both of the imaginary point via an osculating ellipse and of the celestial body in its real orbit must coincide, i.e.

$$\frac{dx}{dt} = \frac{d\xi}{dt} , \quad \frac{dy}{dt} = \frac{d\eta}{dt} , \quad \frac{dz}{dt} = \frac{d\zeta}{dt} . \tag{B}$$

The conditions (A) and (B) are the starting point for the derivation of the differential equations determining the osculating elements as functions of time t, i.e. $\Omega(t)$, $i(t)$ ... $T(t)$.

4 The Expansion of the Perturbation Force

In the most general case the vector of the perturbation force forms an arbitrary angle with the orbital plane as well as with the vector of velocity of orbital motion. Therefore it is convenient to decompose the vector of the perturbation force into three mutually perpendicular components, in order to reveal the action of the perturbation force on various elements of the orbit.

The decomposition of the perturbation force can be realized in various ways. The so-called **orbital** components – the orthogonal Q, the tangential T and the normal N – are rather convenient parameters.

The **orthogonal** component, Q, is directed perpendicularly to the orbital plane, and is assumed to be positive when directed towards the north pole. The **tangential** component, T, is directed towards the tangent to the given point of contact in the orbit, and is assumed to be positive when it is directed along the direction motion. The **normal** component, N, is directed perpendicularly to the vector of velocity V (or T), and is assumed to be positive when is directed inside the orbit. Thus, two components among three,

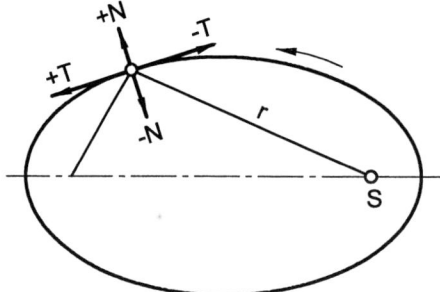

Fig. 11.2. *Tangential (−T and +T) and normal (−N and +N) components of perturbative forces acting on the planet P. The orthogonal component is directed perpendicularly to the plane of the figure.*

i.e. the tangential (T) and the normal (N), are located within the orbital plane (Fig. 11.2).

The introduction of orthogonal, tangential and normal components makes the role of the perturbation action on the various elements of the orbit rather easier to understand; some examples are discussed below. However, for the quantitative analysis of the action of these forces, these components should be projected on the corresponding axes. Thus, the **coordinate** components of the perturbation force, F_x, F_y, and F_z, appear. The latter are useful for derivation of the differential equations including the orbital components Q, T and N, the solution of which will give the desirable functions, i.e. $\Omega(t)$, $i(t)$ ... $T(t)$.

5   The Action of Perturbation Forces

Let us qualitatively consider several examples illustrating the influence of one or another component of the perturbation force on the orbital elements.

1.   The action of the orthogonal component. The orthogonal component Q, being directed perpendicularly to the orbital plane, cannot have any influence on all those elements of the orbit that lie in this plane. The vector of orbital motion V remains invariant with respect to Q. Recalling the dependence of V on the semimajor axis a,

$$V^2 = k^2 M \left(\frac{2}{r} - \frac{1}{a} \right) , \tag{1}$$

one can conclude that the semimajor axis a, as well as the eccentricity e, remain invariant.

The invariance of the vector V in magnitude and direction implies invariance of the position of the orbital plane, and hence, invariance of two more elements, the angular distance of the perihelion from the line of nodes ω, and the time of passage of the body through perihelion T. There are only two elements, the longitude of the line of nodes Ω and the inclination of the orbital plane to the ecliptic i, which are influenced by the orthogonal component.

For greater clarity, consider the motion of an artificial satellite around the Earth, taking into account the Earth's nonsphericity. This problem can be simplified by replacing the flattened Earth by a model of a uniform sphere with a massive surrounding belt or torus as shown in Fig. 11.3. Obviously, this massive belt will be a source of perturbation, particularly of the orthogonal component Q if a satellite is moving above or below it, i.e. on the Earth's equator. Consider now the manner in which the appearance of Q can affect the orientation of the orbital plane, i.e. the elements Ω and i.

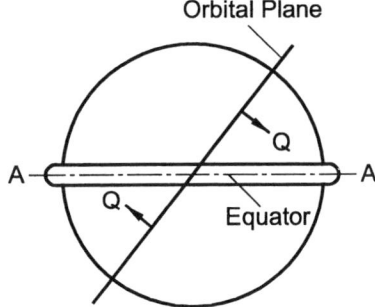

Fig. 11.3. *The problem of the origin of the precession of the orbital plane of an artificial satellite arising from the Earth's flattening.*

Turn to Fig. 11.4, in which AA is the equator, i.e. the mean position of the massive belt, and BB is the plane of the satellite's orbit, which is initially determined by the elements Ω_0 and i_0. Assume that, at the moment when the satellite appears at the point B, below the equator, it is influenced for a moment by a force Q, the gravitation caused by the massive belt. One can easily see that, as a result of such an action, the inclination of the orbital plane will change – the satellite's path will pass through the line BB_1, the orbital inclination will increase, $i_1 > i_0$, and the line of nodes will move to the left, from the point N_0 to the point N_1, i.e. $\Omega_1 > \Omega_0$. Assume that, at the point B_1, located above the equator, symmetrically with respect to B, there is a new instantaneously acting force Q, again perpendicular to the orbital plane, but in this case directed downwards, towards the belt. As a result, the declination of the orbital plane will change by the same magnitude, but now in the opposite direction, and we will have $i_2 = i_0$, so that the initial inclination of the orbital plane is recovered. Regarding the line of nodes, as before it moves to the left, from the point N_1 to the point N_2, insofar as $\Omega_2 < \Omega_1 < \Omega_0$.

Thus, under the action of the orthogonal component Q the inclination of the orbital plane remains invariant, but the line of nodes shifts in the direction opposite to that of the satellite's motion (the horizontal arrow

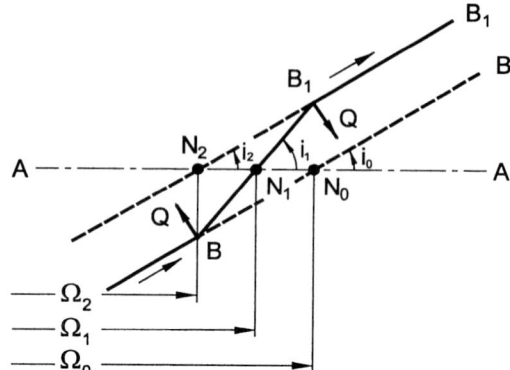

Fig. 11.4. *The influence of the orthogonal component of perturbative force on the orbit's elements Ω and i.*

in Fig. 11.4). This reversal of motion of the line of nodes is also called **precession of the orbital plane**. The speed of this motion is determined by the speed of precession $\dot{\Omega}$ and is measured in degrees or radians per day. The magnitude of $\dot{\Omega}$ depends on many parameters, particularly the declination of the orbital plane i, the height of the satellite above the Earth, etc. The precession vanishes when $i = 90°$, i.e. when the orbit is polar. For an orbit with $i \sim 50°$ the precession amounts to nearly $5°$ per day. Indeed, in the case of an ideal sphere without any additional mass in the equatorial zone, the component Q vanishes completely and so does the precession, and the orbital plane will remain absolutely invariant with respect to the stellar coordinate system.

2. The action of the tangential component. The impact of the action of the tangential component T of the perturbation force on the orbit's elements strongly depends on the sign of T; if it is positive then the velocity V will increase, if negative then V decreases. In both cases, however, the instantaneous direction of the motion remains invariant.

The variation in the velocity V first of all affects the magnitude of the semimajor axis a. As follows from Eq. (1)

$$a = \frac{1}{\dfrac{2}{r} - \dfrac{V^2}{k^2 M}}, \tag{2}$$

the increase in V leads to an increase in a and vice versa; a decrease in the orbital velocity is accompanied by a decrease in the orbit's size. The last circumstance has found practical applications, particularly for the landing of items of spacecraft on the Earth. In this case, the spacecraft is initially oriented in such a manner that the vector of the jet force, T, is directed opposite to the direction of V (Fig. 11.5). As a result, the instantaneous velocity of the spacecraft will decrease, and, according to Eq. (2) the orbit's semimajor axis a will also decrease, and the spacecraft will pass from the initial ellipse E_1 to a new, smaller ellipse E_2. The common focus of both ellipses coincides with the Earth's center O, where the total mass of the Earth is as if concentrated. The ellipse E_2 is smaller than the Earth's radius; it will inevitably intersect with the Earth's surface (the broken line in Fig. 11.5) at some point M. In the case of the absence of this surface, further motion of the spacecraft in the field of the imaginary Earth, centered at the point O, will proceed in the ellipse E_2, denoted by the broken line in Fig. 11.5. Obviously, the landing of real items of spacecraft on the Earth is realized in a somewhat different manner, in order to have zero orbital velocity at the landing point M, by virtue of a less steep orbit – in fact, a parabolic one, to take into account the effects within the Earth's atmosphere, the parachute facilities, etc.

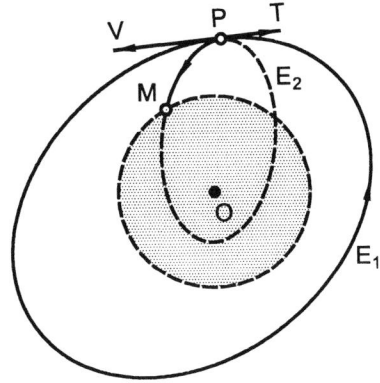

Fig. 11.5. *The transition of an artificial satellite P from an ellipse E_1 to ellipse E_2 as a result of the action of a tangential component of negative sign T leading to the landing of the satellite at the point M on the Earth's surface.*

Completely opposite are the consequences of the action of a tangential component of positive sign, i.e. when the vector of T coincides with the direction of V. In this case, the semimajor axis of the orbit increases, according to Eq. (2), to be as large as possible. In such a way there, is realized in particular the passage of the spacecraft from one ellipse E_1 to another, larger ellipse E_2, and in fact the escape of the spacecraft from the Earth (Fig. 11.6).

Regarding the eccentricity of the orbit, e, the situation is as follows. If $2a$ is the major axis and $OO_1 = 2c$ is the interfocal distance (Fig. 11.6), then the eccentricity of the orbit yields $e = 2c/(2a)$. For positive T, as has been mentioned above, $2a$ will increase. However, $2c$ also increases. In one particular case, namely, at perihelion, both $2a$ and $2c$ increase to one and the same magnitude. Insofar as $2c$ is smaller than $2a$, the relative growth of $2c$ will be greater, and as a result the eccentricity at perihelion will increase. When the body P is at the apogee, $2a$ will increase, whereas $2c$ will decrease by the same magnitude. As a result, the eccentricity at the apogee will

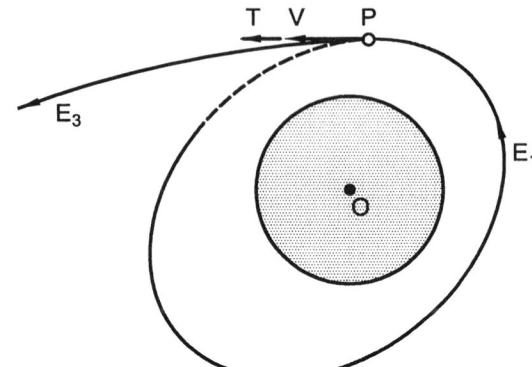

Fig. 11.6. *The transition of an artificial satellite P from an ellipse E_1 to a larger ellipse E_2 as a result of the action of a tangential component of positive sign T leading to the escape of the satellite from the Earth.*

decrease. Hence, a point should exist, or more precisely, two points between the perihelion and apogee, where the eccentricity remains invariant. It is not difficult to ensure that the eccentricity will be invariant at the ends of the small axis of the orbit. Thus, under the action of positive T, the eccentricity varies periodically, oscillating within definite limits.

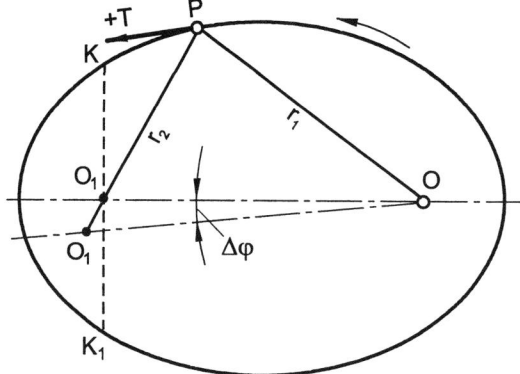

Fig. 11.7. *The influence of the tangential component T of the perturbative force on the line of apsides.*

The behavior of the line of **apsides**, i.e. of the line OO_1 connecting the two focuses of the ellipse (Fig. 11.7), under the action of the tangential component is also of particular interest. A tangential component of positive sign will increase $2a$. For the ellipse we have $2a = r_1 + r_2$, where $r_1 = OP$ is the distance of the body P from the first focus, where the central body is located, and $r_2 = O_1P$ is the distance of P from the second focus O_1 where no object is located. Under the action of $+T$, the position of the focus O of course does not change. r_1 also does not change and the instantaneous position of body P remains invariant. Therefore the increase in $2a$ by an amount $2\,\Delta a$ leads to an increase in r_2 by $2\,\Delta a$ because $r_2' = (2a + 2\,\Delta a) - r_1 = r_2 + 2\,\Delta a$. However, the direction of r_2 remains unchanged; therefore the focus O_1 will move to the point O'_1 and $O_1 O'_1 = 2 \, \Delta a$. As a result, the line of apsides OO_1 occupies a new position OO'_1 performing, in fact, a rotation around the focus O. It is not difficult to see that the picture will be a mirror reflection with respect to the axis OO_1 when the body moves to the lower half of the orbit. The largest remoteness of the line of apsides is reached when the body P is at the points K and K_1.

Thus, under the action of positive T the line of apsides oscillates within the limits of a small angle $\pm \Delta \varphi_0 = \Delta a / (ae)$ with respect to the focus where the central body is located and with a period equal to the period of rotation of the body P.

The behavior of the line of apsides also determines the behavior of the third element of the orbit – ω, the angular distance of the perihelion from the line of nodes. Under the action of the tangential component T, the position of the perihelion will influence cyclic shifts within the limits of the angle $\pm \varphi_0$. Correspondingly, the magnitude of ω will oscillate during a full rotation of the body P within the limits between $\omega + \Delta \varphi_0$ and $\omega - \Delta \varphi_0$.

The variation of the fourth element – the time T of the passage of body P through the perihelion, depends on the variations in the period of rotation P and the position of the major axis. The period P is defined by the well-known formula

$$ P = \frac{2\pi a^{3/2}}{k\sqrt{M}} \, , $$

and depends only on the major axis a which increases when T is positive. Therefore the moment of the passage through the perihelion happens later compared with that in the previous passage T_0. If we denote by P_0 the period in the absence of perturbation ($\Delta a = 0$), then the dependence of the moment of passage T on the magnitude of perturbation of Δa can be presented in the form

$$ T = T_0 + P_0 \left(1 + \frac{\Delta a}{a} \right)^{3/2} . $$

The next two elements – Ω and i, cannot be affected by the action of the tangential component T because the latter lies in the orbital plane.

3. The action of the normal component. The normal component N of perturbation, being perpendicular to the velocity vector, does not change the instantaneous velocity and therefore cannot change the position of the major axis.

However, the normal component can change the direction of the tangent to the given point of the orbit. Insofar as both radius-vectors r_1 and r_2 of both focuses form equal angles with the tangent, in spite of the invariance of r_1 with respect to magnitude and direction, the radius-vector r_2 must vary with respect to direction, remaining invariant with respect to magnitude (given that $r_1 + r_2 = 2a$ and a are invariants). As a result, the second focus O_2 will be displaced from its position and will occupy a position below OO_1). The situation is the same as in the case of the action of the tangential component (Fig. 11.7), i.e. the oscillation of the line of apsides with respect the focus O within the limits of the angle $\pm\psi_0$ depending on the magnitude of the normal component N. The oscillation of the line of apsides under the action of the normal component also leads to the oscillation of ω within the limits between $\omega + \Delta\psi_0$ and $\omega - \Delta\psi_0$.

Insofar as the major axis does not influence any variations, the behavior of the eccentricity will depend only on the behavior of the interfocal distance $2c$ since $e = 2c/(2a)$. However, at the invariant r_1 and r_2 the position of the second focus is changed, and, as a result, the interfocal distance will influence periodic changes both in the smaller and in the larger sides around $2c$, depending on the position of the body on the orbit. The instantaneous changes in the eccentricity vanish upon the appearance of the body at perihelion and at apogee. The other three elements, T, Ω and i, are absolutely not influenced by the normal component N.

Summarizing the results in Table 11.1, we represent the influence of the positive values of Q, T and N on all six elements of the orbit, Ω, i, ω, a, e, and T. The minus sign $(-)$ stands for a decrease, and the sign plus $(+)$ for an increase, in the oscillation of the given element within definite limits.

T a b l e   11.1

Summary of the influence of the components, normal N, tangential T and orthogonal Q, of the perturbative force on six elements of an elliptic orbit

Elements of orbit	Q	T	N
Ω	$-\Delta\Omega$	0	0
i	$\pm\Delta i$	0	0
ω	0	$\pm\Delta\omega$	$\pm\Delta\omega$
a	0	$+\Delta a$	0
e	0	$\pm\Delta e$	$\pm\Delta e$
T	0	$+\Delta P$	0

In a number of cases certain components of the perturbation forces do not affect any of the elements of the orbit. Note, however, that there always exists an element on which one of the components of the perturbation force does have an influence. The range of affected elements is also nonuniform: the largest number of elements are influenced by the tangential component (T) – four elements at once, ω, a, e, and T.

6 Precession of the Spacecraft's Orbital Plane

The orbital plane of a spacecraft tests a perturbation caused by the Earth's non-spherical form. Earth's flattening, i.e. the difference in equatorial and polar radii of 21 km, essentially influences the behavior of the orbital plane of a spacecraft; the qualitative consequences of this fact are examined in the previous section.

Under the action of the perturbation due to the Earth's flattening, the orbital plane of a spacecraft will test a motion opposite to that of the satellite's orbital motion. Due to the resulting precession, the line of nodes of the orbital plane shifts also in the direction opposite to the satellite's motion. Only in one case, when the satellite's orbital plane passes through both the Earth's poles, does precession vanish completely.

The speed of rotation of the orbital plane of the spacecraft is determined by the speed of the precession $\dot{\Omega}$, i.e. $\Delta\Omega = \dot{\Omega}$. In its turn. $\dot{\Omega}$ depends the on the declination of orbital plane φ, the altitude of the spacecraft above the Earth H and on the eccentricity e of the spacecraft's orbit. This gives the following relationship

$$\Delta\Omega = -0.00167 \left(\frac{R_\star}{a}\right)^2 \frac{nt}{(1-e^2)^2} \cos\varphi \; \text{degree day}^{-1} \qquad (3)$$

where $R_\star = 6380$ km is the Earth's equatorial radius, a and e are the semimajor axis and the eccentricity of the satellite's orbit, n is the angular velocity of the satellite through its orbit measured in degree day^{-1}. The minus sign means that the rotation of the orbital plane takes place opposite to the satellite's motion around the Earth.

The orbital period, i.e. the time taken for the satellite to circulate once around the Earth at altitude 300 km, is 1.5 hours. For given altitude H the duration of the orbital period is given by a relationship

$$P = 1.5 \left(\frac{R_\star + H}{R_\star + 300}\right)^{3/2} \text{hour}.$$

Then we have for the angular speed of the spacecraft

$$n = \frac{360°}{P} = 240 \left(\frac{R_\star + H}{R_\star + 300}\right)^{3/2} \text{degree hour}^{-1}. \qquad (4)$$

From (3) and (4) we find for the speed of precession

$$\Delta\Omega = -9.62 \left(\frac{R_\star}{R_\star + 300}\right)^2 \left(\frac{R_\star + 300}{R_\star + H}\right)^{3/2} \frac{\cos\varphi}{(1-e^2)^2} \text{degree day}^{-1}. \qquad (5)$$

For the polar orbit, i.e. when $\varphi = 90°$, we have from (5): $\Delta = 0$, i.e. the precession vanishes completely.

Many Russian spacecraft, for example, are launched from the site in Baykonur into orbits inclined by the angle $\varphi = 53°$ relative to the Earth's equator. Then we have from (5) for the speed of precession, i.e. for the rotation of the orbit's plane in the case of a circular orbit, i.e. when $e = 0$

$$\Delta \Omega = -5.79 \left(\frac{R_\star}{R_\star + H} \right)^2 \left(\frac{R_\star + 300}{R_\star + H} \right)^{3/2} \text{ degree day}^{-1}. \tag{6}$$

For typical orbits of altitude $H = 400$ km, we find from (6)

$$\Delta \Omega = -5.0 \text{ degree day}^{-1},$$

i.e. the satellite's orbital plane will smoothly rotate with a speed of five degrees per day in the direction opposite to the satellite's motion along its orbit. This is quite a high velocity, e.g. in ten days the orbital plane will be turned by 50°. This means that during periods when the orbit's plane is perpendicular to the direction of the Sun, the satellite will be constantly under solar rays.

As follows from Eq. (60), the largest magnitude for the speed of precession occurs at $\varphi = 0$, i.e. when the orbital plane is matched with the Earth's equator. In this case $\Omega = -9.62$ degree day^{-1}. Hence, by means of launching spacecraft into orbital planes under various inclinations φ one can regulate the speed of precession. In particular, we can select the inclination at which the precession will be compensated by daily rotation of the Earth around the Sun, i.e. by $360°/365.25 = 0.985$ degree day^{-1}. Such an orbit is called *Sun–synchronous*; in this case the Sun's rays will be constantly directed perpendicular to the orbit's plane of the spacecraft.

The latter property has been exploited by astrophysicists in space experiments when the stellar telescope has to be directed exactly to the opposite side from the Sun, along the line Earth–Sun. Putting therefore in (6) $\Omega = 0.985$ degree day^{-1}, we obtain for the angle of inclination for such an orbital plane φ_o again at $e = 0$ and $H = 500$ km: $\varphi_o = 97°.1$, i.e. the orbit's plane is inclined slightly relative to the Earth's polar axis but is inclined on the opposite side by $7°.1$. For example, the Dutch astrophysical satellite TD-1A was launched on such an orbit in 1972 to obtain ultraviolet spectra of bright stars. This satellite-observatory was constantly under the Sun's rays for eight months of its functioning in that orbit. The American spacecraft Sert-2, which was testing the ionic engines supplied by energy from a solar battery, was operated under the same regime in interplanetary conditions.

7 Instantaneous Elements

Up to now we have considered the qualitative aspects of the problem of the influence of the perturbation force on the elements of the orbit. Now we will deal with the quantitative treatment of this problem, i.e. with the determination of functions $\Omega(t)$, $i(t)$, ..., $T(t)$. This problem itself includes a number of preliminary steps which should be examined successively.

We start from the classical problem – the two-body problem, i.e. the motion of a body P of mass m and coordinates ξ, η, ζ with respect to the central body of mass m_0 under the action of their mutual gravity. Assume that the body m_0 is located at the center of the coordinate system (ξ, η, ζ). The equations of relative motion of the body P have the following form:

$$\frac{d^2\xi}{dt^2} + k^2 M \frac{\xi}{\rho^3} = 0 \ ,$$

$$\frac{d^2\eta}{dt^2} + k^2 M \frac{\eta}{\rho^3} = 0 \ , \tag{7}$$

$$\frac{d^2\zeta}{dt^2} + k^2 M \frac{\zeta}{\rho^3} = 0 \ ,$$

where $M = m_0 + m$ and ρ is the distance of the body P from the center of the coordinate system:

$$\rho = (\xi^2 + \eta^2 + \zeta^2)^{1/2} \ . \tag{8}$$

The motion determined by the system of equations (7) is called **nonperturbed** or **Keplerian**.

The general solution of Eqs. (7) is known. It is represented by six elements of the orbit and contains the time t, so that it can be rewritten in the form

$$\xi = f_1 \left(\Omega, \ i, \ \ldots, \ T, \ t \right) \ ,$$

$$\eta = f_2 \left(\Omega, \ i, \ \ldots, \ T, \ t \right) \ , \tag{9}$$

$$\zeta = f_3 \left(\Omega, \ i, \ \ldots, \ T, \ t \right) \ .$$

If the body P, in addition to being influenced by the gravity of the central body, is also under the action of external forces of various origin, with the sum of the components, mF_x, mF_y, and mF_z, then its position will no longer correspond to the previous coordinates ξ, η, ζ, but will be determined by new ones x, y, z for the same moment of time t. The equations of motion of P under the simultaneous actions of the gravity of the central body and of the external perturbations will take a form different from Eq. (7):

$$\frac{d^2x}{dt^2} + k^2 M \frac{x}{r^3} = F_x \,,$$

$$\frac{d^2y}{dt^2} + k^2 M \frac{y}{r^3} = F_y \,, \tag{10}$$

$$\frac{d^2z}{dt^2} + k^2 M \frac{z}{r^3} = F_z \,,$$

where

$$r = (x^2 + y^2 + z^2)^{1/2} \,. \tag{11}$$

The motion given by Eq. (10) is called **perturbed motion**. In the particular case in which the perturbation is caused by Newtonian gravity of n bodies, the functions F_x, F_y, and F_z in Eq. (10) should be replaced by the right-hand sides of the system (47) (Chapter VIII), which includes the perturbation function R_{ij}.

The solution of Eqs. (10) in the general case is impossible. However, one can slightly reformulate the problem, i.e. search for the solution of Eq. (10) in the same form as (9) but consider the orbital elements not as constants as before, but as somehow defined functions of time. The elements $\Omega(t)$, $i(t)$, ..., $T(t)$ obtained in such a way are called **instantaneous** in view of the fact that numerical values of these elements are suitable only for the given moment of time t. Correspondingly, the total set of these elements determines the instantaneous orbit of P for the given moment of time. Having the instantaneous elements, one can compute the coordinates of the body P for every moment of time with the help of the formulas derived for Keplerian motion.

8   Osculating Elements

The elements Ω, i, ..., T in the case of unperturbed motion are constants. Therefore we can write from Eq. (9) for the components of the velocity of unperturbed motion

$$\dot{\xi} = \frac{df_1}{dt} \,, \qquad \dot{\eta} = \frac{df_2}{dt} \,, \qquad \dot{\zeta} = \frac{df_3}{dt} \,. \tag{12}$$

If the instantaneous elements $\Omega(t)$, $i(t)$, ..., $T(t)$ also obey the additional conditions

$$\frac{\partial f_i}{\partial \Omega} \frac{d\Omega}{dt} + \frac{\partial f_i}{\partial i} \frac{di}{dt} + \cdots, + \frac{\partial f_i}{\partial T} \frac{dT}{dt} = 0 \,, \tag{13}$$

for $i = 1$, 2 and 3, then the same Eq. (12) can be used for the determination of the velocity of the perturbed motion. The instantaneous elements determined by Eqs. (9) and (10) are called **osculating**. More definitely they can be formulated as follows: osculating elements for a given moment of time t are the elements $\Omega(t)$ ... $T(t)$, which determine the position of the body and its velocity by means of the formulae derived for the unperturbed (Keplerian) motion.

To compare the real motion with the imaginary one via the osculating orbit, which has only one point of contact (osculation) with the real orbit, let us rewrite Eqs. (7) in another form:

$$\frac{d\dot{\xi}}{dt} + k^2 M \frac{\xi}{\rho^3} = 0 \ ,$$

$$\frac{d\dot{\eta}}{dt} + k^2 M \frac{\eta}{\rho^3} = 0 \ , \tag{14}$$

$$\frac{d\dot{\zeta}}{dt} + k^2 M \frac{\zeta}{\rho^3} = 0 \ .$$

Eqs. (10) can also be represented in the following form:

$$\frac{d\dot{x}}{dt} + k^2 M \frac{x}{r^3} = F_x \ ,$$

$$\frac{d\dot{y}}{dt} + k^2 M \frac{y}{r^3} = F_y \ , \tag{15}$$

$$\frac{d\dot{z}}{dt} + k^2 M \frac{z}{r^3} = F_z \ .$$

However, we already know that, at the contacting points (osculations) of the real orbit with the osculating ellipse, two important conditions (1) and (2) are satisfied, i.e. the invariance of the coordinates and components of the velocity:

$$x = \xi \qquad y = \eta \qquad z = \zeta \ ,$$
$$\dot{x} = \dot{\xi} \qquad \dot{y} = \dot{\eta} \qquad \dot{z} = \dot{\zeta} \tag{16}$$

respectively, as well as $\rho = r$ according to Eqs. (8) and (11). Then, from

Eqs. (14) and (15), taking into account also Eq. (16), we obtain

$$\frac{d\dot{x}}{dt} - \frac{d\dot{\xi}}{dt} = F_x \; ,$$

$$\frac{d\dot{y}}{dt} - \frac{d\dot{\eta}}{dt} = F_y \; , \qquad (17)$$

$$\frac{d\dot{z}}{dt} - \frac{d\dot{\zeta}}{dt} = F_z \; ,$$

These equations have clear physical content: the difference in accelerations between the perturbed and unperturbed motions is equal to the acceleration corresponding to the action of the external perturbation forces on the body P.

9   The Principal Operation

Consider the function Ψ including in general all components of the perturbation motion as a function of time t, i.e. six elements of the orbit Ω, i, $\ldots$, T, three coordinates x, y, z, three components of velocity $\dot{x}$, $\dot{y}$, and $\dot{z}$, and time t:

$$\psi\left(\Omega, \; i, \; \ldots, \; T, \; x, \; y, \; z, \; \dot{x}, \; \dot{y}, \; \dot{z}, \; t\right) = 0 \; . \qquad (18)$$

On differentiating this function with respect to time t, we obtain

$$\frac{\partial\psi}{\partial\Omega}\frac{d\Omega}{dt} + \frac{\partial\psi}{\partial i}\frac{di}{dt} + \ldots + \frac{\partial\psi}{\partial x}\frac{dx}{dt} + \ldots$$
$$+ \frac{\partial\psi}{\partial\dot{x}}\frac{d\dot{x}}{dt} + \ldots + \frac{\partial\psi}{\partial t} = 0 \; . \qquad (19)$$

Consider now another function Ψ of the components of nonperturbed motion, again in terms of its dependence on time. Of course, the six elements of the orbit are absent now, insofar as in the case of unperturbed motion they are constants:

$$\psi\left(\xi, \; \eta, \; \zeta, \; \dot{\xi}, \; \dot{\eta}, \; \dot{\zeta}, \; t\right) = 0 \; . \qquad (20)$$

On differentiating this function, we have

$$\frac{\partial\psi}{\partial\xi}\frac{d\xi}{dt} + \ldots + \frac{\partial\psi}{\partial\dot{\xi}}\frac{d\dot{\xi}}{dt} + \ldots + \frac{\partial\psi}{\partial t} = 0 \; . \qquad (21)$$

However, at the moment of the tangential contact (osculation) of the real orbit with the osculating ellipse, Eq. (16) must be satisfied. Therefore, by subtracting Eq. (21) from (19), we obtain

$$
\frac{\partial \psi}{\partial \Omega} \frac{d\Omega}{dt} + \frac{\partial \psi}{\partial i} \frac{di}{dt} + \cdots + \frac{\partial \psi}{\partial T} \frac{dT}{dt} + \frac{\partial \psi}{\partial \dot{x}} \left(\frac{d\dot{x}}{dt} - \frac{d\dot{\xi}}{dt} \right)
$$
$$
+ \frac{\partial \psi}{\partial \dot{y}} \left(\frac{d\dot{y}}{dt} - \frac{d\dot{\eta}}{dt} \right) - \frac{\partial \psi}{\partial \dot{z}} \left(\frac{d\dot{z}}{dt} - \frac{d\dot{\zeta}}{dt} \right) = 0
$$

(22)

or, substituting the values of expressions into the parentheses from Eq. (17), we obtain finally

$$
\frac{\partial \psi}{\partial \Omega} \frac{d\Omega}{dt} + \frac{\partial \psi}{\partial i} \frac{di}{dt} + \cdots + \frac{\partial \psi}{\partial T} \frac{dT}{dt} + \frac{\partial \psi}{\partial \dot{x}} F_x
$$
$$
+ \frac{\partial \psi}{\partial \dot{y}} F_y + \frac{\partial \psi}{\partial \dot{z}} F_z = 0 \ .
$$

(23)

The relationship (23) is called **principal operation**. It gives the dependence between the derivatives of the osculating elements $\Omega(t)$, $i(t)$, ..., $T(t)$ and the components of the perturbation acceleration F_x, F_y, and F_z.

The principal operation, on being applied to six independent relationships of type (18), gives us six differential equations for the osculating elements $\Omega(t)$, $i(t)$, ..., $T(t)$. The solution of those equations will give the sought functions, i.e. the dependence of all six elements of the orbit on time.

10   Euler's Equations

Here our problem will be the derivation of differential equations determining the dependence between the osculating elements, their derivatives with respect to time, and the components of the perturbing acceleration. Obviously, six equations are needed for all six elements. As starting equations we shall use the well-known relationships derived for the case of Keplerian motion that relate various elements to each other and to coordinates, components of the velocity and other parameters. For each such relationship the principal operation (23) will be applied.

1. **The parameter of the ellipse (p), longitude of nodes (Ω), and inclination of orbit (i).** As initial relationships we take the three integrals of areas for the two-body problem, i.e. Eq. (18) from Chapter III:

$$
k\sqrt{p} \, \sin i \, \sin \Omega = y\dot{z} - z\dot{y} \ ,
$$

$$
k\sqrt{p} \, \sin i \, \cos \Omega = x\dot{z} - z\dot{x} \ ,
$$

$$
k\sqrt{p} \, \cos i = x\dot{y} - y\dot{x} \ ,
$$

where $M = 1$.

The application of the principal operation (23) to these equations gives

$$\frac{1}{2p}\sin i \sin \Omega \frac{dp}{dt} + \sin i \cos \Omega \frac{d\Omega}{dt} + \cos i \sin \Omega \frac{di}{dt} = yF'_z - zF'_y \ ,$$

$$\frac{1}{2p}\sin i \cos \Omega \frac{dp}{dt} - \sin i \sin \Omega \frac{d\Omega}{dt} + \cos i \cos \Omega \frac{di}{dt} = xF'_z - zF'_x \ , \qquad (24)$$

$$\frac{1}{2p}\cos i \frac{dp}{dt} - \sin i \frac{di}{dt} = xF'_y - yF'_x \ ,$$

where

$$F'_x = \frac{1}{k\sqrt{p}}F_x \ , \quad F'_y = \frac{1}{k\sqrt{p}}F_y \ , \quad F'_z = \frac{1}{k\sqrt{p}}F_z \ . \qquad (25)$$

On the right-hand side of Eq. (24) we have x, y, and z, which can be replaced by the known expressions

$$x = r \left(\cos \Omega \cos u - \sin \Omega \sin u \cos i\right) \ ,$$

$$y = r \left(\sin \Omega \cos u + \cos \Omega \sin u \cos i\right) \ , \qquad (26)$$

$$z = r \sin u \sin i \ ,$$

where $u = v + \omega$ is the so-called argument of the pericenter.

On solving the system (24) with respect to dp/dt, $d\Omega/dt$ and di/dt, we obtain

$$\frac{dp}{dt} = 2prT' \ , \qquad (27)$$

$$\sin i \frac{d\Omega}{dt} = r \sin u W' \ , \qquad (28)$$

$$\frac{di}{dt} = r \cos u W' \ , \qquad (29)$$

where

$$S' = F'_x \left(\cos u \cos \Omega - \sin u \sin \Omega \cos i\right)$$

$$+F'_y \left(\cos u \sin \Omega + \sin u \cos \Omega \cos i\right) + F'_z \sin u \sin i \ ,$$

$$T' = F'_x(-\sin u \cos \Omega - \cos u \sin \Omega \cos i)$$

$$+F'_y(-\sin u \sin \Omega + \cos u \cos\Omega \cos i) + F'_z \cos u \sin i \ ,$$

$$W' = F'_x \sin \Omega \sin i - F'_y \cos \Omega \sin i + F'_z \cos i \ .$$

In these expressions another form of expansion of the perturbation force in terms of the components is used, i.e. in terms of the radius-vector for S, the orthogonal to the radius-vector in the orbital plane T, and the normal to the orbital plane for W, where

$$S' = \frac{1}{k\sqrt{p}} S \ , \quad T' = \frac{1}{k\sqrt{p}} T \ , \quad W' = \frac{1}{k\sqrt{p}} W \ . \tag{30}$$

2. The major semi-axis (a) and eccentricity (e). We have from the integral of energy

$$V^2 = k^2 M \left(\frac{2}{r} - \frac{1}{a} \right)$$

or, taking $M = 1$ and writing it otherwise:

$$k^2 \left(\frac{2}{r} - \frac{1}{a} \right) = \dot{x}^2 + \dot{y}^2 + \dot{z}^2 \ . \tag{31}$$

By applying the principal operation to this relationship, we obtain

$$k^2 \bar{a}^2 \frac{da}{dt} = 2\dot{x} F_x + 2\dot{y} F_y + 2\dot{z} F_z \ . \tag{32}$$

Differentiating Eq. (26) taking into account the properties of osculating elements and substituting the obtained values of $\dot{x}$, $\dot{y}$, and $\dot{z}$ into Eq. (31), we obtain

$$\frac{da}{dt} = 2a^2 e \sin v S' + 2a^2 \frac{p}{r} T' \ . \tag{33}$$

On differentiating the known expression for the parameter p of the ellipse

$$p = a \left(1 - e^2 \right)$$

we obtain

$$2ae \frac{de}{dt} = (1 - e^2) \frac{da}{dt} - \frac{dp}{dt} \ .$$

By substituting here da/dt from Eq. (33) and using the relationships

$$r = \frac{p}{1 + e \cos v} \ , \qquad r = a \left(1 - e \cos E \right) \ , \tag{34}$$

we obtain for the derivative of the eccentricity

$$\frac{de}{dt} = p \sin(v S') + p \left(\cos v + \cos E \right) T' \ . \tag{35}$$

3. The argument of the pericenter (ω). On multiplying the first of Eqs. (26) by $\cos \Omega$ and the second equation by $\sin \Omega$ and summing we obtain

$$r \cos u = x \cos \Omega + y \sin \Omega \ , \tag{36}$$

where $u = v + \omega$.

We apply the principal operation (36), taking into account the dependence of the true anomaly v not only on x, y, and z but also on $\dot{x}$, $\dot{y}$, and $\dot{z}$ and on the osculating elements and on time. Therefore, after the determination of dv/dt first via the osculating elements, and then from Eq. (36) we obtain

$$(-x \sin \Omega + y \cos \Omega) \frac{d\Omega}{dt} = -r \sin u \left[\frac{d\omega}{dt} + \left(\frac{dv}{dt} \right) \right] , \qquad (37)$$

or, putting in the values of x and y, we obtain

$$\frac{d\omega}{dt} = - \left(\frac{dv}{dt} \right) - \cos i \, \frac{d\Omega}{dt} . \qquad (38)$$

We also have to find dv/dt. We have

$$r^2 = x^2 + y^2 + z^2 ,$$

whence

$$r\dot{r} = x\dot{x} + y\dot{y} + z\dot{z} . \qquad (39)$$

We have from Eq. (34)

$$\dot{r} = \frac{k}{\sqrt{p}} e \sin v$$

or, in slightly modified form

$$\dot{r} \cot v = \frac{k}{\sqrt{p}} e \cos v .$$

On substituting here $\dot{r}$ from (39) and equating to Eq. (34), we find

$$(x\dot{x} + y\dot{y} + z\dot{z}) \cot v = k\sqrt{p} - \frac{k}{\sqrt{p}} r . \qquad (40)$$

After applying our principal operation to this relationship we obtain

$$S \cot v - \frac{\dot{r}}{\sin^2 v} \left(\frac{dv}{dt} \right) = \frac{k}{2r\sqrt{p}} \left(1 + \frac{r}{p} \right) \frac{dp}{dt} , \qquad (41)$$

where dv/dt should be obtained. Substituting into Eq. (41) dp/dt from Eq. (27) we obtain dv/dt after which, from Eq. (38), the final expression for the derivative of the argument of the pericenter ω will be

$$e \frac{d\omega}{dt} = -p \cos v S' + (r + p) \sin v T' - e \cos i \, \frac{d\Omega}{dt} . \qquad (42)$$

4. The mean anomaly (M_0). The mean anomaly at the epoch M_0 is associated with the sixth and last element of the orbit T, the moment of transit of the celestial body through the perihelion, which has the following form:

$$M = n(t - T) = M_0 + n(t - t_0) , \tag{43}$$

where

$$M_0 = n(t_0 - T) \tag{44}$$

is the mean anomaly for the moment t_0, or the **mean anomaly for the epoch** t_0. Therefore, determining the derivative for M_0 with respect to time, i.e. dM_0/dt, we have dT/dt. On differentiating Eq. (43), we obtain

$$\frac{dM}{dt} = \frac{dM_0}{dt} + (t - t_0)\frac{dn}{dt} + n , \tag{45}$$

where n is the mean daily motion and is given by

$$n = \frac{k}{a^{3/2}} , \tag{46}$$

i.e. n depends on the semimajor axis a.

Differentiating Eq. (43) only with respect to the osculating elements, we have

$$\left(\frac{dM}{dt}\right) = \frac{dM_0}{dt} + (t - t_0)\frac{dn}{dt} \tag{47}$$

Both anomalies – eccentric (E) and mean (M) – are dependent on time directly as well as through the osculating elements. Denoting by dE/dt and dM/dt the derivatives taken with respect to the osculating elements, we can obtain from the relationships

$$M = E - e\sin E ,$$
$$r = a(1 - e\cos E)$$

for the sought derivatives

$$\left(\frac{dM}{dt}\right) = (1 - e\cos E)\left(\frac{dE}{dt}\right) - \sin E\,\frac{de}{dt} ,$$
$$\frac{r}{a}\frac{da}{dt} - a\cos E\,\frac{de}{dt} + ae\sin E\left(\frac{dE}{dt}\right) = 0 .$$

By excluding from these equations first dE/dt, we can express dM/dt through da/dt and de/dt. Substituting the values of these derivatives from

Eqs. (33) and (35), we obtain for dM/dt:

$$\frac{e}{(1-e^2)^{1/2}} \left(\frac{dM}{dt} \right) = (p \cos v - 2er) S'$$
$$+ \frac{p}{\sin v} \left(\cos^2 v + \cos v \cos E - 2 \right) T'$$

or, after some transformations,

$$\left(\frac{dM}{dt} \right) = \frac{(1-e^2)^{1/2}}{e} \left[(p \cos v - 2er) S' - (r + p) \sin v T' \right], \qquad (48)$$

which in combination with Eq. (47) enables one to find dM_0/dt, i.e.

$$\frac{dM_0}{dt} = \left(\frac{dM}{dt} \right) - (t - t_0) \frac{dn}{dt} . \qquad (49)$$

One has also to find the explicit form of dn/dt. We have from Eq. (46)

$$\frac{dn}{dt} = -\frac{3}{2} \frac{k}{a^{5/2}} \frac{da}{dt} = -\frac{3}{2} \frac{n}{a} \frac{da}{dt}$$

or, substituting da/dt from Eq. (33), we obtain

$$\frac{dn}{dt} = -3nae \sin v S' - 3na \frac{p}{r} T' , \qquad (50)$$

which completes the determinations of all necessary combinations and derivatives.

Thus, for estimation of the six osculating elements, we have the following set of differential equations:

$$\frac{da}{dt} = 2a^2 [e \sin v S' + pr^{-1} T'] ,$$

$$\frac{de}{dt} = p \sin v S' + p (\cos v + \cos E) T' ,$$

$$\frac{di}{dt} = r \cos u W' ,$$

$$\frac{d\Omega}{dt} = r \sin u \operatorname{cosec} i W' , \qquad (51)$$

$$\frac{d\omega}{dt} = \frac{1}{e} [-p \cos v S' + (r + p) \sin v T'] - \cos i \frac{d\Omega}{dt} ,$$

$$\frac{dM}{dt} = \frac{(1-e^2)^{1/2}}{e} [(p \cos v - 2er) S' - (r + p) \sin v T'] .$$

These equations were discovered by Euler in 1753, and are called **Euler's Equations**. Euler's equations establish the dependence between the osculating elements, their derivatives and the components of the perturbation accelerations S, T, and W. Their joint solution must give the sought functions $m(t), e(t), \ldots, T(t)$, i.e. the dependence of the osculating elements $a(t), e(t), \ldots, T(t)$ on time.

Then, one has to determine for a given moment of time the numerical values of all six elements a, e, $\ldots$, T, and thereafter, to obtain the coordinates x, y, z, which is the final aim, with the help of the usual Keplerian formulas from the classical two-body problem. Euler equations are universal, in that no limitations are posed and no conditions concerning the nature or the properties of the perturbation acceleration were used in their derivation.

Euler wrote his equations in order to explain the motion of planets. Nowadays these equations are "governing" the flights of artificial satellites.

11 The Lagrange Equations

The first important modification of Euler's equations was realized by Lagrange. It concerns the most important case from the point of view of astronomical applications, namely the case of the perturbing acceleration caused by potential forces (fields). Newtonian gravity, for example, belongs to this category of interaction. If the acting perturbation on a given object is caused by n potential forces, then in fact one has an n-body problem. The traditional means of solution of this problem, i.e. derivation and solution of differential equations of motion, as was shown above, is not efficient. However, this problem can be considered within perturbation theory.

Recall the basic property of the potential function: its derivatives with respect to coordinates are components of the force, i.e.

$$F_x = \frac{\partial R}{\partial x} , \qquad F_y = \frac{\partial R}{\partial y} , \qquad F_z = \frac{\partial R}{\partial z} , \qquad (52)$$

where R is the potential function. We will call the latter **the perturbation function**, generalizing, therefore, the previous term.

Lagrange formulated the aim of transforming the Euler equations so that, instead of the components of perturbation accelerations S, T, and W, one can use the partial derivatives of the potential function R with respect to the orbital elements. Consider the partial derivative of R, say, with respect to element a:

$$\frac{\partial R}{\partial a} = \frac{\partial R}{\partial x} \frac{\partial x}{\partial a} + \frac{\partial R}{\partial y} \frac{\partial y}{\partial a} + \frac{\partial R}{\partial z} \frac{\partial z}{\partial a} , \qquad (53)$$

or, taking into account Eqs. (25) and (52), we shall have

$$\frac{1}{k\sqrt{p}} \frac{\partial R}{\partial a} = F'_x \frac{\partial x}{\partial a} + F'_y \frac{\partial y}{\partial a} + F'_z \frac{\partial z}{\partial a} \ , \tag{54}$$

where the transition from F'_x, F'_y, F'_z to S', T', W' is realized with the help of Eq. (29a).

The derivatives of coordinates x, y, z, with respect to element a, can be obtained with the help of Eq. (26). Combining all these manipulations within Eq. (54) we obtain the sought expression for $\partial R/\partial a$. Analogously, we find the derivatives for the potential function R with respect to the next five elements. As a result we obtain the following set of equations:

$$\frac{\partial R}{\partial a} = k\sqrt{p}\,a^{-1}rS' \ ,$$

$$\frac{\partial R}{\partial e} = \frac{ka}{\sqrt{p}}[-p \cos v S' + (1+p) \sin v T'] \ ,$$

$$\frac{\partial R}{\partial i} = k\sqrt{pr} \sin u W' \ ,$$

$$\frac{\partial R}{\partial \Omega} = k\sqrt{p}\,[r \cos i T' - r \sin i \cos u W'] \ , \tag{55}$$

$$\frac{\partial R}{\partial \omega} = k\sqrt{p}\,rT' \ ,$$

$$\frac{\partial R}{\partial M_0} = ka^{3/2}[e \sin v S' + pr^{-1}T'] \ .$$

It is necessary to solve this set jointly with the Euler equations so that the magnitudes S', T', and W' will be excluded completely, and then, as a result, the derivatives of the orbital elements would be represented through the derivatives of the perturbation function R with respect to that element. Then we have finally

$$\frac{da}{dt} = \frac{2}{na} \frac{\partial R}{\partial M_0}$$

$$\frac{de}{dt} = \frac{1-e^2}{na^2} \frac{\partial R}{\partial M_0} - \frac{(1-e^2)^{1/2}}{na^2} \frac{\partial R}{\partial \omega} \ ,$$

$$\frac{di}{dt} = \frac{\cot i}{na^2(1-e^2)^{1/2}} \frac{\partial R}{\partial \Omega} - \frac{\mathrm{cosec}\,i}{na^2(1-e^2)^{1/2}} \frac{\partial R}{\partial \Omega} \ ,$$

$$\frac{d\Omega}{dt} = \frac{\mathrm{cosec}\,i}{na^2(1-e^2)^{1/2}} \frac{\partial R}{\partial i} \ , \tag{56}$$

$$\frac{d\omega}{dt} = \frac{(1-e^2)^{1/2}}{ena^2} \frac{\partial R}{\partial e} - \frac{\cot i}{na^2(1-e^2)^{1/2}} \frac{\partial R}{\partial i} ,$$

$$\frac{dM_0}{dt} = -\frac{2}{na} \frac{\partial R}{\partial a} - \frac{1-e^2}{ena^2} \frac{\partial R}{\partial e} .$$

These equations were discovered by Lagrange in 1778, twenty-five years after Euler; they are called **the Lagrange equations**.

Note two properties of Lagrange equations.

a. The time is present in these equations only through the derivatives of the perturbation function R.

b. The orbital elements are divided into two groups, one including a, e, and i, and the other including ω, Ω, and M_0. The differential equations of the elements for each group contain the partial derivatives of R only with respect to the elements of the other group.

12   Solution of the Lagrange Equations

The Lagrange equations are too complicated even to think about their explicit solution. The only way out is the application of the method of successive approximations. Moreover, it is justified by the fact that, under the real conditions of motion of planets, comets and other celestial bodies, the perturbations are really small.

The essence of the application of the method of successive approximations is as follows. Let us start from the perturbation function R_{ij}, which, as is known, takes into account the interaction of all planets, i.e. the sources of the perturbation. For simplicity, we confine our problem to consideration of a system consisting of only two planets of masses m_1 and m_2, respectively, besides the central body – the Sun. Then the corresponding perturbation functions will be written in the form

$$R_1 = k^2 m_2 \left(\frac{1}{r_{12}} - \frac{x_1 x_2 + y_1 y_2 + z_1 z_2}{r_2^3} \right) ,$$

$$R_2 = k^2 m_1 \left(\frac{1}{r_{12}} - \frac{x_1 x_2 + y_1 y_2 + z_1 z_2}{r_1^3} \right) ,$$

$$(57)$$

where r_{12} is the distance between the planets m_1 and m_2:

$$r_{12} = [(x_1 - x_2)^2 + (y_1 - y_2)^2 + (z_1 - z_2)^2]^{1/2} , \qquad (58)$$

and r_1 and r_2 are the distances of the planets m_1 and m_2 from the Sun, respectively:

$$r_1 = (x_1^2 + y_1^2 + z_1^2)^{1/2} ,$$

$$r_2 = (x_2^2 + y_2^2 + z_2^2)^{1/2} ,$$

$$(59)$$

where x, y, and z are given by Eq. (26).

Our problem is the determination of the osculating elements for the planet m_1. To a first approximation, while solving Eqs. (56), one can assume that $m_0 = 0$. Then, $R_1 = 0$ and the right-hand sides of all six equation in (56) become zero:

$$a = a_0 , e = e_0, \ldots T = T_0 .$$

On substituting these elements into the expression for the perturbation function (57), and into the right-hand side of Eq. (56), we have

$$\frac{da}{dt} = f_1 (a_0, \ e_0, \ \ldots, \ T_0, \ t) ,$$

$$\frac{de}{dt} = f_2 (a_0, \ e_0, \ \ldots, \ T_0, \ t) ,$$

etc.

After the integration of these equations, we find

$$a(t) = a_0 + \int_0^t f_1 (a_0, \ e_0, \ \ldots, \ T_0, \ t) \, dt ,$$

$$e(t) = e_0 + \int_0^t f_2 (a_0, \ e_0, \ \ldots, \ T_0, \ t) \, dt ,$$

(60)

etc.

On integrating numerically the right-hand side of Eq. (60), we obtain

$$a(t) = a_0 + \delta_1 a(t) ,$$

$$e(t) = e_0 + \delta_1 e(t) ,$$

etc. The values $\delta_1 a(t)$, $\delta_1 e(t)$, ... etc. are the first-order perturbations.

With this first set of approximations of elements Eq. (60) we return again to the right-hand sides of Eqs. (56) and (57). Here, from the general terms on the right-hand side of Eq. (56), one can easily select the first approximation terms, i.e. the functions f_1, f_2, ... Then we have

$$\frac{da}{dt} = f_1 (a_0, \ e_0, \ \ldots, \ T_0, \ t) + \varphi_1 (a_0, \ e_0, \ \ldots, \ T_0, \ t) ,$$

$$\frac{de}{dt} = f_2 (a_0, \ e_0, \ \ldots, \ T_0, \ t) + \varphi_2 (a_0, \ e_0, \ \ldots, \ T_0, \ t) ,$$

etc.

After the integration, we obtain

$$a(t) = a_0 + \delta_1 a(t) + \delta_2 a(t) ,$$

$$e(t) = e_0 + \delta_1 e(t) + \delta_2 e(t) ,$$

etc., i.e. the second-order perturbations $\delta_2 a(t)$, $\delta_2 e(t)$, ... are obtained. Obviously, this cycle can be repeated an arbitrary number of times, and as a result, one can obtain the osculating elements with the required accuracy:

$$
\begin{aligned}
a(t) &= a_0 + \delta_1 a(t) + \delta_2 a(t) + \ldots + \delta_n a(t) \ , \\
e(t) &= e_0 + \delta_1 e(t) + \delta_2 e(t) + \ldots + \delta_n e(t) \ ,
\end{aligned}
\tag{61}
$$

etc.

13 Determination of the Motion in a Real Orbit

After obtaining the osculating elements from the Lagrange equations, $a(t)$, $i(t)$, ..., $T(t)$, the estimation of the coordinates x, y, z of the body P is performed with the help of the usual formulas of Keplerian motion (the two-body problem) in the following way.

1. For the given moment of time t we obtain from Eq. (61) the elements a, i, ..., T.

2. The mean anomaly M is obtained for the given moment of time t

$$
M = n(t - T) \ ,
$$

where

$$
n = \frac{k}{a^{3/2}} \ .
$$

3. The eccentric anomaly E is obtained by means of the Kepler equation

$$
E - e \sin E = M \ .
$$

4. The radius-vector r and real anomaly v are obtained from the formulas

$$
r = a(1 - e \cos E) \ ,
$$

$$
\tan \frac{v}{2} = \left(\frac{1+e}{1-e} \right)^{1/2} \tan \frac{E}{2} \ .
$$

5. The Cartesian heliocentric ecliptic coordinates x, y, z are given by the formulas

$$
\begin{aligned}
x &= r \left[\cos(v + \omega) \cos \Omega - \sin(v + \omega) \sin \Omega \cos i \right] \ , \\
y &= r \left[\cos(v + \omega) \sin \Omega + \sin(v + \omega) \cos \Omega \cos i \right] \ , \\
z &= r \sin(v + \omega) \sin i \ .
\end{aligned}
$$

6. The Cartesian heliocentric equatorial coordinates x', y', z' are obtained from Eq. (7) of Chapter V.

7. The geocentric polar coordinates, i.e. the right ascension α, the declination δ, and the distance of the body P from the Earth ρ are obtained from the formulas

$$\tan \alpha = \frac{y' + Y_\odot}{z' + Z_\odot} \, ,$$

$$\tan \delta = \frac{z' + Z_\odot}{x' + X_\odot} \cos \alpha \, ,$$

$$\rho = \frac{z' + Z_\odot}{\sin \delta} \, ,$$

where $X_\odot$, $Y_\odot$, $Z_\odot$ are the coordinates of the Sun at the given moment of time t. Thus, the solution of the problem of the determination of the coordinates of a celestial body P for a given moment of time t is completed.

Along with the analytic methods of integration of the equations of the motions in celestial mechanics that we are discussing, obviously, methods of numerical integration of these equations have been elaborated as well. The analytical methods of integration in celestial mechanics are especially efficient when the problem is to determine the behavior of orbits for a large number of orbitings, for example, of spacecraft or orbital stations around the Earth.

However, when the problem is the determination of the orbits of only a few orbitings around any celestial body, as well for flights to the planets of the Solar system, the numerical methods of integration are most powerful. Numerical methods are also rather convenient for determination of the orbits of comets, small planets and asteroids. The efficiency of the latter method is enhanced sharply in the epoch of computer techniques and currently the matter is about the reasonable combination of both methods, theoretical and numerical, in each particular problem.

14   Classification of Perturbations

Perturbations of arbitrary order, i.e. expressions of the form $\delta_i a(t)$, $\delta_i e(t)$, ..., include the time t. Hence, the time-dependent character of the functions $\delta_i a(t)$, $\delta_i e(t)$, ... is determined by the generalized form of dependence of all orders of perturbations δ_i on time. The latter is itself determined by the dependence of the acceleration of the perturbation force, i.e. of the perturbation function on time.

Below, in order to realize the integration in the formulas of the type (60) for the higher-order perturbations it is necessary to represent the perturbation function explicitly with respect to time. This demand may be realized as a rule by expansion of the perturbation function in infinite series. The problem of finding the proper expansion of the perturbation function is an essential branch of celestial mechanics. Note that, since the position of a planet is a periodic function of the mean anomaly M with a period 2π, the perturbation functions should also be periodic functions. These functions, after integration of Eq. (60), obviously should coincide with the expressions of perturbations of various orders δ_i. As a result, in general, in the expressions of perturbations δ_i, one will have trigonometric functions of the argument of time t.

However, the perturbations can also contain nonperiodic functions proportional, say, to t^p, where p can be positive, negative or even zero. Thus, in general, the perturbation function depends on a number of quantities. It can be shown that a perturbation of arbitrary order can be presented in the form

$$\delta_i = At^p \cos(\kappa t + \beta) , \qquad (62)$$

where A is some coefficient determining the magnitude of perturbation in the case in which $p = 0$, $\kappa = 0$ and $\beta = 0$. Eq. (62) is called the **perturbation** or **inequality** of the corresponding element.

Several types of perturbations are distinguished. Thus, if $p = 0$, we have

$$\delta_i = A \cos(\kappa t + \beta) ; \qquad (63)$$

such perturbation is called **periodic**, where κ is the frequency and $2\pi/\kappa$ is the period of the perturbation. Depending on the absolute magnitude of $2\pi/\kappa$, the perturbations can be **long-periodic** or **short-periodic** ones; in the first case the period $2\pi/\kappa$ is much larger than the period $2\pi/n$ for the given planet, in the second case the two periods are of the same order.

When $p > 1$ and $\kappa = 0$, then

$$\delta_i = A't^p , \qquad (64)$$

and we will have a **secular** perturbation, and, if $p > 1$ and $\kappa \neq 0$, then the perturbation will be **mixed**. In principle, it is possible to have the frequency k very close to zero for periodic perturbations, and then we will have a perturbation with an extremely large period. Such extra-long perturbations can hardly be distinguished from secular perturbation.

15   The Remarkable History of the Shoemaker–Levy 9 Comet

The appearance and disruption of the comet Shoemaker–Levy 9 is a remarkable example of the action of perturbation forces. This event gave astronomers an unprecedented opportunity to learn how cometary nuclei can split and evolve. It marks the first detected event of a cometary breakup by Jupiter.

This periodic comet was accidentally discovered near Jupiter on March 25, 1993 by Shoemaker and Levy on March 24, 1993, on a slide taken with the help of a small Schmidt camera as a bar-like image of one arc minute long, with a faint, wispy tail. A few days later a much more impressive image of this comet was obtained with a high-resolution CCD camera on Hawaii's 2.2-m telescope atop Mauna Kea; it is shown in Plate I (top).

The history of Shoemaker–Levy 9 had no precedent. Just after its discovery it became clear that the string of cosmic pearls probably did not have the near-parabolic orbit so characteristic of most of newly discovered comets. By late May, 1993, computations revealed that Shoemaker–Levy had passed no more than 50 000 kilometers from Jupiter's cloud layers in June, 1992. That was close enough to the Roche limit, i.e. when the tidal forces of the planet had to pull the comet out of its orbit. Such disruption was predicted by Roche, but the process itself had never been observed in the Solar system, especially in such a dramatic manner.

Then, based on one and half hundred observations made between the comet's discovery and its slipping into the Sun's glare four months later, dynamists from the Jet Propulsion Laboratory (D. Yeomans, P. Chodas) concluded that Shoemaker–Levy had been looping around Jupiter much earlier, since at least 1970. They concluded that on July 8, 1992, it had passed 43 000 km, i.e. only one-third of Jupiter's diameter, from the planet, had split apart, and had taken up an orbit with an eccentricity exceeding 0.99. A year later, by July 16, 1993, the fragments were spread over 50 000 000 km, nearly Mercury's distance from the Sun.

The comet ended its life a year and three months later, in July, 1994, by direct impact with Jupiter (Fig. 11.8). An image of this "train" of 21 fragments of the comet just before its impact on Jupiter's atmosphere is shown in Plate I (below) obtained with the help of the Hubble Space Telescope.

The computations by Marsden, however, seem to indicate that this comet was most probably captured by Jupiter earlier, on May 16, 1992, during its transit within 0.007 astronomical units of Jupiter and been ripped apart by tidal forces near this planet.

Presumably the periodic comet with a 11.5-year orbital period was performing its motion via a highly elongated elliptic orbit around the Sun, and quite accidentally entered the zone of influence of Jupiter's gravity.

Such an event could have been a result of an extremely rare combination of the gravitational actions past Jupiter and the Sun. Moreover, the transit past Jupiter occurred so close to the planet's surface that the comet appeared within its Roche lobe and, as a result, was disrupted by its tidal forces. The result was over 20 fragments, all with short tails. Analysis of the Hubble images suggests that the comet's largest fragments are probably no more than six kilometers across.

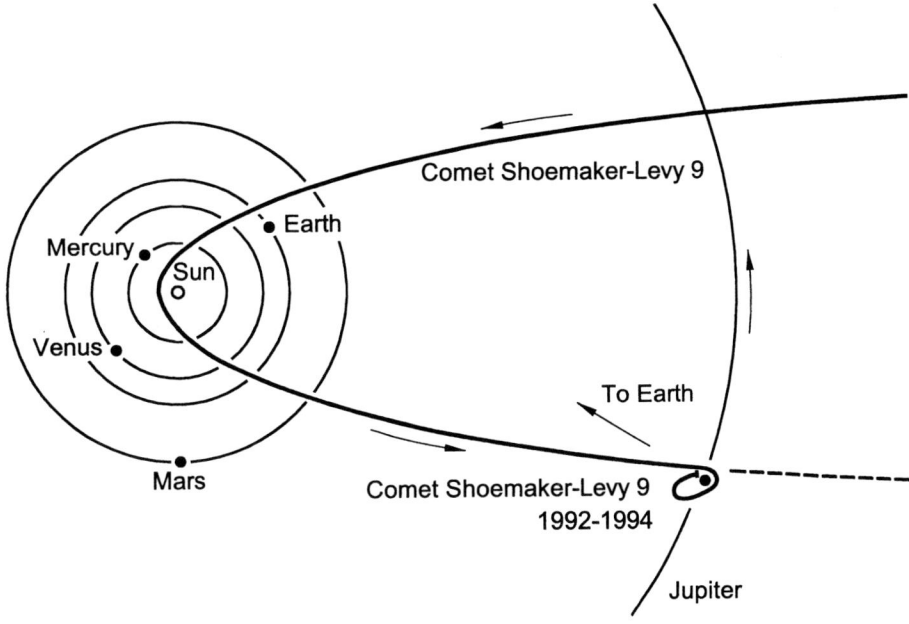

Fig. 11.8. *The journey of the periodic comet Shoemaker–Levy 9 through the Solar system in an elongated elliptical orbit, and its meeting with and capture by Jupiter on July 7, 1992.*

We can estimate the maximum distance of the comet's transit near Jupiter by using Eq. (52), Chapter XI, for the radius of the Roche lobe:

$$R_L = 1.36 \left(\frac{\mathcal{M}}{\rho} \right)^{1/3} .$$

Taking for Jupiter's mass $\mathcal{M} = 0.001 \mathcal{M}_\odot$ and for the density of the comet's matter (ice) $\rho = 1 \ g \ cm^{-3}$, we obtain $R_L = 170\,000 \ km$.

The initial size of the comet is estimated to be about 10 km in diameter, and, equivalently, about $5 \times 10^{17} g$ in mass; the sizes of most large fragments were of order $1 - 2 \ km$ (Sekanina *et al.*, 1994). The different directions of

the velocity vectors of the products of the breakup, even in the case of their
having the same initial values, are enough to induce a noticeable scatter
in their initial orbital velocities. As a result, the fragments formed a **train**
along a line directed towards the central body – Jupiter, as one would expect.
The train's total length was increased from about 50 arc seconds at the time
of discovery to about 2 arc minutes by mid-December 1993. For the same
reason, we will have a family of different elliptic orbits with different major
axes; in Fig. 11.9 the elliptic orbits for boundary nuclei of the "train" are
shown by dashed lines (not to scale).

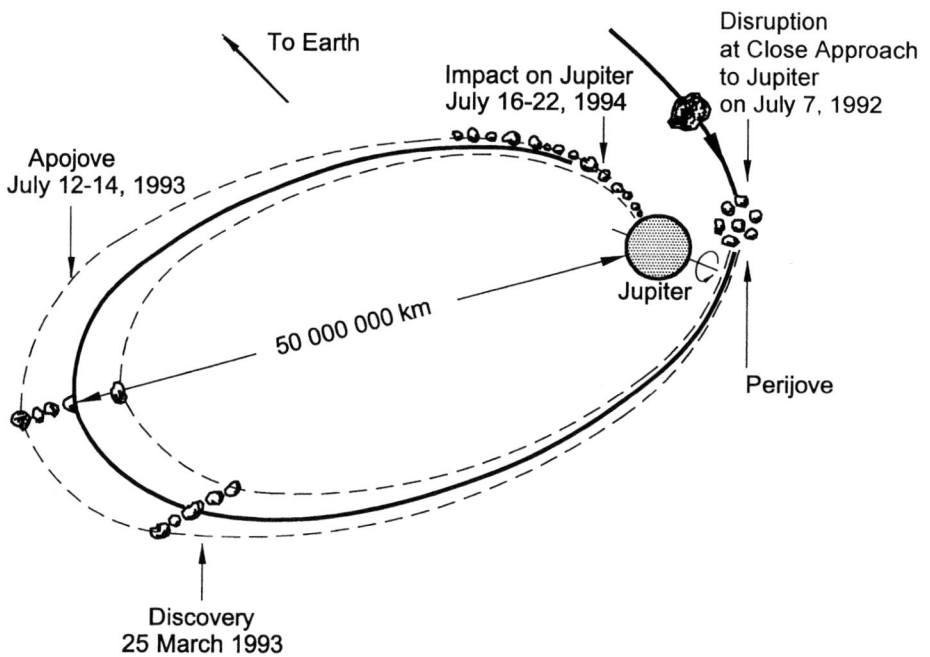

Fig. 11.9. *A schematic representation of the tidal breakup of the parent nucleus of the
comet Shoemaker–Levy 9 in July 1992, the orbital evolution of its fragments, and their
collision with Jupiter during 16–22 July 1994 (not to scale).*

    The train of cometary fragments reached its largest distance from Jupiter,
i.e. the apojove, during July 12–14, 1993, so that the time of flight from the
nearest point (perijove) to the farthest point (apojove) was 361 days. This
should correspond to nearly half of the comet's rotation period P around
Jupiter, i.e. $P = 722$ days. Using Eq. (66) of Chapter III for the major

half-axis a, namely,

$$a = 0.067 \, \mathcal{M}^{1/3} \left(\frac{P}{2\pi} \right)^{2/3}$$

we obtain $a = 0.156 \, a.u. = 23.4 \cdot 10^6 \, km$, i.e. 165 times the diameter of Jupiter.

According to trajectory computations, the ratio of the maximal distance r_a at the apojove of Jupiter to the minimal distance r_p at its perijove is about 520. Hence, by using Eq. (81) of Chapter III, namely,

$$\frac{r_a}{r_p} = \frac{1+e}{1-e} \, ,$$

we obtain for the eccentricity $e = 0.996$, i.e. nearly a parabola. For the minimal distance from Jupiter's center we have

$$r_{min} = r_p = a(1-e) = 90\,000 \, km \, .$$

Comparison of this result with the radius of the Roche lobe obtained above allows one to conclude that, first, the comet has in fact been within the Roche lobe; and, second, its transit took place at the distance ΔH given by

$$\Delta H = 90\,000 - R_J = 19\,000 \, km$$

from Jupiter's surface, where $R_J = 71\,000 \, km$ is Jupiter's radius.

According to trajectory computations in the framework of the two-body problem, the comet, or more precisely the train of its fragments, would pass at a distance of $30\,000 \, km$ from Jupiter's center. Jupiter's radius is $71\,000 \, km$, hence, the collision of cometary nuclei with Jupiter would have been inevitable, which in fact happened on July 16, 1994 for the first fragment and in the next days for the other pieces.

Here the mission of celestial mechanics is finished, and the realm of astrophysics starts. The consequences of the penetration of cometary nuclei into Jupiter's extended atmosphere were recorded by means of a number of ground-based telescopes, including infrared and radio ones, as well as by major space missions, including the Hubble Space Telescope (HST). This exceptional observational event was caused by the impact of a bodies of about one hundred million tons with the largest planet in the Solar system at a relative velocity of $60 \, km \, s^{-1}$, so that the observed giant waves in the form of circles reached sizes of more than $10\,000 \, km$, i.e. that of the Earth's diameter.

Thus, the unique event of the appearance and death of the Shoemaker–Levy 9 comet can be considered as a demonstration of several phenomena of celestial mechanics with one and the same celestial body within the rather short period of time of two years.

a. Drastic modification of the orbit, namely, the transition from the giant ellipse with the Sun at its focus to a smaller ellipse with a planet (Jupiter) at its focus; in other words the capture of a celestial body (comet) by another more massive body (Jupiter) in the presence of a third body (the Sun).

b. The passage of a celestial body (the comet) within the Roche lobe of a planet (Jupiter), and its resulting tidal disruption.

c. The possibility of the destruction of a celestial body, at least of the size of the comet's head, finding itself in the Roche lobe of a massive planet and its transformation into pieces.

d. The group motion of the fragments of the disrupted body in a closed orbit (ellipse) in the form of a "train" of celestial bodies of smaller sizes.

e. The impact of celestial bodies (cometary fragmentation) with a more massive body (Jupiter).

All these events happened with one and the same celestial object, the Shoemaker–Levy comet. Truly, "It was the scientific event of the century."

It remains for us to regret that the classics of celestial mechanics have no opportunity of similar experience, to observe the triumph of their creation, in this case associated with the comet's death.

The retroscopic searches bring us, however, to the discovery of the "train"-type celestial bodies of smaller masses as a result of the disruption of a more massive body, perhaps of a comet or a large meteor, within the Roche lobe of a more massive body, particularly that of the Moon or satellites of more massive planets. Such an example is shown in Plate II, one of the hundreds and thousands of pictures of the surface of Jupiter's largest satellite Ganymede taken with the help of the Galileo spacecraft (see Chapter XV). These pictures show fine details of this icy world, including fresh impact scans and terrain of varying brightness and texture. Among the more intriguing views was of Enke Catena, a 150-kilometer-long chain of craters. Three such features are known to exist on Ganymede, and more than a dozen can be found on the surface of Callisto. Crater chains are believed to arise when a comet or asteroid comes too close to Jupiter and is disrupted by tidal forces produced by the giant planet's tremendous gravity. The resulting pieces travel closely together up to the moment of their fall onto the surface of Ganymede.

One more example is the Apollo image, Plate III, of a fragment of the Moon's surface. Here nearly two dozen small pits, each 1 to 3 km wide, cross the floor and east rim of the near-side lunar crater Davy, in the northeast margin of Mare Nubium. The entire string was formed, again, during a single impact event, most probably due to the fall of a comet onto the Moon's surface. Note the similarity of this image to Plate II.

16 The Stability of the Solar System

From the cosmogonical point of view, the large-scale or secular perturbations are of extreme importance, especially concerning the problem of the stability of motion of the planets as well as that of the stability of the Solar system as a whole. How can one formulate the problem of the stability of motion of planets? In the framework of the two-body problem, finite motion can occur via a precisely elliptic trajectory, performing with absolute accuracy an infinite number of rotations around the central body, unless the system is not under the action of external perturbations. Such motion is stable, since no deviation exists between the parameters of successive orbital elements. The appearance of external perturbations results in deviations in these parameters from one orbit to the next.

However, one can have a situation in which a body, though deviating from some stable state, nevertheless can remain close to that state on an infinitely long time scale. Without going into the details, we mention that this kind of equilibrium is known as **stability in the Lyapunov sense.**

Together with this, one can also define **Lyapunov instability**, i.e. when the smallest initial departures from the equilibrium state exponentially increase with respect to time.

In contrast, there can be cases in which the body that is being removed from the equilibrium state by some external perturbation can return to that state on some time scale independent of the form of the initial perturbations. In such cases of damping of initial perturbations, one speaks of the **asymptotically stable** state of equilibrium.

The deviation from the stable state means the deviation of all six elements of the orbit. However, it is not difficult to ensure that, in the problem of the stability or instability of the motion, only three elements – semimajor axis a, the eccentricity of the elliptic orbit e, and the declination of the orbital plane i play a key role; the variations in the other three elements, i.e. Ω, ω and T, do not influence the problem of the stability of the motion.

Below, in order to ensure the stability of the planet's motion, it suffices to maintain the values of only three elements, a, e and i, within definite limits so that the trajectory of the planet will be confined within a certain torus (Fig. 11.10) with the Sun at its center. Obviously, the position of the body in the real trajectory, e.g. its distance from the Sun r, the height from the plane of the ecliptic z and the declination of the orbital plane i, will be confined within certain limits (within a circle of radius ρ):

$$R - \rho \leq r \leq R + \rho ,$$
$$-\rho \leq z \leq \rho , \tag{65}$$
$$-i_0 \leq i \leq i_0 ,$$

where
$$\tan i_0 = \rho/R, \quad \tan i = z/R, \quad a = R, \quad e = \rho/R \ .$$

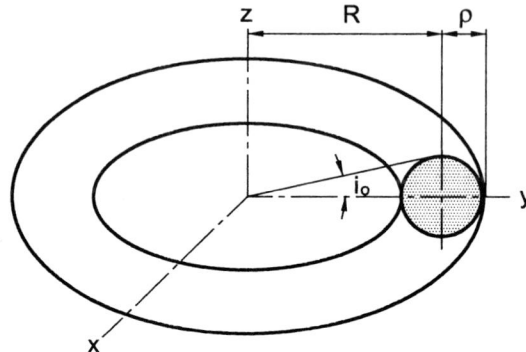

Fig. 11.10. *Tori within which the orbits of the planet should be included when their motion is stable.*

If the motion of a planet satisfies the conditions in Eq. (65), then the motion will be stable regardless of the differences between various real trajectories; e.g. the planet can neither escape from the Solar system nor fall onto the Sun. Every planet has its own torus, with a certain size and section, situated at some angle to the ecliptic. Obviously, their sizes can be defined so that the various tori never intersect with each other. Then collisions between the planets can also be excluded.

Simplifying the situation to concern the Solar system, the problem of the stability of motion can be formulated in the following manner, i.e. a planetary system can be considered as stable if:

i.   the planets do not fall onto the Sun;

ii. the planets do not leave the Solar system; and

iii. the planets do not collide with each other.

Historically, the formulation of the problem of the stability of the Solar system coincides with the period of formation of celestial mechanics itself, or more precisely, that of the analytical methods of celestial mechanics. Already in 1773 Laplace, and soon thereafter Lagrange, set down an important theorem, called the "Laplace–Lagrange theorem", which runs: if the mean motions of planets are incomparable, then the semimajor axes (a) and the average motions (n) do not contain secular (long-scale) perturbations of the first order. Hence, the Solar system seems to be stable, i.e. the motions of all planets take place within their tori. Is it valid? No affirmative answer can be given based on the Laplace–Lagrange theorem since it takes into account only the first-order perturbations. In 1809 Poisson managed to prove that the perturbations of the second order for the semimajor axes again do not contain secular terms. Does that suffice to enable one to claim the stability of the Solar system? Apparently not, since the contributions of perturbations of the third or even fourth order still have to be understood. Numerous attempts in this direction that have been performed during the past century do not clarify the situation, e.g. it seems that, for example, secular perturbations of the third order for the semimajor axes of planetary orbits do exist. What is clear is that the Laplace–Lagrange theorem as well as Poisson's theorem are inadequate to demonstrate the stability of the Solar system in the distant future.

The Laplace–Lagrange theorem ensures the stability of the Solar system, or more precisely, of the semimajor axes of planetary orbits, only within a finite time interval. The fate of the Solar system on cosmogonical time scales remains unspecified.

In the formulation of the Laplace–Lagrange theorem the condition of the "comparability of the mean motion" was mentioned. What does that mean?

An interesting fact concerning the parameters of the mean motions, i.e. the mean angular velocity of the orbital motion of two or more planets, had already been noticed long ago: from time to time the planets, after a certain number of revolutions around the Sun, appear to be situated along a line passing through the Sun. One speaks then about a **resonance**. If the mean motion of a planet is n_i (in angular degrees per day), then for real integers k_i, which are positive, negative and zero, the following condition is known as a **resonance relation**:

$$\sum_i k_i n_i = 0 \; . \tag{66}$$

For example, for Jupiter ($n_J = 0.083091$) and Saturn ($n_S = 0.033460$) this condition is most accurately satisfied at the values of $k_J = 2$ and $k_S = 5$:

$$2n_J - 5n_S = -0°.001118 \; ,$$

i.e. $k_J/k_S = 5/2$. In the cases of Neptune ($n_N = 0.005981$) and Pluto ($n_P = 0.003979$) the condition is almost precisely satisfied at $k_N = -2$ and $k_P = 3$:

$$3n_P - 2n_N = -0°.000025 \; .$$

In the case of the Venus–Mars–Saturn combination, one has

$$n_V - 3n_M - n_S = -0°.002551 \; .$$

One might think that the Solar system is almost completely in resonance; for all nine planets there exist integers k_i for which the condition (66) is fulfilled with negligible errors. It is important to bear in mind that the numbers k_i are not so large – usually 0, +1, +2, rarely above 5, i.e. a circumstance of extreme importance, which cannot be ascribed to chance, and of course, the existence of the resonance relations themselves.

The resonance phenomenon appears to be a privilege not only of planets: similar scaling in mean motions is observed also for natural satellites of the planets and not rarely for asteroids. For example, for three satellites of Jupiter – Io, Europa and Ganymede, the frequency relationships with n_1, n_2 and n_3 were discovered by Laplace in the following form:

$$n_1 - 3n_2 + 2n_3 \approx 0 .$$

As for the questions of how and why the Solar system attained the resonance states during its evolution, the situation is thus far unclear. We know nothing reliable about the real mechanism of the "transition" of the Solar system into the resonance regime. Analogously, no proofs exist concerning the transition of oscillating systems similar to the Solar system to the resonance regime. Obviously, the role of dissipation effects, such as the tidal forces, the resistance of interplanetary matter, etc., which have been neglected within the described celestial dynamical approach, could be rather crucial. Though the role of the dissipation forces is not large, even weaker than the perturbations caused by the planets' mutual influence, on the cosmogonical time scale their contribution might not be negligible. For the time being, it is rather hard to securely evaluate the role of these effects in the fate of the system, particularly with regard to its transition to the resonance state.

However, it seems that the problem of the resonance motions can be considered to be in some way analogous to the singular (libration) points, i.e. as a problem concerning the stability properties of the trajectories. The existence of resonances could be interpreted as an indication of a certain regularity peculiar to the planetary orbits, a consequence of a kind of principle of selection whereby nature reveals a preference for some properties and rejects the others. Incidentally, the careful consideration of resonance effects enabled the prediction the new satellites of Uranus (see Gorkavy and Fridman, 1995), before their discovery by Voyager 2 in 1986.

New insight into this problem has been introduced by the recent development of the ideas of chaos, marginal stability, and other collective dynamical effects concerning the properties and evolution of the Solar system. This problem will be discussed in the next section.

Let us return to the classical problem of the stability of Hamiltonian systems, in particular, the Solar system, i.e. those in which the only interaction is Newtonian gravitation, and the role of dissipation forces and the existence of the resonances are neglected. That problem, with two hundred years of history, has experienced an essential breakthrough in our time, particularly due to the fundamental work by Arnold in the 1960s.

The results obtained by Arnold, based on previous work by Poincaré and Kolmogorov, are of universal character and concern the behavior of perturbed Hamiltonian systems (see Chapter XIV). If the conditions of the Kolmogorov–Arnold–Moser (KAM) theorem are satisfied for the Solar system, and the influence of the planets can be considered as a perturbation to the central field of the Sun, then one can come to the following conclusions: *for the majority of initial conditions the real motions of the planets are conditionally periodical and are close to their unperturbed (Lagrangian) motions. In particular, the eccentricities and declinations will remain small, and the semimajor axes will remain close to their initial values.*

Does that mean that, in the spirit of the Laplace–Lagrange theorem, the Solar system should remain stable forever? Once again, no, it does not. The point is not only the difficulty in checking whether the conditions of the KAM theorem are really satisfied for the Solar system, but also fundamental uncertainties concerning the resonance cases that are not covered by that theorem. As has already been mentioned, the study of chaotic phenomena, including those with cosmogonically large time scales, will be decisive in the understanding of the stability of the Solar system (Laskar, 1994, 1997; Tremaine, 1994; Morbidelli, 2002).

17   Chaos in the Solar System

When the system of differential equations composed for this or that concrete body of the Solar system is written to include the perturbations arising from the other members of the system, for such a system with a few tens or hundreds of thousands of equations, and is being solved by numerical integration on the time scale of hundreds of millions or trillions of years, then, often, it turns out that the solution brings completely inordinate and nontraditional results; in such cases we talk about **chaos**.

Local instability, sensitive dependence on initial conditions, and chaos are known to be the typical properties of any many-dimensional nonlinear system (Ferraz-Mello, 1992; Gurzadyan and Pfenniger, 1994; Gurzadyan and Ruffini 2000). The Solar system, being an example of such a system if described by a given nonlinear Hamiltonian, should obviously reveal these properties, some of which could be crucial in understanding the present structure of the Solar system and especially its distant future. One of the impressive results obtained in this area is the chaotic behavior revealed for the obliquity of planets, i.e. the declination of their rotation axes with respect to the ecliptic (Laskar, 1994). Insofar as the obliquity essentially determines the climate of the planet (e.g. the change of seasons on Earth), irregularity in properties of obliquity can readily lead to catastrophic changes of climatic conditions, and could even modify the parameters of the orbit to such an extent as to lead to its intersection with the orbit of another planet, escape or collision of the planets.

The analysis of the corresponding set of differential equations to describe the long time scale evolution of a planet, asteroid, or comet, aside from powerful computational facilities, obviously requires sophisticated methods of numerical integration of those equations. Just this problem has been overcome while studying the obliquity properties. The situation concerning the Earth–Moon system, i.e. the stability of the rotation axis of the Earth in the presence and absence of the Moon, is quite interesting. The results of computations carried out by Laskar for time scales of 18 million years indicate that the obliquity of the Earth is stable, but, as shown in Fig. 11.11, existence of an essentially chaotic region has been revealed, within the angles 60° and 90°. The reason for the stability lies in the presence of the Moon. Without the Moon, the obliquity of the Earth would suffer significant chaotic variations, by up to 85°, including up to 60° within less than two million years. Thus, the presence of life on the Earth is essentially determined by the existence of the Moon, a conclusion of tremendous heuristic content (Laskar, 1994).

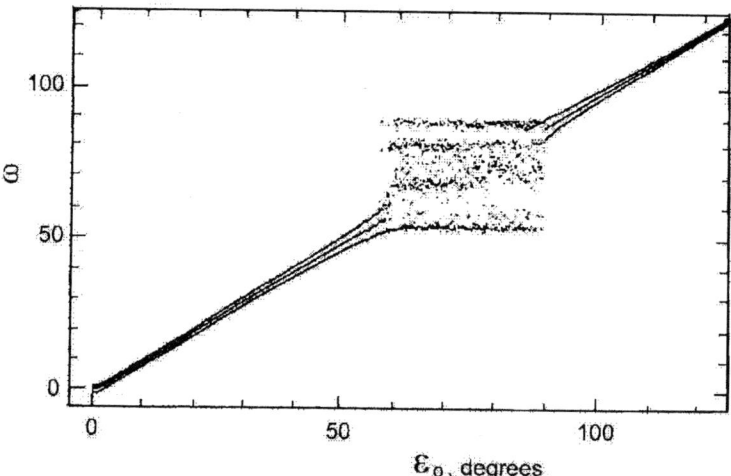

Fig. 11.11. *The stability of the rotation axis of the Earth in the presence of the Moon. Numerical integration of the precession equations is performed over 18 million years for each value of the initial obliquity ε_0 of the Earth ranging from 0 to 125 degrees. An extended chaotic zone is clearly visible between $\varepsilon_0 = 60°$ and $\varepsilon_0 = 90°$ (Laskar, 1994, see the discussion in the text).*

Among other results one can mention the large-scale chaos established for the Solar system: on a time scale of billions of years, the orbits of the planets reveal strong chaotic variations. In the case of Mercury, for example, these variations may even lead to a strong increase in its eccentricity, i.e. of its semimajor axis, leading to the intersection of the orbits of Venus and Mercury. At the same time, the two planets could experience a close encounter, which might lead to the escape of Mercury from the Solar system or its collision with Venus within a period of less than 3.5 billion years. The chaotic behavior of the eccentricity of Mercury could be that dramatic.

The possible role of chaos has been analyzed for the motion of Halley's comet, with loss of predictability occurring after 29 revolutions. Also, it was shown that, due to the perturbation of Jupiter, there exists a large chaotic zone for nearby parabolic comets' orbits, which extends up to the Oort cloud (Fernandez, 1994). Many cometary orbits are chaotic; moreover, the chaos could play a major role in the diffusion of comets from the external regions of the Solar system.

The nontypical obliquity of the rotation axis of Uranus could be chaotic, with $\varepsilon = 98°$ being the character of the rotation of Hyperion, a small satellite of Saturn. The latter fact seems to be confirmed by direct observations by the space missions *Voyager 1* and *Voyager 2* (Klavatter, 1989), see also section 6, Chapter XIII). Within the framework of the concept of chaos one can also explain other well-known facts, e.g. the distribution of the orbits of asteroids within the orbits of Mars and Jupiter and the so-called "Kirkwood gaps" (Ferraz-Mello, 1992).

There is evidence for chaoticity in the obliquity of the orbit of Mars as well, though it is still impossible to make predictions for periods longer than a few million years. Also, the obliquity of Mars is subject to much larger variations, ranging from about 0 to 60 degrees in less than 50 million years. In any case, the study of the past climatic history of Mars should be reconsidered in the light of these results (Jakosky *et al.*, 1993; Touma and Wisdom, 1993).

Thus, the Solar system seems to be strongly associated with chaotic evolution, though revealing marginal stability on time scales comparable to its age. In general, the orbits of planets farther out than Jupiter appear to be more stable than those of the inner planets, and practically no diffusion can occur among the degrees of freedom of those planets. A number of fundamental problems concerning the origin of the Solar system, the origin of planets and their satellites, of the Moon in particular, the general problem of the origin of the axial rotation of planets and many others, await their understanding within the framework of the idea of chaos, of this fruitful concept that has already entered the arsenal of celestial mechanics.

18   On the After-Effects of Large Scale Chaos

The definition of Chaos is given at the beginning of the previous section, from which it follows that Chaos is operated in the time scale of the order of billions of years. However, the characteristic time scale for the divergences of inner planets of the Solar System, from Mercury to Mars, is about five million years, and all solutions of the orbital motions of these planets becomes practically nonperiodic after 100 Myr. On a billion-year time scale, the motions of the planets of the Solar system can be a random process, where the orbits wander erratically (Laskar, 1989).

Numerical integrations of differential equations composed for the concrete body of the Solar System, including the perturbations arising from the other members of the system, lead to extraordinary results. So, the motion of the large planets, Jupiter and farther, is always very regular. The chaotic zones for Venus and the Earth are moderate in size. In contrast with these cases, the chaotic diffusion of Mercury is extremely large; its eccentricity in some periods, as follows from Fig. 11.12, can potentially reach values very close to 1. As a result, the ejection of Mercury from the Solar System and even a close encounter with Venus, as noted in the previous section, is possible in less than 3.5 Gyr.

Even more remarkable is Fig. 11.13, where the variations of distances in a.u. of Mercury and Venus from the Sun of the orbit on the line of nodes, i.e. on the line of intersection of orbital planes both of Venus and Mercury are given. The positive (+) and negative (−) signs in Fig. 11.13 correspond to both directions on the line of nodes relative to the Sun.

As follows from the curves in Fig. 11.13, the variations of the Venus position relative to the Sun are much less, within limits 0.68–0.74 a.u., around its present distance 0.723 a.u., while the distance of Mercury from the Sun is varied within large limits, from 0.10 a.u. up to 0.72 a.u. around its present distance 0.387 a.u.

However, most important is the next result: most of the time, on the line of nodes, the orbit of Mercury crosses the orbit of Venus – three such points are taken into the circles. This phenomenon lasts a few thousand years, and during that time the two planets, Mercury and Venus, can experience a close encounter which could lead to the escape of Mercury, or even to collision.

That is not all. The durations of the periods of interactions between both these objects are long enough as compared with the periods of their orbital revolutions around the Sun, and this circumstance is increasing the probability of their collision.

Thus, in Fig. 11.13, the variations of the planet's distance from the Sun correspond to the time interval from 3.5679 Gyr up to 3.5686 Gyr, and during this period, Mercury undergoes no less 10 sharp changes of its distance, from 0.10 a.u. up to 0.70 a.u., i.e. one sharp change at

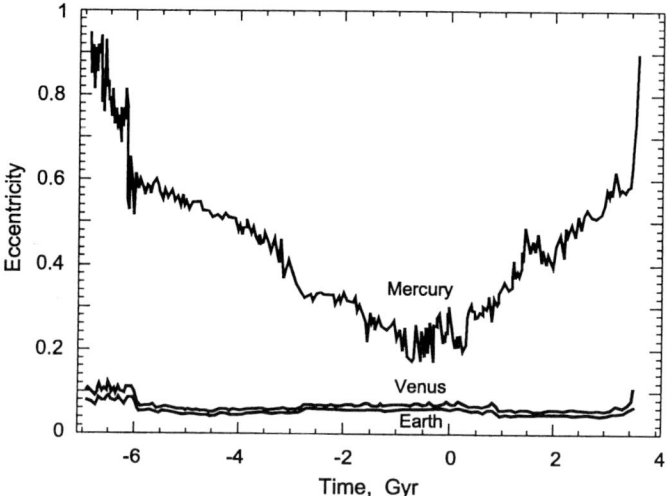

Fig. 11.12. *The evolution of the eccentricity of Mercury's orbit within a billion-years intervals, from 7 Gyr backward (-) to 4 Gyr forward (+). The variations for Mercury, Venus and the Earth are shown. The eccentricities for the Earth and Venus are practically on a constant level during of the time interval 10 Gyr, equal to their present magnitudes. The eccentricity of Mercury's orbit is undergoing strong variations up to the value e = 1, hence, with possible escape from the Solar System at -6.6 Gyr and +3.5 Gyr (Laskar, 1995).*

(3.5686 − 3.5679)/10 = 0.00007 Gyr = 0.07 Myr or 70 000 years. In cosmogonical scale this is extremely short period.

However, even during this short period, 0.0007 Gyr = 0.7 Myr = 700 000 years, three crossings of the orbits of Mercury and Venus occur, which gives for the frequency of crossings with a fatal ending one variation during ∼ 700 000/3 ∼ 250 000 years = 0.25 Myr. During a cosmogonical period, say 5 Gyr = 5 000 Myr, the total number of crossings with a fatal ending will be 5 000 / 0.25 = 20 000! And, in spite of this result, Mercury still exists. Why?

Of course, the crossing of the orbits of Mercury and the Venus still does not mean a fatal end for Mercury – its escape or collision with Venus. But the total number of crossings, 20 000, is too large. It should be assumed, therefore, that the probability of crossings with a fatal ending is much less, say, one fatal crossing among 20 000 normal crossings.

With this connection, a new problem arises, namely, to find ways of computing the probability of the very near transits of these planets, Mercury and Venus, with a fatal ending, the escape of Mercury or its collision with the Venus, at one act of crossing of their orbital plans.

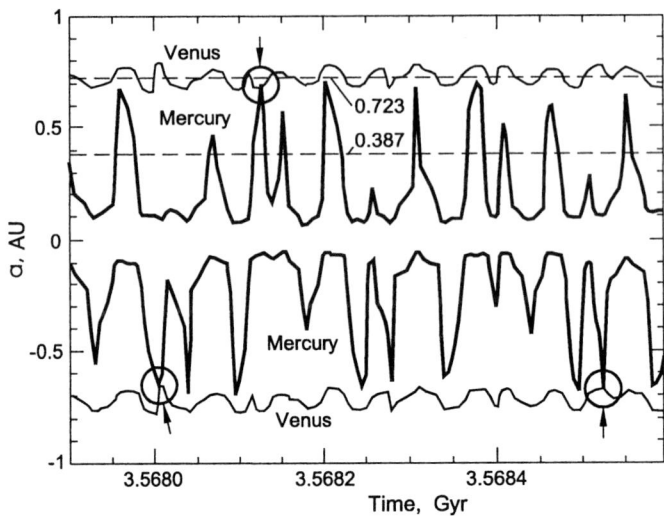

Fig. 11.13. *Relative position of the orbits of Mercury and the Venus computed at the line of intersection of their two orbital planes (line of nodes). The distance, in a.u., from the Sun of the orbit on the line of nodes is plotted against the date in Gyr. Most of the time, the orbit of Mercury stays within the orbit of Venus, but in some periods, it crosses the orbit of Venus: three such cases are circled.*

It should be added that, for very high eccentricity of Mercury, the model used here (Laskar, 1994) is no longer a good approximation for the motion of Mercury, but it should be outlined that in this approximation the chaotic zone allows the escape of a planet from the solar system in a time smaller than the expected life of the Solar system due to diffusion in the chaotic zone. So, these results should present definite theoretical interest.

As to the behavior of the outer planets, Jupiter and farther, up to Uranus, their motions are regular, i.e. their eccentricities during a cosmogonical period of Gyrs remain strongly constant, which is the present situation: the "eccentricity–time" dependencies for these planets in Fig. 11.13 appear as strongly horizontal lines (not shown).

The further solutions, i.e. the numerical integrations of the system of differential equations, mentioned in the previous section, with hundreds and thousands of terms on a time scale of hundreds of millions or trillions of years, brings new and no less interesting results concerning the behavior of the spacing of the inner planets or their distances from the Sun. In particular, as it is shown by Laskar (1997) the orbital configurations of the inner planets, Mercury, Venus, Earth and Mars, could result solely from the presence of large-scale chaos in their orbits, whose wanderings are then only limited by the conservation of angular momentum. With this connection, a new parameter, the so-called *angular momentum deficit* (AMD), is introduced. Calculations carried out for various numerical values of AMD lead to results for the numerical values of these four planets' distances from the Sun coinciding with their present distances almost completely. This leads to the conjecture that dynamical evolution can be responsible for the origin of the spacing of the planets in the inner Solar system.

The difference in behavior between the large planets and the inner planets, as outlined above, is very striking. One reason for this is probably that the large planets are not perturbed mutually or by inner planets.

From these considerations, it appears that the organization of the inner Solar System seems to result from its dynamical evolution. The Solar System was probably progressively driven towards more and more regular states, through collision processes. The angular momentum deficit, which was invoked by Laplace 200 years ago to establish the linear stability of the Solar system, is used now in a different context to understand the limitations of its chaotic behavior.

Thus, the strongest receptivity to Chaos is possessed by Mercury, then Venus. This very important theoretical result is confirmed by an extremely indicative fact, namely, the absence of satellites around Mercury and Venus; it is quite clear that, with such strong receptivity to Chaos, when the fate of the planets themselves is in doubt, it would be hard for them to keep a stationary satellite.

Receptivity to Chaos falls catastrophically with distance from the Sun, starting from the Earth, and completely vanishes for the massive planets, Jupiter and farther. This theoretical deduction is also confirmed by facts: Earth has only one satellite, Mars two, and farther, in view of the complete absence of the intervention of Chaos in the "life" of the planets, the number of natural satellites increases to an avalanche because nothing is preventing them from keeping even a tiny satellite at remote distances from the central planet.

We believe that just this real situation, with the presence or absence of satellites as well as their distribution around the planets of the Solar System, should be considered as the strongest proof of the correctness of the concept of Chaos.

For now, the Solar system is the only possibility for checking the idea of Chaos. Perhaps in the remote future these considerations may be checked for the planetary systems around nearby stars as well. Unpredicted facts, however, might be expected. For example, the existence of planets around stars at distances 10 times closer to the central star than Mercury is to the Sun (see Chapter XI).

Let us to return to Fig. 11.13. It is not difficult to notice that the microsurgery of time variations of the planet's distance, is surprisingly one and the same during a more or less long time period. If so, then from the point of view of final results, the behavior and the action of long time scale chaos may be identified, quite formally, with a strong oscillation – in cosmogonical scale – of the gravity of Mercury, but not of the Sun. Such formal identification of Chaos and "Oscillating Gravity" may simplify radically the calculation process. Note that the sharp variations in Fig. 11.13 of Mercury's distance from the Sun, namely, non-symmetrically relative to the horizontal axis but periodic with more or less constant intervals, can be approximated by a quadratic trigonometric function $a(t)sin^2kt$, where k is the frequency of oscillations, which can be determined or estimated directly from this type of Figure. More problematical is the amplitude of oscillations $a(t)$, which, although not constant, can apparently be estimated also in the same manner.

In all cases, one should not exclude the possibility of a situation where for formal quantitative analysis, both these approaches, Chaos and the Oscillating Gravity Analogy, can be equivalent. Like the "Membrane Analogy" in the Elasticity theory which helps with the solution of complex and difficult problems having nothing common with the membrane itself.

19   Extrasolar Planets

A star cannot be without planets. Independent of the mechanism of its origin. Just because the Sun is a normal star and has planets. That is why the problem of the discovery of so-called extrasolar planets, i.e the planets around neighboring stars is of special interest.

There are two means of detecting unseen satellites, in particular planets around stars. Both are based on one and the same principle, namely, the catching of infinitely small periodic displacements, like periodic vibrations, of the star around the center of the "star–planet" system. In the first case, the star is undergoing microvibrations of its position in the sky, i.e. on the plane **perpendicular** to the line of sight, and the problem is catching those extremely small **angular** vibrations. In the second case, the star is undergoing periodic variations of the Doppler shift, i.e. of the radial velocity **along** the line of sight, and the problem is to measure these extremely small Doppler variations in the **radial velocity** of the star.

As we see, both the methods of discovery and the study of extrasolar planets are linked with the fundamental laws of celestial mechanics.

The first method is the oldest one, it is applied and still may be applied for unseen but relatively massive satellites around a star. In such a way, for example, nearly a century ago Vega's very faint satellite-star, a future white dwarf, was discovered. As to the second method, it usually has been used for the study of double stars. In both cases, astronomers are manipulating the shift in star positions and the radial velocities.

However, when the matter concerns a planet around a star, say, of Jupiter type, both these variations, positional and of radial velocities, should be about 1000 times smaller. At present it is practically beyond the limits of positional astronomy. The situation is better with the fixation of Doppler-type microvariations.

Since 1995 essential results have been achieved concerning the discovery of extrasolar planets around the nearest solar-type stars by a means of the Doppler technique by groups led by Mayor, Marcy and others.

Though the main idea, the Doppler technique, i.e. the measurement of tiny wavelength shifts, is not new, nevertheless the technique of registration is new, namely, sending stellar light into an advanced spectrometer with a CCD detector installed in the focal plane of telescopes, particularly the Lick observatory 3m telescope, the 3.9m Anglo-Australian Telescope, the 9m Hobby-Eborly Telescope in Texas, and the 10m Keck I in Hawaii. The sensitivity of the measurements was better than 3 meters per second (!), which corresponds to a wavelength change of 1 part in 100 million. Until recently, such precision was far beyond reach. This is the essence of planetary detectors.

The fact of discovery of a planet around a star is being confirmed by the detection of variations or wobbles of its radial velocity. The orbital period P of the discovered planet is then determined directly from the wobble's period. Then, from the known orbital period and Kepler's Third Law, i.e. from the relation $a \sim P^{2/3}$, the semimajor axis a, i.e. the planet's average distance from the star, can be determined. If the velocity varies with time like a perfect sine wave, this will mean that the orbit is circular. Otherwise, the skewness of the velocity curve can be analyzed to find the eccentricity of the elliptical orbit e. From two examples, shown in Fig. 11.14, i.e. two orbits, one of which has practically zero eccentricity (47 UMa), the second with very large eccentricity (16 Cyg B), we may be convinced how accurately the type of orbit can be determined.

So, the three parameters of the planet's orbit, P, a and e, are discovered.

The amplitudes of Doppler variations indicate the planet's mass, in fact the magnitude of $m \sin i$, where i is the inclination of the orbit's plane relative to the line of sight which is unknown. If we observe the orbit nearly edge-on, i.e. when $i \approx 90^o$, the star's full orbital velocity is reflected in the Doppler shifts, so in this case the mass derived will be the planet's true mass. Since the angle i is also unknown, the deduced mass of the planet will be its minimum mass, i.e. $m \sin i$. In fact, i is the only parameter which cannot be discovered by direct observation.

The mass of the planet m obtained in such a way is the fourth and last parameter which may be derived from direct Doppler shift measurements.

In this way planets are discovered around many stars, mainly of G type, brighter then $7-8^m$, within 30 pc from the Sun. The list of some of those

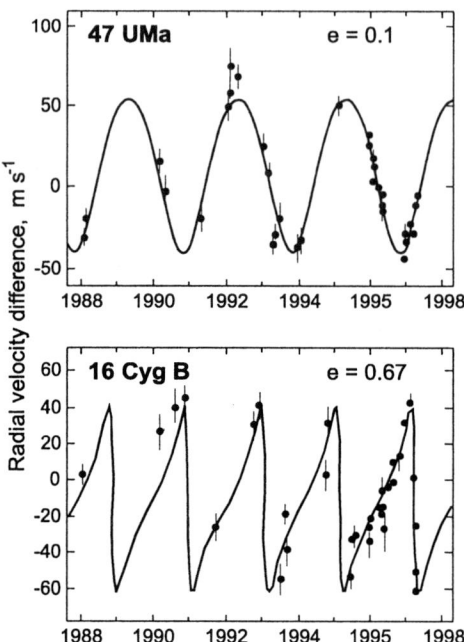

Fig. 11.14. *Velocity curves for the two stars with various degrees of orbit precession. The curve above, for 47 UMa, is nearly sinusoidal, revealing a nearly circular orbit (e = 0.096). The curve below, for 16 Cyg B, shows the distinctive signature of a much more elliptical orbit (e = 0.67).*

stars is given in Table 11.2 along with the above-mentioned four parameters, orbital period P, semimajor axis a, orbit's eccentricity e and planet's mass m in units of Jupiter's mass M_{JUP}. The magnitude of the semimajor axis a has been obtained directly from the Keplerian relationship $a \sim P^{2/3}$, taking, however, the sum mass of the system, i.e. the star plus the planet mass as constant for all stars which is a reasonable assumption beeing in mind that $a \sim M^{-1/2}$.

Two cases are of special interest, i.e. when two planets were discovered, of periods 2.9861 and 29.83 days, around one and the same star, HD 83443, and ϵ And, around which three planets were discovered of periods 4.617, 241.2 and 1266.6 days. In the first case (2.9861) the planet is exactly ten times nearer to its central star than Mercury is to the Sun.

T a b l e 11.2

Extrasolar Planets

Orbital period P in days, semimajor axis a in AU, orbit's eccentricity e and planet's mass m in units of Jupiter's mass M_{JUP} for the planets discovered around 46 neighbor stars. In the last column, the orbital velocities V_{cir} for discovered planets are given. In the last row the data for Mercury are given.

S t a r	Star's spectral type	Orbital period days	Semimajor a x i s AU	Orbital eccentricity e	Planet's mass M_{JUP}	Circular velocity km s^{-1}
ε Eridani	K2 V	2502.	3.3	0.608	0.86	16
14 Her	K0 V	1619	2.5	0.3537	3.3	19
HD 190228	G5 IV	1127	2.31	0.43	4.99	20
47 UMa	G1 V	1095	2.10	0.096	2.41	21
HD 10697	G5IV	1083	2.0	0.12	6.59	21
16 Cyg B	G1.5 Vb	804	1.70	0.67	1.5	23
HD 222582	G5	576	1.35	0.71	5.4	26
HD 19994	F8 V	454	1.3	0.2	2.0	26
HD 82943	G0	442.6	1.16	0.61	2.24	27
HD 210277	G0	437	1.097	0.45	1.28	28
HD 177830	K0	391	1.00	0.43	1.28	30
HD 92788	G5	340	0.94	0.36	3.8	31
HR 810	G0 V	320.1	0.925	0.161	2.26	31
HD 12661	K0	264.5	0.789	0.33	2.83	34
HD 134987	G5 V	260	0.78	0.25	1.58	34
HD 89744	F7 V	256	0.88	0.7	7.2	32
HD 169830		230.4	0.823	0.34	2.96	33
HD 37124	G4 IV-	155	0.585	0.19	1.04	30
GJ3021	G6 V	133.82	0.49	0.505	3.31	43
HD 52265	G0 V	118.96	0.49	0.29	1.13	43
70 Vir	G4 V	116.6	0.43	0.4	6.6	44
HD 80606		112		0.93		
HD 114762	F9 V	84.03	0.3	0.334	11.	50
HD 16141	G5 IV	75.82	0.35	0.28	0.215	51
HD 121504	G2 V	64.6	0.32	0.13	0.89	53
Gliese 876	M4 V	60.85	0.21	0.27	2.1	65
HD 168443	G5	57.9	0.277	0.54	5.04	55
	G5	58	0.3		7.7	55
		1753	2.2		$\sim$ 20	18
ρ CrB	G0 Va	39.645	0.23	0.028	1.1	63
HD 83443	K0 V	29.83	0.174	0.42	0.16	72
HD 192263	K2 V	23.87	0.15	0.03	0.76	78
HD 6434	G3 V	22.09	0.15	0.30	0.48	78
HD 195019	G3 IV-V	18.3	0.14	0.05	3.43	80
Gliese 876	M4	30.1	0.12		0.6	86
		61.0	0.20		0.9	87
HD 13445	K1 V	15.78	0.11	0.046	4	90
55 Cnc	G8 V	14.648	0.11	0.051	0.84	90
HD 38529	G4	14.41	0.1293	0.280	0.81	83
HD 108147	F8/G0 V	10.881	0.098	0.558	0.34	96
HD 130322	K0 III	10.724	0.088	0.048	1.08	101

Continue Table 11.2

HD 217107	G8 IV	7.11	0.07	0.14	1.28	113
HD 168746	G5	6.409	0.066	0.	0.24	117
ε And	F8 V	4.6170	0.059	0.034	0.71	123
		241.2	0.83	0.15	2.11	32
		1266.6	2.50	0.18	4.61	19
51 Peg	G2 Iva	4.2293	0.05	0.0	0.47	134
HD 209458	G0 V	3.52478	0.045	0.0	0.69	141
HD 75289	G0 V	3.51	0.046	0.054	0.42	140
BD-10 3166	G4 V	3.487	0.046	0.	0.48	140
τ Boo	F6 IV	3.3128	0.0462	0.018	3.87	141
HD 187123	G5	3.097	0.042	0.03	0.52	146
HD 46375	K1 IV	3.024	0.041	0.	0.249	148
HD 83443	K0 V	2.9861	0.038	0.08	0.35	154
		29.83	0.174	0.42	0.16	72
Mercury		88	0.39	0.20	0.00017	48

Historically, the first to be discovered was planet around 51 Peg, a G 2–3 main sequence star, in October 1995 by Mayor and Queloz using a spectrograph on the 1.9m telescope at Haute-Provence observatory in France, achieving a velocity resolution of 12 m per second. They discovered a wobbling of the star with a semiamplitude of 51 m per second. This result was immediately confirmed by several teams. Moreover, the data points fit a perfect sine curve, as shown in Fig. 11.15, implying that the planet's orbit is circular. The planet, with a mass at least half of the mass of Jupiter, is a distance from the central star of 0.047 AU, eight times closer than Mercury is to the Sun. At such a distance the planet should be heated to over 1000 K by a stellar disk appearing 10° wide in the sky.

At around the same time a planet around 16 Cyg B was discovered, orbiting with an extremely elongated ellipse; its eccentricity was one of the largest, $e = 0.67$.

The data in Table 11.2 give us a possibility for some conclusions. First, all planets discovered around these stars are of masses $m \sin i < 7$ M_{JUP}, although companions of 20–80 M_{JUP} mass would have been much easier to detect. The planets' mass function rises steeply toward smaller masses but it turns over 0.5 M_{JUP}, presumably because of poor detectability. The lowest known value of $m \sin I$ in Table 11.2 is 0.16 M_{JUP} (HD 83443). To the point, in four cases, HD 16141, 83443, 46375 and 168746, the planet's mass is even less than Saturn's mass (0.30 M_{JUP}). However, for two-thirds of cases the masses of the planets are larger than Jupiter's mass.

Such a distribution of planets' mass, placed at $\sim M_{JUP}$, is not the case for Solar system and, what is important, no existing theories of planet formation predict such a peak (Levison, 1998). Apparently, giant planets that end up closer to their stars than the Solar asteroid belt did not keep their circular orbits.

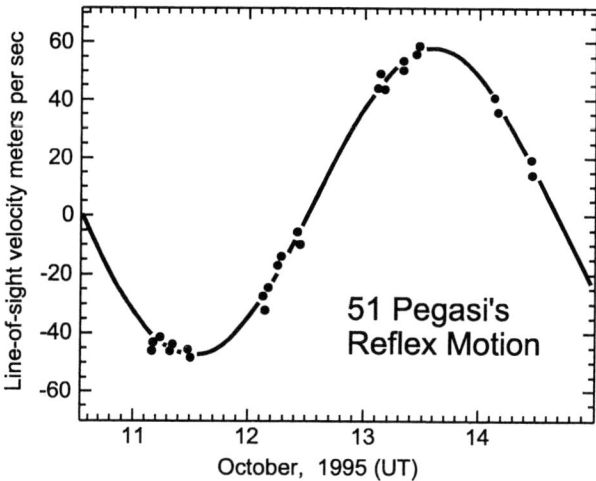

Fig. 11.15. *The radial velocity changes of 51 Peg within the limits of ±51 m per second stipulated by the presence around this G 2 type star of a planet of half the mass of Jupiter at a distance ∼ 0.05 AU from the central star. The observational points fit a perfect sine curve of circular orbit.*

It is also puzzling that a large number of planetary companions have dramatically elliptic orbits, of eccentricities larger than 0.3, up to 0.6 and even 0.7 (!). The distribution of the planet's orbit by eccentricities is as follows:

$$
\begin{array}{lll}
\text{With} & e > 0.3 & 16 \quad \text{orbits} \\
& e > 0.4 & 11 \quad " \\
& e > 0.5 & 8 \quad " \\
& e > 0.6 & 5 \quad " \\
& e > 0.7 & 2 \quad "
\end{array}
$$

For comparison, the largest eccentricities in the Solar system, for Mercury and Pluto, are about 0.2. All other planets have nearly circular orbits. Even with such an eccentricity Mercury's smallest distance is still rather too far from the Sun.

Why do those stars have planets in such strongly elliptical orbits? Especially striking are the orbits of planets around 16 Cyg B, HD 222582 and HD 89744. These highly elongated orbits intersect, for example, all circular orbits of four inner planets of the Solar system, as shown in Fig. 11.16 (above), Mercury, Venus, Earth and Mars. Variations of the distances from the central star in the case of the planet of 16 Cyg B are more than five times – from 0.56 AU up to 2.8 AU. The variation of the radiation flux on the surface of this planet from its central star will yield at least twenty-five fold! It is difficult to imagine the consequences of such dramatic changes on the environment of this planet during an orbital cycle (804 days).

However, most mysterious is the fact of concentration of more than half of the planets – 25 from 46 – within an orbit smaller than Mercury's orbit; the orbits for 12 such planets are shown in Fig. 11.16 (above). That is not all, the radii of the orbits at least of 14 planets are less than 0.1 AU. These are the planets with orbital periods from 3 days up to 15 days, compared to Mercury's 88-day orbit around the Sun. Five to six of these planets are orbiting their central stars at distances ten times closer than Mercury's orbit, hence, the radiation fluxes onto their surfaces will be at least a hundred times larger on Mercury's. The temperature on Mercury's surface is above 400 C, so how high should be the temperatures on the surfaces of these heated planets? Recall that all these planets are supermassive, of Jupiter's mass (see Table 11.2). Moreover, these planets are orbiting their central stars with fantastic circular velocities, up to 150 km s^{-1} (last column of Table 11.2).

How did these giant planets preserve their existence, and how did these supermassive planets end up so close to their central stars? Their existence contradicted existing Solar System formation theories, according to which the giant planets must form at least several AU from the central host star.

The images of the surfaces of planets and their satellites, as well as of asteroids, minor planets, comets etc. obtained with the help of interplanetary missions (Chapter XV), witness with all persuasiveness how diverse the cosmogonical past of the Solar System has been. Eros, for example, a cylindrical minor planet-asteroid (Plate V), in the images obtained by a space probe, reveals a huge signature of an impact with another body.

Even more remarkable is the giant 10 km diameter impact crater on the 20–25 km Phobos (Plate IV), i.e. the crater occupies nearly half of that Martian satellite. No less striking, for example, is the shape and structure of Miranda, one of the Uranus' moons (Plate V), carrying the signs of powerful bombardment.

Thus, the surfaces of planets and their moons without exception, from Mercury to Pluto, are covered by an infinite number of impact craters of all sizes, from tiny ones up to giant ones, of a hundred and more kilometers in size. All these impact formations are old, comparable with the ages of the planets and satellites. These facts indicate the strong active stage of the interplanetary medium in its early history, i.e. when the planets and their satellites were under long and powerful bombardments by large enough bodies.

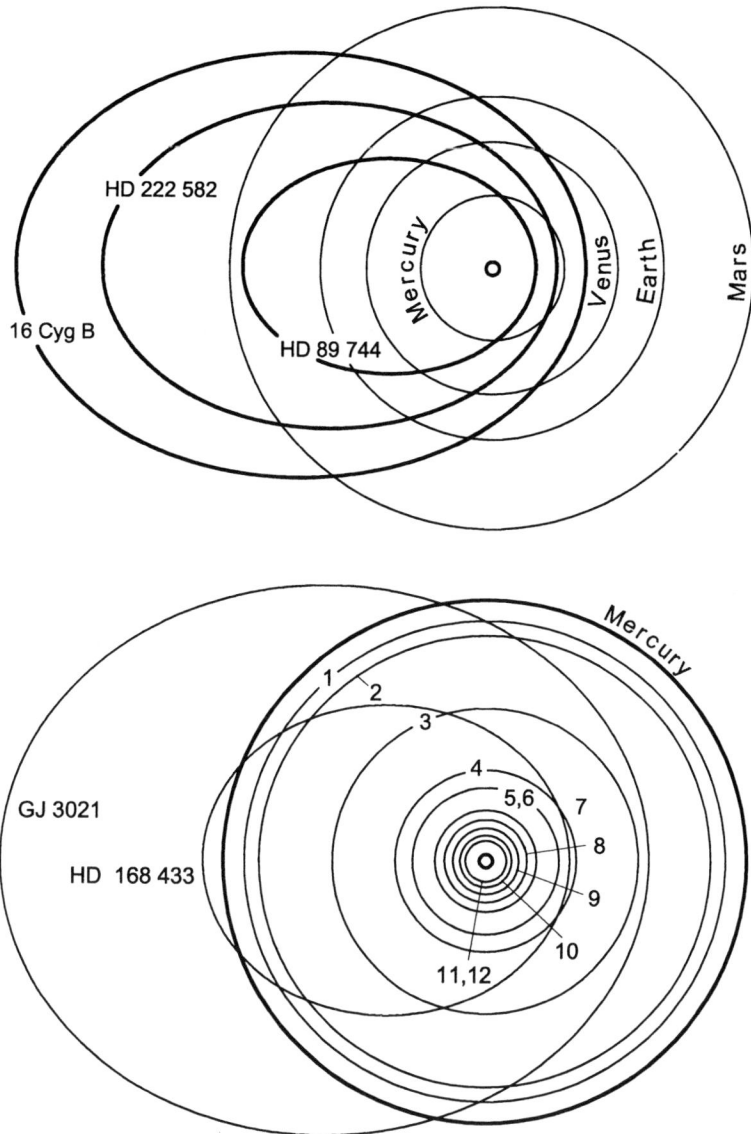

Fig. 11.16. *The orbits of 17 extrasolar planets imposed upon the orbit of four inner planets of the Solar system, up to Mars (above) and Mercury (below). The overwhelming majority of planets are concentrated within Mercury's orbit. The elliptical orbits of three extraordinary planets with large eccentricities, 16 Cyg B, HD 222582 and HD 89744, are shown as well. In the lower figure, some of the planetary orbits from Table 11.2 are drawn: 1. HD 114762, 2. HD 16141, 3. ρ CrB, 4. HD 195019, 5. HD 13445, 6. ρ Cnc, 7. HD 21707, 8. ε And, 9. 51 Peg, 10. τ Boo, 11. HD 46375, 12. HD 187123.*

It is difficult to doubt that a sufficient part of interplanetary matter was expelled from the Solar system not only in the form of diffuse matter but also as large solid bodies. This aspect, the escape or expulsion of large-scale bodies from the Solar system, should by considered as one of the important events of the Solar system in that period.

Undoubtedly, quite soon the number of discovered extrasolar planets will amount to many thousands. and an enormous amount of information will be accumulated about those planetary systems. The next generation of space telescopes will catch objects up to 31–32^m, and hence should obtain their direct images, etc.

A special space mission, Terrestrial Planet Finder (TPF), is also proposed for the study of extrasolar planets, as a cosmic interferometer of varying base from 75 m up to 1000 m of four telescopes of 3.5m mirror each. It will study 250 selected stars in up to 50 light-year distance to discover planets of the mass of Saturn and even of the Earth. TPF is supposed to locate on the orbit around the Earth passing the Earth–Moon libration point L_2. Combined particularly with the efficiency of the Next Generation Space Telescope (NGST), this can indeed lead to an essential breakthrough in the study of extrasolar planets. Apparently, only then the general theory of formation of planetary systems around individual stars can have enough basis for development.

Chapter XII

Trajectories of Interplanetary Flights

1 Peculiarities of Interplanetary Flights

By interplanetary flights we understand uncontrolled journeys of a space probe, launched in particular from the Earth, to a definite point of the Solar system.

From the point of view of celestial mechanics, interplanetary flight implies first of all the passage of a space probe into the zone of Solar gravity after its departure from the Earth and escape from the zone of action, so that the motion of the probe is determined by the influence of the Sun. The probe escaping from the Earth thereby becomes an independent member of the Solar system, and therefore its motion within the interplanetary space can, to a first approximation, be described by the laws of the two-body problem with the Sun as the central body (Celnikier, 1993). As has already been described, that motion around the central body must take place according to a conical section – an ellipse, parabola or hyperbola.

Thus, space probe that has left the Earth can reach any point of the Solar system only via a certain trajectory – by a conical section, and, which is important to outline, without any waste of energy during its journey, unless some means are used to accelerate or decelerate its motion or to change the trajectory.

From all three types of conical sections we have the smallest initial velocity in the case of an elliptic orbit, as was shown in Chapter III. Hence, from the energetic point of view, the elliptic orbit must be accepted as the most economic one for such flights. Therefore, one can conclude that the flights from a given point of the Solar system to another should occur in elliptic orbits, with the Sun at one of the focuses.

The probe launched from the Earth towards any of the planets should in the most general case perform an arc of an ellipse, with the Earth at one end and the destination planet at the other.

The destination of the space probe can be at various distances from the Earth and differently oriented with respect to the Sun. As a result, we will have ellipses of various sizes and spatial orientations. As was shown in Chapter III (Section 8), the sizes and orientations of ellipses are determined particularly by the magnitude and direction of the initial velocity of the space probe.

In view of all these factors, one can formulate the basic properties of interplanetary flights in the following manner.

i. Interplanetary flight from one point of the Solar system to another should be realized in an elliptical orbit with the Sun located at one of the focuses.

ii. The size and the orientation of an elliptic orbit are in every case determined by the magnitude and direction of the initial velocity.

One can ask whether interplanetary flights are possible, say, in parabolic orbits. In principle, yes. Moreover, such orbits possess definite advantages, namely, they give a gain in time, i.e. the duration of the flight will be significantly less. However, a parabolic orbit requires an initial velocity $\sqrt{2}$ times greater than that for the corresponding elliptic orbit, and therefore, from the energetic point of view, parabolic orbits are not preferable, since we need a rocket that is twice as powerful.

2   Hohmann Ellipses

Consider the flight of a space probe from the Earth to an outer planet, e.g. Mars. After leaving the Earth's sphere of influence, the probe will appear within the region of Solar control, so its trajectory must be a Keplerian ellipse. In general, this ellipse can intersect with the Martian orbit at some point M_2 (Fig. 12.1). Formally, the intersection may occur also at the point M_2', after passage through the apogee of the ellipse $E_0 M_2 M_2'$. However, this part of the trajectory is rather long, and, more importantly, the duration of the flight will be incomparably longer, taking into the account the decrease of the velocity in motion far from the Sun.

The elliptic orbit of type $E_0 M_2 M_2'$ (Fig. 12.1) intersecting with the Martian orbit is not the optimal one, first of all from the **energetic point** of view. Note that the encounter of the probe with Mars at the point M_2 (or M_2') takes place with a velocity significantly exceeding the Martian orbital velocity and, hence, if the aim is not a flyby of Mars but rather a landing on Mars, one needs additional energy.

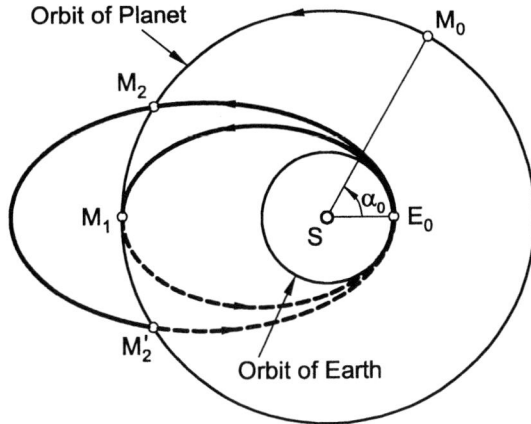

Fig. 12.1. *The flight from the Earth to an outer planet in an intersecting ellipse $E_0 M_2 M_2'$ as well as in a contacting ellipse $E_0 M_1$.*

With decreasing size of the ellipse $E_0 M_2 M_2'$, the initial launch velocity of the probe from E_0 in the Earth orbit should decrease, and the orbit will become more economical. On the other hand, the semimajor axis of the ellipse a, chosen for the orbit, cannot be smaller than half the sum of the radii of the orbits of Earth and Mars, a_0 and a_M, respectively; otherwise, the probe would not reach its destination. It is not difficult to see that the ellipse with the semimajor axis satisfying the condition

$$2a = a_0 + a_M \tag{1}$$

will be the optimal one for the flight from the Earth to Mars. In fact the question concerns only half of that ellipse $E_0 M_1$, as shown in Fig. 12.2, where E_0 is the Earth, occasionally at the perihelion, and M_1 is Mars, at the apogee with an angle of transit 180°.

A transition trajectory of $E_0 M_1$ type from the Earth's orbit to the Martian one and to any outer planet is called a **semi-elliptical** or more often a **Hohmann orbit**, after Hohmann, who in 1925 proposed such orbits for interplanetary flights. The most remarkable property of this ellipse is that it is in contact with the circular orbit of the Earth from the outside and with the circular orbit of Mars from inside, exactly at the points of the perihelion and the apogee. At this, the large semiaxis of this ellipse passes through the center S of both circular orbits, of the Earth and Mars.

It is not difficult to obtain the basic parameters of a Hohmann trajectory. The semimajor axis a is given by the relationship (1), i.e.

$$a = \frac{a_0 + a_M}{2} = \frac{a_0}{2} \left(1 + \frac{a_M}{a_0} \right) . \tag{2}$$

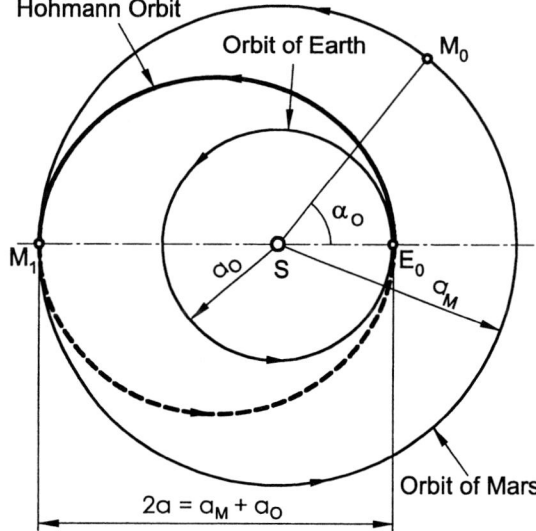

Fig. 12.2. *The Hohmann trajectory $E_0 M_1$ of the flight from the Earth to Mars. M_0 is Mars' position at the moment of the launch of the spacecraft.*

We obtain the eccentricity of the ellipse e from the following obvious relationship:

$$e = \frac{a_M - a_0}{a_M + a_0} = \frac{a_M/a_0 - 1}{a_M/a_0 + 1} \, . \tag{3}$$

One of the important parameters, the duration of the flight in the Hohmann trajectory, T_f, is determined as half the full period of rotation of the probe in this ellipse, namely,

$$T_f = \frac{P}{2} = \frac{\pi a^{3/2}}{k\sqrt{M}} \, , \tag{4}$$

or, on substituting the value of a from Eq. (2), we have

$$T_f = \frac{\pi a_0^{3/2}}{2^{3/2} k\sqrt{M}} \left(1 + \frac{a_M}{a_0} \right)^{3/2} \, . \tag{5}$$

Bearing in mind that

$$\frac{2\pi a_0^{3/2}}{k\sqrt{M}} = P = 1 \; sidereal \; time \, ,$$

we obtain from Eq. (5)

$$T_f = \frac{1}{4\sqrt{2}} \left(1 + \frac{a_M}{a_0} \right)^{3/2} \; stellar \; years \, . \tag{6}$$

Note that, as one might expect, the duration of the flight T_f does not depend on the eccentricity of the Hohmann ellipse.

The motions of all outer planets in their orbits take place more slowly than does the Earth's motion. Therefore the encounter of the probe with Mars at the angular distance of flight 180° can be performed only with a particular mutual initial configuration of the Earth and Mars on their orbits; the latter is determined by the angle α_0 (Fig. 12.3):

$$\alpha_0 = 180° - \omega_\star T_f , \qquad (7)$$

where $\omega_\star$ is the mean angular velocity of the planet (Mars) in its orbit. Obviously $\omega_\star T_f$ is the arc of the orbit passing by the planet during the flight period T_f of the probe in a Hohmann orbit. It follows from Eq. (7), that at the moment of departure from the Earth, $\alpha_0 < 180°$, i.e. Mars, and in general, any other outer planet must be in front of the Earth.

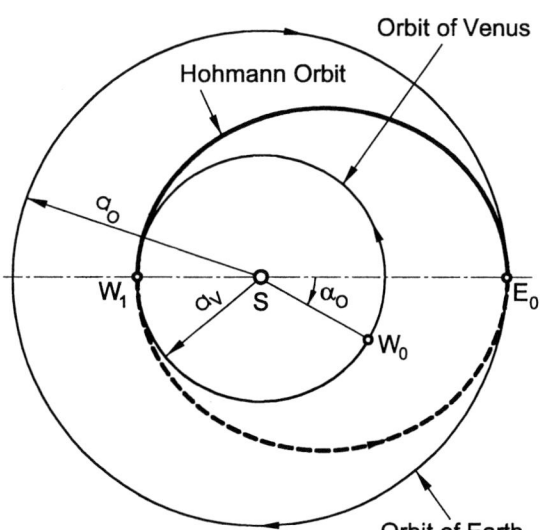

Fig. 12.3. *The Hohmann trajectory E_0W_1 of a flight from the Earth to Venus. W_0 is the position of Venus at the moment of launch of the space-craft.*

One of the important characteristics of the Hohmann orbit is the magnitude of the initial velocity V_0, i.e. the velocity of departure of the probe from the Earth. Hohmann velocity V_0 is formed by two components, namely, the velocity of the Earth's orbital motion and an additional velocity for the transference of the spacecraft from Earth's orbit into the Hohmann ellipse. The magnitude of V_0 determines by the formula

$$V_0 = (V_{ad}^2 + V_\star^2)^{1/2} , \qquad (8)$$

where $V_\star = 11.186 \ km \ s^{-1}$ is the second space velocity, and V_{ad} is the additional velocity, which, superimposed upon the Earth's own velocity, ensures the attainment of the Hohmann orbit of given parameters for the probe. Bearing in mind that both orbits, the Hohmann one and the Earth's, are in contact at the starting point, V_{ad} can be represented as a simple difference between the orbital velocities of the two orbits, i.e.

$$V_{ad} = V_{Ho} - V_{Er} . \qquad (9)$$

For the Earth $V_{Er} = 29.785 \ km \, s^{-1}$, and V_{Ho} is determined from the well-known relationship

$$V_{Ho}^2 = k^2 M \left(\frac{2}{r} - \frac{1}{a_0} \right) , \qquad (10)$$

where $a_0 = r$ should be replaced by its value from Eq. (2). Then we have

$$V_{Ho}^2 = 2 \, \frac{k^2 M}{a_0} \, \frac{a_v/a_0}{1 + a_v/a_0} . \qquad (11)$$

However, the expression $k^2 M / a_0$ is nothing other than the square of the circular velocity of the Earth on its orbit, i.e.

$$V_{Er}^2 = \frac{k^2 M}{a_0} . \qquad (12)$$

Therefore we have for the velocity in the Hohmann orbit

$$V_{Ho} = V_{Er} \sqrt{2} \left(\frac{a_v/a_0}{1 + a_v/a_0} \right)^{1/2} . \qquad (13)$$

On substituting this into Eq. (9), we obtain for the additional velocity

$$V_{ad} = V_{Er} \left[\sqrt{2} \left(\frac{a_v/a_0}{1 + a_v/a_0} \right)^{1/2} - 1 \right] . \qquad (14)$$

Eqs. (2), (3), (6), (7) and (14) may be used for determination of the parameters of Hohmann trajectories of the flights to any of the outer planets, i.e. to Mars and farther, with the substitution of appropriate values of $a_\star$ and $\omega_\star$ for the given planet.

3   Trajectories of Flights to Mars

Let us apply the formulas that were derived above for determination of the parameters of the flight of a space probe to Mars in the Hohmann trajectory.

The common configuration of all three orbits, those of the Earth (circular), Mars (circular) and Hohmann (ellipse), is shown in Fig. 12.2. Mutual positions of the Earth, E_0, and Mars, M_0, at the moment of the probe leaving the Earth are shown. The encounter of the probe with Mars occurs at M_1. One of the focuses f of the Hohmann ellipse coincides with the Sun.

We have for the radius of the orbit of Mars $a_m = 1.524$ (in units of the radius of the Earth's orbit $a_0 = 1$ $a.u.$). Then, from Eqs. (2) and (3) we obtain for the semimajor axis and eccentricity of the Hohmann ellipse

$$a = 1.262 \ a.u. \ ,$$

$$e = 0.208 \ .$$

For the duration of the flight T_f of the probe from the Earth to Mars we have

$$T_f = 0.71 \ year = 259 \ days \ ,$$

i.e. nearly nine months. Correspondingly, for the angle of the initial configuration α_0 we have from Eq. (7), taking for Mars $\omega_\star = 0.524$ $degree \ day^{-1}$; that

$$\alpha_0 = +44°.3 \ .$$

The Hohmann orbit with these parameters is reached whit an initial velocity V_0 of the probe launched from the Earth. This velocity is determined by Eq. (8) and represents the sum of two velocities, namely, of the second space velocity $V_\star$ and of an additional velocity V_{ad}. For the additional velocity V_{ad} we have from Eq. (14):

$$V_{ad} = 2.945 \ km \ s^{-1} \ .$$

On substituting this into Eq. (8), we obtain for the initial velocity

$$V_0 = 11.567 \ km \ s^{-1} \ .$$

The most important elements of the Hohmann trajectory, a, T_f and e, increase in accordance with the above formulas with increasing radius of the planet of destination. Therefore, the largest values for these parameters are those that we will have in the case of Pluto: $V_0 = 16.27 \ km \ s^{-1}$ and $T_f = 45.6 \ years(!)$.

The heliocentric velocity of the spacecraft at the moment of arrival at Mars by Hohmann trajectory is 21.54 $km \ s^{-1}$. The Mars' orbital velocity is 24.11 $km \ s^{-1}$. Hence, at the moment of arrival, Mars is moving faster than the spacecraft on an amount $V_{rel} = 2.57 \ km \ s^{-1}$. Thus, in order to arrive at Mars it is still necessary for the spacecraft to gain an additional velocity. Besides this, one has to overcome the gravity of Mars for soft landing on its surface. Because of the strong rarefaction of the Martian atmosphere it is not possible to rely on atmospheric resistance. The solution of both these problems is impossible without jet engines.

4   Flights to Mars

There are many reasons making flights to Mars a kind of standard, among the infinitive number of varieties of space routes. In particularly, in Fig. 12.2, we analyzed the Hohmann trajectory. One can think that, Mars can indeed become a "utilized" planet, the most likely planet for future space programs up to the formation of "Martian Civilization"....

The minimum velocity V_0 of the flight from the Earth to Mars by a Hohmann trajectory is equal to 11.567 $km\,s^{-1}$ (the trajectory shown in Fig. 12.4). At this trajectory we have the largest duration of the flight, 259 days, nearly nine months. It definitely cannot be considered as the most economic version, bearing in mind the expenditures of energy on inevitable corrections of the trajectory.//[6mm]

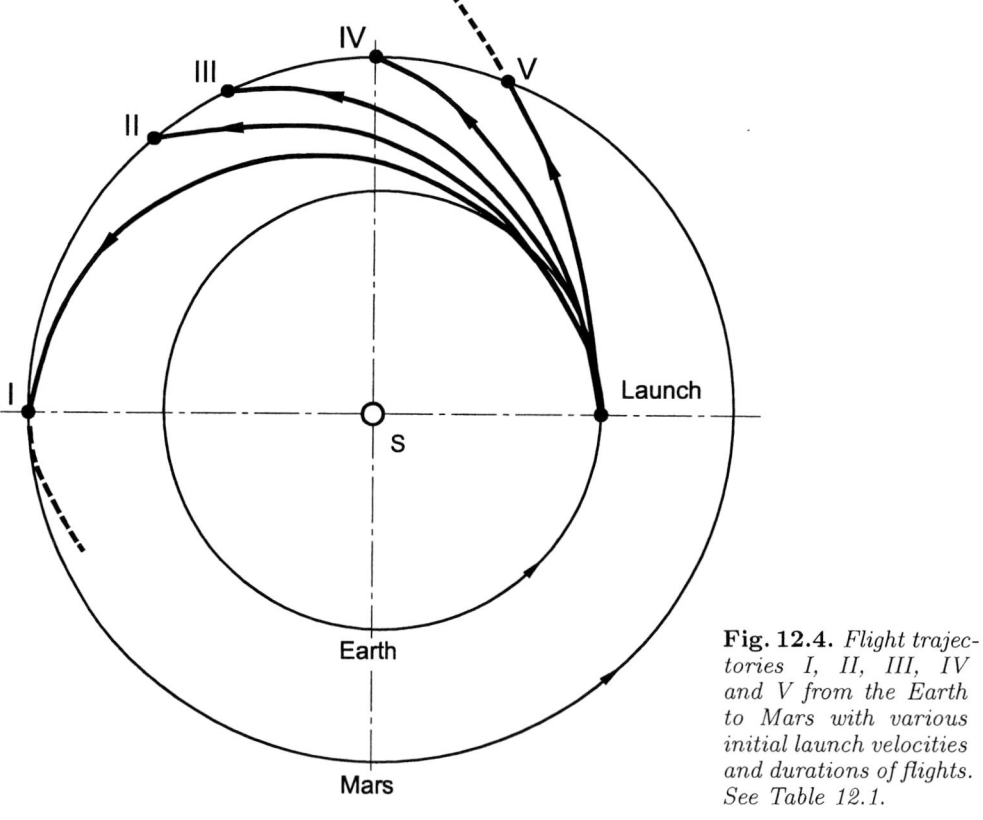

Fig. 12.4. *Flight trajectories I, II, III, IV and V from the Earth to Mars with various initial launch velocities and durations of flights. See Table 12.1.*

T a b l e   12.1

Parameters of the flight to Mars by trajectories I, II, III, IV and V

	I (*)	II	III	IV	V (**)
Initial velocity of launch, $km\,s^{-1}$	11.56	11.80	12.00	13.00	16.53
Semimajor axis of flight orbit, a, *a.u.*	1.26	1.37	1.46	1.98	∞
Eccentricity, e	0.208	0.268	0.313	0.494	1
Distance from the aphelion, *a.u.*	1.52	1.73	1.91	2.96	∞
Duration of flight, *days*	259	164	144	105	70

(*)   Hohmann trajectory
(**)   Parabolic trajectory

In Table 12.1 we present the results of calculations for flight trajectories by crossing ellipses II, III and IV with progressive initial velocities of launch. As we see, by increasing the launch velocity the flight duration is shortened strongly. For example, with an increase of the initial velocity to 0.23 $km\,s^{-1}$ (trajectory II), the spacecraft will arrive at Mars in 164 days, i.e. three months faster. For a flight by parabolic orbit (trajectory V), i.e. at the third space velocity $V_0 = 16.65\ km\,s^{-1}$, the flight duration will be only 70 days, i.e. the spacecraft will arrive at Mars half a year earlier than on the flight by Hohmann trajectory (I). However, then the energy spent will be twice as large.

Flights by any of these trajectories, I–V, are possible at an interval of 26 months, i.e. two years and four months. The same concerns the intervals for return flights from Mars. Thus, the duration of a single return flight to Mars will be nearly four and half years; faster is impossible, at least theoretically. The favorable period for the start of flights from the Earth by these orbits are within only 20 days. If this period is missed, one must wait 2.17 years for the next favorable moment for launch.

Any possible flight from the Earth to Mars should be on a plane passing through three points: the Earth at the moment of launch, the Sun and Mars during the flight. Flight by a Hohmann trajectory, i.e. diametrically in the opposite point on Mars' orbit with an angular distance of 180° is possible only if the flight begins and ends on the line of nodes. However, satisfying this and other demands is practically impossible, therefore periodic corrections of the trajectory and the flight's regime on the Earth–Mars route is inevitable.

The diversity of problems and the purposes of flights to Mars strongly depend on the combination of the required trajectory and the dates of the launch from the Earth and arrival at Mars. An example of such an interesting trajectory is shown in Fig. 12.5. The spacecraft is dispatched not by a Hohmann trajectory but by a trajectory providing the return flight from

Mars without any waiting near Mars. Upon the arrival of the spacecraft at Mars, a probe will be separated from the spacecraft which will then be braked and captured by the gravity field of Mars. After performing its task in the neighborhood of Mars, the probe will return to the circum-Martian orbit to join the main spacecraft with a hyperbolic velocity, for their joint return to the Earth. The essence of this version is the guarantee of a return to Earth at an exactly fixed date.

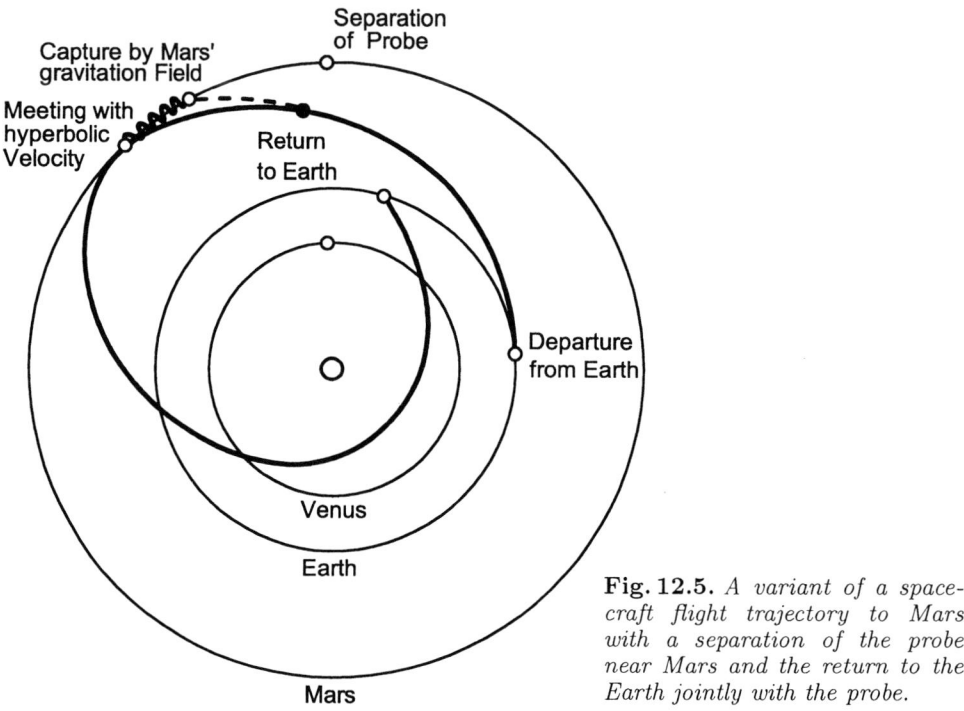

Fig. 12.5. *A variant of a space-craft flight trajectory to Mars with a separation of the probe near Mars and the return to the Earth jointly with the probe.*

The possibilities of flight trajectories to Mars are too broad. In Fig. 12.6 a calculated trajectory of the flight by a short route Earth–Mars–Earth without landing on Mars is given, along with the concrete data of various operations. A 10 day delay is provided near Mars, for the realization of the operation of the passage of the spacecraft into a circular orbit around Mars at an altitude of 1000 km and with the return to and landing on the Earth. Total duration of the flight is 400 days. The spacecraft arrives at the Earth with a velocity 12.008 $km\,s^{-1}$, 0.825 $km\,s^{-1}$ larger than the parabolic velocity of escape. For the landing, the velocity has to be reduced to 8 $km\,s^{-1}$ or lower.

Another example of the calculated trajectory by the route Earth–Mars–Earth for a flight period 1981–83 is presented in Fig. 12.7. The complete duration of the flight in this case is 456 days including a 20-day delay in the

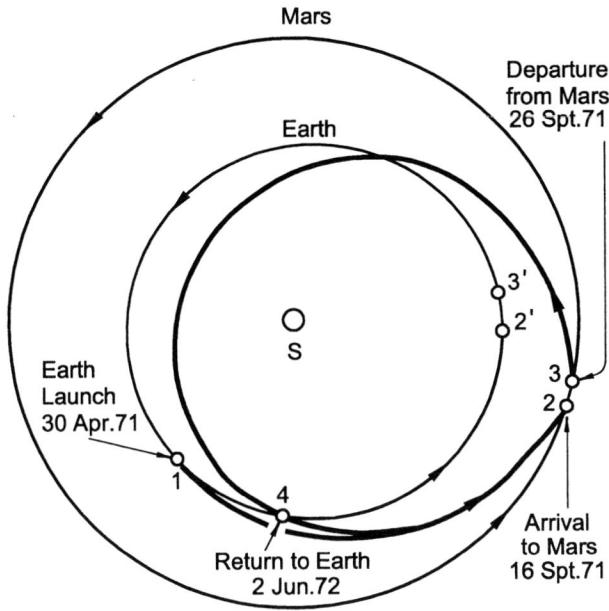

Fig. 12.6. *Theoretical flight trajectory by a route Earth–Mars–Earth. Duration of the flight is 400 days with a 10-day delay in the environment of Mars. 1. Earth launch, 2. Arrival at Mars, 3. Departure from Mars after 10-day delay, 4. Arrival to Earth. 2' and 3' are the Earth's location at Mars' positions 2 and 3.*

neighborhoods of Mars. The trajectory in the Earth–Mars part is exactly of Hohmann type. Although this trajectory twice crosses the Venus orbit, a meeting with Venus is not provided. However, in this case the entry of the spacecraft into the Earth's atmosphere takes place with a velocity larger than 14 $km\,s^{-1}$, i.e. essentially larger than in the previous case.

Another remarkable trajectory for Earth–Mars–Earth flight is shown in Fig. 12.8. Again, in the first part of this route, Earth–Mars, the trajectory is exactly of Hohmann type. The spacecraft's stay near Mars in this case is longer, 110 days. After departure from Mars, the spacecraft continues its flight by a trajectory which crosses the Venus orbit, thus encountering Venus as well. Here, in the gravity field of the Venus, the spacecraft undergoes a gravity assist (see Section 4, Chapter XIV), which noticeably reduces the total duration of the flight, making it 642 days.

The physical and atmospheric conditions on Mars strongly differ from those of the Earth. The atmosphere pressure on its surface is smaller, by a hundred times, than that on the Earth. The daytime temperature on its surface near the equator is 10–20° C, at night – 60–70° C, in polar regions up to −130° C. Winds of velocity up to 15 $m\,s^{-1}$ are common near the surface, and up to 40–50 $m\,s^{-1}$ at altitudes of 200 m. The composition of Mars' atmosphere also differs essentially from the Earth's, thus, 95% is CO_2, nitrogen (N) 2–3%, argon (Ar) 1–2%. Oxygen is only about 0.1–0.4%.

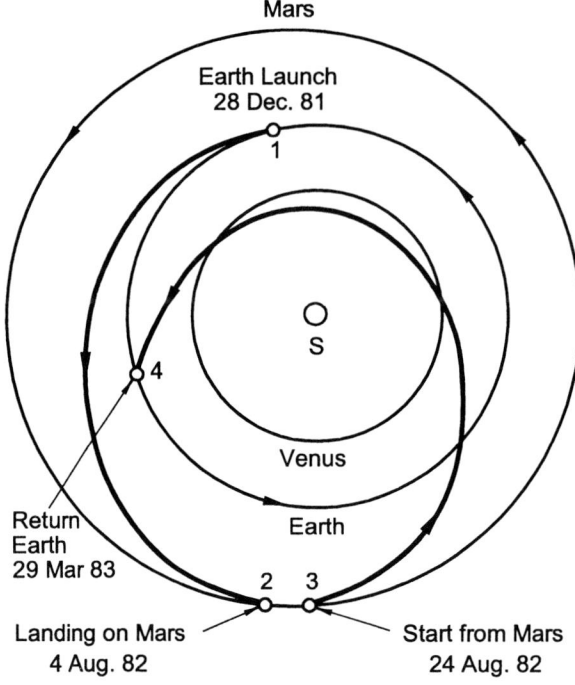

Fig. 12.7. *Calculated trajectory for 456-day Earth–Mars–Earth flight, including a 20-day delay in the vicinity of Mars. 1. Launch from Earth, 2. Landing on Mars, 3. Departure from Mars, 4. Return to Earth.*

Martian color images obtained by Voyager, Viking and other interplanetary probes from various distances are rather impressive. The surface is covered by craters formed as a result of meteorite impacts as well as of ancient volcanic activity. An enormous number of geological formations, huge canyons, are discovered, e.g. one of a depth of 6 km, up to 100–250 km in width and 4000 km in length. Numerous dried-up rivers were discovered, which undoubtedly could have been formed as a result of water flows or flows of some liquid many millions of years ago, since apparently water disappeared from the Martian surface millions of years ago. Broad valleys, mountains up to 5 km high, huge craters of diameter up to 65 km were discovered. A truly giant mountain, Olympus Moons, of altitude 26.4 km above the mean surface, was discovered as well. These facts provide a unique insight in to Martian ancient history.

However, the main feature should be considered to be the low gravity on the surface of Mars, namely, 2.6 times smaller than at the Earth's surface. This means that future Martians in principle cannot return to Earth and survive its gravity.

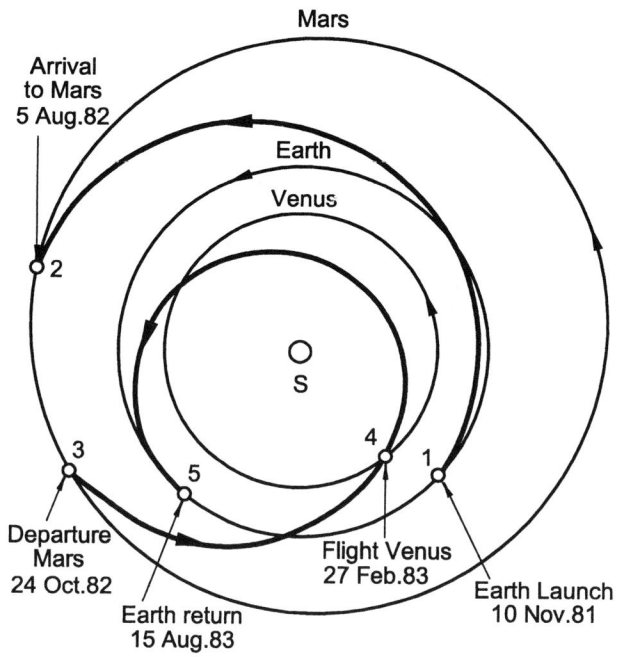

Mars

Arrival
to Mars
5 Aug.82

Earth

Venus

2

S

3

5

4

1

Departure
Mars
24 Oct.82

Flight Venus
27 Feb.83

Earth Launch
10 Nov.81

Earth return
15 Aug.83

Fig. 12.8. *Calculated 642-day duration cruise by a route Earth–Mars–Venus–Earth. 1. Launch from Earth, 2. Arrival at Mars exactly by a Hohmann ellipse, 3. Departure from Mars after 110-day delay, 4. Flyby near Venus from a distance 4700 km, 5. Return to Earth.*

5 Flights to Venus and Mercury

In the case of flights to the inner planets of the Solar system, for which $a_v/a_0 < 1$, all the above-mentioned formulas are applicable with minor changes, namely,

$$e = \frac{1 - a_v/a_0}{1 + a_v/a_0} \,,$$

$$T = \frac{1}{4\sqrt{2}} \left(1 + \frac{a_v}{a_0} \right)^{3/2} \; years,$$

$$\alpha_0 = \omega_\star T_f - 180° \,,$$

$$V_{ad} = V_{Er} - V_{IIo} = V_{Er} \left[1 - \sqrt{2} \left(\frac{a_v/a_0}{1 + a_v/a_0} \right)^{1/2} \right] \,.$$

(15)

Using these formulas for the flight to Venus for which $a_v/a_0 = 0.723$ and $\omega_\star = 1.602$ *degree day*$^{-1}$, we obtain (see Fig. 12.3)

$$a_0 = 0.861 \,, a.e. \,,$$

$$e = 0.163 \,,$$

$$T_f = 146 \; days \,,$$

$$\alpha_0 = -54° \,,$$

$$V_{ad} = 2.496 \; km \, s^{-1} \,,$$

$$V_0 = 11.461 \; km \, s^{-1} \,.$$

Note that the initial launch velocity V_0 in the case of Venus hardly differs from that in the case of Mars ($11.567 \; km \, s^{-1}$), and both differ only sightly from the second space velocity $11.186 \; km \, s^{-1}$. That is why Venus and Mars are more easily reachable for space flights. As for the duration of the flight to Venus, it is essentially shorter – nearly five months – than that to Mars. The reason is clear: the velocity of motion of any object moving in a Keplerian ellipse will increase on approaching the Sun. In general, the ratio of durations of flights in Hohmann orbits to outer planets T_{ou} and to inner planets T_{in} can be determined with the help of the following formula

$$\frac{T_0}{T_{in}} = \frac{(1 + a_v/a_0)_0^{3/2}}{(1 + a_v/a_0)_i^{3/2}} \,, \tag{16}$$

so that for Mars (T_M) and Venus (T_V) we have

$$\frac{T_0}{T_{in}} = 1.77$$

Note that, for the inner planets, particularly Venus, when the orbital velocities are larger than the Earth's velocity, the angle of initial configuration α_0 is negative insofar as, at the moment of departure of the probe, these planets are situated behind the Earth.

Venus and Mercury are inner planets and therefore the spacecraft leaving the Earth's sphere of action should move in the direction opposite to the Earth's orbital motion, as shown in Fig. 12.3. This is simply explained: the orbital velocities of inner planets are larger – $47.9 \; km \, s^{-1}$ for Mercury, $35.0 \; km \, s^{-1}$ for Venus, as compared with the Earth's velocity on its own orbit – $29.8 \; km \, s^{-1}$. Correspondingly, it is necessary to pass to the spacecraft an additional velocity which can be realized at the opposite motions through the orbits. This is contrary to the situation in the case of outer planets, which possess orbital velocities smaller as compared with the Earth's velocity.

The largest number of spacecraft have been sent to Venus, for example, up to ten spacecraft of the Venera series, also Mariner 5 and Mariner 10. In the majority of cases, the trajectories to Venus are of Hohmann type, i.e. an ellipse which contacts both the Earth's and Venus' orbits. One such trajectory realized in the flights of Venera 9 and Venera 10 is shown in Fig. 12.9. The trajectories of both flights were absolutely identical, with almost the same duration of the flights: 133–135 days, about ten days faster than predicted by the exact Hohmann trajectory – 146 days. The launch from the Earth was realized with an initial velocity 11.461 $km\,s^{-1}$.

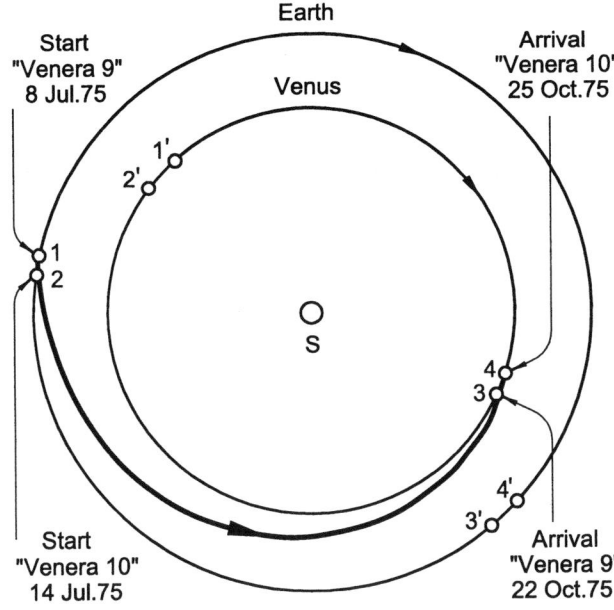

Fig. 12.9. *Heliocentric Hohmann-type trajectories of Venera 9 and Venera 10 flights from the Earth to Venus.*

Venus is considered as a planet with a large enough mass to be used for the realization of a so-called perturbative maneuver in its gravity field during a close flyby (Chapter XIV, Section 4); such a flyby can essentially reduce the duration of further flight of the spacecraft. Venus was used quite often for such a goal, as we shall see in Chapter XV while discussing flights to the outer planets, Jupiter and farther.

In Fig. 12.10 one such trajectory is shown, definitely not of Hohmann type, for the flight from the Earth to Mars through the Venus orbit where the gravitation maneuver, i.e. the so-called gravity assist, should be realized. The spacecraft arrives at Venus after 158 days and, undergoing a gravity assist in its zone, continues its flyby to Mars. After a 10-day stay in the vicinity of Mars, the spacecraft continues its return flight to the Earth.

In contrast with flights to Mars, the spacecraft arriving at Venus is leaving the planet with a velocity $\Delta V = 2.5\ km\,s^{-1}$. The escape velocity on the Venus' surface is 10.23 $km\,s^{-1}$ (circular velocity is 7.24 $km\,s^{-1}$). At an altitude of 1000 km the escape velocity is reduced to $V_{es} = 9.49\ km\,s^{-1}$. Hence, the spacecraft arrives at this planet on this altitude with a surplus of hyperbolic velocity $V_h = (\Delta V^2 + V_{es}^2)^{1/2} = 9.814\ km\,s^{-1}$. The circular velocity at this altitude is $V_c = 6.71\ km\,s^{-1}$. So, in order to transfer the

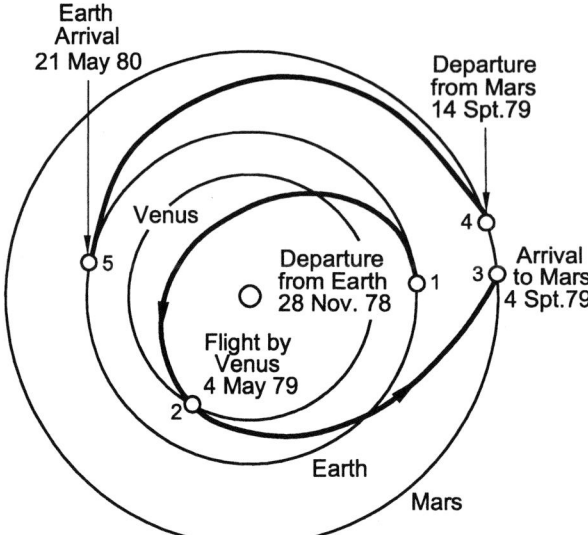

Fig. 12.10. *Trajectory of a flyby to Mars via Venus in whose gravitation field a gravity assist should be realized. The return is after a 10-day stay near Mars.*

spacecraft into a circular orbit around Venus it is necessary to diminish the velocity by the amount $\Delta V' = V_h - V_c = 8.104 \; km \, s^{-1}$, using jet engines.

As we see, for flights to both Mars and Venus, the transfer of the spacecraft into a circular orbit around those planets needs extra energy expenditure.

Venus possesses a dense atmosphere, and in contrast to Mars, the operation of landing needs no jet engines and a parachute brake is enough even at large weights of the probe.

Thus, there are two options for the realization of the landing of spacecraft on the planet's surface:

i. Landing from the circular orbit on the given point of the planet's surface, realized with a high accuracy.

ii. The spacecraft first carries out observations of the surface of the planet, and then the landing is performed.

In all cases for landing from a circular orbit it is necessary first to get rid of the momentum, and then, at the following step to perform aerodynamic braking if the planet possesses an atmosphere (Mars, Venus, Jupiter). The parachute systems and the braking and landing systems begin to operate, as a rule, automatically when they detect signals from their sensors of atmospheric pressure depending on the goals and content of the spacecraft.

The physical conditions of Venus as a planet are rather complex. Venus possesses a powerful and hot atmosphere; the pressure on its surface is 100 bar, temperature $480°$ C. The density of its atmosphere is only 14 times lower than the density of water (for comparison, Earth's atmosphere is 450 times less dense). At an altitude of 40 km the pressure is equal to 3.5 bar, and at 51 km, it is nearly one bar (in the Earth's atmosphere at this altitude the pressure is 0.001 bar). The high density of the atmosphere is the reason for poor visibility of Venus' surface; it is covered by powerful clouds. The speed of winds near its surface is small, near 1 $m\,s^{-1}$, while at higher altitudes it reaches up to $100m\,s^{-1}$. The composition of Venus' atmosphere is as follows: CO_2 is 98%, CO nearly 2%, small quantities of oxygen, carbon, water vapors.

The next and the innermost planet is Mercury. It possesses the shortest orbital period, 88 days, the smallest escape velocity, 4.2 $km\,s^{-1}$, and the largest velocity of orbital motion, 47.9 $km\,s^{-1}$, which creates definite difficulties for sending a spacecraft.

The theoretical duration of a flight to Mercury by a Hohmann orbit is 105 days at initial velocity of launch from the Earth of $13.5ofkm\,s^{-1}$. However, to send a spacecraft to Mercury without an acceleration at least via Venus' gravity field is practically impossible. Such a trajectory is shown in Fig. 12.11, which permits the spacecraft to reach Mercury after undergoing a gravity assist in Venus' gravity field. To the point, just in that way Mariner 10 realized the first gravitation maneuver in Venus' field, i.e. a flyby at velocity 10 $km\,s^{-1}$ at a distance of 5700 km from the surface of Venus. At that flyby the spacecraft velocity increased by 4.5 $km\,s^{-1}$, while deviation from this velocity by 1 $km\,s^{-1}$ from the calculated point near Venus would lead to a shift of the distance from Mercury of up to 1000 km.

The first and only spacecraft, Mariner 10, was sent to Mercury in 1973. It passed near Mercury on February 4, 1974 at a distance of 720 km from planet's surface, obtaining over two thousand color pictures of the planet's surface. After correction of the flight trajectory, the second transit of Mercury was realized on September 21, 1974 at a distance of 48 000 km, and after a year, on April 16, 1975, the third and the final meeting with this interesting planet took place..

6 Recurrence of the Primary Configuration

In order to ensure the meeting of a space probe launched in a Hohmann ellipse with a given planet at an angular separation of $180°$, it is necessary, as has been mentioned above, to launch it from a certain configuration of the Earth with respect to the planet of destination. This configuration is determined by a **configuration angle** α_0 of positive sign in the case of the outer planets, and of negative sign for the inner planets. The appearance of this configuration is determined by the launch time T_0 of the probe from the Earth. Possible deviations from the configuration angle by a magnitude

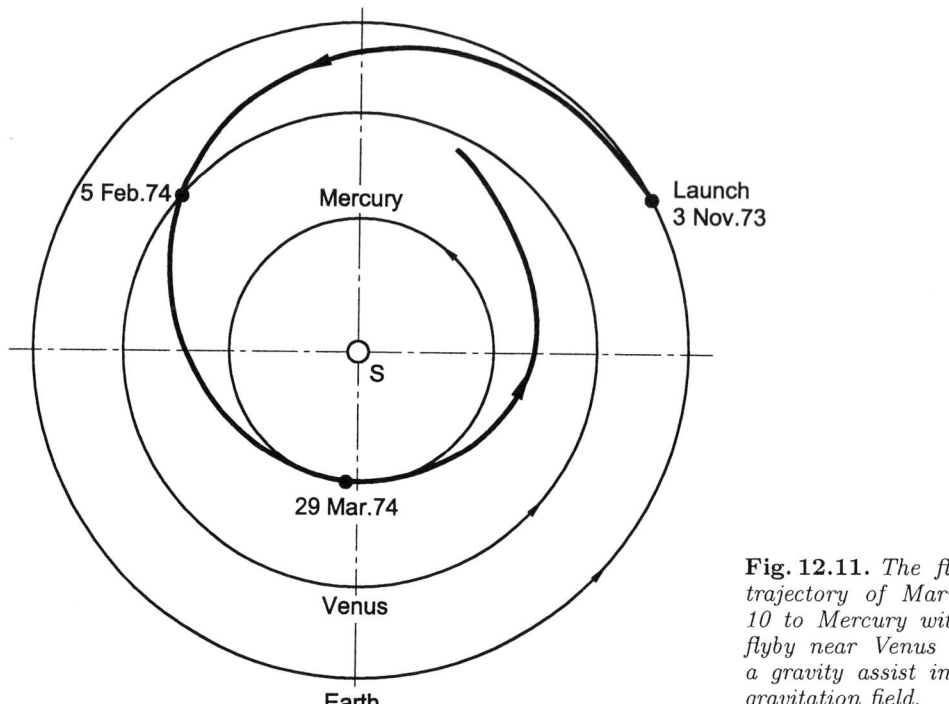

5 Feb.74

Mercury

Launch
3 Nov.73

S

29 Mar.74

Venus

Earth

Fig. 12.11. *The flight trajectory of Mariner 10 to Mercury with a flyby near Venus and a gravity assist in its gravitation field.*

$\pm\Delta\alpha_0$ and hence of time of launch T of $\pm\Delta T_0$ can modify the initial velocity from the Earth as well as having an effect on the correction of the orbit during the flight.

The question of the frequency of appearance of the required configuration and, hence, the convenient periods of time for two successive launches of the probe in the direction of a given planet, therefore arises.

The initial configuration obviously appears a certain period Δt before the moment at which the inner planet, in one case, reaching the Earth finds itself on the Sun–Earth line, and, in the second case, before the Earth reaches the outer planet on the same Sun–Planet line. This interval Δt can be determined with the help of the following formula:

$$\Delta t = \frac{\alpha_0}{\omega_0 - \omega_\star}\, T_f \,, \tag{17}$$

both for outer and for inner planets; here ω_0 and $\omega_\star$ are the mean angular velocities of the Earth and the planet, respectively; T_f is the duration of the flight. From Eq. (17) we obtain for the synodic period P_C of recurrence of the given configuration

$$P_c = \frac{P_0 P_\star}{P_\star - P_0} \tag{18}$$

for the outer planets, and

$$P_c = \frac{P_0 P_\star}{P_0 - P_\star} \tag{19}$$

for the inner ones. In these expressions, P_0 and $P_\star$ are the periods of rotation of the Earth and the given planet around the Sun, respectively.

With the help of these formula, we arrive, in particular, at the following conclusion: two successive launches of a space probe may be realized with an interval of time of 2.12 years in the case of Mars, and of 1.60 years in the case of Venus.

For the outer planets $P_\star$ is significantly larger than P_0. Then we shall have from Eq. (18), for those planets

$$P_c \approx P_0 \ ,$$

i.e. the launch of the probe in the direction of distant planets may be realized annually: e.g. $P_C = 399 \ days$ for Jupiter, and $P_C = 367 \ days$ for Pluto.

7 Flight from Mars to Venus

The search of the optimal flight trajectories from the Earth to the planets of the solar system represents first of all practical interest. As to the flights between other planets of solar system, in the present state of exploration of interplanetary space it may present at least a theoretical interest. However, since in this case the possibility of unpredicted results is not excluded, we shall examine here some of those problems.

We start from the examination of a flight from Mars to Venus by Hohmann trajectory; it is shown in Fig. 12.12. For the semimajor axis a of this semi-elliptic trajectory we have

$$a = \frac{a_V + a_M}{2} \tag{20}$$

where a_V and a_M are the semimajor axes of the orbits of Venus and Mars, respectively. Substituting their numerical values $a_V = 0.723 \ a.u.$ and $a_M = 1.523 \ a.u.$ we have: $a = 1.123 \ a.u.$

The eccentricity of this ellipse contacting with Mars' orbit from the inside and Venus' orbit from the outside, as shown in Fig. 12.12, we find from the above relationship Eq. (3):

$$e = \frac{a_M - a_V}{a_M + a_V} = 0.355 \ . \tag{21}$$

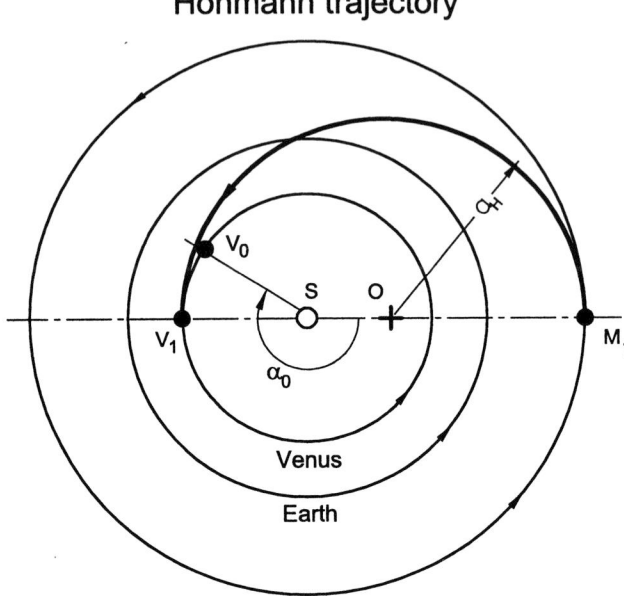

Fig. 12.12. *The Hohmann trajectory $M_1 V_1$ of a flight from Mars to Venus. V_0 is the position of Venus at the moment of launch of the spacecraft from Mars.*

For the duration of the flight T_f from Mars to Venus by Hohmann ellipse we have from the relationship Eq. (6)

$$T_f = \frac{1}{4\sqrt{2}} \left(1 + \frac{a_M}{a_V} \right) = 0.971 \; year = 354 \; days \,. \tag{22}$$

At the angular velocity of the orbit of Venus $\omega_\star = 360/P_V = 1.602 \; degree/day$, and at angular distance of flight $180°$ for the mutual initial configuration α_0 of Mars and Venus on their orbits we have

$$\alpha_0 = \omega_\star T_f - 180° = 207.°1 \,. \tag{23}$$

An important characteristics of the flight Mars–Venus is the magnitude of the initial velocity, i.e. the velocity V_0 of departure of the spacecraft from Mars to Venus; it is given by the above relationship Eq. (8)

$$V_0 = (V_e^2 + V_{ad})^2 \tag{24}$$

where $V_e = 5.0 \; km \; s^{-1}$ is the velocity of escape from Mars' surface. As to V_{ad}, the additional velocity which is superimposed upon Mars' own velocity to ensure the attainment of the Hohmann orbit at given parameters for the spacecraft, it is determined by the relationship

$$V_{ad} = V_M \left[1 - \sqrt{2} \left(\frac{a_V/a_M}{1 + a_V/a_M} \right)^{1/2} \right] = 4.83 \; km \; s^{-1} \qquad (25)$$

where we have used the value $V_M = 24.1 \; km \; s^{-1}$ for the orbital velocity of Mars. Then we have from (24)

$$V_0 = 6.95 \; km \; s^{-1} . \qquad (26)$$

With such an initial velocity the spacecraft is leaving Mars at the point M_1 (Fig. 12.12) and via the Hohmann orbit will encounter Venus exactly at the point V_1.

Comparing the result (26) with that obtained above in Section 4, $V_0 = 11.567 \; km \; s^{-1}$ for the Earth–Mars flight by Hohmann orbit, we find that for the taking off of a spacecraft from the Mars' surface it is necessary to have a velocity almost twice as small as in the case of the Earth–Mars flight. From energetic point of view, the gain is nearly quadruple.

8   Flight from Mars to Earth

As a second example of flights from the outer planets to the orbits of inner planets, we shall examine here the conditions of flight from Mars to Earth by Hohmann trajectory. For many reasons, this mission, Mars–Earth, must present a special interest.

In this case, as follows from Fig. 12.13, we have for the main semimajor axis of Hohmann orbit: $a_M = 1.52 \; a.u.$ for the Mars and $a_E = 1.00$ for the Earth. Then we find from (1) for the semimajor axis of Hohmann ellipse: $a = 1.262 \; a.u.$ with an eccentricity $e = 0.206$ according to relationship (2). The duration of the Mars–Earth flight T_f by such an orbit will be, in accordance with (3)

$$T_f = \frac{1}{4\sqrt{2}} \left(1 + \frac{a_M}{a_E} \right)^{3/2} = 0.709 \; year = 259 \; days \qquad (27)$$

i.e. nearly 100 days shorter than we had in the case of Mars–Venus flight.

During such a flight time and with an angular velocity of Earth's orbital motion $\omega_\star = 360/P_E = 0.985 \; degree/day$, the spacecraft will cover the arc $\omega_\star T_f = 255°.1$.

Therefore the meeting of the spacecraft sent from Mars to the Earth at the angular distance of flight 180° can occur only at the maintenance of the initial configuration both of Mars and the Earth on their own orbits determined by an angle

$$\alpha_0 = \omega_\star T_f - 180° = 75°.1 .$$

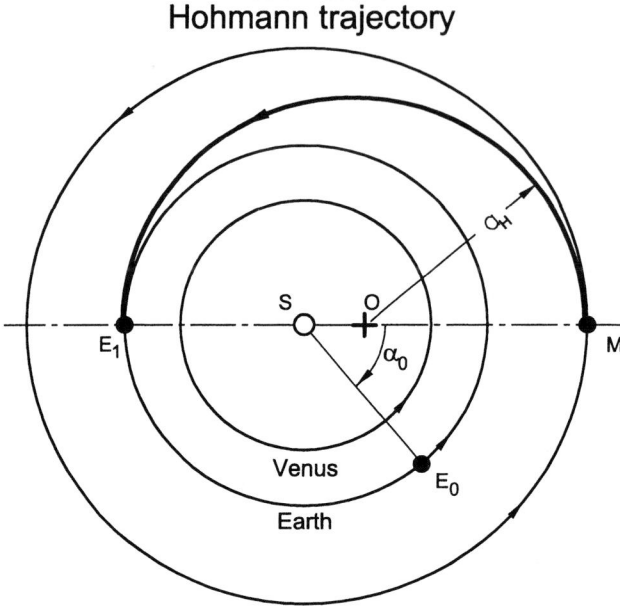

**Mars - Earth
Hohmann trajectory**

Fig. 12.13. *The Hohmann trajectory M_1E_1 of a flight from Mars to the Earth. E_0 is the position of the Earth at the moment of the launch of the spacecraft from Mars.*

For additional velocity V_{ad} which must be passed to the spacecraft in order to ensure that it enters the Hohmann orbit, we have

$$V_{ad} = V_M \left[1 - 2 \left(\frac{a_E/a_M}{1 + a_E/a_M} \right)^{1/2} \right] = 2.70 \ km \ s^{-1} \qquad (28)$$

where again $V_M = 24.1 \ km \ s^{-1}$ is Mars' orbital velocity, and $a_E/a_M = 0.658$. From here we find, for the second space velocity of escape of the spacecraft from Mars' surface which is necessary for the landing of the spacecraft on the Earth, again at $V_E = 5.0 \ km \ s_1$,

$$V_0 = (V_e^2 + V_{ad})^{1/2} = 5.68 \ km \ s^{-1}, \qquad (29)$$

That is, in the case of the flight from Mars to the Earth we have an initial velocity noticeably, by 20%, smaller as compared with the Mars–Earth flight.

Thus, the smallest escape velocity $V_0 = 5.68 \ km \ s^{-1}$ corresponds to the flight from Mars to the Earth. This somewhat unexpected result has far going consequences. First of all, it means that from the energetic point of view the Mars–Earth flight will be $(6.95/5.68)^2 = 1.50$ times more economical as compared with the Mars–Venus flight. This remarkable conclusion finds some unexpected applications.

Assume that a piece of rock is thrown from Mars with a velocity 5.7 $km\,s^{-1}$ in the direction of its, Mars', orbital motion. This velocity is far to be enough for the flyby of the rock, say, to the Venus. Such a rock can arrive exclusively on the Earth. If this rock leaves Mars with a velocity 5.7 $km\,s^{-1}$ in the conditions noticed above, then it will have a Hohmann orbit and after 259 days will appear exactly on the Earth's orbit at the opposite point, i.e. on an angular distance of 180°, and if at that moment the Earth happens to be at this point then the rock will fall onto the Earth.

Now consider another scenario, namely, assume that on Mars sometimes volcanoes are active during the whole year and eject rocks continuously. If some of the rock pieces gain velocity 5.7 $km\,s^{-1}$, and if its vector is directed tangentially to Mars' orbit, then there will be a train of rocks moving by Hohmann trajectory up to the Earth's orbit. Since the rocks arrive continuously during the whole year, then sooner or later the Earth will appear exactly at that point and the fall of rocks onto the Earth will become an inevitable event.

Though such an event of course is less probable, the conclusion that on a scale of millions of years some rocks ejected from Mars will sooner or later appear on the Earth seems quite reasonable.

The probability of the fall of Martian rocks onto Venus is small because in this case it is necessary to have a noticeably larger outburst velocity for the rocks, nearly 7 $km\,s^{-1}$.

It is not difficult to ensure that the inverse process, namely, the transfer of rocks from the Earth to Mars by Hohmann trajectory, is impossible even theoretically. In that case the escape velocity from the Earth would have to be 11.7 $km\,s^{-1}$, i.e. twice as larger as the escape velocity from Mars, or four times more powerful from the energetic point of view.

9 Flight from Mars to Earth by Crossing Trajectory

Assume that the ejection of rocks from the surface of Mars takes place with a velocity equal to its escape velocity $V_0 = 5,68\ km\,s^{-1}$, but now in the opposite direction to its orbital motion around the Sun, as shown in Fig. 12.14. That velocity is smaller by an amount of 0.68 $km\,s^{-1}$ than the magnitude obtained above for the second space velocity $V_0 = 5.68\ km\,s^{-1}$ for the flight from Mars to the Earth by a Hohmann trajectory. Apparently, in this case the space probe cannot arrive at the point E_1 to meet the Earth. In this case, the initial velocity will be $V_H - V_e = 24.1 - 5.0 = 19.1\ km\,s^{-1}$, i.e. smaller than we had in the previous case and, hence, now we shall have a new elliptic orbit with a smaller target distance than in the first case, i.e. $a_M + a_0$. However, this new ellipse will certainly intersect with the Earth's orbit with any conceivable consequences. Our problem is to find out this new, so-called "crossing" trajectory.

We have for the orbital velocity

$$V_M^2 = k^2 M \left(\frac{2}{r} - \frac{1}{a_H} \right) . \tag{30}$$

Writing this relationship for the point of the start from Mars, i.e. at $r = a_M$ and $V = V_H$, we have

$$V_M^2 = k^2 M \left(\frac{2}{a_M} - \frac{1}{a_H} \right) = \frac{k^2 M}{a_M} \left(2 - \frac{a_M}{a_H} \right) , \tag{31}$$

where a_H is the semimajor axis of the sought "crossing" trajectory, a_M is the semimajor axis of Mars' orbit. However, the left-hand side of this expression contains the square of the circular velocity of Mars, and the multiplier before the parentheses is nothing other than the square of Mars' orbital velocity

$$V_M^2 = \frac{k^2 M}{a_M} . \tag{32}$$

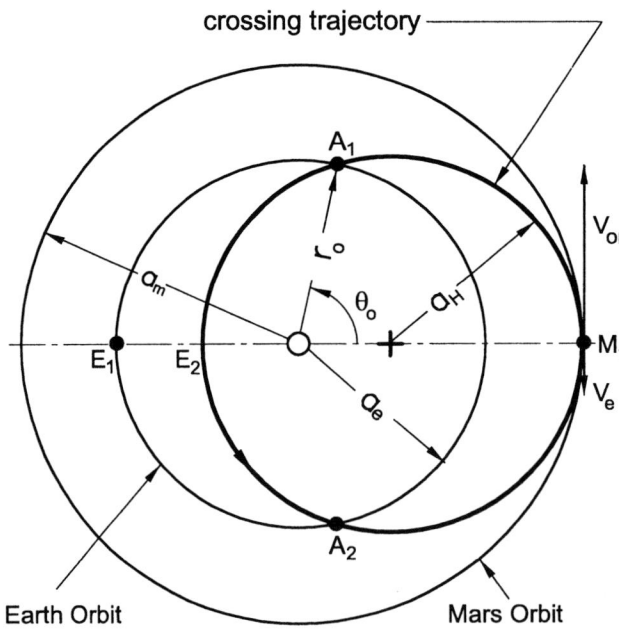

Mars - Earth crossing trajectory

Fig. 12.14. *The "crossing" trajectory $M_1 A_1 E_2$ of a flight from Mars to the Earth. Intersection of the crossing trajectory with the Earth's orbit occurs twice, at the points A_1 and A_2.*

Therefore we have from the unknown semimajor axis of crossing orbit a_H

$$\frac{a_M}{a_H} = 2 - \frac{V_H^2}{V_M^2} . \qquad (33)$$

Substituting for $V_H = 19.1 \; km \; s^{-1}$ and $V_M = 24.1 \; km \; s^{-1}$, we find

$$a_H = 0.728 \, a_M = 1.06 \; a.u. \qquad (34)$$

or from the point M_1 up to the extreme position E_2 on this crossing orbit

$$2a_H = 2.16 \; a.u.$$

that is less than the distance $M_1 E_1 = a_E + a_M = 2.52 \; a.u.$. Hence, this case really will be a short way up to the point E_1 and, hence, the intersection with the Earth's orbit in the point A_1 will be inevitable.

By known a_H and a_M we obtain for the eccentricity of this new ellipse

$$e = \frac{a_M - a_H}{a_M + a_H} = 0.178 \qquad (35)$$

which gives for the small axis b of the crossing ellipse

$$b = a\sqrt{1 - e^2} = 1.04 \; a.u. . \qquad (36)$$

We have for the orbital period for any body through the crossing orbit

$$P = \frac{2\pi a_M^{3/2}}{k\sqrt{M}} \left(\frac{a_H}{a_M}\right)^{3/2} \left[2 - \left(\frac{V_H}{V_M}\right)^2\right] . \qquad (37)$$

However, the first term in the right-hand side is nothing other than the orbital period of Mars, i.e.

$$\frac{2\pi a_M^{3/2}}{k\sqrt{M}} = P_M . \qquad (38)$$

Therefore we have for the orbital period through the crossing orbit

$$P = P_M \left(\frac{a_H}{a_M}\right)^2 \left[2 - \left(\frac{V_H}{V_M}\right)^2\right] . \qquad (39)$$

Putting $a_H/a_M = 0.728$, $V_H/V_M = 19.1/24.1 = 0,792$, we find finally

$$P = 364 \; days \qquad (40)$$

or for the flight duration T_f from the point of start M_1 up to point E_2

$$T_f = P/2 = 182 \; days \; . \tag{41}$$

The intersection of this crossing orbit with the Earth's orbit takes place nearly at the halfway point, i.e. in about three months.

However, finding the moment of intersection of the crossing orbit with Earth's orbit more precisely is possible if the coordinates of the point of intersection A_1 are determined. We have for the Earth's orbit

$$r_1 = \frac{a_1(1 - e_1^2)}{1 + e_1 \cos \vartheta_1} \tag{42}$$

and for the crossing orbit

$$r_2 = \frac{a_2(1 - e_2^2)}{1 + e_2 \cos \vartheta_2} \; . \tag{43}$$

At the point of intersection of both orbits we have: $r_1 = r_2$ and $\vartheta_1 = \vartheta_2 = \vartheta_0$. Then we obtain for the angle ϑ_0 (Fig. 12.14)

$$\cos \vartheta_0 = \frac{K - 1}{e_1 - e_2 K} \; , \tag{44}$$

where we have denoted

$$K = \frac{a_1}{a_2} \frac{1 - e_1^2}{1 - e_2^2} \; . \tag{45}$$

Putting the values of necessary parameters: $a_1 = 1.00 \; a.u.$, $a_2 = 1.06 \; a.u.$, $e_1 = 0.0167$, $e_2 = 0.172$, we obtain from (45): $K = 0.973$ or from (44)

$$\vartheta_0 = 80°.3 \; . \tag{46}$$

Accordingly, we have for the radius-vector r_0

$$r_0 = \frac{a_1(1 - e_1^2)}{1 + e_1 \cos \vartheta_0} = \frac{a_2(1 - e_2^2)}{1 + e_2 \cos \vartheta_0} = 0.970 \; a.u. \; . \tag{47}$$

The distinguishing property of the crossing trajectory is as follows: its intersection with Earth's orbit takes place twice, in the symmetrically situated points A_1 and A_2 (Fig. 12. 14).

Speaking formally, in this case, i.e. at crossing orbit, the probability of a direct hit of a rock thrown from the Mars onto the Earth will be twice as larger as we had in the case of a Hohmann flight orbit with a meeting at point E_1, i.e. only once during each flight from Mars to the Earth.

Continuing this abstraction, we note that two intersections with Earth's orbit occur also in the imaginary case when the rock is thrown in the opposite direction to the orbital motion of Mars with a velocity 14.1 $km\,s^{-1}$. In that case the rock will "fall" onto the Sun by a line M_1S.

Speaking about the "crossing" trajectory, i.e. the one crossing the Earth's orbit, it is necessary to take into account also the difference in spatial positions of the planes of the Earth and Mars orbits. Of course, the spatial orientation of the plane of motion of a rock thrown from Mars may differ from the orbital planes both of the Earth and of Mars. However, one circumstance can to some degree weaken this demand: namely, the existence of the zone of gravity around the Earth of radius equal to 932 000 km, i.e. nearly one million km or three times the Earth–Moon distance. So, if the rock thrown from Mars, during its travel to the Earth via such a crossing trajectory, finds itself within a circle of diameter two million km with the Earth in its center, its fall onto the Earth turns out to be inevitable.

Now something deductive. In 1994 and 1996 small meteorites were discovered in Antarctica and California, both of rock type, with assumed traces of organic matter. They were found to be of Martian origin, with the far reaching consequences. As suggested, those pieces were ejected from Mars during a large impact, probably making their way to Earth in less than a million years. Without touching the biological aspect of the problem, in the light of the above statements, we see that from the viewpoint of celestial mechanics the appearance of Martian rocks on the Earth is not a myth.

10   An Alternative to Hohmann Trajectories

Up to now the Hohmann orbit has been considered as optimal purely from the geometrical point of view. However, there is also the energetic aspect of the problem. More careful examination of the problem brings a conclusion unexpected at first glance, namely, that the Hohmann orbit may turn out to be advantageous from the energetic point of view if the eccentricity of the orbits is within the limits

$$0 \le e \le 0.8794$$

or if the ratio of both circular orbits a_M/a_0 is within the limits

$$1 \le a_M/a_0 \le 15.6 \ .$$

In other words, for the Hohmann passage from one circular orbit of a radius a_0 to another circular orbit of a radius a_m that ratio is definitely not larger than 12.14 (Ladwin, 1963). Moreover, more economical is shown to be the so-called three-momentum version, shown in Fig. 12.15, according to which, by the first momentum V_1 applied at point A, i.e. tangentially to the initial orbit, the probe is put on an elliptic orbit with a semimajor axis larger than the semimajor axis of a Hohmann orbit. Then, at the second point B, the second momentum V_2 is applied with the help of which the spacecraft is transferred into the new elliptic orbit with a contact at point C. The third momentum V_3 is applied just at this point but in this case opposite to the direction of the orbital motion. It is interesting to notice that the larger the distance OB, the more economic will be the maneuver. In the limiting case when the point B is at infinity, the ellipse transfers into the parabolic orbit and the magnitude of the momentum V_2 at this point B falls to zero.

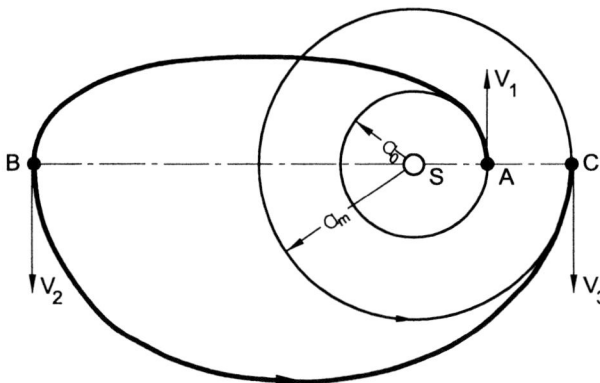

Fig. 12.15. *The flight from the point A of an inner planet to the point C of an outer planet by three-momentum trajectory.*

The essence of this version is in the following: at $a_M/a_0 > 15.6$ the three-momentum maneuver is more favorable from an energetic point of view as compared with the Hohmann maneuver if the point B lies outside the large circle (a_M).

In fact Hohmann's maneuver may be accepted as a particular case of the three-momentum version when the point lies on the large circle and the magnitude of the momentum applied on the point is zero. In principle, the three-momentum maneuver should be presented as a development of Hohmann's maneuver and, hence, Hohmann's transition should be estimated as not optimal.

The cases with $a_M/a_0 > 15.6$ are too rare, therefore three-momentum trajectories, although quite interesting, should be considered as having mostly theoretical interest. The Hohmann ellipses should be accepted as more reasonable ones. The flights to remote planets, farther than Saturn for which the ratio a_M/a_0 far exceeds 15, are realized based on rather different principles and requirements, particularly connected with the necessity to realize gravity assists in the gravitation fields of more massive planets, especially of Jupiter's and Saturn's, with smallish corrections during the rest of the flight.

11 The Hohmann Orbit for the Sun

In principle, a space probe may be launched in the Hohmann ellipse to the Sun, or more precisely, to a point on the opposite surface of the Sun. In this case, the Hohmann ellipse will come into contact with the inner side of the Earth's orbit at the aphelion, and with the Sun's surface at the perihelion.

Consider a more or less real case, i.e. when the probe is crossing the Solar corona of a distance about $\sim 1.07\,R_0$ from the Sun's center, i.e. when $a_\star = 7.5 \cdot 10^{10}\ cm = 0.005\ a.u.$ The parameters of the Hohmann orbit obtained from Eq. (15) in this case are

$a = 0.5025\ a.u.$

$e = 0.990$

i.e. the Hohmann orbit of the Sun manifests itself as an extremely elongated ellipse; its minor semi-axis is equal to $b = a(1 - e^2)^{1/2} = 0.07\ a.u.$, i.e. 7.5 times the diameter of the Sun.

The duration of flight from the Earth to the Sun is $T_f = 65.1\ days$.

The additional velocity is $V_{ad} = 26.82\ km\ s^{-1}$, and the initial velocity is $V_0 = 29.062\ km\ s^{-1}$.

In Chapter III (Section 11) the problem of the launch of a probe to the Sun not in a Hohmann orbit but in a straight line, i.e. free-fall onto the Sun, was considered. In that case, the duration of the flight, i.e. the free-fall time, is

$$T_0 = \frac{\sqrt{2}}{3}\,\frac{a^{3/2}}{R_0(2g_0)^{1/2}}\,, \tag{48}$$

where a is the distance to the Sun. On substituting in Eq. (48)$\cdot a = 1\ a.u. = 1.5 \cdot 10^{13}\ cm$, $R_0 = 7 \cdot 10^{10}\ cm$, and $g_0 = 2.74 \cdot 10^4\ cm\ s^{-2}$, we obtain

$$T_0 = 27.3\ days$$

i.e. nearly two and half times less than the duration of the flight in a Hohmann ellipse.

However, in terms of the basic parameter, the initial velocity of the probe, the Hohmann orbit does not provide any remarkable gain – the difference in initial velocities amounts to only $31.816 - 29.062 = 2.754\ km\ s^{-1}$. The conclusion is clear: sending a probe to the Sun is, in both versions, i.e. via the fourth space velocity or in a Hohmann orbit, still an impossible problem for the present level of rocket technique.

12   Flight in a Parabolic Orbit

One has the least initial launch velocities of the space probe in the case of the Hohmann ellipses, therefore, these trajectories should be accepted as the most economic ones from an energetic point of view. For the same reason, the largest flight durations are again for Hohmann trajectories. In the case, e.g. of Pluto, this duration, as was shown above (Section 3), amounts to nearly half a century.

A reduction in the duration of the flight is possible via an increase in initial velocity. Unless this increase is not large, the trajectory will remain elliptic; in this case we shall have an ellipse intersecting with the orbit of the planet of destination (ellipse $E_0 M_2$ in Fig. 12.1).

However, a noticeable increase in initial velocity, say, by a factor of one and a half, is possible if one moves to parabolic orbits. Thus, we arrive at the problem of interplanetary flights via parabolic orbits (Fig. 12.16) with the Sun at the focus, which is also the center of the circular orbits both of the Earth and of the planet of destination.

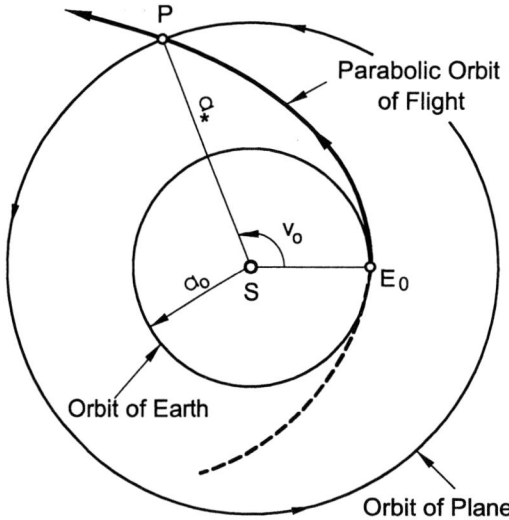

Fig. 12.16. *Flight to the outer planets in a parabolic orbit $E_0 P$.*

The basic parameter of a parabolic orbit is the duration of the flight T_p from the Earth to a given planet. The value of T_p is not difficult to determine using the equation of motion of a celestial body in a parabolic orbit (Chapter IV). We have

$$\frac{k(t-T)}{q^{3/2}\sqrt{2}} = \tan\frac{v}{2} + \frac{1}{3}\tan^3\frac{v}{2},\qquad(49)$$

where T is the time of launch from the Earth, i.e. from the perihelion of the parabola, and v is the true anomaly, i.e. the angular distance of the position of the body on the parabola from the perihelion E_0. At the point P of the intersection of this parabola with the planet's orbit (Fig. 12.16), we have

$$T_p = t - T, \qquad r = a_\star, \qquad v = v_0 .$$

The linear distance of the perihelion from the focus of the parabola q is equal to the Earth's radius, i.e. $q = a_0$. Therefore, we obtain from Eq. (49) for the duration of the flight

$$T_p = \frac{\sqrt{2}}{k\sqrt{M}} a_0^{3/2} \left[\tan \frac{v_0}{2} + \frac{1}{3} \tan^3 \frac{v_0}{2} \right] . \tag{50}$$

We obtain the value of $\tan v_0/2$ from the equation of the parabola, namely

$$r = \frac{q}{\cos^2 \dfrac{v}{2}} .$$

On substituting here $r = a_\star$, $q = a_0$, and $v = v_0$, we obtain

$$1 + \tan^2 \frac{v_0}{2} = \left(\frac{a_\star}{a_0} \right) ,$$

and from Eq. (50)

$$T_p = \frac{\sqrt{2}}{k\sqrt{M}} a_0^{3/2} \left(\frac{a_\star}{a_0} - 1 \right)^{1/2} \left[1 + \frac{1}{3} \left(\frac{a_\star}{a_0} - 1 \right) \right] . \tag{51}$$

But

$$\frac{\sqrt{2}}{k\sqrt{M}} a_0^{3/2} = \frac{\sqrt{2}}{V_0} a_0 ,$$

where V_0 is the circular velocity on the Earth's orbit. Then we have

$$T_p = a_0 \frac{\sqrt{2}}{V_0} \left(\frac{a_\star}{a_0} - 1 \right)^{1/2} \left[1 + \frac{1}{3} \left(\frac{a_\star}{a_0} - 1 \right) \right] , \tag{52}$$

or, if the distance is presented in a.u.   and we take the value $V_0 = 29.8\ km\ s^{-1}$,

$$T_p = 82.212(a_\star - 1)^{1/2} \left(1 + \frac{1}{3}(a_\star - 1) \right)\ days . \tag{53}$$

From this formula we can estimate the durations of flights T_p in parabolic orbit from the Earth to any planet. The results for the outer planets are given in Table 12.2. For comparison the durations T_H for Hohmann trajectories are given as well.

T a b l e   12.2

Durations (in years) of interplanetary flight from the
Earth to the outer planets in the case of two trajectories, the parabolic, T_P, and Hohmann, T_H

Planet	T_P	T_H
Mars	0.19	0.71
Jupiter	1.11	2.73
Saturn	2.53	6.05
Uranus	6.78	16.04
Neptune	12.97	30.62
Pluto	19.33	45.60

As we see, the difference between the durations of flights for the two
trajectories, parabolic and Hohmann, is rather significant – a factor of two–
three. Therefore, in every case the problem of the choice of the type of
trajectory to the given planet should be solved taking into account all the
relevant factors – the aims of the flight, the local resources of the probe,
energy resources, etc.

The initial velocity of launching V_0 is determined by Eq. (8) with the
only difference being that the additional velocity V_{ad} should be taken as
the difference between the parabolic velocity, V_p, and the Earth's circular
velocity, V_0:

$$V_{ad} = V_p - V_0 = 42.122 - 29.785 = 12.337 \ km \, s^{-1}$$

on condition that the vector V_{ad} will be parallel to the vector of the Earth's
orbital velocity V_E and directed in the same direction. On substituting this
value of V_{ad} into Eq. (8) we obtain for the initial velocity in the case of a
parabolic trajectory of flight to Mars

$$V_0 = 16.67 \ km \, s^{-1} \, .$$

This is larger by 44% than the initial velocity of the Hohmann trajectory.
From an energetic point of view this implies that one needs a rocket twice
as powerful as the one required for the Hohmann trajectory to Mars.

In certain cases, orbits intermediate between parabolic and Hohmann
orbits, i.e. elliptic orbits with various initial velocities at perihelion, can be
more convenient. In Table 12.3 the results of computations are presented,
in particular, for the durations of flights to Mars in the case of four elliptic
(including one of Hohmann type (I)) orbits and a parabolic orbit. In the
last row the angle of initial configuration α_0, according to Eq. (7), is given,
which, of course, differs from that for the Hohmann ellipse (I).

T a b l e   12.3

Parameters of flight to Mars in the case of four elliptic orbits, including one of Hohmann type (I), as well as by parabolic orbit

Flight parameters	Elliptic				Parabolic
	I	II	III	IV	
Initial velocity V_0, $km\,s^{-1}$	11.57	11.8	12.0	13.0	16.67
Duration of flight, days	259	164	144	105	70
Angular distance, degrees	180	129	116	92	72
Angle α_o, degrees	44	43	41	37	35

The following curious circumstance is worthy of particular attention. According to computations, the initial velocity for flight to Pluto in a Hohmann elliptical trajectory equals 16.270 $km\,s^{-1}$, which is slightly less than the initial velocity of a parabolic orbit (16.67 $km\,s^{-1}$). This insignificant difference, only 0.40 $km\,s^{-1}$, leads to an essential difference, more than twofold, in the duration of flight, i.e. 19.3 years instead of 45.6 years in the case of the Hohmann trajectory. The explanation is that for Pluto, we have an extremely elongated Hohmann trajectory, namely an ellipse of eccentricity $e = 0.95$. As a result, the ratio of initial velocity V_{max} at perihelion (i.e. on the Earth) to the velocity V_{min} at aphelion, i.e., in the point of contact with Pluto's orbit, is very large:

$$\frac{V_{max}}{V_{min}} = \frac{1+e}{1-e} = 39 \,,$$

which gives $V_{min} = V_{max}/39 = 29.785/39 = 0.765 \; km\,s^{-1}$. In other words, the velocity of motion of the space probe in a Hohmann trajectory near aphelion, i.e. at the meeting point with Pluto, is even less than 1 $km\,s^{-1}$, which is much less than the parabolic flyby velocity of 6.7 $km\,s^{-1}$.

13   Flight by the Shortest Route

Flights from the Earth to the outer planets by the shortest route, i.e. along the radius-vector Sun–Earth–Planet, theoretically cannot be excluded. Such flights cannot be of economical interest, insofar as the initial velocity of launch will be incomparably larger than in the case of the Hohmann trajectory. Moreover, such a flight cannot be realized in a typical Keplerian trajectory even though the probe should constantly be within the Solar sphere of influence, and therefore cannot be *a priori* predicted as the shortest route in terms of the duration of the flight.

Consider the concrete case of a flight from the Earth to Mars by the shortest route. The mutual disposition of the Sun (S), Earth (E_1) and Mars (M_1) at the moment of departure of the space probe from the Earth is shown in Fig. 12.17. The statement of the problem is as follows: we wish to find the minimal starting velocity V_0 at which the probe moving strongly along radius-vector SE_1 will meet Mars at the point M_2. We also have to find the moment of launch, i.e. the position of Mars in its orbit relative to the Earth determined by the angle Ψ_M, as well as the duration of flight t_0 from the Earth to Mars via the shortest route E_1M_2. At the moment of encounter of the probe with Mars at the point M_2, the Earth will be at the point E_2, i.e. at an angular distance Ψ_2 from its position E_1 at the starting moment.

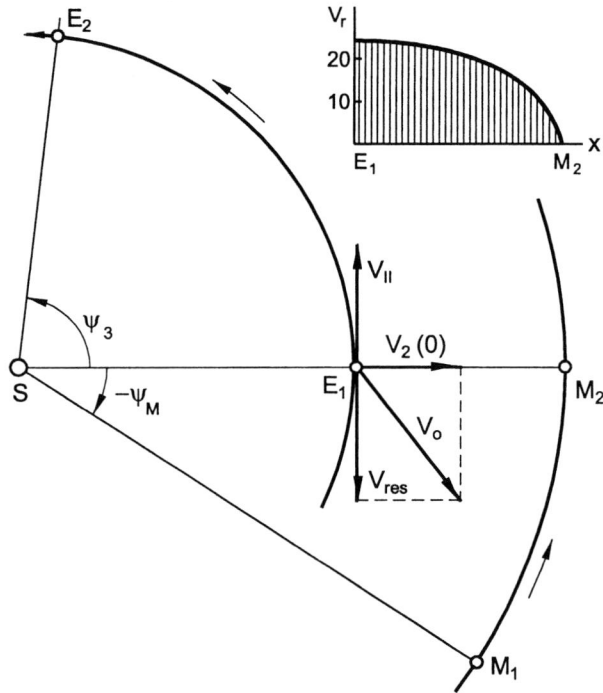

Fig. 12.17. *Flight from the Earth to Mars in a straight line E_1M_2. E_1 and M_1 are the positions of the Earth and Mars at the moment of launch of the spacecraft, E_2 and M_2 their positions at the moment of the meeting of the spacecraft with Mars. The upper diagram shows the law of variation of the spacecraft's velocity along the path E_1M_2.*

Realization of the flight by such a straight line route is possible if the following three operations are implemented.

a. The escape of the probe from the Earth: the second space velocity $V_{II} = 11.186\ km\ s^{-1}$ should be reached.

b. The instantaneous stopping of the probe with respect to the Sun at the point E_1, i.e. compensation of the heliocentric velocity of the Earth $V_{Er} = 29.785\ km\ s^{-1}$. For that, the probe should again

gain the same velocity, V_{Er}, but in the opposite direction. Then the resulting velocity V_{Res} required for the probe to escape from the Earth's influence and to be at rest with respect to the Sun will be

$$V_{Res} = (V_{II}^2 + V_{Er}^2)^{1/2} = 31.82 \; km \; s^{-1} \;.$$

c. The motion of the probe via the radius-vector SE_1, in the direction of point M_2 with an initial radial velocity $V_r(0)$, so that the meeting of the probe with Mars at the point M_2 will occur with zero radial velocity.

On the route E_1M_2, the probe will be under the action only of Solar gravitation. In fact we have "thrown a stone" from the point E_1 strongly vertically upwards, so that at the "highest" point M_2, where the velocity of the stone vanishes, we have Mars. If the acceleration due to Solar gravity g is assumed constant within the interval E_1M_2, then the initial radial velocity $V_r(0)$ can be obtained readily from the following relationship:

$$V_r(0) = (2g\Delta s)^{1/2} \;, \tag{54}$$

where $\Delta s = E_1M_2 = a_M - a_0$ is the minimal distance between Mars and the Earth.

However, the ratio of the accelerations at the points E_2 and M_2 due to Solar action, is $g_\oplus/g_M = (a_0/a_M)^2 = (1.52)^2 = 2.31$ and, hence, we cannot assume $g = constant$. Therefore, we have for the differential equation of motion in the field of Solar gravity

$$\frac{d^2x}{dt^2} = -g_0 \left(\frac{R_0}{x}\right)^2 \;, \tag{55}$$

where x is the distance of the probe from the center of the Sun, g_0 is the gravity at the Sun's surface, and R_0 is the radius of the Sun.

Eq. (55) can be rewritten in the following form

$$\frac{dV_r^2}{dx} = -2g_0 \left(\frac{R_0}{x}\right)^2 \;. \tag{56}$$

After integration, we find

$$V_r(x) = (2g_0R_0)^{1/2} \left(\frac{R_0}{x} - \frac{R_0}{a_M}\right)^{1/2} \;, \tag{57}$$

where the constant of integration has been obtained from the condition $V_r(a_M) = 0$ for the probe at the moment of its meeting with Mars, i.e. at the point $x = a_M$.

The term within the second set of parentheses in Eq. (57) is nothing other than the parabolic velocity of take-off from the surface of the Sun. Therefore, we can write instead of Eq. (57)

$$V_r(x) = V_p \left(\frac{R_0}{a_M} \right)^{1/2} \left(\frac{a_M}{x} - 1 \right)^{1/2} . \tag{58}$$

This formula is applicable within the interval $a_0 < x < a_M$. By substituting $x = a_0$ and $V_p = 617.7 \ km \ s^{-1}$ in Eq. (58), we obtain for the initial radial launch velocity

$$V_r(0) = 24.68 \ km \ s^{-1} .$$

This is much higher than we had even in the case of a parabolic orbit ($V_r = 16.67 \ km \ s^{-1}$).

For the variation of the radial velocity $V_r(x)$ within the interval from $x = a_0$ to $x = a_M$, we obtain

$$V_r(x) = 34.23 \left(\frac{a_M}{x} - 1 \right)^{1/2} \ km \ s^{-1} , \tag{59}$$

as is shown in Fig. 12.17 above. To obtain the resulting initial velocity V_0, at which we can realize simultaneously all three operations, (a), (b) and (c), one should add vectorially to the initial radial velocity $V_r(0)$ also the velocity of "stopping" V_{st}:

$$V_0 = [V_r^2(0) + V_{st}^2]^{1/2} = 40.27 \ km \ s^{-1} .$$

The direction of the vector V_0 is defined by the relationship

$$\tan \varphi = \frac{V_r(0)}{V_{st}} , \tag{60}$$

where φ is the angle between the vector and the tangent to the Earth's orbit at the point E_1 (Fig. 12.17). Numerically, $\varphi = 37°.8$.

The next problem is connected with the duration of the flight t_0 from the Earth to Mars. We have from Eq. (58)

$$\frac{dx}{dt} = V_p \left(\frac{R_0}{a_M} \right)^{1/2} \left(\frac{a_M}{x} - 1 \right)^{1/2} , \tag{61}$$

which gives

$$t_0 = \frac{(a_M/R_0)^{1/2}}{V_p} \int_{a_0}^{a_M} \frac{dx}{\left(\frac{a_M}{x} - 1 \right)^{1/2}} , \tag{62}$$

or, after integration,

$$t_0 = \frac{a_0}{V_p} \left(\frac{a_M}{R_0}\right)^{1/2} \left[\left(\frac{a_M}{a_0} - 1\right)^{1/2} + \frac{a_M}{a_0} \arctan\left(\frac{a_M}{a_0} - 1\right)^{1/2}\right]. \quad (63)$$

On substituting here the numerical values of all parameters, we obtain

$$t_0 = 84.7 \; days \; ,$$

which is not much longer than we had (70 *days*, Table 12.3) for the duration of the flight to Mars in a parabolic orbit, i.e. not by the shortest route.

This result can be checked by control computations. Thus, assuming that the flight from the Earth to Mars takes place with a constant value of g, we have for the duration of the flight

$$t = \left(\frac{2\Delta s}{g}\right)^{1/2}. \quad (64)$$

Taking $g = g_0(R_0/a_0)^2 = 0.60 \; cm \; s^{-2} = constant$, which corresponds to the acceleration due to the Sun's gravity at the Earth's distance, we have for the minimal value of t

$$t_{min} = 58.9 \; days \; .$$

Adopting now $g = g_0(R_0/a_M)^2 = 0.26 \; cm \; s^{-2} = constant$, which corresponds to the acceleration due to the Sun's gravity at Mars' distance, we shall have for the maximal value of t

$$t_{max} = 89.6 \; days \; .$$

The above obtained exact value of t_0 (84.7 *days*) lies between t_{min} and t_{max}, but is essentially close to t_{max}, which shows that the most time is spent overcoming the second half of the route.

In Fig. 12.17, the Earth–Mars configuration at the moment of launch (E_1 and M_1) as well as at the moment of arrival at Mars (E_2 and M_2) are shown; the latter is determined by angles Ψ_3 and Ψ_M:

$$\Psi_M = -\omega_M t_0 = -0°.524 \cdot 84.7 = -44°.4 \; ,$$

$$\Psi_3 = \omega_3 t_0 = 0°.986 \cdot 84.7 = 83°.5 \; .$$

At first glance we find ourselves in a paradoxical situation: during the flight by the short route from the Earth to Mars both the initial velocity, V_0, and the flight duration, t_0, turned out to be much larger than in the case of a parabolic trajectory; see the table below:

Type of Flight	Initial velocity V_0 $(km\,s^{-1})$	Duration of flight (days)
Parabolic trajectory	16.67	70
By shortest route	40.27	84.7

Of the two parameters V_0 and t_0, the more crucial is, of course, the initial velocity V_0, in that the value of V_0 "by shortest route" is two and a half times greater, and energetically 6–7 times larger, than in the case of a parabolic trajectory. It is surprising that, even at such a high velocity, the shortest route does not lead to any essential advantages in terms of the duration of the flight.

Note that, though the encounter with Mars takes place with zero radial velocity, the difference $24.13\ km\,s^{-1}$ in the direction of orbital motion of Mars is preserved. Hence, it is necessary to compensate this value with additional momentum. This problem can be simplified if the probe at the moment of launching is made also to gain tangential velocity $24.13\ km\,s^{-1}$ at the expense of the orbital velocity of the Earth. Then at the moment of departure, not all the velocity $V_3 = 29.785\ km\,s^{-1}$, but only $\Delta V = V_3 - V_M = 29.875 - 24.129 = 5.656\ km\,s^{-1}$ need be compensated. As for the radial velocity, it remains unchanged, i.e. $V_r(0) = 24.68\ km\,s^{-1}$. Then we shall have for the escape velocity $V_{re} = 11.186 + 5.656 = 16.842\ km\,s^{-1}$, and for the final starting velocity

$$V_0 = [(16.842)^2 + (24.68)^2]^{1/2} = 29.88\ km\,s^{-1}\ ,$$

which is significantly lower than the earlier result ($40.27\ km\,s^{-1}$).

Earth–Mars flight by the shortest route is completely impossible from a practical point of view. The goal of our analysis was to demonstrate by a concrete example the illusory character of the Euclidean definition of "the least route" concerning interplanetary fights controlled by the laws of celestial mechanics. From the energetic point of view the best solution is a flight in a Hohmann trajectory, but from the point of view of least duration of the journey, the parabolic trajectory would be best.

14   Flights to the Planets with Minimum Energy Spend

The flight from the surface of a planet to the surface of another planet can be divided to three phases.

i. The flight from the surface of a planet up to the boundary of its sphere of action.

ii. The flight in heliocentric space up to the boundary of action sphere of the destination planet.

iii. The landing on the surface of that planet.

Phase I includes passage on the waiting orbit around the planet of origin, before getting on a Hohmann orbit with necessary hyperbolic excess of velocity at the point from where the spacecraft is leaving the sphere of action of the initial planet.

Phase II consists of the period of passive flight under the dominating action of the Sun's gravity by elliptic trajectory, including the inevitable corrections during the flight. This phase occupies most of the flight time.

Phase III presents a repetition of phase I, though in the reverse sequence, including the operations for the passage from planetocentric orbit of rapprochement into the waiting orbit around the destination planet and the landing on its surface.

The return flight has to be realized with the same three phases with an addition of a very important element, namely, the delay time up to the return launch which is determined by the elements of orbits of both planets. In other words, the start of the spacecraft's return journey can occur only when both planets have a favorable position with respect to the Sun.

As was noticed above, the most economic are the Hohmann orbits, i.e. in that case the energy spent is minimal. The duration of the flight T_r in this case is determined by Eq. (6). As to waiting time T_w, it is determined by the formulas given in Section 5. Obviously the total duration of the flight T_0 will be: $T_0 = 2T_r + T_w$. In Table 12.4 the magnitudes of T_w, T_r and T_0 are presented for the planets of the Solar system. In the last column, the eccentricities for Hohmann flight orbits are given.

Table 12.4 enables us to draw some conclusions.

i. Manned flights to planets more distant than Mars are excluded because their durations is to long.

ii. Return flights to Venus, Mercury, and Mars are not excluded in principle from the point of view of flight durations. However, in the case of Mars and Venus we have the largest waiting times, two–three times larger than the durations of the flights.

iii. Extremely high temperatures, 400–500° C, on Venus and Mercury, as well as extremely high pressures up to 100 bar in Venus' atmosphere will create essential difficulties for the stay on these planets.

iv. Mars is the only planet of the Solar System on which it is possible to stay with suitable equipment.

T a b l e   12.4

Duration of flight from the Sun to planets of the Solar system T_r, the time of delay on the planet of destination T_w and total time of the flight T_0 are in years

Destination planet	Flight duration T_f	Duration of waiting T_w	Total time T_0	Eccentric. of flight's e	Additional velocity $V\ km\ s^{-1}$
Mercury	0.289	0.183	0.76	0.44	7.54
Venus	0.400	1.278	2.08	0.16	2.50
Mars	0.709	1.242	2.66	0.21	2.95
Jupiter	2.731	0.588	6.05	0.68	8.80
Saturn	6.048	0.936	13.03	0.81	10.30
Uranus	16.04	0.932	33.01	0.91	11.29
Neptune	30.62	0.766	62.01	0.94	11.66
Pluto	45.47	0.061	91.00	0.95	11.82
Interstellar Medium	–	–	–	–	12.34

How large should be the real velocity for the realization of these flights? The problem in fact comes down to the determination of the excess of velocity necessary to send the spacecraft onto the required heliocentric orbit. For an interplanetary flight the departure from the sphere of the earth's action should be realized with a hyperbolic velocity.

The magnitude of hyperbolic excess of the velocity V with which the spacecraft leaves the sphere of Earth's action of radius ρ, is given by an expression

$$V = \left[\frac{2Gm}{\rho} + v_e(2V_e + v_e)\right]^{1/2} \tag{65}$$

where V_e is the velocity of escape, i.e. the parabolic velocity

$$V_e = V_c \sqrt{2} \tag{66}$$

where V_c is the circular velocity of the orbit of radius ρ.

In order to move a spacecraft that is on the orbit of radius a_1 around the body of mass M (Sun) to a circular orbit of radius a_2, it is necessary to increase the velocity by about v_e which is determined by a relationship

$$v_e = \left(\frac{GM}{a_1}\right)^{1/2} \left[\left(\frac{2a_2/a_1}{1 + a_2/a_1}\right)^{1/2} - 1\right]. \tag{67}$$

This relationship determines the magnitude of v_e in (65).

If the spacecraft is leaving the Earth's sphere of action with velocity V in the direction of the Earth's orbital motion, then the heliocentric orbital velocity of this spacecraft will be

$$V_n = V + v_0 \qquad (68)$$

where $v_0 = 29.8 \; km \, s_{-1}$ is the velocity of the Earth's motion in its orbit around the Sun.

In the last column of Table 12.4 the magnitudes of V are given for the departure of spacecraft from the Earth up to the orbit of one of the planets of the Solar system. In the last line the velocity V is given for the spacecraft escaping forever from the solar system. As should be expected, we have the largest velocity in this case. Because of these velocities the probes Pioneer 10 and Pioneer 11 as well as Voyager 1 and Voyager 2 are leaving the Solar system. Any spacecraft or any other object possessing velocity smaller than $12.34 \; km \, s^{-1}$ at such distances will be a permanent member of the Solar System.

15 Factors Influencing the Orbital Motion of Satellites

The examples of interplanetary flights examined above are of course simplified, since e.g. the orbits of the planets are assumed to be circular and therefore the eccentricities of real orbits, their spatial orientation, non-coplanarity of orbital planes, etc., have been neglected. However, the methods of exact computations taking into account the factors mentioned do not differ from the picture already described; except in that, in the case of exact computations, the formulas, are much more cumbersome (Anderson, 1989).

As for the real trajectories of the flights, they will differ for many reasons from the computed trajectories – Hohmann, elliptic and parabolic. Hence, it is inevitably necessary to perform periodic corrections of the trajectory of the space probe during flight.

Flights around the Earth also have their peculiarities. The basic motion of the satellites via various elliptic orbits of course occur under the action of the Earth's gravity. This motion is completely described in the framework of the two-body problem, since both the satellite and the Earth can be represented as material points, with the Earth at one of the focuses. Together with this, the following factors can also influence the motion:

a. Nonsphericity, i.e. the flattening of the Earth;

b. The resistance of the Earth's atmosphere;

c. The contribution of perturbations by the Sun, the Moon and other planets;

d. The magnetic field of the Earth;

e. The Solar radiation;

f. The Solar wind.

The effect of nonsphericity of the Earth is taken into account through the potential function U, which can be represented in the form

$$U = G\frac{M}{r} - G\frac{C_z - A_x}{2r^3}\left(3\sin^2\phi - 1\right),\qquad(69)$$

where M is the Earth's mass, C_z and A_x are the moments of inertia of the Earth along the axes OZ and OX, correspondingly, r is the radius-vector of the satellite, and ϕ is the angular height of the satellite above the Earth's equator. The first term on the right-hand side in Eq. (37) characterizes the basic motion, i.e. the ellipse; the second term is the perturbed potential conditioned by the Earth's flattening. Depending on the required accuracy, terms of higher order could be added to the right-hand side of Eq. (69) as well. Very often the contribution of the third term turns out to be essential; this term has the form

$$\frac{1}{2}\omega^2 r^2 \cos^2\phi$$

and is considered to be a perturbed potential caused by the Earth's rotation. One of the important consequences of the flattening of the Earth is the precession of the orbital plane as discussed in Chapter X.

Atmospheric resistance has an essential influence on satellite motion; it is noticeable even at large heights. Physically this resistance is expressed in the form of continuous collisions with a velocity of about $\sim 8\,000\ m\,s^{-1}$ of the satellite with molecules and atoms of the atmosphere. The resulting effect, i.e. the deceleration of satellite motion, depends on many factors, and first of all on the form, dimensions, mass and orientation of the satellite. All these factors are usually taken into consideration through special coefficients determined by empirical methods. Macroscopically, the effect of air resistance is proportional to the square of the orbital velocity (V^2) and to the density of the atmosphere at a given height, as well as being dependent on the geographic latitude.

The influence of the Solar gravity and that of the Moon on satellite motion is not large when the satellite is in an orbit with apogee less than $1500\ km$. The influence of other planets, including Jupiter and Saturn, is also negligible. Any satellite obviously carries metallic parts in it construction, and therefore, the excitation of vertical currents within the magnetic field of the Earth is inevitable. As a consequence, the satellite can be influenced by a weak additional resistance force. In general this effect is negligible, though it can play some role in the motion of satellites in polar orbits.

In the case of relatively light satellites of large dimensions (pneumatic balloons, for example) a noticeable influence may be expected from the Solar radiation pressure. In the case of the USA's satellite balloon *Echo–I*, which was launched into an orbit with apogee $3\,000$ km, this effect has indeed been found. The influence of this effect on ordinary satellites is negligible.

The influence of collisions with the Solar wind on a satellite is analogous to the effect of atmospheric resistance. The formal differences are, first, that the particle collisions take place at high velocities, over $1\,000$ $km\,s^{-1}$; second, the Solar wind particles are ionized and therefore interact with the electromagnetic field at the satellite's surface. These processes are rather complicated and have not been adequately studied, but in all cases their influence on the satellite's motion, in general, is negligible.

Note that accounting for the effect of any of these factors on the satellite's motion can be considered as perturbations to the Keplerian motion in the two-body problem.

Thus, the basic perturbations of the Keplerian orbit of a satellite's motion are related to two effects: the flattening of the Earth and the resistance of the Earth's atmosphere. In practice the corrections to the elements of elliptic orbits of satellites are usually realized by accounting for these two factors only, and under their action the elliptic orbit normally continuously approaches the circular one.

16   Estimation of Coordinates of Space Probes

Without going into the details of estimation of the orbits of space probes, we will confine ourselves to representation of the sequence of steps for calculation of the coordinates of the probe. We assume that, for the given moment of time t, all six elements of the orbit are known.

1. The eccentric anomaly E should be obtained with the help of Kepler's equation

$$E - e\sin E = M = n(t - T) , \qquad (70)$$

where n is the mean angular motion by the orbit.

2. The mean anomaly v is obtained from the formula

$$\tan\frac{v}{2} = \left(\frac{1+e}{1-e}\right)^{1/2}\tan\frac{E}{2} . \qquad (71)$$

3. The determination of polar coordinates r and v of the probe is done by means of the equations

$$r\sin v = a\,(1 - e^2)^{1/2}\sin E ,$$
$$r\cos v = a\,(\cos E - e) . \qquad (72)$$

4. The latitude argument u of the space probe is obtained from

$$u = v + w \; .$$

5. The declination δ and the increase in the longitude of the ascending node $\Delta\Omega$ is obtained with the help of the relationships

$$\sin \delta = \sin u \, \sin i \; ,$$
$$\tan \Delta\Omega = \tan u \, \cos i \; . \tag{73}$$

6. The geographic coordinates of the probe, i.e. the projection of its position onto the Earth's surface, are obtained from $\varphi = \delta$ and $\lambda = \alpha = \Omega + \Delta\Omega$.

7. The Cartesian heliocentric elliptic coordinates of the probe are obtained from the equations

$$x = r \left(\cos u \, \cos \Omega - \sin u \, \sin \Omega \, \cos i \right) ,$$
$$y = r \left(\cos u \, \sin \Omega + \sin u \, \cos \Omega \, \cos i \right) , \tag{74}$$
$$z = r \sin u \, \sin i \; .$$

These coordinates can be used for the determination of the equatorial geocentric coordinates of a space probe realizing interplanetary flight.

The above formulas are also useful for determination of the position of a probe in orbit around the Earth, with the only difference being that now one should take into account also the variations in the orbit's elements occurring within appropriately small time intervals, say, during a day or a single spiral of the orbit. Some idea concerning these variations is given by the following formulas describing the variations of the elements a, e, T, Ω, w and H_n (orbital heights correspond to the perigee) during a spiral:

$$\Delta a = \frac{2 c_x s}{m} \rho_n a^2 \left(\frac{2\pi}{\nu} \right)^{1/2} \left(1 + \frac{1}{8\nu} + 2e + \frac{3}{2}e^2 - \frac{3}{4}\frac{e}{\nu} \right) ,$$

$$\Delta e = \frac{2 c_x s}{m} \rho_n a \left(\frac{2\pi}{\nu} \right)^{1/2} \left(1 - \frac{3}{8\nu} + e \right) ,$$

$$\Delta T = \frac{3 \Delta a}{2a} T \; ,$$

$$\Delta H_n = \frac{2 c_x s}{m} \rho_n a^2 \left(\frac{2\pi}{\nu} \right)^{1/2} \left(\frac{1}{2\nu} - \frac{3}{16\nu^2} \right) , \tag{75}$$

$$\Delta\Omega = -j \left(\frac{R}{a} \right)^2 \frac{\cos i}{(1 - e^2)^2} \, nt \; ,$$

$$\Delta w = -0.5j \left(\frac{R}{a} \right)^2 \frac{5 \cos^2 i - 1}{(1 - e^2)^2} nt \; .$$

In these formulas ρ_n is the density of the atmosphere at the height H_n, $\nu = ae/k_n$ (k_n is the gradient of decrease of air density), m is the mass of the probe, c_x is the coefficient characterizing the aerodynamic properties of the probe, s is the area of the largest cross-section of the probe, a is the major semi-axis of the orbit, $f = C - 0.0017287$ (C is the flattening of the Earth), R is the equatorial radius of the Earth, and n is the mean daily motion (angular velocity) of the probe.

17   The Sphere of Attraction, the Sphere of Action, and Hill's Sphere

The motion in a Hohmann or parabolic orbit is determined by the gravity of the Sun, so that the latter is located at the focus of these orbits. If the probe moving in the Hohmann ellipse somehow misses Mars or Venus, then the probe should return, following the second half of the same Hohmann ellipse, to the point of the Earth's position at the moment of launch of the probe (these halves of the orbit are shown by dashed lines in Figs. 12.2 and 12.3). In these cases, the Keplerian motion is performed completely within the **sphere of action** of the Sun. However, if the Sun, as a center of gravity, has its sphere of action, then the planets should also have their own spheres of action. Thus, the question concerning the exact definition of the sphere of action and, in general, the interrelation of spheres of action of various gravitational centers, arises.

Obviously, the Keplerian motion of a space probe with respect to an isolated body as a center of gravity can occur within a sphere of radius theoretically equal to infinity. Therefore, strictly speaking, one has an action sphere of infinite radius.

Consider now two celestial bodies of masses m and M, respectively. It is clear that the probe will be influenced by forces from both these sources of gravity. If the probe is at rest with respect to both of these bodies, located, say, between these bodies on the line mM, then there always exists a distance R_g (from m) at which the gravitational actions (accelerations) exerted by both bodies on the probe will be equal to each other; in this case we speak of a **gravitational sphere** of radius R_g.

However, if the probe is performing a Keplerian motion and, hence, has a definite acceleration, then it is clear that each of these bodies will have its **sphere of action** of finite radius R_k. The problem, therefore, is to determine those radii, R_g and R_k.

To obtain R_g is easy. If we denote by A the distance between m and M, then the radius of the gravitation sphere R_g around the mass m should be obtained from the condition

$$\frac{m}{R_g^2} = \frac{M}{(A - R_g)^2} \, ,$$

which gives

$$R_g = \frac{A}{1 + (M/m)^{1/2}} \, . \tag{76}$$

In the case e.g. of the Moon (m) this formula gives for the radius of the gravitational sphere with respect the Earth (M) $R_g = 38\,000\ km$ if $M/m = 81$.

Quite different is the means of determination of the radius of sphere of action R_k, if one considers the Keplerian motion around a given body. In this case, the gravitational acceleration of the probe caused by the two bodies m and M can be estimated in two ways.

a) When the motion of the probe around the bigger mass (M) is under examination, then it is necessary to estimate the perturbations as the difference between the accelerations caused by the body M with respect to the probe and with respect to the body m. This difference equals

$$\Delta g_M = G \frac{m}{x^2} - G \frac{M}{(A - x)^2} \, , \tag{77}$$

where x is the distance of the probe from m.

b) Considering the motion of the probe around the smaller mass (m), one has to estimate the perturbations as the difference between the accelerations in the probe caused by the body m and the body M:

$$\Delta g_M = G \frac{M}{(A - x)^2} - G \frac{m}{A^2} \, . \tag{78}$$

It is clear that for given m, M and distance x of the probe from the body m, one of these differences, Eq. (77) or Eq. (78), will prevail. Correspondingly, one can speak of the sphere of **action** of the body m or M (for fixed values of A and M/m). In both cases the probe is performing a Keplerian motion either within the action sphere of body m or within that of body M.

One should make two comments. First, the shape of the action region is not strongly spherical but is slightly flattened. Second, the boundary of the action sphere is not clearly distinguished as in the case of the definition of R_g. Correspondingly, its radius R_k may be defined only approximately. As was shown by Tisserand, the surface of the action region is close to a sphere and the radius of this sphere R_k surrounding the body of smaller mass with respect to the body of larger mass M can be represented with high enough accuracy by the formula

$$R_k = a \left(\frac{m}{M}\right)^{2/5} , \qquad (79)$$

where a is the large semiaxis of the orbit, m is the mass of the planet, M is the mass of the Sun. In the case of the Earth–Sun system the radius of this sphere around the Earth is 925 000 km.

For the motion of a comet or spacecraft near a planet the concept of the **sphere of action** is clear. Accordingly, one can define the radius r_k of the sphere of action for the motion of the spacecraft around a given body. In the case of the Earth–Moon system, the radius of the sphere of action around the Moon is determined by the relationship

$$r_K(Moon) = \rho_m \left(\frac{m_0}{M_\oplus}\right)^{2/5} \qquad (80)$$

where $/rho_m$ is the distance of the Moon from the Earth (380 400 km), m_0 and $M_\oplus$ are the mass of the Moon and the Earth, respectively. Using $m_0/M_{/oplus} = 1/82$, we have, from (80), $r_k(Moon) = 66\,000\ km$.

In the cases of both the Earth and the Moon the radius of the sphere of action R_k is larger than the radius of the gravitational sphere R_g.

In Table 12.5 the radii of the sphere of action R_k for the combination of the system Planet–Sun are given in millions of kilometers. As follows from Eq. (79), this radius is increasing not only with the growth of the mass m, but also with the growth of the distance from the Sun. This factor explains the rapid increase of this radius, R_k, with the distance off from the Sun. The privileged role of the Sun's mass M means that, at distances of few million kilometers from any planet, the spacecraft is moving in a central gravitation field depending only on the square of the distance from the Sun.

T a b l e   12.5

Radii of the sphere of action R_k for various combinations of the system Planet–Sun

Planet	R_k million km	Planet	R_k million km
Mercury	0.112	Jupiter	48.1
Venus	0.615	Saturn	54.6
Earth	0.625	Uranus	52.0
Mars	0.579	Neptune	85.9
		Pluto	34.0

The sequence of the falling of the Sun's field of gravity becomes especially clear, if we compare the sphere of the action of the Earth with that for Pluto. Pluto's mass is very small, nearly six times smaller from the Earth's mass. However, the radius of the sphere of action in the case of Pluto is 35 times larger than that of the Earth and is comparable with the radius of the sphere of action of Jupiter (!), although Jupiter's mass is 2000 times larger than the mass of Pluto.

The spacecraft's motion within the sphere of action of a given planet is controlled completely by the gravity of that planet. As a result the conditions of the maneuver for a spacecraft are much more favorable in the spheres of the action of outer planets than those in of inner planets.

Within the limits of the sphere of action the space probe moves in one of the Keplerian orbits, namely according to an ellipse, parabola or hyperbola. When passing to the sphere of action of another central body, the probe again moves in a Keplerian orbit, be it ellipse, parabola or hyperbola, but now around that body. It is clear that, at the boundary between the two spheres of action, a certain **conjugation** of the two Keplerian sections should take place; they may be of different types. Because of the inaccuracy in the definition of the boundary of the sphere of action, the estimation of the trajectories of the space probe can hardly be made with sufficient confidence, and usually the main errors accumulate during the computation of the motion just in the vicinity of the boundary of the sphere of action.

Together with the mentioned types of spheres, another one, namely, the **Hill sphere**, can also be distinguished. This concept appeared in connection with the restricted three-body problem, i.e. the motion of an infinitely small body (for example, the space probe) under the gravity of finite bodies m and M (see Chapter IX). It was shown by Hill that, at given values of M/m and A, there is a definite zone (sphere) with a radius R_H around m (or M) within which the third body of small mass can stay for an infinitely long time, initially with a closed (elliptic) orbit around the given body m (or M). The means of definition of this sphere and, particularly, the estimation of its radius R_H are described in Chapter IX. Though not complicated, those definitions are somewhat cumbersome, so that the radius of Hill's sphere R_H cannot be represented by a simple formula depending on the configuration and the masses m and M, as in the case of R_g and R_k (Eqs. (76) and (79)). However, the computations lead to an important conclusion: the radius of Hill's sphere exceeds the radii of both spheres – those of gravitation and of action, at all values of m and M and for any distance between them. Thus, e.g. in the case of the Moon Hill's sphere lies at $R_H = 700\,000$ km from the Earth, whereas the distance between the Moon and the Earth varies between the limits $364\,000$ km and $402\,000$ km, being $384\,000$ km on average. Hence, the Moon lies deeply within Hill's sphere surrounding the Earth, and cannot leave this sphere. Moreover, it has been proved that no perturbations (perturbations by planets, the Sun, the eccentricity of the Earth's orbit, etc.) can change the main character of this result, namely, the existence of an upper limit of geocentric distance for the Moon. In such a manner, Hill arrived at the conclusion of the stability of the Moon's motion.

The concept of one more sphere, the *sphere of influence*, was introduced by Kislik relatively recently, in 1964. It defines a sphere on the boundary of which the action, even though in nonactive form, of the given celestial body on the motion of the space probe, is **beginning**. The corresponding radius of the influence sphere R_I, is given by the formula

$$R_I = 1.15a \left(\frac{m}{M} \right)^{1/3} , \tag{81}$$

where a is the distance between the bodies.

The radius of the sphere of influence R_I is slightly larger than the radius of the sphere of action R_k. Their ratio is

$$\frac{R_I}{R_K} = 1.15 \left(\frac{m}{M} \right)^{-1/15} . \tag{82}$$

In Table 12.6, as an illustration, the values of the radii of the mentioned spheres for three cases of motion are given: that of the Moon with respect to the Earth ($a = 380\,000\ km$), that of the Earth with respect to the Sun ($a = 150 \cdot 10^6\ km$), and of the Sun with respect to the Galaxy ($a = 10\,000\ pc$) assuming that all the mass of the Galaxy is concentrated at its center.

T a b l e   12.6

Radii of various spheres – gravitation (R_g), action (R_k), Hill's (R_H) and influence (R_I) for three combinations of the motion of the Moon relative to the Earth (km), Earth relative to the Sun (km) and of the Sun relative to the Galaxy (pc)

Combination	M/m	Radius of Sphere			
		R_g	R_k	R_H	R_I
Moon–Earth	81.3	38 000	65 000	700 000	102 000
Earth–Sun	$3.29 \cdot 10^5$	260 000	932 000	2 500 000	1 500 000
Sun–Galaxy	$1.4 \cdot 10^{11}$	0.1	1.15	3.65	6.84

Note the following circumstance. The radius of the Earth's gravitation sphere is roughly equal to $\sim 260\,000\ km$, so that the Moon is outside this sphere. Then we have to conclude that the Moon were be attracted by the Sun more strongly than by the Earth. Indeed, if the Moon were suddenly to stop its motion, it would fall onto the Sun and not onto the Earth. It is only owing to the fact that the Moon is in a state of **geocentric** motion around the Earth that this Solar domination remains without any consequences. This example reveals the practical uselessness of the concept of the gravitational sphere, since neither bodies at absolute rest exist nor can they be suddenly stopped. The idea of the sphere of action is of more practical interest. Its radius is always larger than the radius of the gravitational sphere and smaller than the radius of Hill's sphere and characterizes the degree of stability of the motion of a body within this sphere.

In Table 12.7 we give the orbital and kinematical elements for the planets required in interplanetary flight calculations.

T a b l e   12.7

Orbital and kinematical elements of planets necessary for interplanetary flight calculations

Planet	Semimajor axis of orbit *a.u.*	Sidereal period *Days*	Orbital velocity kms^{-1}	Escape velocity kms^{-1}	Eccentric. *e*
Mercury	0.387	87.97	47.9	4.2	0.2056
Venus	0.723	224.701	35.0	10.3	0.0067
Earth	1.000	365.256	29.8	11.2	0.0167
Mars	1.523	686.98	24.1	5.0	0.0934
Jupiter	5.20	4332.6	13.06	61	0.0484
Saturn	9.54	10759	9.64	37	0.0556
Uranus	19.18	30685	6.81	22	0.0472
Neptune	30.06	60189	5.43	25	0.0086
Pluto	39.44	90465	4.74	–	0.250

Thus, in practical cosmodynamics we have to deal with at least four types of spheres that a space probe can cross during its flight from one body to another. They can be summarized in the following manner.

The gravitational sphere (R_g) is the region within which the gravity from both celestial bodies would be the same if they could be imagined to be motionless.

The sphere of action (R_k) is the region within which the probe realizes a Keplerian motion. The boundary between two spheres of action is at the same time the boundary separating two Keplerian trajectories. Hill's sphere (R_H) is the region within the limits of which the given planet can keep its satellite forever.

The sphere of influence (R_I) is the region within the boundary of which the action of a body on the motion of the probe starts to come into effect.

All these spheres are unified by a common property, i.e. a small body in the system always remains undistorted. From this point of view the situation with the Roche sphere is completely different.

18 Roche Limit

Consider Eq. (66) (Chapter III) for the rotation period P through a circular orbit in the following form:

$$\frac{2\pi}{P} = \frac{kM^{1/2}}{a^{3/2}} = \omega , \tag{83}$$

where ω is the angular velocity of the body m realizing a motion in a circular orbit of radius a around a central body of mass M. We have from Eq. (83)

$$\omega^2 = \frac{G\mathcal{M}}{a^3} , \tag{84}$$

where $k^2 = G$ is the gravitational constant, and $\mathcal{M} = m + M$. On dividing both parts of Eq. (50) by $\pi G\rho$, where ρ is the density of the body m, we will have

$$\frac{\omega^2}{\pi G\rho} = \frac{M}{\pi \rho a^3} = C . \tag{85}$$

Roche showed as early as in 1847 that, for an infinitesimal, homogeneous satellite of density ρ, rotating around a solid spherical planet of mass M in a circular Keplerian orbit, there is a critical radius $a = R_R$, i.e. there exists some value of the dimensionless constant C in Eq. (85) for which the mass m can remain undistorted if

$$\frac{M}{\pi \rho R_R^3} \leq C . \tag{86}$$

The lower limit to R_R is obtained from the equality in Eq. (86) and is called the **Roche limit**. Correspondingly, it defines the **Roche lobe (zone)**. Numerically, $C = 0.090093$ according to Roche and 0.12554 according to Jeans and Chandrasekhar; in the latter case, only the tidal forces between a satellite and its planet are taken into account. If the body m enters the Roche limit, i.e. the sphere of radius R_R, then it will be disrupted under the action of tidal forces of the central planet M. The origin of Saturn's rings is explained by this phenomenon; for this planet the Roche limit equals 2.44 times the radius of the planet, and any satellite that entered this sphere would be torn into small pieces by the tidal forces of the central planet. This, however, is not the only possible explanation of the origin of Saturn's rings.

The Roche limit can be defined, as it turned out, also for binary stellar systems, with the only difference being that, first, now we have not one but two Roche zones, and, second, these zones are not exactly spherical in shape; they are pear–shaped surrounding both masses m_1 and m_2 of the system, contacting at acute angles the libration point L_2, as shown in Figs. 9.3 and 9.5 for the value of the Jacobi constant $C = C_2$.

Binary stellar systems can always be determined by the equipotential surfaces (Chapter IX) corresponding to definite values of the Jacobi constant C which will coincide with the contours of the Roche lobe. Of special interest is the figure-eight-shaped equipotential surface shown, for example, for values of C_3 and C_4 in Fig. 9.3 or for $C = 4$ in Fig. 9.9. The existence of stable gaseous configurations within this figure-eight-shaped zone is impossible; they would be quickly disrupted under the action of tidal forces. As a result, the transfer of gaseous matter from one component of the system to another occurs with numerous interesting consequences.

The fraction of so-called close binary systems among the binary stars is rather high. At the same time, the formation of a common Roche zone surrounding both components of a close binary system is inevitable. As a result the space between the components in such systems turns out to be a zone of violent processes of transfer and redistribution of gaseous matter.

Incidentally, together with this one can have another extreme case, in which gravity has almost no role. It appears that the internal dynamics, the structure and the stability of planetary nebulae are determined by forces of a non-Newtonian nature – electromagnetic, hydrodynamic, radiation pressure, etc.

One of the fundamental discoveries of celestial mechanics, namely the Roche lobe, has therefore found even wider application in astrophysics, particularly of close binary stellar systems, including relativistic systems, than for planetary systems in which its role is only episodic.

Chapter XIII

Flights to the Moon

1 Trajectories of Minimal Velocity

Flights to the Moon are of special interest among interplanetary flights. First, the Moon is the nearest celestial body to the Earth. Second, in contrast to the planets, which perform Keplerian motions in heliocentric trajectories, the motion of the Moon, though again Keplerian or almost Keplerian, occurs with the Earth as the central body.

The aims and problems of flights to the Moon may be different and the trajectories of the flights will be correspondingly different. One such aim could be the delivery of space probe to the Moon's zone of action, i.e. within the sphere of a radius $33\,000\ km$ around the Moon with a subsequent impact or soft landing of the probe onto its surface. Other aims are possible as well: a flyby of the Moon past its opposite side with or without return to the Earth; the transformation of the probe into a satellite of the Moon moving in a selenocentric orbit; the transfer of space probe into a special orbit for periodic flights around the Moon or flights periodically to the Moon and the Earth, etc.

It appears that flights to the Moon can be realized via different trajectories – elliptic, parabolic, hyperbolic, etc., even if the problems and aims are the same. Each of these trajectories has its versions differing, in particular, in the magnitude of the initial starting velocity from the Earth. For example, in the case of elliptic trajectorics, flights to the Moon can be realized by the following trajectories.

a. **By a "straight" ellipse,** i.e. by a route A (Fig. 13.1) in which the space probe is launched from the Earth absolutely vertically and directly to the Moon in a line EM_A with encounter with the Moon at the point M_A. The minimal initial velocity of launch is determined from the condition that the velocity of the probe will be zero at the point of its meeting with the Moon M_A. In this case the initial velocity obtained by the method described in the Section 8, Chapter XI, is equal to 14.13 $km\ s^{-1}$, which, as we shall see below, is too large.

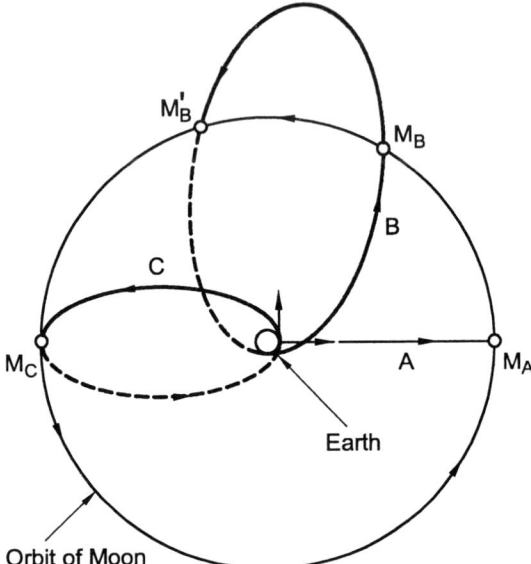

Fig. 13.1. *Three flight trajectories to the Moon: A, in a straight line (meeting with zero velocity at the point M_A); B, in an intersecting ellipse (at points M_B and M'_B); C, in a Hohmann ellipse (touching at the point M_C).*

b. **By the ellipse B** intersecting with the Moon's orbit twice – either at M_B or at the point M'_B, after passing the apogee. Bearing in mind the extremely small velocity of motion of the probe in the vicinity of the apogee, a flight with its destination at point M'_B will be very prolonged. The relative size of the arc of the ellipse exceeds the Earth–Moon distance and therefore this route cannot be considered as optimal from an energetic point of view.

c. **By the ellipse C.** In this case we shall have an ellipse of minimal size because it will come into contact at perigee with the surface of the Earth, or more precisely, with the circular orbit around the Earth from which the escape of the probe has been realized; and at apogee, with the Moon's orbit. The encounter of the probe with the Moon takes place at the point M_C.

It is not difficult to ensure, however, that, in the case of the ellipse C, we have a typical Hohmann orbit. If we assume the minimal use of energy as a basic property of a Hohmann trajectory, then the ellipse C satisfies this condition fully. In this case the space probe realizes a flight from one circular orbit – around the Earth, to another circular orbit – the Moon's orbit; the

Earth is at the center of both circles. The difference is terminological: in the case of the planets we have **heliocentric** Hohmann trajectories, whereas in the case of the Earth–Moon flight we have a **geocentric** Hohmann trajectory.

There is one more reason to consider the trajectory C to be of Hohmann type, i.e. the initial velocity for the ellipse C is minimal among all ellipses of type B. In our quantitative analysis of this trajectory, aiming in particular to derive the basic parameters of the flight, we will assume that the flight itself takes place in the Moon's orbital plane.

The radius of the Earth's action sphere was shown above to be equal to $900\,000\ km$, i.e. much bigger than the Earth–Moon distance. Therefore, after the launch of the space probe from the Earth, its further Keplerian motion in the direction of the Moon must take place completely under the action of the Earth's gravitation. In other words, the focus of the ellipse C must be at the center of the Earth. Then, from the formula for the radius-vector

$$r = a\left(1 - e\,\cos E\right), \tag{1}$$

we have for the distance of perigee $r_p(E = 0)$ and apogee $r_a(E = 180°)$, respectively

$$
\begin{aligned}
r_p &= a\left(1 - e\right), \\
r_a &= a\left(1 + e\right),
\end{aligned}
\tag{2}
$$

where a and e are the semimajor axis and the eccentricity of the Hohmann trajectory of the flight C. The peculiarity of the situation is that the numerical magnitudes both of r_p and of r_a are known:

$$
\begin{aligned}
r_p &= R_\oplus + H\,, \\
r_a &= a_0\,,
\end{aligned}
\tag{3}
$$

where $R_\oplus$ is the Earth's radius, H is the height of the circular orbit from the surface of the Earth, from which the escape of the probe occurs, and a_0 is the geocentric distance of the Moon.

From Eqs. (2) and (3) we have for the numerical value of the eccentricity

$$e = \frac{a_0/(R_\oplus + H) - 1}{a_0/(R_\oplus + H) + 1}. \tag{4}$$

Having e, we obtain easily the value of a from one of the relationships in Eq. (2):

$$a = a_0\,\frac{1}{1 + e}. \tag{5}$$

Let us illustrate the further steps of the computation by a concrete example, i.e. when $H = 350\ km$. Then, substituting $a_0 = 384\,000\ km$, and $R_\oplus = 6\,370\ km$, we obtain

$$a = 195\,524\ km$$

$$e = 0.966\,.$$

To move via a circular orbit at height H, the space probe should have the circular velocity V_c given by

$$V_c = \left(\frac{GM}{r_p}\right)^{1/2} = \left(\frac{GM}{R_\oplus + H}\right)^{1/2}, \tag{6}$$

where G is the gravitational constant, M is the total mass of the Earth and the Moon. On substituting the numerical values ($M_E/M_M = 81.3$, $GM = 3.986 \cdot 10^{20}$ in CGS units) we obtain from Eq. (6) at $H = 350\ km$

$$V_c = 7.70\ km\ s^{-1}\,.$$

It is necessary to know the magnitude of the additional velocity ΔV in order to put the probe into the Hohmann orbit C; it yields

$$\Delta V = V_p - V_C\,, \tag{7}$$

where V_p is the velocity at perigee of the ellipse C. We have

$$V = \left[GM\left(\frac{2}{r} - \frac{1}{a}\right)\right]^{1/2} \tag{8}$$

or, because $r = r_p = a(1 - e)$, we obtain for V_p

$$V_p = \left(\frac{GM}{a}\frac{1 + e}{1 - e}\right)^{1/2}. \tag{9}$$

On substituting here the values of a and e just derived, we obtain

$$V_p = 10.80\ km\ s^{-1}\,,$$

which gives for the additional velocity in accordance with Eq. (7)

$$\Delta V = 3.10\ km\ s^{-1}\,.$$

Indeed, with smaller values of ΔV the probe cannot reach the Moon at the point M_C. For larger values of ΔV we shall have a trajectory similar to the ellipse B. Hence, the obtained value of ΔV is minimal for realization of flight to the Moon. Therefore, the trajectory C is also called the "trajectory of minimal velocity" or the "trajectory of minimal energy".

Among the most important parameters, the duration of the flight T_0 is of particular importance. It is defined as half of the total duration P_0 of the flight of a hypothetical body in an ellipse C, i.e.

$$T_0 = \frac{P_0}{2} = \frac{\pi a^{3/2}}{(GM)^{1/2}},$$

or, substituting the value of a from Eq. (5), we have

$$T_0 = \frac{\pi}{(GM)^{1/2}} \left(\frac{a_0}{1+e} \right)^{3/2}. \tag{10}$$

Substituting the numerical values, we obtain

$$T_0 = 119.5\ h = 4.98\ days \approx 5\ days\,.$$

Thus, the Earth–Moon flight in a Hohmann trajectory C, i.e. the trajectory of minimal velocity, can be realized in five days. This is the largest duration of the flight. Every shortening of this duration will strongly affect the economic aspect of the flight.

If the encounter with the Moon does not occur, the space probe will return, again in *five days*, to the starting point, i.e. to the Earth, in the second half of the ellipse C, as shown by the dashed line in Fig. 13.2.

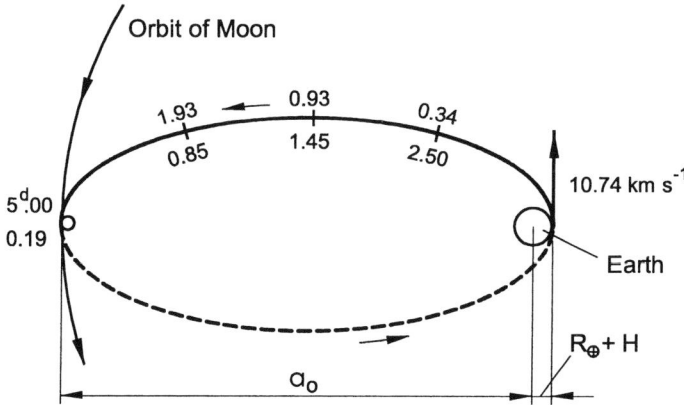

Fig. 13.2. *The structure of a Hohmann trajectory (ellipse) for a flight to the Moon. Numbers denote: velocity (lower row) and days (top row) during different stages of the flight.*

The magnitude of the velocity of the space probe V_a at the moment of its contact with the Moon's orbit at the point $M - C$ is also of interest. We have from Eq. (8), substituting $r_a = a(1 + e)$, that

$$V_a = \left(\frac{GM}{a} \frac{1 - e}{1 + e}\right)^{1/2}. \tag{11}$$

From that we obtain $V_a = 0.19\ km\ s^{-1}$, i.e. nearly 58 times smaller than the velocity of the probe at perigee. Such a large difference is explained by an extremely large value of eccentricity, i.e. by extremely strong elongation of the ellipse C.

These formulas are enough also for computation of the parameters of the motion of the space probe, in particular, the velocity and the duration of its flight at each point of the flight trajectory. For the duration of the flight, $T(r)$, up to the distance r from the Earth, we have from Kepler's equation

$$T(r) = \frac{1}{n}(E - e\sin E), \tag{12}$$

where the magnitude of the eccentric anomaly E is obtained from Eq. (1) to be

$$\cos E = \frac{1}{e}\left(1 - \frac{r}{a}\right) = \frac{1}{e}\left(1 - \frac{r}{a_0}(1 + e)\right). \tag{13}$$

We find the value of n, the mean daily motion of the probe through the orbit C, from the formula

$$n = \frac{2\pi}{2T_0} = (GM)^{1/2}\left(\frac{1 + e}{a}\right)^{3/2}. \tag{14}$$

In our case $n = 36\ degree\ day^{-1} = 0.631\ rad\ day^{-1}$. As for the velocity of motion of the probe at the point r, we have

$$V_r = \left(\frac{GM}{a_0}\right)^{1/2}\left(2\frac{a_0}{r} - (1 + e)\right)^{1/2} = 1.02\left(2\frac{a_0}{r} - (1 + e)\right)^{1/2}\ km\ s^{-1} \tag{15}$$

With the help of Eqs. (12) and (15) the duration of the flight $T(r)$ and the velocity of the motion are obtained at relative distances from the Earth r/a_0 equal to 0.25, 0.50, 0.75 and 1, as well at perigee for which $r/a_0 = 0.0175$. The results are presented in Table 13.1 and in graphic form in Fig. 13.2.

T a b l e 13.1

The duration of the flight, $T(r)$, and the velocity of the motion, $V(r)$, at various relative distances from the Earth, r/a_0, during flight from the Earth to the Moon in a Hohmann trajectory

Relative distance r/a_0	Flight duration $T(r)$, (days)	Velocity $V(r)$ (km s^{-1})
0.0175	0	10.74
0.25	0.34	2.50
0.50	0.93	1.45
0.75	1.93	0.85
1.00	5.00	0.19

It follows from these data that the velocity of the motion of the space probe decreases rapidly almost immediately after starting from the Earth and, as a result, the main flight time is spent on the second half of the route; its duration is five times longer than that of flight on the first half.

The initial escape velocity 10.80 $km\ s^{-1}$ of the probe in the case of an ellipse C is noticeably smaller than the velocity of flight in a straight trajectory A (Fig. 13.1), i.e. 11.09 $km\ s^{-1}$. This fact justifies the orbit C being called a flight with minimal velocity.

2 Parabolic Trajectory

Flight to the Moon is possible also in a parabolic orbit, the focus of which is the Earth (Fig. 13.3). In this case the initial velocity of escape is given by

$$V_p = \sqrt{2}\,V_c = \left(\frac{2GM}{R_\oplus + H}\right)^{1/2}, \tag{16}$$

which gives $V_p = 10.86\ km\ s^{-1}$ at $H = 350\ km$, i.e. it is only 0.060 $km\ s^{-1}$ larger than that within an elliptic orbit B. This insignificant difference turns out to lead to a significant gain in the duration of the flight.

We have for the motion of the probe in a parabolic trajectory

$$\left(\frac{GM}{2}\right)^{1/2}\frac{T}{q^{3/2}} = \tan\frac{v}{2} + \frac{1}{3}\tan^3\frac{v}{2}, \tag{17}$$

where q is the orbital parameter, i.e. the distance of the perigee from the focus, and T is the time of flight calculated from the moment of the start (at

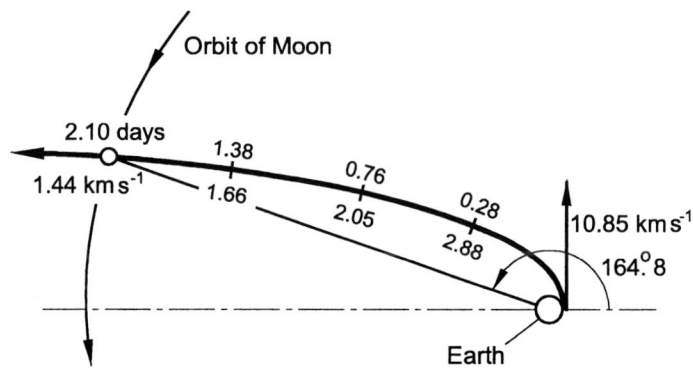

Fig. 13.3. *Flight to the Moon in a parabolic trajectory. Numbers denote velocity (lower row) and days (top row) during different stages of the flight.*

perigee). The magnitude of the real anomaly v corresponding to a given distance of the probe r from the focus, i.e. from the Earth's center, is obtained with the help of the parabola equation

$$\cos \frac{v}{2} = \left(\frac{q}{r}\right)^{1/2} . \tag{18}$$

In our case $q = R_\oplus + H$. Therefore we shall have instead of Eqs. (18) and (17)

$$\cos \frac{v}{2} = \left(\frac{R_\oplus + H}{r}\right)^{1/2} , \tag{19}$$

$$T(r) = \frac{(R_\oplus + H)^{3/2}}{(GM/2)^{1/2}} \left(\tan \frac{v}{2} + \frac{1}{3} \tan^3 \frac{v}{2}\right) . \tag{20}$$

These relationships simply determine the magnitude of the duration of the flight $T(r)$ when the probe is at a distance r from the center of the Earth.

We obtain the velocity of the motion from Eq. (8) if $a = \infty$:

$$V(r) = \left(\frac{2GM}{r}\right)^{1/2} . \tag{21}$$

With the help of these formulas we have obtained the most important parameters of the flight in a parabolic orbit, i.e. v, $T(r)$ and $V(r)$; they are presented in Table 13.2 and shown graphically in Fig. 13.3. The calculations are carried out for the same initial parameters, in particular, for $H = 350 \ km$.

T a b l e   13.2

Parameters of flight in a parabolic orbit to the Moon: the
true anomaly v, the launch velocity $V(r)$ and the duration
of the flight $T(r)$ for various distances r/a_0 from the Earth

r/a_0	v	$V(r)$ $(km\ s^{-1})$	$T(r)$ (days)
0.0175	0	10.86	0
2.25	149°.2	2.88	0.28
0.50	158.4	2.05	0.76
0.75	162.4	1.66	1.38
1.00	164.8	1.44	2.10

The most important result should be considered the duration of the flight
to the Moon – in the case of a parabolic orbit it turns out to be always two
days, which is 20% of the elliptic trajectory (five days). However, the velocity
of the encounter of the probe with the Moon in the case of the parabolic orbit
is significantly larger, $1.44\ km\ s^{-1}$, which is always an order of magnitude
larger than those obtained for the elliptic orbit, i.e. $0.19\ km\ s^{-1}$ (Table 13.1).

3   Intermediate Trajectories

The extreme sensitivity of the duration of flights to the Moon to the initial velocity of the start forces us to broaden our analysis in at least two
directions: to examine flights in an elliptic trajectory of type B, i.e. with
intersection with the Moon's orbit, and to consider flights with superhy-
perbolic velocity. Let us start from the elliptic trajectory of type B (Fig.
13.1). In contrast with the ellipse C for which the linear distances of perigee
and apogee from the focus are known precisely, in the case of an ellipse of
type B we know only the distance of the perigee; it is equal, as before, to
$r_p = a(1 - e)$, and remains unchanged for all ellipses B with various semimajor axes a and eccentricities e. As for the apogee, its position is different
for ellipses with various semimajor axes $a > a_0$, and is unknown. Therefore,
in this case, the semimajor axis a and eccentricity e of the ellipse B cannot
be determined in the same simple manner as in the case of the ellipse C.
This is the first point.

Second, we will change in some degree the statement of the problem;
namely, we wish to find the duration of flight T_0 in an ellipse B in terms of
its dependence on the initial velocity V_0. In this case the other parameters
of the ellipse, particularly a and e, should also be presented through V_0.

For the duration of the flight T_0 from the Earth to the Moon we have from Eq. (12)

$$T_0 = \frac{a^{3/2}}{(GM)^{1/2}} \left(E_0 - e \sin E\right), \qquad (22)$$

where $n = (GM)^{1/2}/a^{3/2}$, and the mean angular motion of the probe through the ellipse B is also used. The semimajor axis a of this ellipse as a function of V_0 may be obtained with the help of the formula

$$a = \left(\frac{2}{R_\oplus + H} - \frac{V_0^2}{GM}\right)^{-1}. \qquad (23)$$

As for the eccentricity anomaly E_0 in Eq. (22), its numerical value can be obtained from the relationship

$$\cos E_0 = \frac{1}{e} \left(1 - \frac{a_0}{a}\right), \qquad (24)$$

which has been derived from Eq. (1) bearing in mind the necessity to have $r = a_0$ at the point of intersection of the ellipse B with the Moon's orbit. Finally, at perigee we have $a(1 - e) = R_\oplus + H$ and

$$e = 1 - \frac{R_\oplus + H}{a}. \qquad (25)$$

The sequence of steps for the determination of T_0 from these formulas is as follows. For the given value of V_0 we obtain a from Eq. (23) and then e from Eq. (25). Having both a and e, it is easy to find E_0 from Eq. (24), after which T_0 is available from Eq. (22). At the end T_0 will be presented through V_0 – the initial velocity of escape of the probe from its circular orbit around the Earth. The analysis of all these formulas makes clear that a rather small increase in V_0 (of the order of a few $m\ s^{-1}$) is enough to give a significant shortening of the flight duration.

In principle the same is true for the analysis of a flight with hyperbolic initial velocity of escape; one could proceed from the formulas given in the previous section. Without going into details of the computations, we will confine ourselves to representation of the final results, i.e the dependence of T_0 on V_0 (for the case $H = 350\ km$ and with preservation of the remaining parameters as invariable). These results are as follows.

$V_0,\ (km\ s^{-1})$	$T_0,\ (days)$
10.92	5.0
10.97	2.5
11.02	2
11.52	1

The result $T_0 = 2\ days$ corresponds to the parabolic velocity. It appears that the duration of flight may be shortened by up to a day on increasing the initial velocity only by 0.5 $km\ s^{-1}$ with respect to the parabolic velocity.

These results are not only of theoretical interest. In extreme situations, say for rescue operations on the Moon, the hyperbolic orbit can be rather useful: only a slight increase in the starting weight is required. Presumably, flights with hyperbolic velocity will be realized sometime in the future.

The problems examined in the last three sections pursue primarily practical aims, i.e. revealing the conditions of flights to the Moon. The methods of computation of the trajectories are presented. They are rather simple and quite accessible even bearing in mind that these computations are of qualitative character, and the results can be used for further detailed surveys.

The next step is the realization of more accurate trajectory computations taking into account the following factors.

1. The orbital plane passes through the Earth's center, whereas the launching facility, i.e. the starting point, is on the Earth's surface, above this plane. As a result, the flight of the space probe cannot be realized in the plane of the Moon's orbit. In fact, the flight takes place in the plane perpendicular to the plane of the Moon's orbit. The main defect of such a flight is the impossibility of using additional velocity for the space probe conditioned by the daily rotation of the Earth. In fact, the problem is reduced to the examination of the spatial problem of reaching the Moon when the orbital plane of the flight is inclined by a nonzero angle with respect to the Moon's orbital plane.

2. The Moon's orbit is not circular as it was assumed to be in the above computations. At the eccentricity of the Moon's elliptical orbit $e = 0.0549$, the distance of the Moon at perigee will be 363 300 km and at apogee 405 500 km, i.e. the distance differs from the mean distance 384 400 km) by 21 100 km which is not negligible. The difference is about ($\sim$ 6 %), and therefore cannot be ignored in precise trajectory computations. This difference affects the duration of the flight – it can be shortened by a few hours.

3. The finiteness of the Earth's size has been taken into account in the trajectory computations described above – it was not assumed to be a mathematical point, but the Moon was assumed to be a mathematical point. Obviously, in concrete computations the finiteness of the Moon should be taken into account, especially for those concerning landing on the Moon's surface. Accounting for the finiteness of the Moon leads to a shortening of the flight duration by about 30 min for two reasons: due to the accelerating action of the Moon, as well as because the meeting takes place not at the Moon's center but on its surface or, in other words, because of the reduction in the whole flight path by the amount of the Moon's radius (1 740 km).

4. In all the cases considered above, the Keplerian trajectory during the entire flight is determined primarily by the Earth's gravitation, whereas, at the end of the flight, nearly five–sixths of the way to the Moon, the motion of the space probe occurs under the action of the Moon. As a result, the parameters of motion of the probe should be modified; in particular, its velocity will be increased. As a result, the duration of the flight will be shortened. Quantitatively this effect will be considered in the next section.

5. During flyby of the Moon, the space probe should experience some influence of the gravity of the Sun. The perturbation originates as a result of the different accelerations caused by the Sun upon the space probe on one side and the Earth on the other. This difference vanishes when the flight takes place on the plane perpendicular to the Sun–Earth radius-vector, i.e. when the Moon is in the first or in the last quarter. The influence of the Sun is relatively large when both the Moon and the Earth are situated along this radius-vector, i.e. either at the new Moon or at the full Moon. Quantitatively this effect (about 1%) influences the duration of the flight – it will slightly decrease. However, as the computations indicate, the expected shortening is small – only three minutes.

6. The influence of the nonsphericity (oblateness) of the Earth should also be taken into account in precise trajectory computations. The oblateness brings about a shortening in the duration of the flight. According to the computations, the shortening is about 1 000 *min*, i.e. five–six times larger than the shortening of the flight duration caused by the Sun's gravitation.

4   The Trajectory of Impact with the Moon

The trajectories considered above – elliptic (Hohmann), parabolic, and hyperbolic – can be used for any aim connected either with the direct impact (landing) of the probe on the Moon or its flyby, or the transfer of the probe at the speeding-up trajectory. The strong influence of starting conditions, in particular, of the magnitude and the direction of initial velocity of escape from the Earth on the parameters of the flight, and first of all, on its duration, is becoming clear.

In principle, the landing on the Moon may be realized by two versions, namely, directly from the Earth–Moon trajectory as well as after a preliminary going out into a closed, circular or elliptical, orbit around the Moon. In the first case, the trajectories of the direct landing on the Moon may be realized, depending on the inclination angle to the Moon's surface, which may be vertical as well as inclined when the trajectory crosses the Moon's surface under an acute angle. To the landing may foregoes one, two or three orbitings of the spacecraft from the closed orbit around the Moon during of some time which may be used for the realization of various programs – photography along the flight trace, survey and investigation of the opposite side of the Moon etc.

The Moon does not have an atmosphere, hence, landing on its surface cannot be realized except by the use of rocket engines.

In fact, there are an infinite number of trajectories and conditions for the probe to enter the sphere of action of the Moon. The direct impact or landing of the probe is possible on any point on the Moon, either on its visible side or on the opposite one. A small change is enough – a fraction of a degree, in the direction of the initial velocity vector, and one will have a quite different trajectory with an exact impact or landing point of the probe on the Moon's surface. Similarly, a small change in the magnitude of the initial velocity of escape – only within the limits of a few meters per second(!), and the duration of the flight may become shorter or longer by whole days.

In the most general case, computation of the impact trajectory of the space probe on the Moon should be realized taking into account the influence of the Moon's gravity; the probe in the last part of its motion will be within the sphere of action of the Moon. In other words, the problem connected with the determination of the trajectory of an Earth–Moon flight should be approached as a three-body problem. The trajectory estimated in such a way will of course differ, especially while approaching the Moon, from the Hohmann trajectory computed in the scheme of the two-body problem, i.e. ignoring the presence of the Moon and assuming the Earth to be the central body.

In the scheme of the three-body problem, the system of differential equations of motion of space probe in the coordinate system xyz with the center situated at the Earth's center and a plane XOY coinciding with the Moon's orbit, is written in the following form:

$$\frac{d^2x}{dt^2} = -\mu_0 \frac{x}{r^3} - \mu_\star \frac{x_\star}{a^3} + \mu_\star \frac{x_\star - x}{\rho^3},$$

$$\frac{d^2y}{dt^2} = -\mu_0 \frac{y}{r^3} - \mu_\star \frac{y_\star}{a^3} + \mu_\star \frac{y_\star - y}{\rho^3}, \qquad (26)$$

$$\frac{d^2z}{dt^2} = -\mu_0 \frac{z}{r^3} - \mu_\star \frac{z}{\rho^3}.$$

One should also add to this system the equations of spatial velocities of the probe (dr/dt) and of the Moon $(d\rho/dt)$ represented through their projections

$(dx/dt \ldots, dx_\star/dt \ldots)$ on the corresponding coordinate axes

$$\frac{dr}{dt} = \frac{1}{r} \left(x\frac{dx}{dt} + y\frac{dy}{dt} + z\frac{dz}{dt} \right),$$

$$\frac{d\rho}{dt} = \frac{1}{\rho} \left[(x_\star - x)\left(\frac{dx_\star}{dt} - \frac{dx}{dt}\right) + (y_\star - y)\left(\frac{dy_\star}{dt} - \frac{dy}{dt}\right) + z\frac{dz}{dt} \right].$$

(27)

In these equations $x_\star$ and $y_\star$ are the coordinates of the Moon ($z_\star = 0$), μ_0 and $\mu_\star$ are the gravitational potentials of the Earth and the Moon, respectively, a is the mean distance between the Earth and the Moon (384 000 km), and r and ρ are the distances of the space probe from the centers of the Earth and the Moon, respectively:

$$r = (x^2 + y^2 + z^2)^{1/2}.$$

$$\rho = [(x - x_\star)^2 + (y - y_\star)^2 + (z - z_\star)^2]^{1/2}.$$

The integration of Eqs. (26) and (27) can be realized only by numerical methods. As a result, the equations of motion of the probe will be obtained, i.e. the dependence of the geocentric coordinates of the probe on time:

$$x = f_1(t),$$
$$y = f_2(t),$$
$$z = f_3(t),$$

and also the trajectories of flights with arbitrary initial conditions. Owing to the diversity of initial conditions, i.e. the starting velocity from the Earth, the direction of its vector, the coordinate of the starting point on the surface of the Earth, the height of the circular orbit around the Earth from which the escape of the probe takes place, etc., one will have enormous diversity of flight trajectories.

About two hundred years ago, Laplace proposed an original and very simple method of computation of the trajectory of a celestial body of small mass in the gravitational field of two bodies. The essence of this method lies in the splitting of the trajectory of motion of an infinitely small mass (space probe) into two parts, i.e. the trajectory within the Earth's sphere of action, for which the influence of the Moon can be ignored, and the trajectory within the Moon's sphere of action, for which the Earth's gravity can be ignored. Those local trajectories were later called *"rapprochement trajectories"*.

Owing to such a division of the trajectory into two parts, the restricted three-body problem is reduced to two independent two-body problems, i.e. the probe–Earth and probe–Moon ones. The Laplace method was revived in the early days of the space flight epoch (Egorov, 1965), for computations of flight trajectories to the Moon. The sequence of computation steps of Earth–Moon flight trajectories is as follows.

I. The motion of the probe on the first part of the trajectory from the Earth to the Moon's action sphere is considered to be purely geocentric and should be computed as a Keplerian trajectory in a two-body scheme, i.e. Earth-probe.

II. On the boundary of the transition from the Earth's sphere of action to the Moon's sphere of action, the parameters of geocentric motion should be recomputed in terms of the parameters of selenocentric motion.

III. The selenocentric trajectory of the motion should be computed as Keplerian motion with the Moon as the central body, ignoring the Earth's influence.

The method that has just been described is extremely simple and easily accessible for wide application. However, in all cases, the joint solution of Eqs. (26) and (27) through numerical integration should be considered the most precise solution. Obviously, the results obtained may be used first of all for checking the results obtained via the Laplace method. It appears that the trajectories of the Moon's flybys, and their most important parameters obtained by the Laplace method, are practically the same as the parameters obtained by numerical integration of differential equations of motion.

A series of lunar trajectories have been computed by the method just described, including various initial conditions of the flights. As an example, in Fig. 13.4 six flight trajectories from the Earth to the Moon are shown (in the rotating system of coordinates with the X axis directed along the Earth–Moon line). The curve III corresponds to the parabolic trajectory, i.e. when $V = V_p = 10.86 \ km \ s^{-1}$. These trajectories are computed for an angle of declination of initial velocity vector $\alpha_1 = \pi/2$ but for six different values of the starting velocity V. As a result, essentially different durations T_0 are obtained for the Earth–Moon flight; they are given in Table 13.3 together with the differences in $\Delta V = V - V_p$, where V_p is the escape velocity in the parabolic trajectory.

It should be noted that the duration of the flight in a hyperbolic (I) orbit when the initial velocity exceeds the parabolic velocity by 0.48 $km \ s^{-1}$ is shortened by a factor of two, from 2.070 to 1.084 $days$. Now we see that the initial velocities of escape for flights in the trajectories V and VI differ only by 10 $m \ s^{-1}$, and this extremely small difference leads to differences in the durations of the flights by a factor of one and a half.

The trajectories ensuring the impact of the probe with the Moon without intersection with the lunar orbit are called **ascending** ones. For example, all the six trajectories in Fig. 13.4 are ascending ones; they are distinguished not only by the durations of the flights but also by the selenocentric coordinates of the points of intersection with the lunar surface. There is an enormous diversity of ascending orbits, ensuring impact with any point of the lunar surface.

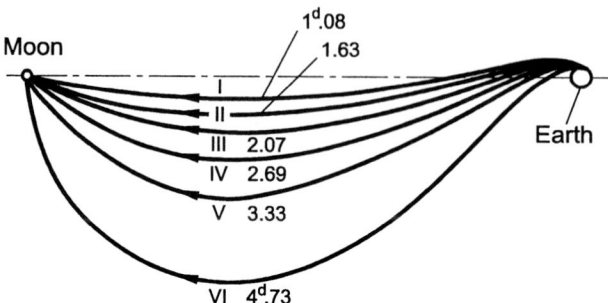

Fig. 13.4. *Ascending trajectories of capture by the Moon in the rotating (with the Earth–Moon line) coordinate system. Numbers on curves denote the duration of the Earth–Moon flight: I and II the launch of the spacecraft at hyperbolic velocity; III, with parabolic velocity; IV, V and VI, with elliptic velocities (see the text).*

T a b l e   13.3

Earth–Moon flight parameters for six trajectories with six initial velocities V and six various durations T_0 of the flight. The trajectory III is parabolic

Trajectory	Δv $(km\ s^{-1})$	V $(km\ s^{-1})$	T_0 $(days)$
I	+0.482510	11.3425	1.084
II	+0.10609	10.9661	1.627
III	0	10.86	2.070
IV	-0.057828	10.8022	2.648
V	-0.082828	10.7772	3.333
VI	-0.092828	10.7672	4.731

Together with this, there are also trajectories, again of enormous number, ensuring the impact of the probe with the lunar surface via descending trajectories, i.e. after intersection with the lunar orbit.

These examples indicate the necessity for classification of the lunar impact trajectories. Such a classification system, rather detailed and complete, was proposed by Ehricke in 1962; in Fig. 13.5 four examples are shown, i.e. four classes of flight trajectories for impact with the Moon. These trajectories are estimated for two cases, $\alpha_1 > 0$ and $\alpha_1 < 0$, where the sign of

α_1 indicates the direction of the flyby of the Earth in the initial stage of the trajectory. The limiting trajectories of each class, i.e. the trajectories corresponding to the angles $\alpha_1 = +\pi/2$ and $\alpha_1 = -\pi/2$, are shown. Two types of trajectories are shown in Fig. 13.5 for each of these four classes, one of which, sidereal, is related to the coordinate system connected with the Earth (trajectories EM), and other one, synodic, is related to the rotational coordinates (trajectories EM'). The circle MM' is the Moon's orbit around the Earth. Finally, all these orbits are nominal, i.e. they intersect with the Moon's center.

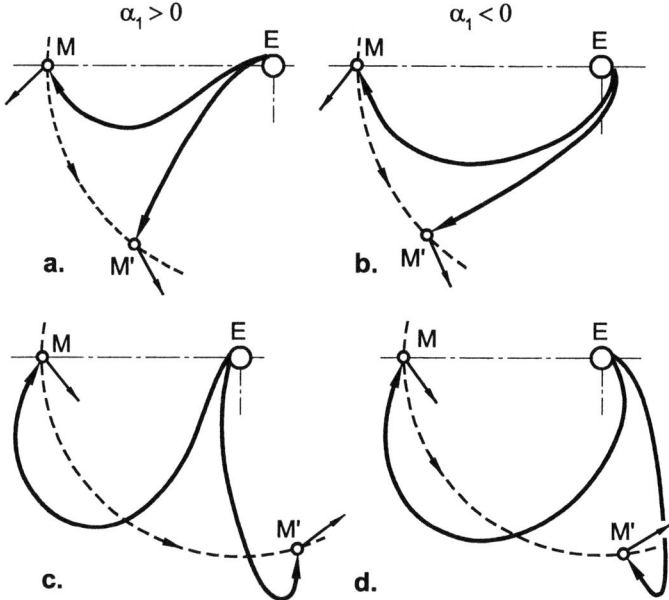

Fig. 13.5. *Four classes of trajectories of capture by the Moon. Solid lines, trajectories in the motionless (geocentric) system of coordinates; broken lines, the same trajectories in the rotating system of coordinates. The sign of α_1 denotes the direction of the trajectory around the Earth in the initial part of the trajectory.*

All trajectories shown in Fig. 13.4 are related to class (a) in Fig. 13.5, i.e. to the class of ascending branch trajectories.

At hyperbolic initial velocities we can obviously have only trajectories of the ascending branch (trajectories I and II in Fig. 13.4). The trajectory III is also an ascending one, i.e. it has a parabolic initial velocity. At velocities smaller than the parabolic one, together with the trajectories of the ascending branch, the appearance of trajectories of the descending branch is also possible. With the subsequent decrease in initial velocity, the trajectories of both classes, corresponding to one and the same value of angle α_1, become close to each other, and at initial velocity equal to its minimal value, the corresponding trajectories are coincident.

The trajectories of direct impact on the Moon, as is shown in Fig. 13.5, practically do not differ from flyby trajectories that run into the sphere of action of the Moon. The distinction relates to the appearance of a small-ish correction near the Moon. The characteristic peculiarity of these flyby trajectories is that after the flyby into the Moon's sphere of action the magnitude of the going out geocentric velocity in fact is always hyperbolic, independent of the initial take off velocity from the Earth. As a result, the spacecraft, after such a flyby with the Moon, leaves in the infinity. Such a momentum without expenditure of fuel in principle may be used for the realization of interplanetary flights in any direction (the arrows near the Moon in Fig. 13.5) dependent on the initial conditions of launch from the Earth.

5   Flyby Trajectories

Among the aims of astronautics and interplanetary flights there are projects for flybys of, say, the Moon, with or without subsequent return to the Earth. In such cases, one speaks of lunar flyby trajectories.

There are an infinite number of lunar flyby trajectories. An example of such a trajectory, a figure-eight-shaped one, is shown in Fig. 13.6; in this case the probe performs the flyby around the Moon via its opposite side and returns again to the Earth. In this case, the probe intersects the lunar orbit "in front" of the Moon, i.e. from the direction of its orbital motion, and this circumstance should be counted as the most characteristic feature of this trajectory. Below we shall see that flyby from the opposite (back) side is excluded in principle. This trajectory is synodic, i.e. it is shown in the coordinate system with the axis X rotating along the Earth–Moon line.

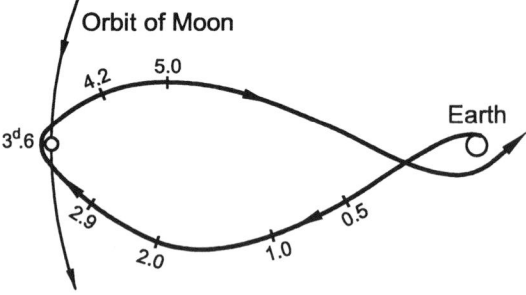

Fig. 13.6. *A typical trajectory of the flight of the Moon in the rotating (with the Earth–Moon line) system of coordinates. The flight always takes place from the front of the Moon.*

In Fig. 13.6 a lunar flyby trajectory is shown in which the probe, however, will not return exactly to the Earth, but will pass at a distance that is not

large. There are also a variety of flyby trajectories ensuring the return and
landing of the probe exactly on the Earth. Examples of such trajectories,
one from each of four types of flyby trajectories, are shown in Fig. 13.7.
These trajectories, at the same time, are nominal, i.e. they pass through the
Earth's center. For those of classes *a* and *b*, the flyby takes place very close
to the Moon.

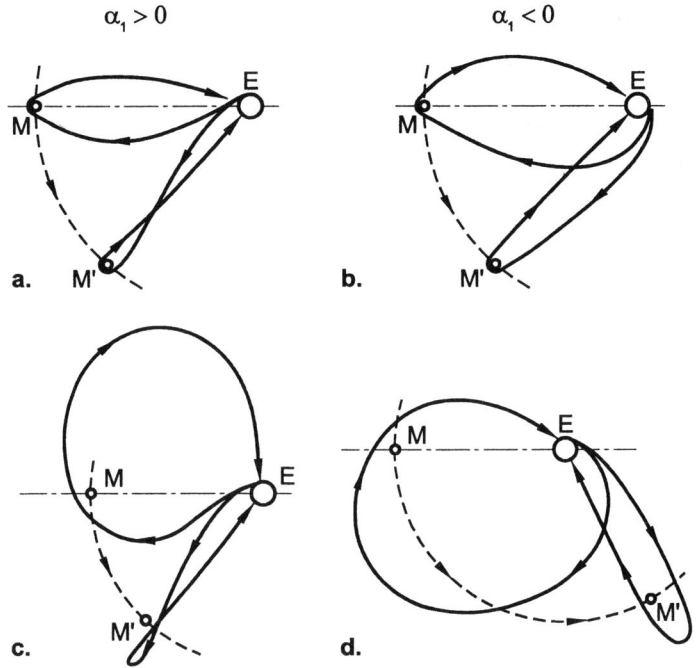

Fig. 13.7. *Four classes of flight trajectories for the Moon (see Fig. 13.5).*

Each of these four classes of trajectories is shown in Fig. 13.7 analogously
with Fig. 13.5, by two trajectories, one of which is geocentric or sidereal,
i.e. in the coordinate system related to the Earth (trajectories EM), the
other synodic, i.e. related to the frame rotating along the Earth–Moon line
(trajectories EM'). The arc MM' corresponds, as before, to the lunar orbit
around the Earth.

In all trajectories shown in Fig. 13.7, the flyby of the Moon takes place in
clockwise direction, i.e. "in front" of the lunar motion. Flyby from the back
side, i.e. in the direction of the orbital motion of the Moon, is impossible for
the following reason. The point is that a sufficient part of the lunar sphere of

action from its back side is a forbidden zone for entry of space probe. This is explained rather simply: the apogee geocentric velocity of space probe near the lunar orbit is about 0.2 $km\ s^{-1}$, i.e. five times smaller than the orbital geocentric velocity of the Moon, equal to 1 $km\ s^{-1}$. It is clear that, under such conditions, the probe simply cannot reach the Moon, i.e. in other words, the sphere of action will simply run away from the former. Only in the cases in which the apogee is far behind the lunar orbit, or when the launch of the probe from the Earth was performed with rather large hyperbolic velocity, can the entry of the probe into the back side of the sphere of action be possible.

Thus, the normal flyby of the Moon is always possible from the front, in a clockwise direction. As for the return, it can occur from the back side of the Moon (trajectories *a*, *b*, and *d*) as well from the front (trajectory *c*).

In which manner does the motion of the probe near the Moon occur?

The appearance of the space probe in the coordinate system connected with the Moon in the lunar sphere of action means that the probe enters this sphere of action from infinity with hyperbolic velocity. The motion of the probe within the sphere of action of any planet, including the Moon, thus takes place by a hyperbolic velocity at the focus of which the Moon is situated. During flyby of the Moon at the point of intersection (Fig. 13.7), the probe will continue its motion in a hyperbolic orbit and will leave the sphere of action with selenocentric velocity. At the same time this seleno-centric hyperbolic velocity is a normal elliptic geocentric velocity. At two points, namely, at the moment of entry into the lunar sphere of action (Fig. 13.7) and at the moment of leaving it, the selenocentric trajectory will be "sewn", i.e. it will smoothly continue to the geocentric elliptical orbit in its ascending and descending halves, from the side of apogee where the Moon is. From that one can understand the apparently amazingly shaped trajectories of flyby orbits.

Thus, the entry velocity of the probe into the lunar sphere of action must always be larger than the selenocentric parabolic velocity on the boundary of this sphere. The latter is $V_p(Moon) = 0.383\ km\ s^{-1}$.

Thus, the space probe entering the lunar sphere of action has two possibilities: to undergo impact with the Moon, or to leave the sphere of action of the Moon, escaping from the Moon with hyperbolic velocity. The capture of the probe by the Moon, i.e. its transformation into a stationary satellite moving in a closed (elliptical) orbit around the Moon, is impossible. The capture of the probe with its subsequent transformation into a satellite of the Moon is possible only if the probe can be gained while it is in the lu-nar sphere of action by decelerating momentum in such a manner that its selenocentric velocity will be smaller than the parabolic velocity of escape from the Moon, i.e. 2.38 $km\ s^{-1}$.

6   Space Probe at Libration Points

One of the fundamental results obtained in the analysis of the restricted three-body problem (Chapter IX) was the establishment of five libration points, i.e. the points of relative equilibrium. These points are situated within the configuration of finite bodies M and m in a remarkable way: three points, L_1, L_2 and L_3, are situated on a straight line connecting the two bodies, M and m; the other two points, L_4 and L_5, form the vertices of two equilateral triangles with a common side coinciding with the line Mm.

The libration points of the Earth–Moon system are of particular interest: their localization is shown in Fig. 13.8 to scale (relative to the Earth–Moon distance, $384\,400\ km$). Numerically, the parameter $1/\mu$ for the Earth–Moon system is large – 82.2 ($[1/\mu = (M + m)/m]$), therefore, the center of the system practically coincides with the position of the Earth (more precisely, it is $4\,700\ km$ from its center). The points L_2 and L_1 are located in the vicinity of the Moon at distances $0.15d = 58\,000\ km$ and $0.17d = 65\,000\ km$ from its center (d is the Earth–Moon distance), i.e. nearly on the boundary of its sphere of action, the radius of which is equal to $R_k = 66\,000\ km$. Point L_3 is located on the opposite side, far from the Earth and nearly within the Moon's orbit. The libration points L_4 and L_5 are situated, again because of the large value of the parameter $1/\mu$, practically within the lunar orbit, at the angular distance of $60°$ from each side of the Earth–Moon line (Fig. 13.8).

These five libration points possess a common property, i.e. they are all rotating with the same angular velocity around the center of the system Mm (around the Earth) with an angular velocity equal to that of the Moon's rotation around the Earth. In other words, this configuration of the points L_1, L_2, L_3 L_4 and L_5 preserves the invariability of their mutual orientation, as if they were fastened to the Earth–Moon line, and therefore rotating as a whole together with this line. However, the linear velocities of these points are different: for the points, e.g. L_4 and L_5 and, in part, for L_3, these velocities are almost the same and are equal to the linear velocity of the orbital motion of the Moon, i.e. $1.02\ km\ s^{-1}$. For the nearest point, L_2, and the farthest one, L_1, these velocities are equal to $0.87\ km\ s^{-1}$ and $1.19\ km\ s^{-1}$, respectively.

The points L_1, L_2 and L_3, located on a line, are called **collinear**. The points L_4 and L_5 are called **triangular**, or **triangle points of libration**.

The problem of the stability of libration points was a matter of special concern even from the time of their discovery. As a result, definite conclusions on the degree and the character of stability of these points were made. Thus, it is considered that the triangle libration points L_4 and L_5 are more stable than the collinear libration points L_1, L_2, and L_3. More precisely, the situation is as follows: any body, natural or artificial, if appearing not

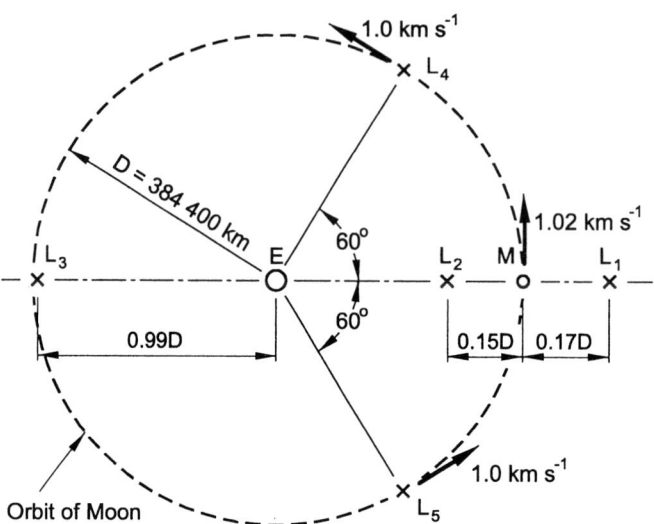

Fig. 13.8. *The positions of the libration points of the Earth–Moon system: collinear, L_1, L_2 and L_3, and triangular L_4 and L_5.*

exactly at the point, say, L_4, but a little away from it, with a small nonzero velocity (in the system of coordinates rotating with the Earth–Moon line), will remain in the future in the vicinity of the point L_4.

The situation is rather different when the probe is situated in the vicinity of the points L_1, L_2 or L_3; in this case even a very small velocity will be enough to remove the probe far from the point. However, the collinear libration points possess a property that is not characteristic for the triangle libration points: around the points L_1, L_2 and L_3 there are, as was shown in Chapter IX, so called "halo-orbits", through which the space probe can move far as long a time as possible. Moreover, the sizes of these halo-orbits turn out to be not small, they are about several thousand kilometers, i.e. even larger than the Moon's diameter. This means that, if in halo-orbit around the libration point L_2 there appears an probe, say, a radio-communication satellite, then this satellite will, at any of its halo-orbit, remain in the zone of direct visibility from the Earth indefinitely; of course, this is only true if the plane of this halo-orbit is perpendicular to the Earth–Moon-point L_2 line. According to computations, the period of rotation of such a satellite-station is about two weeks for a halo-orbit of radius equal to 3500 *km*. If such a station were placed exactly at the libration point L_2, then its signals would not be received on the Earth because of screening by the Moon.

The practical absolute immobility of the libration points, with respect to the Earth and with respect to the Moon, and also with respect to each other, opens wide possibilities from the point of view of their exploitation as future space stations, industrial objects, scientific research centers, etc. In this connection, a problem concerning the substitution of the space probe or transport ships at the libration point arises.

How could a space probe be substituted at the libration point? The answer to this question, as we shall see later, cannot be common for all five libration points. Each of these libration points has its individual properties and, therefore, the optimal solution of the problem is different for each case.

Most probably the main hopes should be connected, first, with the points L_1 and L_2 as the locations of future stationary space centers and stations. Therefore, let us confine ourselves to the problem of substitution of space probe at the points L_1 and L_2.

Let us start from a general analysis of the dynamics of the points L_1 and L_2. The period of their rotation around the Earth is 27.32 *days* (a sidereal month), and these three libration points, L_1, L_2 and L_3, are rotating as rigidly connected points with an angular velocity $\omega = 2.662 \cdot 10^{-6}$ *rad* s^{-1}. Therefore, we have for their linear velocities

$$V(L_1) = (D + 0.17D)\omega = 1.17D\omega = 1.20\,km\,s^{-1},$$

$$V(L_2) = (D - 0.15D)\omega = 0.85D\omega = 0.87\,km{\cdot}s^{-1},$$

$$V(L_3) = 0.99D\omega = 1.01\,km\,s^{-1}$$

which are at the same time the heliocentric velocities of points L_1 and L_2. The velocity of the Moon relative to the Earth is equal to 1.02 *km* s^{-1}. Obviously, the ends of all three velocity vectors must be situated on a straight line passing through the center of the Earth (Fig. 13.9).

Because the geocentric velocity of the point L_2 (0.87 *km* s^{-1}) is smaller than the geocentric velocity of the Moon (1.02 *km* s^{-1}), the point L_2 should lag behind the Moon with a velocity $1.02 - 0.87 = 0.15$ *km* s^{-1}. Hence, when the Moon is moving ahead in its own circular orbit with velocity 1.02 *km* s^{-1}, the point L_2 shifts backward relative to the Moon with a velocity of 0.15 *km* s^{-1}, staying at an invariant distance from the Moon (58 000 *km*). In other words, the point L_2 is performing a circular motion around the Moon. It is clear that, during a lunar month when the Moon accomplishes a complete circuit above the Earth, the point L_2 will also accomplish a complete cycle around the Moon in a circular orbit of radius 58 000 *km*.

We will have an analogous picture also in the case of the libration point L_1. The geocentric velocity of the point L_1 is larger than the orbital velocity of the Moon, and, hence, the point L_1 will move ahead of the Moon with a velocity $1.19 - 1.02 = 0.17$ *km* s^{-1}, and, correspondingly, will make a

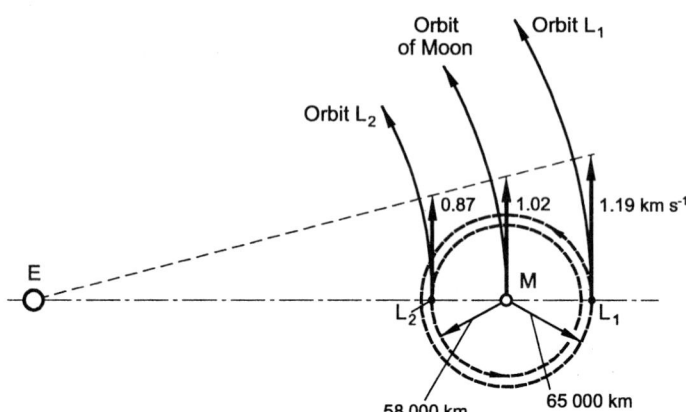

Fig. 13.9. *A diagram showing the heliocentric velocities of the libration points L_1 and L_2 and of the Moon M. E is the Earth. The orbits, both geocentric (solid arcs) and selenocentric (broken circles), of the points L_1 and L_2 are also shown.*

complete cycle in a circular orbit with a radius of 65 000 km around the Moon during a lunar month. Now we will turn to the question concerning the means of substitution of space probe at the libration points L_1 and L_2.

The substitution of space probe at the libration point L_2. The radius of the Moon's sphere of action is equal to 66 000 km, and the libration point L_2 is located between the Moon and the Earth, but nearer to the Moon – at a distance of 58 000 km, not far from the inner boundary of its sphere of action (Fig. 13.10). Therefore, the flight from the Earth to the point L_2 can be fully realized in a Hohmann ellipse, because the probe will be within the Earth's zone of action practically throughout its flight.

Let us determine the elements of the Hohmann trajectory, i.e. of the transition ellipse to the point L_2. Because the focus of this ellipse lies at the center of the Earth, and the point L_2 is at its apogee, we shall have for the distance r_a of the point L_2 from the apogee (see Fig. 13.8):

$$r_a = d - 0.15d = 0.85d = 326\,740 \ km$$

and for the distance from the perigee r_p, taking the height of the circular orbit $H = 350 \ km$ from the Earth's surface (see Fig. 13.10)

$$r_p = R_\oplus + H = 6\,370 + 350 = 6\,720 \ km \ .$$

Then we obtain for the semimajor axis a of the transition ellipse

$$a = \frac{r_p + r_a}{2} = 166\,730 \ km \ .$$

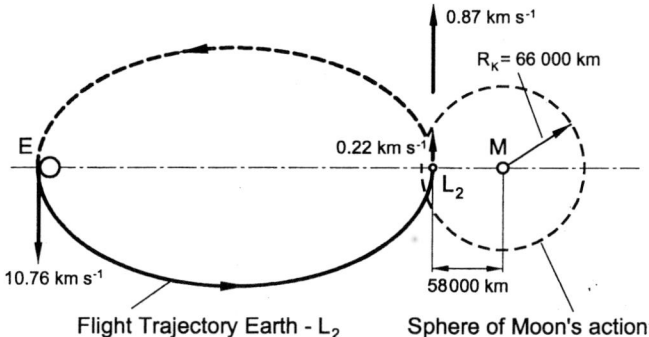

Fig. 13.10. *The motion of a spacecraft towards the libration point L_2 in a Hohmann flight trajectory (in the rotating system of coordinates). The initial velocity is 10.76 km s^{-1}, the arrival velocity at the point L_2 is 0.22 km s^{-1}. The broken line is the trajectory of the return from the point L_2 to the Earth.*

We obtain the eccentricity of this ellipse from the following relationship:

$$\frac{r_p}{r_a} = \frac{1-e}{1+e},$$

which gives $e = 0.960$.

An important parameter is the duration of the flight T_0 from the Earth to the point L_2:

$$T_0 = \frac{\pi a^{3/2}}{(GM)^{1/2}} = 93.97 \ h = 3.91 \ days.$$

For the escape velocity of the probe from a circular orbit around the Earth, i.e. at the perigee of the Hohmann ellipse, we have

$$V_p = \left[\frac{GM}{a}\left(2\frac{a}{r_p} - 1\right)\right]^{1/2} = 10.76 \ km \ s^{-1}.$$

A space probe launched from the Earth with such an initial velocity should reach the point L_2 with a velocity V_a equal to

$$V_a = V_p \frac{r_p}{r_a} = 0.22 \ km \ s^{-1}.$$

This is by a factor of four smaller than the heliocentric velocity of the libration point L_2 (0.87 $km \ s^{-1}$). Hence, to make the probe reach the point

L_2, it is necessary to give it a **momentum** equal to $0.87 - 0.22 = 0.65\ km\ s^{-1}$ in the direction of the lunar orbital motion, or more precisely in the direction of the velocity vector at the apogee of the Hohmann ellipse. Possessing such a velocity, the space probe will leave the Hohmann ellipse, passing to the circular orbit with a radius of $r_a = 326\,400\ km$ around the Earth. Then by transition maneuvers, the probe will be moved into an orbit around point L_2 with the help of a series of minor momentum corrections.

Thus, the operations concerning the substitution of space probe at the libration point L_2 are completed. One also has to keep the probe permanently at the libration point L_2, or more precisely, within the halo–orbit, under the action of various perturbations, by means of correcting maneuvers. Note that the peculiarity of the situation at the libration point enables complete automation of this operation, i.e. keeping the probe in its place.

Bearing in mind the possibility of sending people or equipment to the libration point L_2, we should think about their return to the Earth. In this case it is necessary to impart to the probe or space ship an impulse of magnitude exactly $0.65\ km\ s^{-1}$ but now in the opposite direction. As a result, the heliocentric velocity of the probe will be reduced to $0.22\ km\ s^{-1}$, and it will automatically leave the circular orbit around the Earth, immediately passing into the Hohmann orbit. This operation will be a mirror image of the Hohmann ellipse in the case of the Earth–point L_2 flight, as is shown in Fig. 13.10 by a broken line. The duration of the flight during the return journey will be the same – 3.91 *days*.

The substitution of space probe at the libration point L_1. In principle, the flight of space probe from the Earth to the libration point L_1 can also be realized in a Hohmann trajectory, an ellipse.

For the perigee and apogee $r_p = 6\,720\ km$ and $r_a = 449\,400\ km$, the semimajor axis of the Hohmann ellipse will be $a = 228\,060\ km$, the eccentricity $e = 0.961$ and the semiminor axis $b = 54\,900\ km$. The initial velocity of departure from the Earth will be slightly larger than before, i.e. $V_p = 10.79\ km\ s^{-1}$. The velocity at the apogee, i.e. on approaching the libration point L_1, is equal to $V_k = 0.16\ km\ s^{-1}$. With an additional momentum $1.19 - 0.16 = 1.03\ km\,s^{-1}$, the probe may be driven to the point L_1 and then into the halo-orbit around L_1.

However, as shown by the analysis, flight by the described completely Hohmann trajectory from the start to the destination is not optimal. In the final stage, about the final third of the way, the flight is performed within the sphere of action of the Moon (Fig. 13.11). As a result, the space probe accelerated by the Moon will reach the point L_1 earlier. This disadvantage can be corrected by giving an additional momentum to the probe. To avoid frequent use of jet engines for such corrections, the search for more economic possibilities of Earth–point L_1 flight seems of definite importance. An example of such an orbit is shown by the upper solid line in Fig. 13.11 (Farquhar,

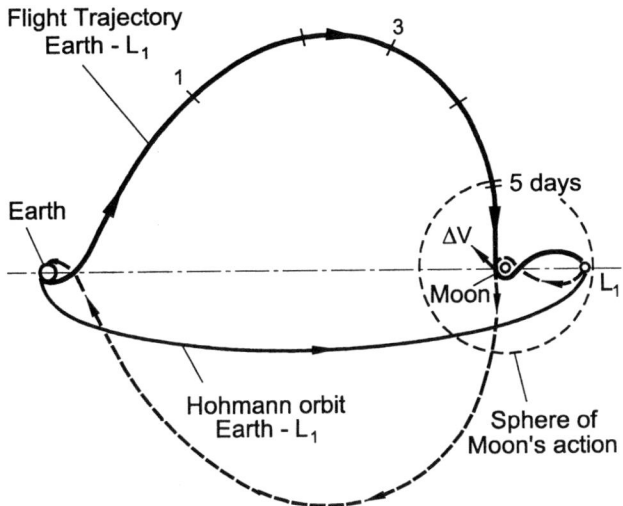

Fig. 13.11. *The leading of a spacecraft to the libration point L_1 in a combined trajectory. At the point A the spacecraft undergoes a 0.18 km s^{-1} braking, and at the point L_1 it receives an impulse of 0.15 km s^{-1}. The return trajectory from the point L_1 to the Earth (broken line) is also shown.*

1972).

First of all, the departure of the spaceship is realized from a very low orbit around the Earth, at altitude 185 km. As a result, the ship will perform a hook-like trajectory shown by a solid line in Fig. 13.11. At the nearest point to the Moon (altitude 110 km) the probe undergoes a braking of 0.18 km s^{-1}, and another one, 0.15 km s^{-1}, close to the point L_1. As we see, the momentum, 0.33 km s^{-1}, is three times smaller than we had in the first case (1.03 km s^{-1}, plus the effect of lunar gravitation); thus the gain in energetics amounts to an order of magnitude. The return trip from the point L_1 to the Earth is realized again by a two-momentum scheme and in a trajectory symmetrical to the trajectory of the Earth–point L_1 flight, as shown in Fig. 13.11 by a broken line.

7 The Satellite on the Libration Points

Thus, in the Earth–Moon system there are five motionless points – L_1, L_2, L_3, L_4 and L_5, called Lagrangian or libration points. Under the joint action of the Earth and the Moon these motionless points are moving in such a manner that their initial locations remain constantly invariable. On other words, in the system of coordinates rotating with the Earth–Moon line, these five points are immovable or stationary. The points L_1, L_2 and L_3 are called **collinear** or **rectilinear** libration points, and the points L_4 and L_5 **triangle** libration points.

In connection with the appearance with the libration points, a problem connected with the behavior of a particle in one or another libration point arises. If the particle quickly moves off from the libration point then such a state of stationarity should be accepted as **instable**. If the particle is in a state of oscillation near the point of libration, then such a state will be **stationary**.

The problem of the stability of the libration points has been forwarded as far back as by Lagrange. Later on, this problem become the subject of more detailed study.

As it quickly turned out, the rectilinear libration points L_1, L_2 and L_3 are **stable** in the two-body problem and **unstable** in the real three-body problem. This means that any spacecraft that finds itself in one of these points should maneuver permanently in order to stay in this point.

The situation with the triangle libration points L_4 and L_5 is different. At present, it may be accepted as a finally argued result that these two points are in fact **stable** at all values of the ratio $m_1/(m_1 + m_2)$. This means that if the spacecraft has too small a velocity, then it should be farther into the neighborhoods of these points, L_4 or L_5, in fact drifting in the limits of some closed space.

The property of the stability of the points L_4 and L_5 forces one to suggest the possibility of the accumulation of interplanetary dust in the libration points L_4 and L_5. The special observations undertaken by Polish astronomers in Sahara desert apparently confirmed, the presence of a diffuse cloud at these points. This means that even in the presence of solar perturbations, the triangle libration points L_4 and L_5 possessed this capability – to hold matter or discrete bodies in their neighborhoods.

There is no data confirming the presence of diffuse matter in the rectilinear libration points L_1, L_2 and L_3.

8   The Moon's Sphere of Influence

The Moon, being the Earth's satellite, has the full complement of spheres mentioned in Chapter XII, namely, a sphere of attraction with a radius of R_g, a sphere of action, R_k, Hill's sphere, R_H, and a sphere of influence, R_I. The numerical magnitudes of these radii were presented in Table 12.6 of Chapter XII. Here we repeat the magnitudes of the radii of these spheres:

Attraction	$R_g =$	$38\,000 km,$
Action	$R_k =$	$65\,000 km,$
Hill's	$R_H =$	$700\,000 km,$
Influence	$R_I =$	$102\,000 km.$

In Fig. 13.12, the Moon's sphere of influence is shown with a radius of $R_I = 102\,000\,km$, with four types of orbits schematically marked: stable, circular, stable–elliptical, non-stable–elliptical, as well as a hyperbolic trajectory of the rapprochement, with an extinguishing of hyperbolic velocity, to a point not very far from the Moon's surface, by means of an momentum acting opposite to the motion. Depending on the power of this action, the spacecraft may pass into the one of above-mentioned orbits.

Two of these four spheres, namely, attraction and action, are joined into one - into the sphere of influence. Hill's sphere is seven times as extended as the sphere of influence. In a given case, however, much more important is the sphere of influence: during a period when the spacecraft is in this sphere, its orbit will be stable.

If the brake-engine fails, the spacecraft continues its flight by the same hyperbolic trajectory (broken line in Fig. 13.12), strongly symmetrical to the trajectory of its ascent part (solid part of the trajectory in Fig. 13.12).

9   Saturn's Libration Points

In the case of more massive planets when the ratio m_1/m_2 is much larger compared with Earth–Moon pair, the locations of the five libration points differ slightly from those of the Earth–Moon system.

As an illustration, in Fig. 13.13, an arrangement of libration points for Saturn is presented for these five points, L_1, L_2, L_3, L_4 and L_5, related to an object orbiting Saturn. If compared with Fig. 13.11 for the Earth–Moon system, we see that the similarity is obvious. However, the essential differences may be noticed as well. The first,in the case of Saturn, the points L_1 and L_2 are both located on the same side relative to the Saturn. The second, both these points, L_1 and L_2, are located very close to one another. The third, the symmetrical triangle points L_4 and L_5, the third rectilinear point L_3 and the geometrical center of both rectilinear points L_1 and L_2 are arranged in a circle with a diameter only three times larger than the diameter of Saturn itself, in contrast with the system of libration points for the Earth–Moon pair where the relative distances of all five libration points are much larger, by at least 30 times, compared with the size of the Earth or the Moon.

Analogically with Fig. 13.8, the rectilinear libration points L_1, L_2 and L_3 are located on a line, the triangle points L_4 and L_5 symmetrically on both sides relative to this line. However, in this case only the points L_4 and L_5 are considered to have some stability.

Saturn's libration points may find an interesting application in future interplanetary experiments concerned, particularly, with the creation of stationary observation bases around Saturn.

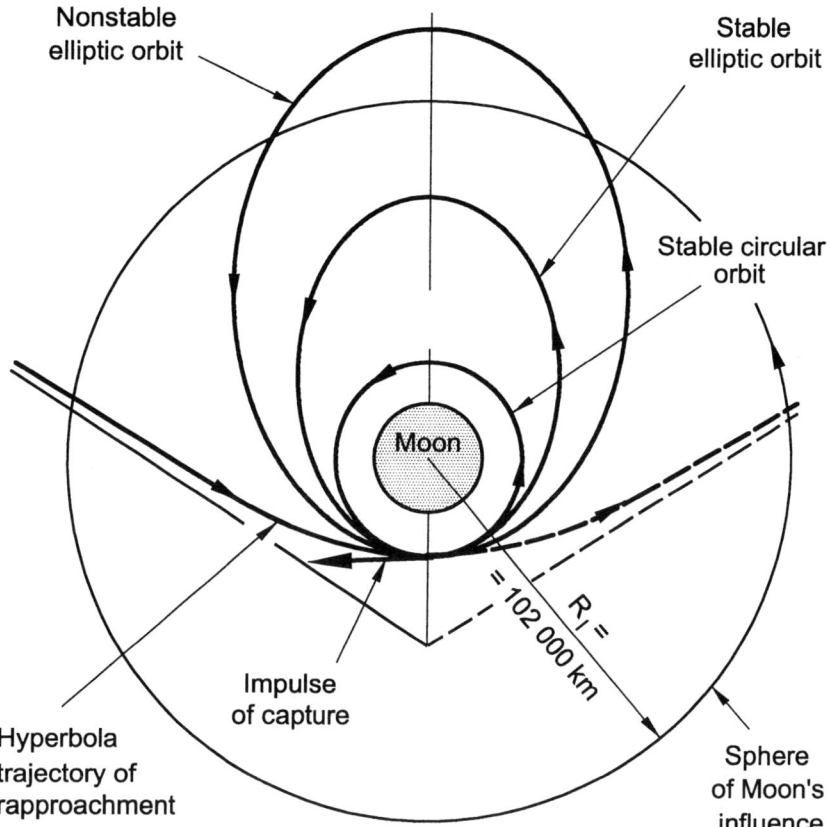

Fig. 13.12. *The motion of a spacecraft through the Moon's sphere of Influence with a radius of $R_I = 102\,000km$. The stable and instable orbits around the Moon are shown, as well as the trajectory of rapprochement of the spacecraft.*

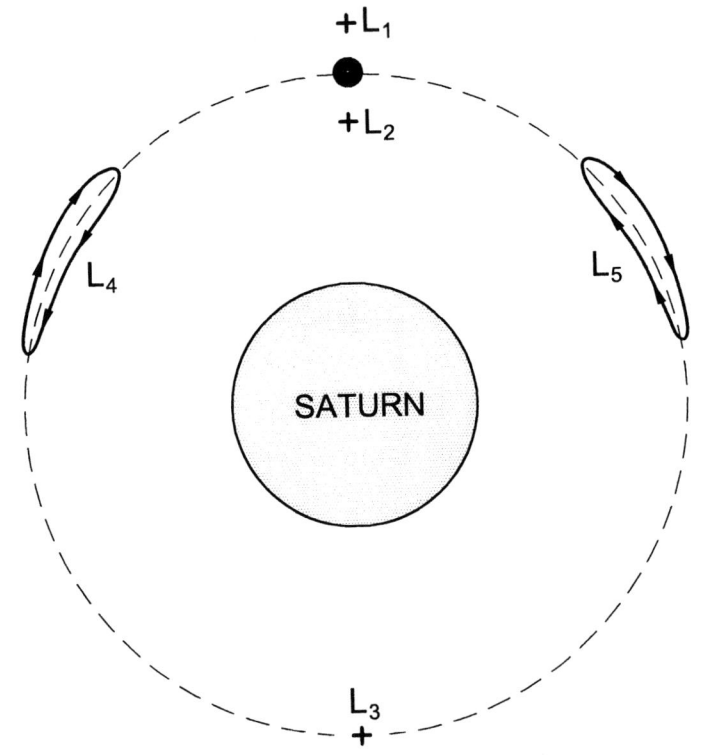

Fig. 13.13. *The five Lagrangian points for an object orbiting Saturn. They are arranged around a circle with a diameter only three times larger than Saturn's diameter.*

Chapter XIV

Flights to the Planets

1 Flyby Trajectories to Planets

In the previous chapter, the problem of flights to the Moon was examined; in particular, the possible trajectories and orbits aiming at flyby of the Moon as well as landing on it were described. The main parameters of the flights and, of course, the starting conditions and the duration of the flight, were derived.

In an analogous way one can also analyze the flyby heliocentric trajectories to other planets – to Mars, Venus, etc. The types of such trajectories are numerous; however, of a practical interest are those that ensure the return of the space probe to the Earth after flyby of the planet and departure from its sphere of action. In other words, one should initially consider those motions of the probe for which the second half of the heliocentric trajectory intersects the Earth's orbit just at the point of their encounter.

The ideal Hohmann trajectories to ensure the return of the probe and its meeting with the Earth are useless since these trajectories do not take into account the perturbations, a kind of "gravitational shock", which the probe will undergo during its passage through the sphere of action of a given planet. The part of the Hohmann heliocentric trajectory close to aphelion where the planet of destination is situated is a hyperbola in the coordinate system connected with the planet. With respect to that planetocentric system, the entrance of the probe into the sphere of action of the planet takes place with hyperbolic velocity. The escape of the probe from this sphere occurs with even larger hyperbolic velocity. In fact, the motion of the probe through the planet's sphere of action occurs with a speeding-up trajectory. As a result, the duration of the Earth–planet flight will be shorter than the duration in an ideal Hohmann trajectory, i.e. in the absence of gravitational action of the planet of destination. The computations indicate a shortening in the case of a two-year flight of the probe in the route Earth–Mars–Earth with close flyby of Mars of about two months.

Return flights from Mars with a longer duration are possible – three years or more. In principle the duration of such flights can be shortened by means of active maneuvers to gain additional momentum. However, the shortening of the flight duration in such a way will be doubly expensive: the velocity of entrance of the probe into the Earth's atmosphere in this case will be very large – 22.5 $km\ s^{-1}$, instead of the usual velocity 12 $km\ s^{-1}$; to dissipate the former momentum obviously requires greater energetic resources.

The situation is different for flyby of Venus. The dependence of the radius of the sphere of action R_k on the planet's mass is weak: $R_k \simeq m^{0.4}$. In addition $R_k \simeq a$, where a is the distance between the Sun and the planet. Therefore, though the mass of Venus is eight times that of Mars, the distance of Venus from the Sun is half that of Mars, and as a result the radius R_k in the case of Venus is of the same order as that for Mars. This means that the "corridor" for Venus flyby trajectories will be extremely broad: the minimal distance of the probe from the surface of Venus can vary from a few hundred kilometers to a few dozen thousands of kilometers. Correspondingly, the duration of the flight of the probe through the sphere of action of Venus can also vary within wide limits; this fact can essentially modify the return Hohmann trajectory, as well as the total duration of the Earth–Venus–Earth flight.

Among the large class of flyby trajectories of Venus there are also numerous trajectories that ensure the return of the probe to the Earth within a year. There are trajectories that are more economical from the energetic point of view that ensure small velocities of entry of the probe into the Earth's atmosphere. However, in this case the total duration of the flight is larger – two years or more.

2   Trajectories of Multiplanetary Flights

Rapid progress in long-distance broadcasting techniques for radio and television information makes rather efficient the creation of space probe capable of flying by several planets successively, performing necessary observations at each flyby and transmitting the results of observations telemetrically to the Earth. Thus, a space probe equipped with appropriate facilities can transmit first the data of Mars, then, continuing the journey, those of Jupiter and its satellites, etc. One can assume a configuration of the planets such that Saturn also will be in the path of the journey. This tour can be continued. Of course, in this case the problem of the return of the probe to the Earth is not raised.

A return tour of a probe having flown by Mars, Venus and Mercury as well as the Earth is possible (Mercury Orbiter. Flights with other sequences are also possible, e.g. first to Venus, and then to Mercury, flyby of the Sun at the perihelion, return to Mercury or to Venus and, after intersecting the Earth's orbit, continuing the tour to Mars.

In the first case, in which no return of the space probe back to the Earth is required, the main problem is to find the necessary initial configuration of planets, and taking into account the duration of flight intervals, to determine the exact starting conditions from the Earth. To find the necessary configuration in the case of flights to, say, Mars and Jupiter, without return of the probe, is not too difficult. However, if we add the third planet, Saturn, then the estimation of the necessary configuration will become complicated. The difficulties are increased incredibly if flyby of four planets is sought. Recall that in all cases, the trajectory of items of space probe is controlled only by the action of the Sun's gravitation, i.e. without any perturbation maneuvers.

However, if the return of the space probe to the Earth is intended after the realization of its mission, then the choice of necessary configuration is much more complicated even in the case of two planets. In this case the important condition of equality of two times should be fulfilled: of the return time of the flight of the space probe, i.e. the spent time up to its intersection with the Earth's orbit, on the one hand, and the time required for the Earth to reach the same point of its orbit, on the other hand.

Although the complexities concerning computations of the trajectory of multiplanetary flights and the estimation of the exact configuration of the planets increase greatly with an increasing number of planets to be visited, the approach to the solution of this problem and the principles of corresponding computations are the same for all. Therefore, we will confine ourselves to the examination of one particular case: starting from the Earth, a close flyby of Mars, then of Venus, with subsequent return to the Earth. Obviously, we will describe the approximate computations without taking into account the perturbations of the planets themselves since our aim is to determine the basic structure of the trajectory and the approximate geometric configuration of the planets. For simplicity we will assume that all three planets, the Earth, Mars and Venus, are moving in circular orbits. We will also ignore the influence of the gravitation of the planets even during the passage of the space probe through their sphere of action.

Computations of the trajectory for any multiplanetary flight, including the Earth–Mars–Venus–Earth flight, always start from the choice of one or even two parameters of the flight's orbit, such as the angular distance ω of perihelion in some direction SX. The magnitude of the semimajor axis a of the trajectory can also be chosen; the choice is determined mainly, say, by the condition of the shortening of the flight duration or to realize the flight with minimal velocity.

We will consider here the trajectory of a multiplanetary flight under the following conditions. a) The argument of the latitude ω is given. b) The launch from the Earth should be realized with minimal velocity.

The flight orbit, an ellipse, as well as the configurations of the Earth E_1, of Mars M and of Venus at the moments corresponding to the two points of intersection with the flight ellipse, W_1 and W_2, are shown in Fig. 14.1. The angular distance of perihelion of the ellipse ω from the axis SX passing through the Earth's position E_1 is also indicated.

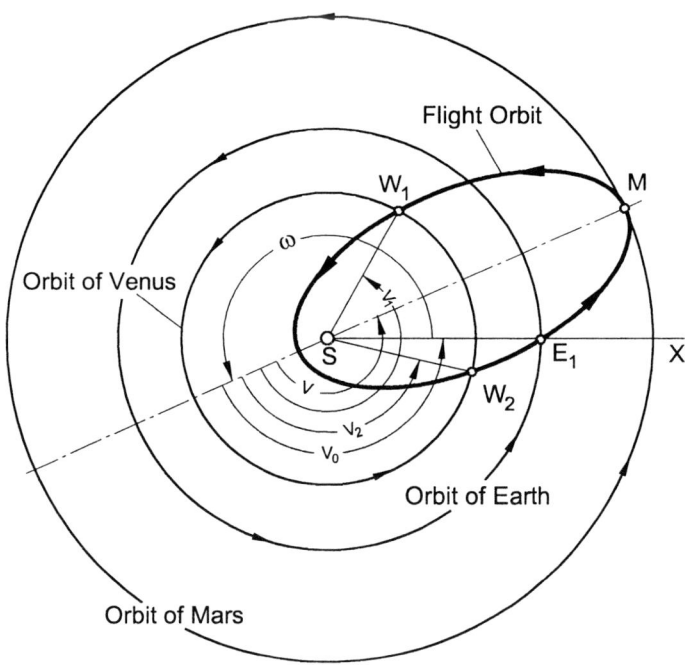

Fig. 14.1. *The elliptical orbit of successive approaches to Mars (at the point M) and Venus (at the points W_1 and W_2) for the flight with minimal velocity. E_1 is the position of the Earth at the moment of launch.*

The second of these conditions will be realized if the aphelion of the flight ellipse will smoothly come into contact with the orbit of Mars at the point M. This means that the aphelion distance of the flight ellipse is known; it is equal to the distance of Mars from the Sun, r_M. Moreover, the true anomaly of the aphelion, i.e. of the point M, is also known: it is equal to $v = 180°$. Therefore we have

$$r_M = \frac{a(1 - e^2)}{1 + e \cos v} = a\,(1 + e)\,. \tag{1}$$

The parameters of the point E_1, i.e. of the point of intersection of the flight ellipse with the Earth's orbit, are known too:

$$r = r_E,$$
$$v_0 = 360° - \omega.$$

(2)

Then we can write

$$r_E = \frac{a\left(1 - e^2\right)}{1 + e \cos v_0} = \frac{a\left(1 - e^2\right)}{1 + e \cos \omega}.$$

(3)

Eqs. (1) and (3) enable one to find, first, the eccentricity of the flight ellipse e, and then, the semimajor axis a:

$$e = \frac{r_M/r_E - 1}{r_M/r_E + \cos \omega},$$

(4)

$$a = \frac{r_M}{1 + e}.$$

(5)

We can determine also the real anomalies v_1 and v_2 at the points W_1 and W_2 of intersection of the flight ellipse with the orbit of Venus, i.e. at the points where $r = r_V$. We have

$$\cos v_1 = \frac{1}{e}\left(\frac{a}{r_V}\left(1 - e^2\right) - 1\right),$$

(6)

and, bearing in mind the symmetrical location of the points W_1 and W_2 relative to the axis of the ellipse, we have

$$v_2 = 360° - v_1.$$

(7)

The calculated duration P of the whole flight of the space probe through the flight ellipse will be

$$P = \frac{2\pi a^{3/2}}{(f M_\odot)^{1/2}}.$$

(8)

Among the parameters of the flight ellipse is also the heliocentric initial velocity of the probe V_R. We have from the integral of energy

$$V_R = \left[f M_\odot \left(\frac{2}{r} - \frac{1}{a}\right)\right]^{1/2},$$

(9)

where $f = 6.670 \cdot 10^{-23} \ km^3 \ g^{-1} \ s^{-2}$ when the velocity is expressed in $km \ s^{-1}$ and $M_\odot = 2 \cdot 10^{33} \ g$.

It is also necessary to know the geocentric velocity of the probe W_G at the initial moment as well as the angle Θ composed by two vectors of the heliocentric velocities – those of the probe V_R and of the Earth V_T.

We determine the magnitude of initial geocentric velocity of the probe with the help of the following formula:

$$W_G = (V_T^2 + V_R^2 - 2V_T V_R \cos \Theta)^{1/2}, \tag{10}$$

where $V_T = 29.76 \ km \ s^{-1}$, and Θ is obtained via the known a and e from the following relationship:

$$\cos \Theta = \left(\frac{a\,(1 - e^2)}{2 - 1/a} \right)^{1/2}, \tag{11}$$

where the semimajor axis a is expressed in $a.u.$ The vector of the heliocentric (orbital) velocity of the Earth V_T at the starting point is perpendicular to the axis SX; therefore, we shall have for the direction of the initial heliocentric velocity of the probe V_R

$$\beta = \Theta + \frac{\pi}{2}. \tag{12}$$

This, basically, completes the computation of the flight trajectory. It remains to obtain the initial configuration of the system Earth–Mars–Venus at the moment of launch. In order to find this, it is necessary to know the durations of the flight via segments of the ellipse $E_1 M$, $E_1 W_1$ (through M) and $E_1 W_2$ (through M). We can find the time of passage of the mentioned segments of the ellipse in the following manner.

We have from Kepler's equation

$$(1 - e \cos E)\,dE = \frac{(f M_\odot)^{1/2}}{a^{3/2}}\,dt.$$

By integrating the left-hand side with the given values of the eccentric anomaly – from E_1 to E_2, and the right-hand side from t_1 to t_2, we obtain

$$(E_2 - E_1) - e(\sin E_2 - \sin E_1) = \frac{(f M_\odot)^{1/2}}{a^{3/2}}\,\tau, \tag{13}$$

where $\tau = t_2 - t_1$ is the time of passage by the probe of the segment of the ellipse from the point with an eccentric anomaly E_1 to the point with an eccentric anomaly E_2. It will be convenient, however, to replace E by the true anomaly v, with the help of the known relationships (see Chapter IV)

$$\tan \frac{E}{2} = \left(\frac{1 - e}{1 + e} \right)^{1/2} \tan \frac{v}{2}, \tag{14}$$

$$\sin E = (1 - e^2)^{1/2} \frac{\sin v}{1 + e \cos v}. \tag{15}$$

Then we find finally from Eq. (13) for the time of passage τ by the probe of the segment of the ellipse corresponding to the interval of the true anomaly from v_1 to v_2

$$\tau = \frac{a^{3/2}}{(fM_\odot)^{1/2}} (1 - e^2)^{1/2} \left\{ \frac{2}{(1 - e^2)^{1/2}} \left\{ \arctan\left[\left(\frac{1-e}{1+e}\right)^{1/2} \tan\frac{v_2}{2} \right] \right. \right.$$

$$\left. \left. - \arctan\left[\left(\frac{1-e}{1+e}\right)^{1/2} \tan\frac{v_1}{2} \right] \right\} - \frac{e \sin v_2}{1 + e \cos v_2} + \frac{e \sin v_1}{1 + e \cos v_1} \right\}. \tag{16}$$

The discovery of the initial configuration of the planets is reduced to the determination of the initial angular distance φ according to the formula

$$\varphi = v + \omega - \tau n, \tag{17}$$

where v is the true anomaly of the probe, τ is the flight time from the Earth to a given planet, determined by Eq. (16), and n is the angular velocity of the motion of the planet in its orbit.

The essence of the application of Eq. (17) is in the determination of the angular distances between the planets of destination and then, by comparison of all the results with the Annual Astronomical Almanac, finding the date on which the required mutual configuration of the Earth, Mars and Venus occurs, hence determining the starting date.

We will illustrate all these considerations by the numerical example of an Earth–Mars–Venus–Earth flight, for which $\omega = 210°$. Then we find from (2) that $v_0 = 150°$.

Substituting into Eqs. (5) and (4) for the distances of Mars and the Earth from the Sun, $r_M = 228 \cdot 10^6$ km and $r_E = 150 \cdot 10^6$ km, we find for the eccentricity and semimajor axis of the flight ellipse $e = 0.795$ and $a = 127 \cdot 10^6$ km.

The magnitudes of the true anomalies v_1 and v_2 for Venus ($r_V = 108 \cdot 10^6$ km) at the points W_1 and W_2 are determined from Eq. (7):

$$v_1 = 135°.5, \qquad v_2 = 224°.5.$$

Finally, we determine the durations of the flight of the space probe in various intervals of the trajectory from Eq. (16):

$$\tau(E_1 M) = 95.9 \ days$$
$$\tau(E_1 W_1) = 210.0 \ days$$
$$\tau(E_1 W_2) = 265.5 \ days$$
$$\tau(W_2 E_1) = 19.4 \ days$$

The total period of motion of the space probe in the elliptic orbit E MW_1W_2E (Fig. 14.1), obtained from Eq. (8), is equal to $P = 284.9\ days$.

An important parameter of the flight ellipse is the initial geocentric velocity W, determined by Eq. (10). However, first, it is necessary to know the heliocentric velocity, V_R; it is equal, according to Eq. (9), to

$$V_R = 26.99\ km\ s^{-1}\,.$$

Then we have from Eq. (10), substituting for the orbital velocity of the Earth $V_T = 29.76\ km\ s^{-1}$,

$$W = 24.95\ km\ s^{-1}\,.$$

and $\Theta = 51.9$ according to Eq. (11). Finally, we have for the direction of the geocentric velocity β from Eq. (12)

$$\beta = \theta + \pi/2 = 141°.9\,.$$

The vectorial diagram of all three types of velocities at the moment of launch of the space probe, i.e. at the point A of the intersection of the Earth's orbit with the flight ellipse, is shown in Fig. 14.2.

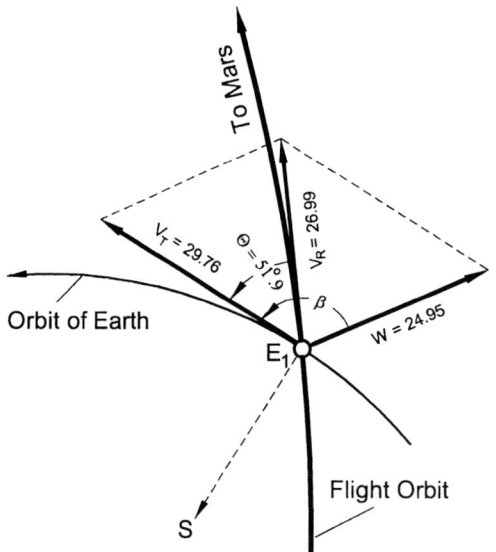

Fig. 14.2. *The vectorial diagram of velocities at the moment of launch from the Earth E_1 of a space probe in the Mars–Venus flight trajectory.*

We now have all the necessary data for determination of the initial angular distances for both Mars and Venus at the moment of launch of the space probe. On substituting into Eq. (17) $n(M) = 0°31'30''$ for Mars and $n(V) = 1°36'$ for Venus, we obtain

$$\varphi(Mars) = v_2 + \omega - \tau(E_1 M) \times n(M)$$
$$= 135°.5 + 210° - 95.9 \times 31'30" = 295°.2$$

$$\varphi(Venus) = v_1 + \omega - \tau(E_1 W_1) \times n(V)$$
$$= 224°.5 + 210° - 210 \times 1°36' = 98°5.$$

Now one has only to find in the Annual Astronomical Almanac the exact date on which the obtained mutual configuration of the Earth, Mars and Venus will be realized, which will be the date of launch of the space probe.

Concerning the mutual location of these three planets, the Earth, Mars and Venus, one should note that this particular configuration relative to the Sun is repeated nearly every 6.4 years; it equals the sum of the synodic periods of Venus and Mars. Therefore, the launch of a space probe to fly by Mars and Venus can be realized every 6.4 years.

The considered trajectory of flight Earth–Mars–Venus–Earth corresponds to a minimal initial velocity and, hence, to minimal energetic resources due to the fact that the flight ellipse smoothly comes into contact with the orbit of Mars (this ellipse, however, is not of Hohmann type). In principle, other ellipses are possible, too, for example with large aphelion distances compared with the Sun–Mars distance, i.e. ellipses intersecting the orbit of Mars at two points. Large aphelion distances imply large initial launch velocities. From the energetic point of view such ellipses are not optimal; however, in this case, the duration of the flight for certain parts of the route can be significantly shortened.

3   The Influence of Starting Errors on the Flight Orbit

The realization of the launch of the space probe with the required accuracy is practically impossible. There are numerous reasons – the divergence of the actual power of the engines from the calculated one, imprecise periods of the action and stopping of various engines of the rocket, inaccuracy in the functioning of the automatic systems, of service, control and measuring facilities, etc. All these can lead to a flight with initial errors, as a result of which the initial parameters of the launch will diverge from their calculated values. This concerns, in particular, the magnitude of the initial velocity, its direction, the moment of launch of the space probe, i.e. the moment of its departure from the Earth, etc. Normally these errors are small; however,

they accumulate during the long journey of the space probe so that, by the
time of flyby of the planet of destination, the parameters of the calculated
trajectory and the real position can differ drastically.

Moreover, as we will see below for concrete examples, the considerable
discrepancy between the real trajectory and the calculated one in the final
stage of the flight can be caused even by extremely small errors in the launch
parameters. These errors are within the limits of the technical possibilities
of modern technique and technology, so that it is not expected that they
will be decreased significantly in the foreseeable future. Hence, we arrive at
the inevitable conclusion that any space flight will require correction of its
trajectory by means of engines on board the spacecraft.

Consider a few examples illustrating the character of the dependence of
the flight trajectory on the errors in the launch parameters. We start from
an error ΔV in the magnitude of the initial departure velocity V.

By differentiating the relationship

$$V^2 = fM_\odot \left(\frac{2}{r} - \frac{1}{a} \right) , \tag{18}$$

we obtain

$$\frac{\Delta a}{a} = 2a \left[\left(\frac{2}{r} - \frac{1}{a} \right) \frac{\Delta V}{V} + \frac{\Delta r}{r^2} \right] , \tag{19}$$

where Δa and Δr are the divergences between the real and calculated values
of the semimajor axis of the flight ellipse a and the radius-vector of probe
r. In order to demonstrate the consequences of the error in the velocity of
departure ΔV for the magnitude of the semimajor axis a, we assume that
$\Delta r = 0$. Then we obtain from Eq. (19)

$$\frac{\Delta a}{a} = 2a \left(\frac{2}{r} - \frac{1}{a} \right) \frac{\Delta r}{V} . \tag{20}$$

We shall have the largest divergence between the real and calculated
trajectories in the final stage of the flight, i.e. at aphelion of the orbit, be it
Hohmann or a contacting ellipse, where the meeting of the spacecraft with
the planet of destination should occur. However, the distance of the aphelion
r_A is given by

$$r_A = a (1 + e). \tag{21}$$

On differentiating the logarithm of this relationship, we obtain

$$\frac{\Delta r_A}{r_A} = \frac{\Delta a}{a} + \frac{\Delta e}{1 + e} . \tag{22}$$

The value of the first term on the right-hand side is known; it is given by Eq. (20). As for the second term in Eq. (22), its value can be obtained using the well-known formula

$$e = \left(1 - \frac{r}{a^2}(2a - r)\sin^2\phi\right)^{1/2},$$ (23)

where ϕ is the angle between the velocity vector V and the radius-vector r. On differentiating this relationship, we obtain

$$\frac{\Delta e}{e} = \frac{r}{a}\left(\frac{r}{a} - 2\right)\frac{\sin\phi}{e^2}\left[\left(1 - \frac{a}{r}\right)\left(\frac{2\Delta V}{V} + \frac{\Delta r}{r}\right)\sin\phi + \Delta\phi\cos\phi\right].$$ (24)

The launch is realized from the Earth, i.e. from the pericenter where

$$r = a(1 - e), \qquad \phi = \pi/2.$$

Then we shall have from Eq. (24), substituting also $\Delta r = 0$ and $\Delta\phi = 0$

$$\frac{\Delta e}{e} = 2\frac{1 + e}{e}\frac{\Delta V}{V}.$$ (25)

By substituting Eqs. (20) and (25) into Eq. (22) we obtain finally

$$\frac{\Delta r_A}{r_A} = \frac{4}{1 - e}\frac{\Delta V}{V}.$$ (26)

It follows from this interesting formula that the divergence in the position of the farthest point of the trajectory, the aphelion, from its calculated point is large when the eccentricity of the flight ellipse is close to unity, other conditions being equal. This rather important conclusion can be used for the choice and computation of flight trajectories; in this case an ellipse with smallest possible eccentricity is therefore preferable.

Let us apply Eq. (26) to the determination of the errors in the apogee distance Δr_A in terms of their dependence on the errors in the initial velocity of the space probe ΔV during the flight to Mars. We obtained above for the multiplanetary flight trajectory to Mars $e = 0.795$, $W = 24.95\ km\ s^{-1}$, and $r_A = 228 \cdot 10^6\ km$. With these data we obtain from Eq. (26)

$$\Delta r_A = 178 \cdot 10^6\ \Delta V\ km.$$

Let the error in the starting velocity be $\Delta V = 1\ m\ s^{-1}$, i.e. 0.4% of the calculated initial magnitude of the velocity V. Then the error in the apogee distance will be

$$r_A = 178\,000\ km,$$

which exceeds the diameter of Mars by a factor of more than 20. This means that, if the initial velocity of the start turns out to be, e.g. smaller than the calculated velocity only by 0.4%, then the spacecraft cannot reach Mars – it will move via the aphelion at an approximate distance 180 000 km from Mars and will escape from it forever.

Consider Eq. (26) in relation to the Moon. We have for the flight to the Moon in a Hohmann trajectory $V = 10.80$ km s^{-1}, $e = 0.966$, and $r_A = 384\,000$ km. From these data we obtain, again with $\Delta V = 1$ m s^{-1}, about 4 200 km for the error in the apogee distance of the lunar trajectory, which greatly exceeds the lunar diameter.

Consider now the influence of the error in the direction of the initial velocity vector on the parameters of the flight of the spacecraft. Let the real direction of the initial velocity vector be declined from the calculated one by $\Delta\phi$ in angular units, and the other errors be equal to zero, including the errors in the magnitude of the initial velocity V.

We have to obtain the dependence of the error in the apogee Δr_A on $\Delta\phi$. This problem can be solved by means of the obvious fact that the divergence in the direction of the velocity vector from the calculated one implies a decrease in the **magnitude** of the initial velocity in the calculated direction by $\omega V = V \Delta\phi$. Hence, to obtain the error at apogee Δr_A, one can use the same Eq. (26), except that in $\Delta V / V$ it should be replaced by Δr_A. Then we obtain the dependence of $\Delta\phi$ in the form

$$\frac{\Delta r_A}{r_A} = \frac{4}{1-e} \Delta\phi. \tag{27}$$

When applied to Mars and the Moon, this formula gives

$$\Delta r_A\,(Mars) = 7,2 \cdot 10^5 \Delta\phi\ km\,,$$

$$\Delta r_A\,(Moon) = 1.32 \cdot 10^4 \Delta\phi\ km\,.$$

where $\Delta\phi$ is in arc minutes.

Let the angular deviation in the initial velocity vector be negligibly small, $\Delta\phi = 1'$. Then the deviation of the final point of the trajectory of the spacecraft from the calculated one will be $\Delta r_A = 13\,200$ km, i.e. four times larger than the Moon's diameter.

As for the influence of the error in the time of the start Δt, this problem can be solved in various ways, in particular, by transformation of Δt into the error in the direction of the initial velocity vector of departure, i.e. into the angle $\Delta\phi$. Then it suffices to obtain the dependence of $\Delta\phi$ on Δt and then substitute the value $\Delta\phi$ into Eq. (27).

The departure of the space probe takes place from a circular orbit around the Earth and at the altitude H from the Earth's surface. Then any delay or other errors in the moment of the start Δt compared with the calculated one will be equivalent to a deviation in the direction of the velocity vector V:

$$\Delta \phi = \frac{V_0}{R_\oplus + H} \Delta t, \qquad (28)$$

where V_0 is the first cosmic velocity, and $R_\oplus$ is the radius of the Earth. On substituting Eq. (28) into Eq. (27), we obtain

$$\frac{\Delta r_A}{r_A} = \frac{4}{1 - e} \frac{V_0}{R_\oplus + H} \Delta t.$$

By substituting $V_0 \approx 8\ km\ s^{-1}$, $H = 350\ km$, and $R_\oplus = 6\,370\ km$, we obtain in the case of flights to Mars and the Moon

$$\Delta r_A\ (Mars) = 2.95 \cdot 10^6\ \Delta t\ km,$$

$$\Delta r_A\ (Moon) = 5.40 \cdot 10^4\ \Delta t\ km.$$

Assume that the launch of the spaceship from the geocentric orbit takes place with an error of $\Delta t = 0.1\ s$. Then the error in the apogee distance will be $\Delta r_A = 295\,000\ km$ in the case of the flight to Mars and $\Delta r_A = 5\,400\ km$ in the case of the Moon.

This analysis can be continued. However, it is already clear that the corrections to the trajectory of the flight should be considered an essential part of projected space probe or ships intended for interplanetary journeys. As for the celestial mechanical aspect, the most optimal and economic scheme of corrections should, as a rule, be predicted.

4 The Perturbation Maneuver. Speeding-Up Trajectories

In all the interplanetary flight orbits considered above, in particular, Hohmann-type ellipses and multiplanetary flight trajectories, the question concerns the orbits determined essentially by the gravitational action of the Sun. However, during close flybys of the planets, the space probe can for some period be within the sphere of action of a planet. As a result, the trajectory of the flight will inevitably undergo certain changes. For example, the motion of the probe in the planetocentric system of coordinates will now take place – within the limits of the sphere of action – in a hyperbolic orbit. Of course, the spacecraft enters the sphere of action of a given planet and leaves it with the same hyperbolic velocity, i.e. $V_{in} = V_{out}$ (see Fig. 14.3) in the same planetocentric system of coordinates. However, the change **direction** of the velocity vector of the motion of the probe yields an essential increase in its heliocentric velocity. In fact, a speeding up of the spacecraft takes place as a result of its flight through the sphere of action.

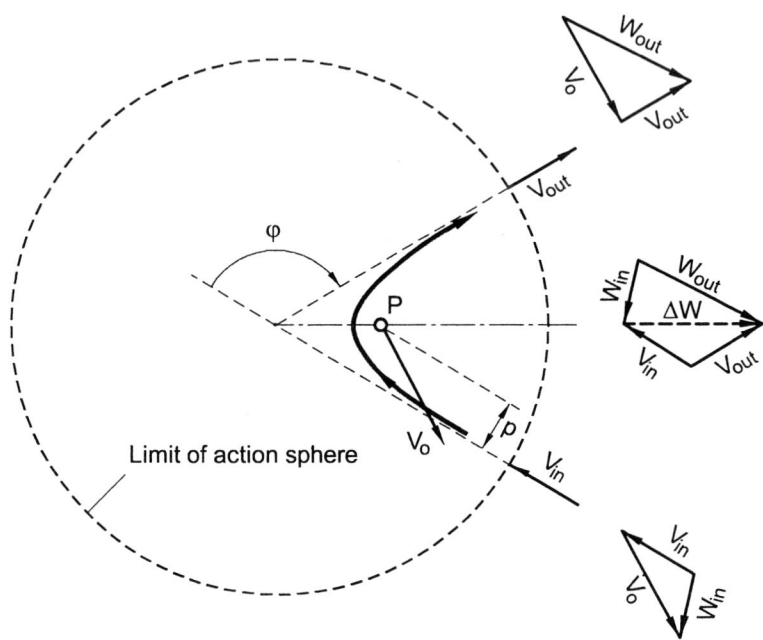

Fig. 14.3. *The perturbation maneuver in the sphere of action of a planet P. V_0 is the heliocentric velocity of the planet, V_{in} and V_{out} are the entry and departure velocities of the probe in the sphere of action of the planet. $V_{in} = V_{out}$ in the planetocentral system of coordinates. W_{in} and W_{out} are the entry and departure velocities of the spacecraft in the heliocentric system of coordinates. ΔW is the increase in the velocity as a result of the perturbation maneuver, p is the intended distance, φ is the angle of the change in the vector of planetocentral velocity during the flight through the sphere of action.*

The degree of acceleration of the spacecraft through the sphere of action is characterized by the curvature of its trajectory, i.e. by the magnitude of φ. The latter depends on the intended distance p, as well as on the velocity of entry v_{in} into the sphere of action:

$$\tan\frac{\varphi}{2} = \frac{fM}{pv_{in}^2}, \tag{29}$$

where fM is the gravitational parameter for the given planet.

What is the essence of the acceleration, and, in general, how should the additional velocity gained by a spacecraft as a result of this operation be determined, i.e. after passage through the sphere of action of a given planet?

When we speak of the acceleration, we have in mind a process with respect to the heliocentric system of coordinates. Hence, in our quantitative calculations, the heliocentric velocity of the planet should itself be involved; it is denoted in Fig. 14.3 by the vector V_0. Obviously, insofar as the probe flies through the sphere of action, the heliocentric velocity of the planet will undergo certain changes both in magnitude and in direction. However, the flight of the probe through the sphere of action takes place too quickly; in all cases, the corresponding time scale is usually negligible compared with the total duration of the flight itself. Therefore, for simplicity we will assume V_0 to be constant.

Although the heliocentric velocity of the planet V_0 is not changed, the planetocentric velocity of the spacecraft v undergoes changes at every point of the trajectory within the sphere of action. The maximal change occurs between the points of entry to and exit from the sphere of action, i.e. between $\bar{V}_{in}$ and $\bar{V}_{out}$. By means of the vector sum of heliocentric velocity V_0 and these limiting vectors of the planetocentral velocity of the spacecraft, we obtain the change in the heliocentric velocity of the probe. This process is illustrated in more detail by the vectorial diagrams in Fig. 14.3.

In the lower vectorial diagram, near the entry of the sphere of action, the addition of two vectors, those of the heliocentric velocity of the planet V_0 and of the planetocentric velocity of the probe V_{in}, is represented. As a result, we have a new vector of heliocentric velocity of the probe W_{in} in front of the sphere of action.

In the upper vectorial diagram the same operation is shown in the case of exit from the sphere of action. Because the direction of V_{out} undergoes a change relative to V_{in}, and V_0 remains unchanged, the sum effect, i.e. the absolute magnitude of the heliocentric velocity of probe W_{out} after leaving the sphere of action, turns out to be larger than that before entry: $W_{out} > W_{in}$. That is readily seen from the middle vectorial diagram in Fig. 14.3, in which the increase in the velocity, ΔW, of the spacecraft is also shown. The magnitude of ΔW can be obtained by means of the following obvious formula:

$$\Delta W = 2 v_{in} \sin \frac{\varphi}{2} \,. \tag{30}$$

Insofar as φ depends on the distance p, one can ensure any magnitude of acceleration by its proper choice, i.e. any increase in the initial velocity of the spacecraft.

However, this increase has its theoretical limit, it cannot exceed $\Delta W_{out} = 2 V_{in}$. This can happen when $\varphi \to \pi$, i.e. when $p \to 0$; in this case, the spacecraft can even change its direction almost to the opposite one with respect to the initial one. However, $p \to 0$ together with flyby of the planet is impossible, insofar as the planet has a size and therefore p cannot be smaller than the effective radius of the planet R_{eff}. If $p = R_{eff}$ in (29), then, combined with the gravitational parameter of the planet $(fM)^{1/2}$, the escape velocity V_{es}, i.e. the parabolic velocity on the planet's surface, will appear. On substituting the obtained value of φ_{max} into Eq. (30), we have finally

$$\Delta W_{max} = \frac{v_{es}^2}{\dfrac{1}{2}\dfrac{v_{es}^2}{v_{in}} + v_{in}}. \tag{31}$$

It follows from this formula that a definite role in the formation of the maximal increase in momentum velocity is played by the velocity V_{es}, the escape velocity at the surface of the planet. As for V_{in}, it appears that the planetocentric velocity of entry of the probe into the sphere of action also has its maximal value, i.e. when it is equal to the velocity at the planet's surface. In this case the increase in the velocity ΔW_{max} gained due to the gravity of the central planet will be maximal.

The largest escape velocity V_{es} that we have is for the four planets of the Jupiter group. Therefore, these planets may be used as powerful generators to accelerate spacecraft. The largest possible increase in the velocity vector of a space probe within the sphere of action of Jupiter could be about 42.7 $km\ s^{-1}$. Venus and the Earth have an intermediate value, only 7–8 $km\ s^{-1}$.

The other planets, as generators of additional velocity, are Mars (3.5 $km\ s^{-1}$), Mercury (3.0 $km\ s^{-1}$) and the Moon (1.7 $km\ s^{-1}$). In the last place is Pluto, with about 1 $km\ s^{-1}$ (Asker, 1993).

In the particular case in which the vector of the heliocentric velocity of a planet V_0 is perpendicular to the axis of the hyperbola, the modulus of the heliocentric velocity of the probe will be unchanged, i.e. $W_{in} = W_{out}$. In other words, one can have a situation in which the probe does not undergo any increase in absolute magnitude of its heliocentric velocity in the sphere of action. However, such an operation certainly leads to a change in the **direction** of the heliocentric velocity of the spacecraft; it is impossible to avoid this effect. However, a new direction of the velocity of motion always means a new heliocentric trajectory. Therefore, the final conclusion can be formulated in the following manner: the perturbation maneuver within the sphere of action of a planet always leads to motion of the probe in a new heliocentric orbit independently from the change in the absolute magnitude of the heliocentric velocity of the probe.

The spacecraft, after accelerating within the sphere of action of a planet and leaving it, continues its flight in a **new** heliocentric trajectory, which is completely determined by the gravity of the Sun. However, even in this case one can conditionally call those trajectories **speeding-up** ones (Morrison, 1982).

5 Periodic Orbits

A broad category among the interplanetary trajectories are the so-called *rapprochement* periodic orbits. The essence of these orbits is as follows: a spacecraft launched in such an orbit can approach a given planet successively in periodic orbits, of course without application of jet engines. Moreover, for certain trajectories even an approach towards the given planet and the Earth one after the other is possible.

These orbits cannot be represented via analytic expressions. Instead, numerical integration of the differential equations of motion of an infinitely small mass within the framework of the restricted three-body problem is required.

Because there are some parallels between experiments in physics and computations in celestial mechanics, the latter are also attributed to the category of experiments (Szebehely, 1967). Therefore the computations of the periodic orbits and trajectories have to be included in the category of experimental celestial mechanics.

The first results concerning the realization of the numerical integration to find the periodic orbits for a parameter $\mu = 10/11$ are associated with the name of Darwin (1897, 1910). Later, analogous computations for $\mu = 0.5$ and $\mu = 0.2$ were realized by Moulton (1914, 1920). However, the appearance of the first complete system of numerical results was due to studies in the period 1913–1939, performed by the astronomers of Copenhagen observatory under Strömgrén (1934); these so-called "Copenhagen category" calculations were carried out for $0.1 < \mu < 0.5$.

The next important stage was connected with the realization of systematic investigations of periodic orbits for the Earth–Moon system, i.e. of the value of the parameter $\mu = 0.01215$ by the same methods of numerical integration, but now carried out with the help of computers. That extremely rich and busy period of the beginning of the computer epoch of experimental celestial mechanics is associated with the names of Egorov (1965), Message (1958), Newton (1959), Broucke (1962), Huang (1962), Arenstorf (1963), Ehricke (1962), Deprit and Henrard (1965) developed later by many others, up to appearance of powerful supercomputers like GRAPES of Makino and collaborators. An enormous number of orbits for broad interval of the parameter μ, periodic and nonperiodic ones, asymptotic ones, orbits with *rapprochement* both with a large mass and with a small mass, orbits ensuring the periodic flight of libration points etc., have been studied.

The shapes of periodic orbits and trajectories essentially differ in their dependence on the system of coordinates in relation to which these orbits are represented, e.g. to a system with a rotating Earth–Moon line (a synodic system) or to the heliocentric system connected with the Earth (a sidereal system). As an example, in Figs. 14.4 and 14.5 the periodic orbits are represented for two different classes and for both systems of coordinates.

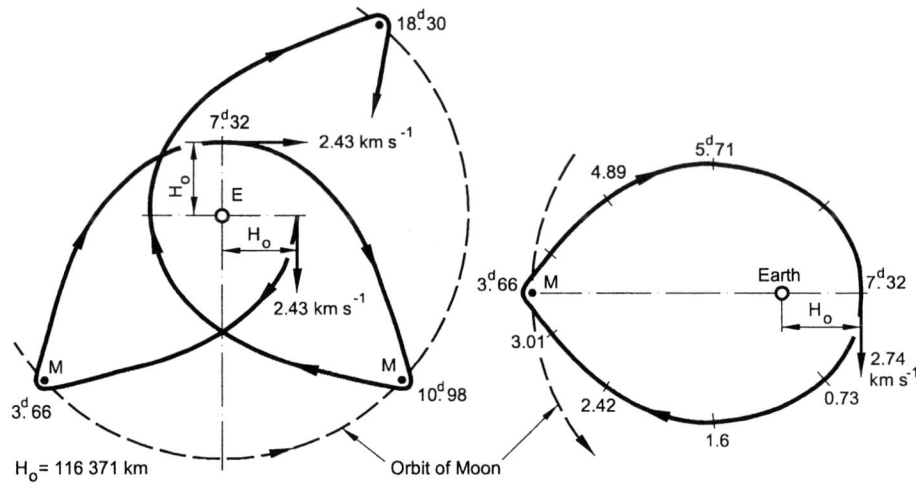

Fig. 14.4. *A periodic flyby trajectory for the Moon. On the left the trajectory is represented in the geocentric system of coordinates; on to the right, in the rotating (around the Earth–Moon line) system of coordinates. The period equals 7.32 days.*

Of special interest is the trajectory in Fig. 14.4 with periodic flybys of the Moon and returns to the Earth. In this case, the spacecraft is launched from an altitude $116\,400\ km$ from the Earth's surface and with an initial horizontal heliocentric velocity of $2.43\ km\ s^{-1}$ (or with the velocity of $2.74\ km\ s^{-1}$ in the system of coordinates that is rotating together with the Earth–Moon line). After 3.66 days the spacecraft will pass via the opposite side of the Moon, at a distance of $260\ km$ from its surface. After passing the lunar sphere of action, the spacecraft will continue its journey in a trajectory symmetrical with respect to the initial one, and will approach the starting point after 7.32 days, exactly at the same distance from the Earth and with the same velocity as at the moment of its launch. Then the spacecraft will perform the same trajectory but now with another orientation of the major axis, and move to a new meeting with the Moon; all this will be repeated with a period of 7.32 days.

The second example of a periodic trajectory in Fig. 14.5 is distinguished by the elegance of its shape in the rotating system of coordinates (to the right). In this case, the duration of the flight from the moment of the encounter of the spacecraft with the Moon until its return to the Earth is nearly one and half months, i.e. a complete flight period from the Earth and back takes nearly three months.

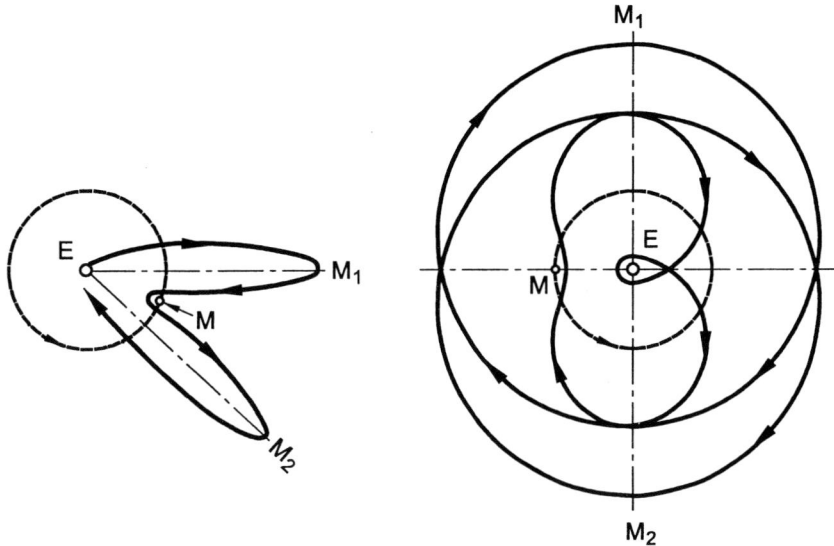

Fig. 14.5. *Periodic flyby trajectory for the Moon. On the left the trajectory is represented in the geocentric system of coordinates; on the right, in the rotating system of coordinates. The period amounts to nearly one and half months.*

6 Periodic Orbits with Rapprochements

The diversity of the shapes and configurations of periodic orbits is enormous. Slight changes in the initial data, in the magnitude and the direction of the initial velocity, in the altitude of the starting point on the Earth, in the required duration of the flight, in the conditions of *rapprochement* with the Moon, etc., are enough to modify the trajectory completely. All these are for one and the same value of the parameter $\mu = 0.01215$ corresponding to the Earth–Moon system. Under such conditions it seems extremely difficult to formulate any unified and complete system of classification of periodic orbits. In all cases, the attempts that have been performed in this area (in particular, the Copenhagen classification) have hardly been worthwhile. In the majority of cases, the forms of the periodic trajectories are symmetrical with respect to the axis OX (Earth–Moon) and sometimes with respect to the axis OY.

We shall confine ourselves to three more examples of periodic "lunar" orbits, in the synodic (rotating) system of coordinates (Broucke, 1962; Arenstorf, 1963). An example of a synodic trajectory of periodic *rapprochement* or flyby by the spacecraft of the main body, i.e. of the Earth, is shown in Fig.

14.6. The *rapprochement* with the Moon, with a doubled period (compared with the period of flyby of the Earth), is far from being close. This is only one of the possible trajectories among the broad class of the orbits of close flybys of the Earth. The crucial factor here is the numerical value of the Jacobi constant; in the case of the described trajectory we have $C = 0.8450$. Even a negligible change in C leads to a completely different trajectory.

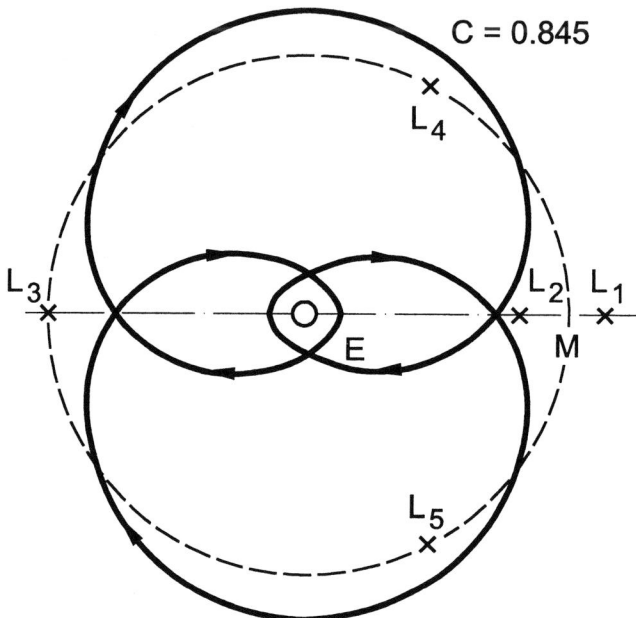

Fig. 14.6. *A periodic rapprochement lunar orbit in the rotating (synodic) system coordinates (with respect to the Earth–Moon line). The Jacobi constant $C = 0.8450$.*

The periodic orbits are characterized, in addition, by the ratio of the so-called "commencorability" T/τ, where T is the period of the orbiting of the planet (Earth) and τ is the orbiting time of the spacecraft. Typical is the situation when $T/\tau = 1/2$, i.e. when for one Moon's orbit around the Earth the spacecraft realizes two orbits around, again, the Earth. We have such a trajectory, for example, in Fig. 14.7 with a ratio $1/2$; in this case for one Moon's orbit around the Earth, the spacecraft realizes two orbits around the Earth. Another example of a periodic rapprochement trajectory with a commencorability $T/\tau = 3/4$ is shown in Fig. 14.8; in this case during three orbits of the Moon around the Earth the spacecraft realizes four orbits.

There is a term, so-called "Hollow Loop", which designates the closed

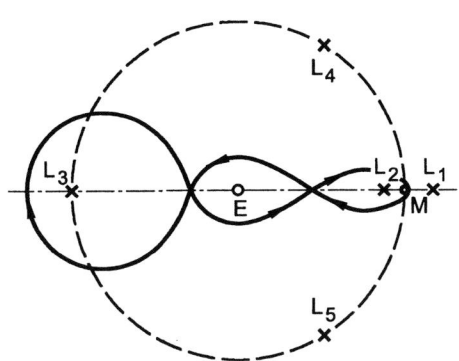

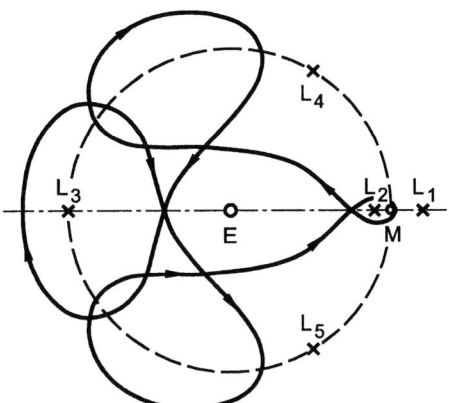

Fig. 14.7. *A periodic lunar orbit of rapprochement with a ratio 3/4 in the rotation (synodic) system of coordinates.*

Fig. 14.8. *An asymptotical periodic orbit for outward flight through the libration point L_1 of Earth–Moon system in the rotating system of coordinates.*

part or parts of the orbit not enveloped by one of these two bodies. In the case of an orbit shown in Fig. 14,8 we have a very interesting trajectory; here three loops among the four are hollow, and only in one case is the Moon enveloped by a small loop. In Fig. 14.12, a unique trajectory for a spacecraft is shown, in fact with a close transit near the Moon; here all four loops are hollow.

However, we have trajectories–orbits without any hollow loops as well; one such example is shown in Fig. 14.11; here all three loops are drawn around the Earth.

One may compute the approximate initial conditions, and construct periodic orbits with successive approaches to a main body in the following manner. Consider the motion of the Earth in an elliptic orbit. The Earth's mass is $1 - \mu$; therefore, for the period of rotation of the spacecraft we have

$$T = \frac{2\pi a^{3/2}}{[f(1-\mu)]^{1/2}} \, .$$

or, for a mean angular motion n

$$n = \frac{2\pi}{T} = \frac{[f(1-\mu)]^{1/2}}{a^{3/2}} \, . \tag{32}$$

The angular period of the Moon in the fixed (geocentric) system is equal to 2π, and the mean motion equals unity.

If, within a certain period of time the ellipse performs l rotations (in the rotating system), and the spacecraft accomplishes k spirals, and both l and k are integers, then the condition for existence of a periodic orbit may be written in the form

$$\frac{l}{k} = \frac{1}{n}, \tag{33}$$

or, bearing in mind (32), we have

$$a = \left(\frac{l}{k}\right)^{2/3} [f(1-\mu)]^{1/3}. \tag{34}$$

This result should be understood in the following manner. During the time in which an ellipse of semimajor axis a given by Eq. (34) makes l rotations in the rotating coordinate system, the spacecraft performs k spirals through this ellipse. In this case the ratio l/k is equal to the ratio of the period of the spacecraft in the motionless (geocentric) system of coordinates and the period of the Moon.

Obviously, given the integer values of l and k, we can find with the help of Eq. (34) the size of the orbit a, and, hence, the elliptic orbit itself, which performs l rotations, while the ellipse covers l revolutions. An example of such an orbit for the *rapprochement* with the Moon's orbit constructed in the rotating system of coordinates, corresponding to the values $l = 1$ and $k = 2$, i.e. with ratio $1/2$, is shown in Fig. 14.7; in this case the spacecraft performs two rotations ($k = 2$) through its elliptic orbit during a time between two encounters with the Moon ($l = 1$). The semimajor axis obtained by Eq. (34) is equal to $a = 0.6274$ in units of the Earth–Moon distance.

An orbit of an amazing shape, for *rapprochement* with the Moon, was shown in Fig. 14.8 in the rotating system of coordinates obtained for the ratio $3/4$ ($l = 3$, $k = 4$ and $C = 2.7598$). However, in this case the minimal distance of the orbit from the lunar center is $1\,531$ km, i.e. less than the lunar radius ($1\,730$ km); hence, it is necessary to make a correction to the orbit during the flyby of the Moon. The major axis of the orbit found from Eq. (34) is equal to $a = 0.8821$.

The orbits of the periodic *rapprochement* of the spacecraft with the libration points of the Earth–Moon system form a special group. Two such examples of asymptotic periodic orbits are shown in Figs. 14.10 and 14.11, both in the rotating system of coordinates (Deprit and Henrard, 1965). In the first case the orbit allows the flight of the spacecraft through the collinear libration point L_1 ($C_1 = 3.10469$); in the second case, through the point L_2 ($C_2 = 3.30690$).

7 Orbits with Flybys Through Libration Points

Of special interest are trajectories with flybys through libration points of the Earth–Moon system. The triangle libration points L_4 and L_5 are of less interest, therefore we shall concentrate on the collinear libration points L_1, L_2 and L_3 only, presenting, as an illustration, two typical examples of trajectories for each of these points. The orbits of the periodic rapprochements of the spacecraft with these libration points form a special group.

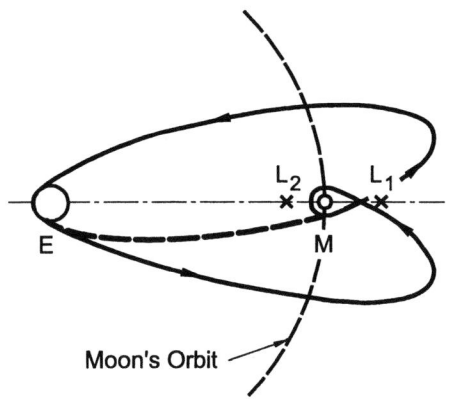

Fig. 14.9. *Periodic flyby orbit for the Moon through the libration point L_1. The trajectory is represented in the geocentric (sidereal) system of coordinates. The period of one cycle is nearly half a month.*

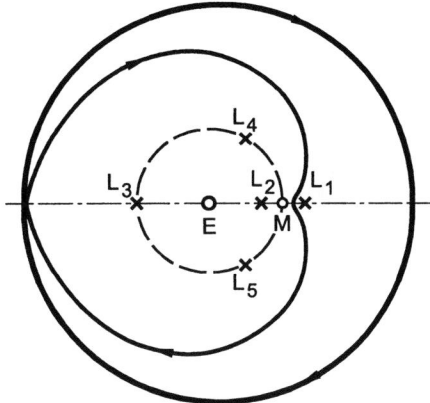

Fig. 14.10. *The asymmetrical periodic orbit for outward flight through the libration point L_1 ($C = 3.10469$) of the Earth–Moon system in the rotating system of coordinates.*

We start from Fig. 14.9, in which a periodic trajectory of one of the classes of spacecraft launched from the Earth is shown. On the right-hand side in this picture is shown a flight trajectory in a rotational system of coordinates (synodic), and on the left-hand side is shown the same trajectory in the stationary, i.e. geocentric (synodic), system of coordinates. Initially the Moon is located at the point M_0 in its orbit. The spacecraft is launched in an elliptic orbit, reaches apogee at the point M_1, intersects with the Moon's orbit and returns to the Earth, again approaching the lunar orbit. At this time, after nearly half a month the Moon will appear at the point M_1, and the spacecraft enters its zone of action and, performing a loop around the Moon, will leave this sphere towards the outer direction of the Moon's orbit. Further motion of the spacecraft takes place in an elliptic orbit that is visually identical to the first ellipse; however, it is of another orientation (another orientation of the major axis). Nearly a month after launch, the

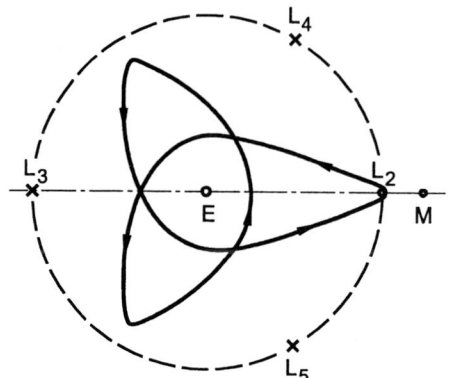

Fig. 14.11. *An asymptotic-periodic orbit for flyby of the libration point L_2 of the Earth–Moon system in the rotating system of coordinates.*

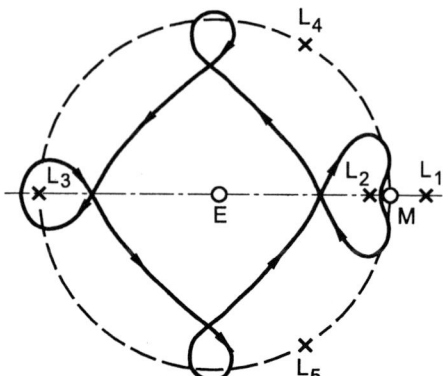

Fig. 14.12. *An asymptotic-periodic lunar orbit for flyby both libration points L_2 and L_3 of the Earth–Moon system in the rotational system of coordinates.*

spacecraft will again return to the Earth and, flying over the Earth's surface at exactly the same height and with the same velocity as at the moment of its launch, will again move towards the Moon, now in a new ellipse (the broken line in Fig. 14.9, left). On reaching its new apogee M_2, it will return to the Moon's orbit, where the second encounter will take place at the point M_3. Performing a new loop around the Moon, the spacecraft will move via a new spiral of the ellipse. This sequence will be repeated periodically; in our case, with a period of a little more than a month.

This trajectory, Fig. 14.9, is especially interesting; it gives a possibility to realize as well a close flyby through the libration point L_1 with a return to the Earth.

The second example of an asymptotic periodic orbit with a flyby through the libration point L_1 is presented in Fig. 14.10 (Deprit and Henrard, 1965); in this case, although periodic, there is no return to the Earth.

In Figs. 14.11 and 14.12, two examples of periodic trajectories with the spacecraft's transit through the libration point L_2 are shown. Of special interest is the trajectory shown in Fig. 14.12; in this case, the flight is realized with a transit of the libration point L_3 as well.

In Figs. 14.13 and 14.14 two interesting periodic asymptotic trajectories are presented, both with flyby of libration point L_3 of the Earth–Moon system. The orbit in Fig. 14.13 is interesting for its figure-eight-shaped form which is unique among interplanetary, particularly Earth–Moon, flight trajectories and which we usually meet in the restricted two-body problem.

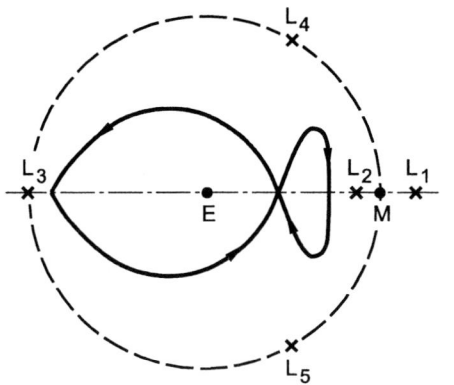

Fig. 14.13. *An asymptotical-periodic orbit with a flyby of the libration point L_3 of the Earth–Moon system in the rotating system of coordinates.*

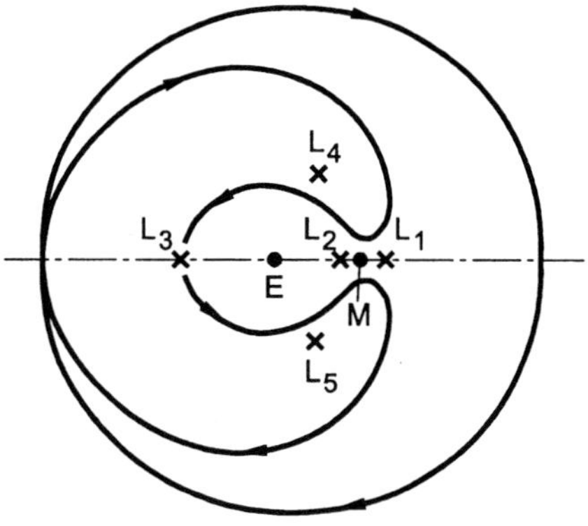

Fig. 14.14. *An asymptotical-periodic orbit with a flyby of the libration point L_3 of the Earth–Moon system in the rotating coordinate system.*

In the case of orbits in Figs. 14.10–14.14, the orbital periods are infinity, and the orbits themselves are asymptotic relative as a minimum to one of the collinear libration points L_1, L_2 or L_3.

All these orbit-trajectories, with an exception of Fig. 14.9, have a shortcoming, namely, none of the flights return to the Earth.

These orbits may be interpreted as follows. If, at the point L_1 or L_2 or L_3, an interplanetary long-duration space station, or a space astrophysical observatory with a superpower telescope, is situated and if we need, say, to take something away from this station by means of jet engines, then we can use one of these periodic orbits, since the spacecraft will return to the point L_1 or L_2 or L_3. Astrophysicists should think about suitable scientific applications of this interesting property of interplanetary periodic orbits.

In all mentioned and unmentioned cases of periodic trajectories we are still dealing, unfortunately, with an idealized problem. Many factors have not been taken into account: noncoplanarity of the orbits, the influence of Solar gravitation, the ellipticity of the lunar orbit, etc., and, the most important factor, namely, the very stringent requirements imposed on the initial conditions at the launch of the spacecraft. For example, an error in the initial velocity on the order of one millimeter per second is enough to destroy each of these periodic and exceptionally beautiful orbits.

The periodic orbits and periodic trajectories constitute yet another graceful demonstration of the amazing possibilities of celestial mechanics.

8   Flights with Small Traction

The launch of a spacecraft or artificial satellite into orbit around the Earth can be realized only with the help of powerful jet engines of large traction working on the basis of chemical fuel and developing an acceleration much larger than the acceleration due to the Earth's gravity on its surface. Practical astronautics also uses so-called ionic jet engines, i.e. engines based on a jet of charged particles in an electrostatic field. At present there are various types of such engines – ionic, plasma type, etc. All these engines are characterized, however, by an extremely small value of jet traction and, correspondingly, by extremely small accelerations, on the order of several $mm\ s^{-2}$, i.e. about a thousand times smaller than the acceleration due to the Earth's gravity. Of course, with such an acceleration departure from the Earth is not possible.

Quite different is the situation in which the spacecraft or satellite is already in orbit around the Earth. The ionic or plasma engine can work continuously and for a long time. Owing to this, the total amount of impulse gained by the spacecraft or satellite during, say, several days, months or even years, can even be comparable with the efficiency of powerful chemical reactive engines, which function only for some few seconds or minutes. The mechanics of flight with such exceptionally small but also prolonged traction can find efficient application in various fields of practical astronautics, in particular the following:

a. For the flights of spacecraft or satellites in the outer layers of a planetary atmosphere when the problem of compensation of the resistance of the medium (atmosphere) by a small but constantly acting impulse arises.

b. For maneuvering the spaceship to perform small changes in the parameters of the orbit, e.g. compensation of relatively small perturbations, enabling rendezvous with another spaceship and maneuvering for docking.

c. For transitions of the spaceship from a closed (circular or elliptic) orbit around a planet to a parabolic escape orbit.

d. For interplanetary flights (planetocentric or heliocentric) with the aim of shortening the duration of the flight.

To this list one can add the possible efficiency of using ionic engines in the case of scientific satellites and space observatories for special purposes for which the use of chemical engines for maneuvering could disturb the required conditions of cleanliness in the immediate vicinity of the satellite or observatory. The application of engines of small traction during the transition of a spacecraft from a closed orbit to a parabolic one can be advantageous in terms of economic considerations; in this case the cost of working with a smaller mass will be smaller than that for a flight using chemical engines. An engine of small traction working continuously and in a stable regime will enable one to realize a slow but constantly increasing long-time modification of the orbit. In this case, the parameters of the orbit are changing continuously: the spacecraft or spaceship executing the successive spiral reaches the parabolic velocity and, on leaving the Earth, will be directed towards the planet of destination. Of course, the loss of time will be much larger than that with the transition from a closed orbit to a parabolic one with the help of chemical engines. However, if the further flight is to be realized under conditions of permanent functioning of the engines of small traction, then this loss will be compensated to some degree.

Thus we will always have some transition trajectory – from a closed orbit to the point at which the spacecraft reaches its parabolic velocity: it is called a **speeding-up trajectory** or an **untwist trajectory**. Some properties of such trajectories will be examined below.

The application of engines of small traction, however, hides a difficulty, namely, that the vector of the traction f_p of an ionic engine must preserve its direction for a long period of time. This means that the spacecraft itself must be stabilized during this period. Moreover, during interplanetary flights there can appear situations when the direction of this vector should be changed in accordance with the flight schedule. Hence, additional energetic losses will be inevitable.

We confine ourselves here, however, to examination of a particular but important case in which the traction of an ionic engine is directed constantly at a tangent to the trajectory at the given point, and the acceleration of this traction f_p is constant in magnitude throughout the flight. The equations of motion of the spacecraft, written in the traditional form, in the Cartesian system of coordinates with the Earth at its center, have the following form:

$$\frac{d^2x}{dt^2} = -k^2 m \frac{x}{r^3} - f_p \cos \alpha \,,$$

$$\frac{d^2y}{dt^2} = -k^2 m \frac{y}{r^3} + f_p \sin \alpha \,,$$

(35)

where x and y are the coordinates of the spacecraft, r is the radius-vector, i.e. the distance from the Earth's center, m is the mass of the Earth, α is the angle between the vector of the traction f_p and the normal at a given point of the trajectory. The latter angle α is not constant and changes from point to point. Therefore in the Eq. (35) we have

$$\cos \alpha = \left[1 + \left(\frac{dy}{dx}\right)^2\right]^{-1/2},$$

$$\sin \alpha = \frac{dy}{dx}\left[1 + \left(\frac{dy}{dx}\right)^2\right]^{-1/2}.$$

(36)

The solution of Eqs. (35) and (36) gives the equations of motion of the spacecraft $x(t)$ and $y(t)$ and, hence, the equation of the trajectory (orbit):

$$\psi\,(x,\,y,\,f_p) = 0\,,$$

(37)

and this curve, due to the fact that $f_p \neq 0$, will not be a closed curve; it represents a spiral with a complete opening after n rotations around the central planet – the Earth. In the limit when $f_p = 0$, we shall have the usual solution, $\Psi(x, y, 0) = 0$, i.e. an ellipse.

The solution of the system of Eqs. (35) and (36) cannot be realized by analytical methods, and the numerical integration is not very simple. However, this procedure may be simplified in some degree by transiting to the polar system of coordinates as shown in Fig. 14.15.

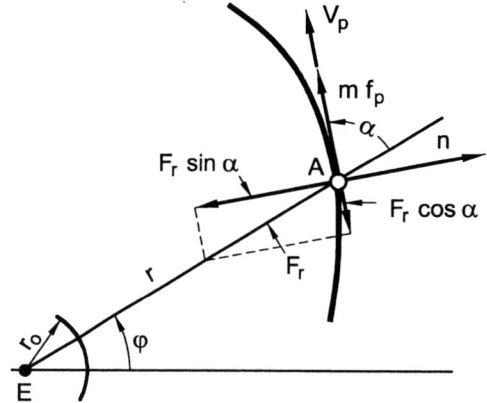

Fig. 14.15. *The problem of the momentum of a spacecraft with small traction. The solid curve denotes the trajectory of the spacecraft A around the Earth's center E. For explanation of notation see the text.*

In Fig. 14.15 r_0 is the Earth's radius, r is the distance of the spacecraft A from the Earth's center E, f_p is the traction-vector of the ionic traction, directed always at a tangent to the given point, i.e. coinciding with the direction of the velocity vector V_p, α is the angle between the radius-vector r and the vector f_p, F_r is the Earth's gravitational force with components normal, $F_r \cos \alpha$, and tangential, $F_r \sin \alpha$, to the normal of the orbit, respectively.

The equation of motion in the projection on the tangent to the orbit is written in the form

$$\frac{dV_p}{dt} = f_p - g_0 \left(\frac{r_0}{r}\right)^2 \cos\alpha, \tag{38}$$

where g_0 is the acceleration due to the Earth's gravitation at its surface. For the equation of motion in the projection normal to the trajectory we have

$$\frac{d}{dt}\cos\alpha = \frac{\sin^2\alpha}{rV_p}\left(V_p^2 - g_0\frac{r_0^2}{r}\right). \tag{39}$$

Bearing in mind that $\alpha \approx 90°$, we also have for the radial component, dr/dt, of the spacecraft velocity V_p that

$$\frac{dr}{dt} = V_p \cos\alpha. \tag{40}$$

However, it is convenient to pass to the dimensionless variables, distance ρ, velocity V, and time τ:

$$\rho = \frac{r}{r_0}, \quad V = \frac{V_p}{(g_0 r_0)^{1/2}}, \quad \tau = \left(\frac{g_0}{r_0}\right)^{1/2} t, \tag{41}$$

Then Eqs. (38)–(40) should be written in the following form:

$$\frac{dV}{d\tau} = f - \frac{\cos\alpha}{\rho^2},$$

$$\frac{d}{d\tau}\cos\alpha = \frac{\sin^2\alpha}{\rho V}\left(V^2 - \frac{1}{\rho}\right), \tag{42}$$

$$\frac{d\rho}{d\tau} = V\cos\alpha,$$

where f is the dimensionless magnitude of the traction of the ionic engine

$$f = \frac{f_p}{g_0}. \tag{43}$$

The unknown quantities in Eqs. (42) and (43) are the parameters ρ and α, which determine the traction trajectory as well as the spacecraft's velocity V at the given point. For complete determination of the trajectory it is necessary to add to Eq. (42) also the equation for the polar angle φ:

$$\frac{d}{d\tau}(\rho^2\varphi) = f\rho\sin\alpha. \tag{44}$$

The solution of Eqs. (42) and (44) is realized by numerical integration. From this solution the equation of the trajectory is derived in the following form:

$$\varphi\left(\rho,\ \alpha,\ f\right) = 0\,, \qquad\qquad (45)$$

The trajectory is a nonclosed curve, an untwisted spiral; an example of such a trajectory computed for the value $f = 0.01$, which is a very large value for ionic engines, is shown in Fig. 14.16 (Beletzky, 1977). At small values of f, say, when $f \simeq 10^{-4}$, which is close to reality, a rapid increase in the number of spirals takes place. In fact, the number of spirals varies inversely proportionally to f, i.e. $n \propto f^{-1}$. This means, for example, that the number of spirals in the case $f = 10^{-4}$ will be a hundred times greater than we have discovered in the case $f = 10^{-2}$ and in all cases the graphic representation of such a trajectory will not be obvious.

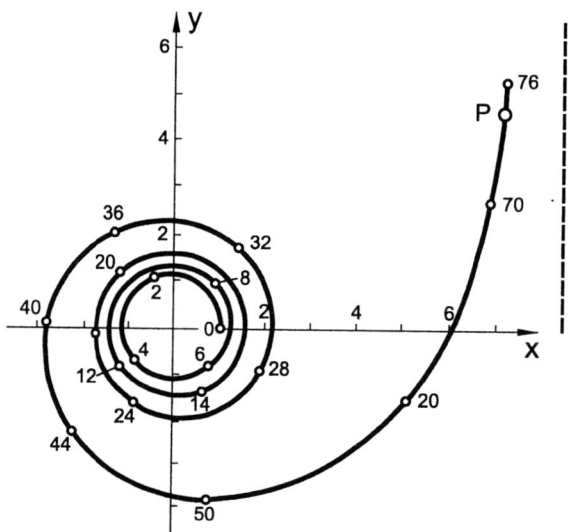

Fig. 14.16. *A speeding-up trajectory with small traction and at constant tangential acceleration* $f_p = 0.01 g_0$. *The acceleration starts with a circular orbit at altitude* 300 *km from the Earth's surface. Numbers on the trajectory denote dimensionless time, and, along of axis* x, *the distance. P is the point of attainment of parabolic velocity. The broken line to the right is the approximate position of the asymptote of the trajectory at large distances.*

As for the behavior of the trajectory after untwisting of the spiral, i.e. at $\alpha \approx 0$ when the component of the velocity in the direction of the normal at the given point of the trajectory vanishes completely $(dr/dt = 0)$, the reaction force acting practically along the trajectory tends to an asymptotical line. It is shown to the right in Fig. 14.16 by a vertical broken line.

The characteristic feature of the solution (45), as well as of the relationship $V = V(\rho, f)$, is as follows. The velocity V, although it is decreasing with increasing distance ρ, is at the same time approaching ever closer to the parabolic velocity at the given point of trajectory. Of course, it is of interest to know the moment of time $\tau_\star$ at which the spacecraft attains this limiting velocity at a distance $r_\star$ from the Earth's center.

The parabolic velocity will be reached, of course, when further change in the velocity ceases, i.e. when $dV/dt = 0$. Then we find from the first equation in Eq. (42) that

$$\rho_\star = \left(\frac{f}{\cos \alpha_\star} \right)^{-1/2} \qquad \rho_\star = \left(\frac{\cos \alpha_\star}{f} \right)^{1/2} . \tag{46}$$

The law of dependence of parabolic velocity on the distance is known to be $V_\star^2 = 2/\rho_\star$. Therefore we will have from Eq. (46) for the parabolic velocity at the distance $\rho_\star$

$$V_\star = \left(\frac{4f}{\cos \alpha_\star} \right)^{1/4} . \tag{47}$$

As for the time $\tau_\star$ of acceleration required to the reach the parabolic velocity, it can be obtained by simple fixation of the time in the results of a numerical integration of the type $\rho = \rho(\tau)$, substituting $\rho = \rho_\star$ at $\tau = \tau_\star$. However, it can be obtained with the help of an approximate formula derived by means of the integral of the energy:

$$\tau_\star = \frac{1}{(1 + V_\star)f} , \tag{48}$$

where $V_\star$ is given by Eq. (47). As for the angle $\alpha_\star$, it is not difficult to see that $\cos \alpha_\star \to 1$ as $\tau_\star \to \infty$. Then we obtain from Eqs. (46)–(48), moving again to usual dimensions, the following approximate formulas, which suffice for qualitative estimations:

$$r_\star = r_0 \left(\frac{g_0}{f_p} \right)^{1/2} , \tag{49}$$

$$V_{\star p} = \sqrt{r_0} \, (4f_p g_0)^{1/4} , \tag{50}$$

$$t_\star = \frac{(r_0 g_0)^{1/2}}{f_p} [1 + \sqrt{r_0} \, (4f_p g_0)^{1/4}]^{-1} . \tag{51}$$

The calculated parameters are the limiting distance $r_\star$ at which the trajectory becomes a straight line, the velocity of the spacecraft $V_{\star p}$ at this

distance and the time $t_\star$ at which the spacecraft reaches the parabolic velocity $V_{\star p}$, for a number of values of f and when the acceleration starts from the
altitude $300\ km$ of the Earth's surface. Their value are presented in Table
14.1.

T a b l e   14.1

Acceleration characteristics for a spacecraft starting from the Earth
and equipped with ionic jet engines of small traction for three values
of relative traction f: 10^{-2}, 10^{-3}, and 10^{-4}

f	f_p $mm\ s^{-2}$	$t_\star$ days	$r_\star$ km	$V_{\star p}$ $km\ s^{-1}$	n
10^{-2}	89.5	0.75	56 630	3.65	4
10^{-3}	8.95	8.5	156 760	2.07	39
10^{-4}	0.9	92	585 150	1.17	398

In the last column the number of spirals, n, up to the straightening of
the trajectory is given. It increases sharply with decreasing traction f of the
ionic engine. The ionic traction engines may also find applications in the
realization of interplanetary flights (Wiessel, 1989).

9   Solar Sailing

The light pressure caused by the radiation of the Sun at the distance of the
Earth's orbit is equal to $p = 4.5 \cdot 10^{-8}\ Pa$. This is of course a very small
quantity. However, acting permanently and constantly in one and the same
direction, this negligible magnitude may turn out to be the equivalent of the
action of some ion jet driver of even smaller traction. However, there is an
important difference, namely that, in the case of using the Solar radiation
we are free from energetic problems, though the problems of the orientation
and stabilization of the spaceship obviously remain. Sailing on radiation
from the Sun or a beam of electromagnetic radiation are forms of photon
propulsion (Wright, 1992).

The action of light pressure on a perfect sail is obvious from Fig. 14.17.
From the incident sunlight, at the angle θ with respect to the sail's normal,
there results a light pressure F_I. However, the reflection of this light at the
same angle θ will also result in a light pressure F_R. The resulting pressure
F_{total} should be the vectorial sum of the two pressures, F_I and F_R. However,
the sail does not reflect all of the incident light; some part of the radiation
will be lost due to imperfect reflection and the processes of absorption and

scattering. With the absolute reflection (100%) we would have $F_I = F_R$, and this vector would be directed exactly normal to the sail, and the effect of the light pressure would be the largest. There follows, therefore, the first important conclusion, namely that the sail should be covered by a reflecting layer.

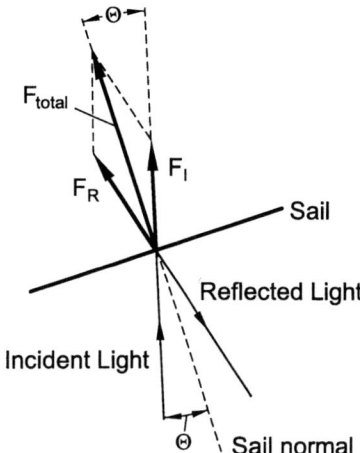

Fig. 14.17. *The force vector for an ideal sail under the action of solar radiation pressure.*

Solar sailing can be used to reach any body within the Solar system, both outer planets (Mars and farther out) and inner ones (Venus and farther in). This is realized by a simple change in the sail's orientation. If the sail is turned in such a way that the lateral component of the force – both gravitation and light pressure of the Sun – is against the direction of motion, then the spaceship will lose energy and angular momentum, resulting in a spiral towards the Sun, as illustrated in Fig. 14.18 (top). If the orientation of the sail results in the motion gaining lateral acceleration, then the sail will spiral outward from the Sun (Fig. 14.18, bottom). In other words, in the case of solar sailing the same manipulations take place as in the case of usual sailing, for which the role of light pressure is played by the pressure of the wind, and the role of gravitational attraction is played by the pressure of the water on the ship.

An attitude control system must exist to execute a steering profile that will cause the sail to follow a trajectory to its intended destination. Of course, it is hard to reject the attractiveness of this idea. Therefore, we will examine here the basic features of these phenomena.

If the sail is of absolutely black material, i.e. its reflection is zero, then the light pressure on the sail will be conditioned only by the pressure of the

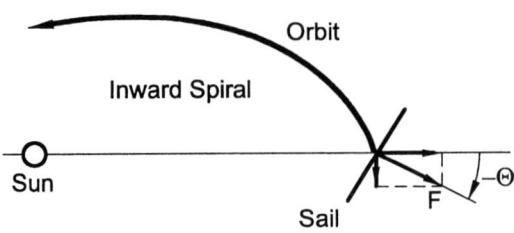

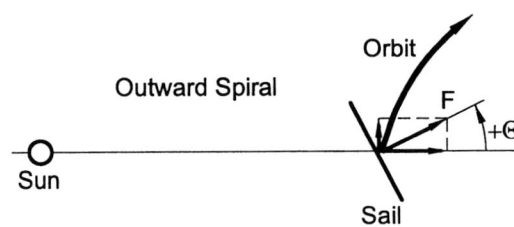

Fig. 14.18. *The maneuvers of a spaceship under the action of solar radiation pressure. Top, an inward spiral, in the direction of inner planets; below, an outward spiral, in the direction of the outer planets.*

absorbed light. The magnitude of this pressure P will be

$$P = p \left(\frac{R_0}{R} \right)^2 S \cos \theta, \tag{52}$$

where S is the area of the sail, R_0 and R_* are the distances of the sail from the Sun and from the Earth, respectively, and θ is the angle between the sail's normal and the incident radiation.

However, the sail may also be made of perfectly reflective material, for example, an aluminum foil. Then, the resulting light pressure will be doubled to

$$P = 2p \left(\frac{R_0}{R} \right)^2 S \cos \theta. \tag{53}$$

For sailing orbits not far from the Earth we have $(R_0/R_*) = 1$.

Of course, when the problem is to achieve parabolic velocity within the shortest possible period, then the construction of the sail as well the regime of its control will be chosen correspondingly. However, insofar as we are interested in a treatment of the problem concerned with the acceleration of a spaceship with the help of solar sailing purely from a theoretical point of view, we omit these questions.

The equations of motion of a sail in polar coordinates φ and r can be written in the following form:

$$\frac{du}{dt} = \frac{v^2}{r} - \frac{\mu}{r^2} + \frac{a_0}{2} \left(\cos \gamma + \cos \varphi \right) f(r, \varphi),$$

$$\frac{dv}{dt} = -\frac{uv}{r} + \frac{a_0}{2} \left(\sin \gamma - \sin \varphi \right) f(r, \varphi),$$

$$\frac{dr}{dt} = u, \tag{54}$$

$$\frac{d\varphi}{dt} = \frac{v}{r},$$

where u and v are the radial and transversal components of the spaceship's velocity, φ is the angle between the directions of radius-vector r and solar rays, γ is the angle between the velocity vector V and radius-vector r, μ is the gravitational parameter for the Earth, and $a_0 = P_\star/m$ is the maximum acceleration developed by light pressure on the sail of given construction and mass m.

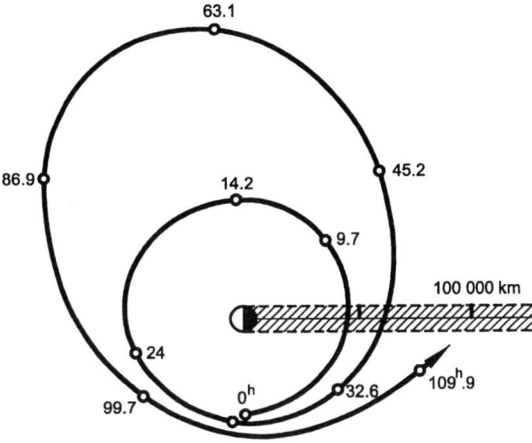

Fig. 14.19. *Acceleration of a spaceship with the help of a solar sail. One of the versions of the computed trajectory is represented. Numbers along the trajectory denote the time periods in hours after the start (0^h). The shaded area is the Earth's shadow (not to scale).*

In Eq. (54) the so-called "shadow function" $f(r, \varphi)$ is also included; it equals zero when the ship enters the Earth's shadow, and unity outside the shadow. However, because neither entry into the shadow not exit from it

takes place sharply, the form of this function is chosen accordingly as a result
of various manipulations.

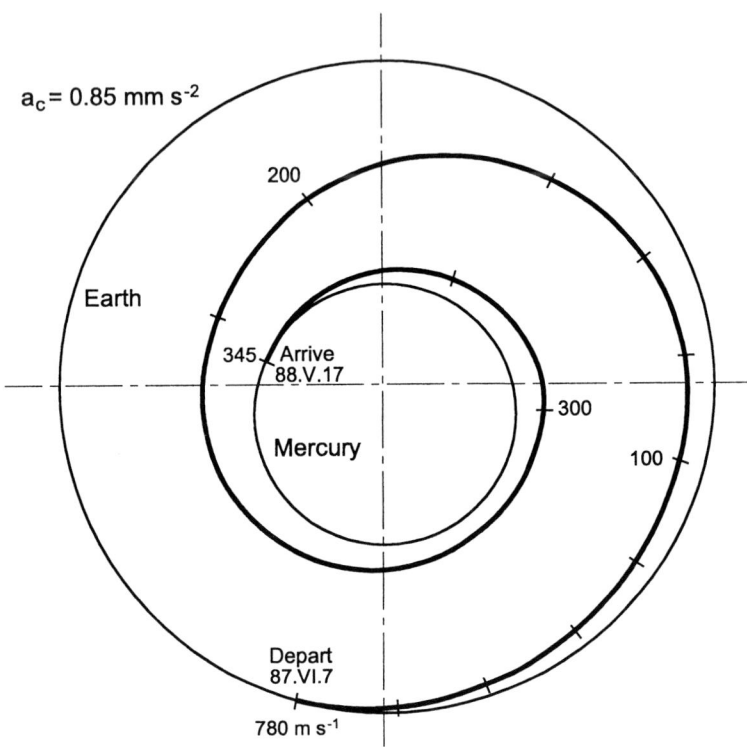

Fig. 14.20. *The trajectory to Mercury from the Earth for a sailing spaceship with a departure velocity from the Earth of 780 m s⁻¹, with zero arrival velocity at Mercury and a journey duration of 345 days.*

One should add to Eq. (54) also the following relationships:

$$V = (u^2 + v^2)^{1/2}, \quad \sin\gamma = \frac{v}{V}, \quad \cos\gamma = \frac{v}{U}. \tag{55}$$

The solution of Eqs. (54) and (55) can be realized only by numerical methods. An example of a momentum trajectory of a spaceship under the action of solar radiation is shown in Fig. 14.19, computed under the following assumptions: the sail is an aluminum foil of depth 10^{-4} mm, area $S = 4.65$ km^2 (a circular sail with a diameter of 2.4 km) and weight about 1.5 ton (spaceship plus sail). In this case the acceleration of the spaceship will be of the order of $0.001g_0$. The acceleration itself starts from the stationary orbit, i.e. at the conditions

$$r_0 = 42\,188\ km\,, \quad u_0 = 0\,, \quad v_0 = (\mu/r_0)^{1/2}\,.$$

The last point on the trajectory (Fig. 14.19) corresponds to on duration of the flight from the start, 109.9 h, nearly five days. The first spiral is twisted for about 30 h, i.e. for six hours longer than the duration of nonaccelerated motion in a stationary orbit.

Another example of a computed trajectory for a spaceship sailing from the Earth to Mercury is shown in Fig. 14.20. The computations were carried out for concrete data concerning the start (July 7, 1987) with departure velocity 780 $m\ s^{-1}$, zero arrival velocity at Mercury and a journey duration of 345 days. The conditions for flight to Venus by means of solar sailing are less demanding (Wright, 1992).

Chapter XV

Interplanetary Missions

1 Flights to the Sun

In the present chapter we will give a brief account of the key interplanetary missions already performed and scheduled. On one hand, this will serve as a "live" illustration for the theory of orbits given in previous chapters, and on the other, will provide definite insight into the intriguing properties of the planets and their satellites, comets, and asteroids of the Solar system.

The investigation of the profound properties of the Solar system has also acquired a special significance in view of the progress in the search for exrasolar planets around nearby bright stars, as discussed in Chapter XI. In particular, precession Doppler measurements via the Keck High-Resolution Echelle Spectrograph reveal periodic Keplerian velocity variations in some G-type stars. In some cases companions of $M \sin i = 0.22 M_{Jup}$, even $M \sin i = 0.46 M_{Jup}$ have been discovered. Moreover, the first system of **multiple planets** was detected in 1999 by Butler and collaborators, followed by the first detection of a **transiting planet**. Such distribution of masses, peaking at $\sim 1 M_{Jup}$, is not the case in our Solar system, and no present theory of planet formation predicts such a peak.

In the survey of interplanetary flights, let us start from the Sun. We have already seen the practical impossibility of sending a spacecraft from the Earth directly to the Sun, either in a Hohmann trajectory or in a straight line (after stopping the probe in the orbit). In both cases the initial launch velocities are rather large – 29.15 and 31.82 $km\ s^{-1}$, respectively.

However, the astrophysics community has stated the importance of realization of such an experiment – the probing of the outer regions of the solar corona with the help of flyby of special probes. In some sense, this fact inspired the search for and, as a result, the discovery of a convincing solution, i.e. the possibility of using the perturbation maneuver in the sphere of action of Jupiter.

As was outlined above, we have the largest increase in the velocity vector during the flight of the probe through the sphere of action of Jupiter – 42.7 $km\ s^{-1}$; for comparison, the increase in the velocity vector in the case of the sphere of action of Mars is only 3.5 $km\ s^{-1}$. This circumstance, i.e. the possibility of the realization of a powerful perturbation maneuver in Jupiter's sphere of action, can be used to send a space probe towards the Sun.

A possible version of such a flight trajectory, Earth–Jupiter–Sun–Earth, is shown in Fig. 15.1. The space probe starts from the Earth with an initial velocity of 16.5 $km\ s^{-1}$ in the direction of Jupiter, and after nearly a year and three months it will reach Jupiter's sphere of action. At the radius of Jupiter's sphere of action, $48.2 \cdot 10^6\ km$, the probe will fly by Jupiter at a very small distance – 378 000 km from its center (5.3 times Jupiter's radius) in a loop trajectory, which we have usually had for the Moon. Leaving Jupiter's sphere of action with an increased velocity, the probe will now be directed towards the Sun. Roughly three years after launch from the Earth, the probe will pass the nearest point to the Sun at a distance of approximately $30 \cdot 10^6\ km$ from its center (nearly 40 solar radii). In principle, an even closer flyby is possible; however, it could lead to extreme heating of the probe, thus disrupting, for example, its telemetric broadcasting facilities.

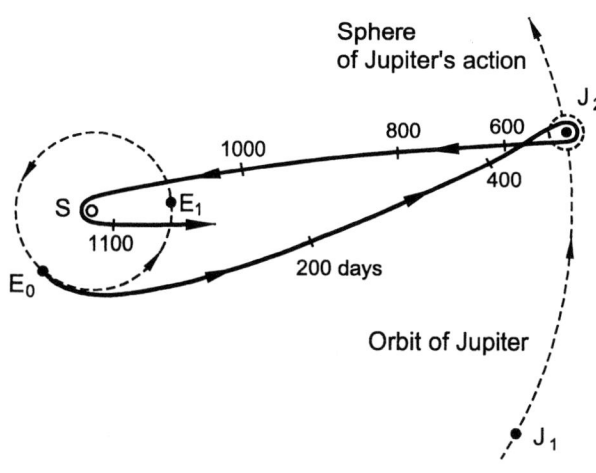

Fig. 15.1. *Flight to the Sun (S) with the help of Jupiter. The perturbative maneuver is realized within Jupiter's sphere of action (J_2). Numbers on the trajectory denote the days after the launch of the space probe from the Earth (E_0). After the flyby of the Sun, the probe will be directed towards the Earth (E_1).*

In Fig. 15.1, the configurations of the Earth and Jupiter at the moment of launch of the probe (J_1 and E_0), as well as at the flyby of Jupiter (J_2 and E_1) are shown. At the moment of flyby of the Sun, the Earth is at the point E_1, not far from the point of intersection of the trajectory of the probe with the Earth's orbit.

The duration of the flight to the Sun with "Jupiter's help" is about three years. In this connection a question arises: can some other planet, say, Mars, be used for the same aim? The answer is simple: no! The perturbative maneuver with the help of Mars in the period of flight of the probe through its sphere of action is insufficient in power to direct the probe towards the center of the Solar system in general. This is not only because the absolute value of the increase in the velocity is small, only 3.5 $km\ s^{-1}$. The largest angle of the rotation φ_{max} of the velocity vector in the planetocentric system of coordinates, caused by the gravity of Mars, is only 90°, i.e. much smaller than the required magnitude, 180°. At such an angle the probe will not be directed towards the Sun, whereas, in the case of Jupiter, this angle of rotation is large, 159°.

Directing the space probe towards the Sun is only one of the applications of the perturbative maneuver within Jupiter's sphere of action. In fact, Jupiter's gravitation can direct the probe to any point of the Solar system. In particular, Jupiter has been used for the realization of extremely interesting tours by the *Voyagers* with multiplanetary passages, which we will discuss in the next sections.

2 Ulysses: Flight Perpendicular to the Ecliptic, the First Round

The study of the physical conditions above the Sun's polar caps represents an astrophysical problem of extreme importance. More precisely, the question concerns the outer atmosphere of the Sun, the corona, which extends out to several solar radii and has a temperature above $10^6\ K$. Neither the heating of the corona nor its role in the properties of Solar wind are well known. Of particular importance are the mechanisms of acceleration of particles up to high energies, the problem of the Solar flares, unpredictable short-duration outbursts with release of huge amounts of energy in various forms: radiation, extending from gamma and X-rays up to radio waves, and energetic particles. It is assumed that all these processes are essentially connected with the active regions, i.e. the centers of the large coronal holes covering the polar caps, but the role of axial rotation of the Sun, the longitude dependence, is unknown. The solar magnetic field at the polar caps is clearly an important component of the Sun's field and is poorly observed from the Earth. Thus the major aim is to determine the field strengths at the polar caps, i.e. of the regions that are free from the influence of the Sun's axial rotation.

To solve these problems, it is necessary to send a spacecraft equipped with measuring facilities in a trajectory passing through the polar regions of the Sun. In other words, the question is to realize the flight of a spacecraft in the plane **perpendicular** to the ecliptic.

The main problem is obviously to gain flight velocity perpendicular to the ecliptic. As has been mentioned above, it is necessary to have an initial launch velocity not less than 30 $km\ s^{-1}$.

The only solution is the use of Jupiter as a powerful tool for radically changing the flight trajectory of the spacecraft, i.e. by realization of a perturbative maneuver. It appears that such an operation can be realized in a manner such that the spacecraft, without any waste of additional energy, can move exactly within the plane perpendicular to the ecliptic, so that, after a certain period, it can reach the polar regions of the Sun. Such a project, *Ulysses*, was successfully realized.

Ulysses was a spacecraft weighing nearly 400 kg, equipped with nine scientific instruments for the registration of elementary particles, the magnetic field of the Sun and the interplanetary medium, Solar wind, cosmic rays, etc., in the interplanetary medium and in the vicinity of the Sun.

Ulysses was launched in October 1990 in the direction of Jupiter in a parabolic trajectory, and in February 1992, i.e. after a year and four months, it reached Jupiter; had it flown in a Hohmann ellipse, the duration would have been much longer – two years and nine months (see Table 12.2). During this period, the *Ulysses* trajectory lay within the plane of Jupiter's orbit around the Sun, i.e. nearly on the ecliptic.

The perturbation maneuver close to Jupiter was organized in a such manner that *Ulysses* took the plane perpendicular to the ecliptic as shown by A in Fig. 15.2. In two and half years, in early June 1994, *Ulysses* reached the south polar region of the Sun, performing measurements throughout a period of four–five months.

We will not discuss the astrophysical aspects of this flight here. For us it is important that *Ulysses* turned out to be, after the *Voyagers* (see Section 4), the second case of application of the perturbative maneuver in Jupiter's sphere of action; however, with the essential novelty that it was the first spacecraft to move perpendicular to the ecliptic. Thus *Ulysses* is an impressive example demonstrating the exceptional possibilities of perturbative maneuvers.

3   Ulysses: Second Round

In April 1998, Ulysses, from the orbit's aphelion, started its second turn over the Sun's polar region. In this case, the probe will be within 7° of the south pole – several hundred million km away – from September 8, 2000 until January 16, 2001. For astrophysicists this should be an especially rewarding pass as the Sun must go through the most dynamic period of its 11-year activity cycle. In particular, during this period the aim was to study the magnetic fields in the Sun's polar regions, as well as the Solar wind, variations of cosmic rays propagating through the Solar system. The

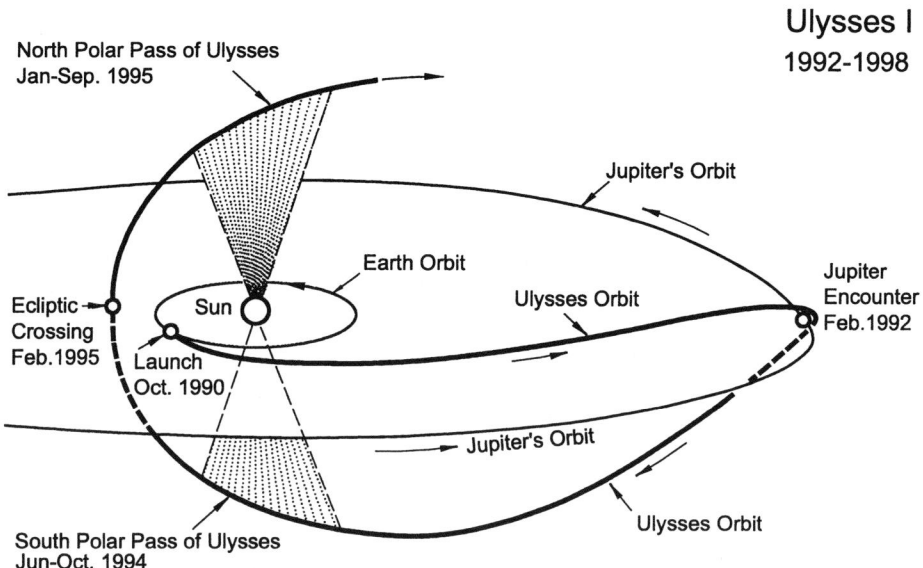

Fig. 15.2. *The Ulysses trajectory from the Earth to Jupiter on the plane of Jupiter's orbit, then, after a perturbative maneuver within Jupiter's sphere of action, the flight through a trajectory perpendicular to the ecliptic and above both of the Sun's poles.*

probe's north-polar pass would occur from September to October 2001.

T a b l e   15.1

The phases of the first and the second traverses of Ulysses through the plane perpendicular to the ecliptic

T r a n s i t s	First cycle	Second cycle
Launch	Oct. 1990	–
Jupiter encounter, aphelion	Feb. 1992	April 1998
South Polar Pass	Jan. – Oct. 1994	Sept. 2000 – Jan. 2001
Ecliptic Crossing, perihelion	Feb. 1995	May 2001
North Polar Pass	Sept.1995–Jan.1996	Sept. – Dec. 2001
Period of a cycle	6 year 2 months	6 year 8 months

In Fig. 15.3 the main structure of Ulysses' second Solar orbit is shown, and in Table 15.1 the main periods of both cycles are presented. The duration of the first cycle was 6 years 2 months, of the second cycle, nearly 6

years 8 months, i.e. the difference, although not too large, is noticeable and
is due to the fact that in the first cycle a gravitation maneuver has been
realized within Jupiter's sphere of action, while the second cycle contains no
such maneuver.

The impression is that both the Ulysses orbits, in the first and the second cycles, are practically the same or too stable, that is, without strong
perturbations.

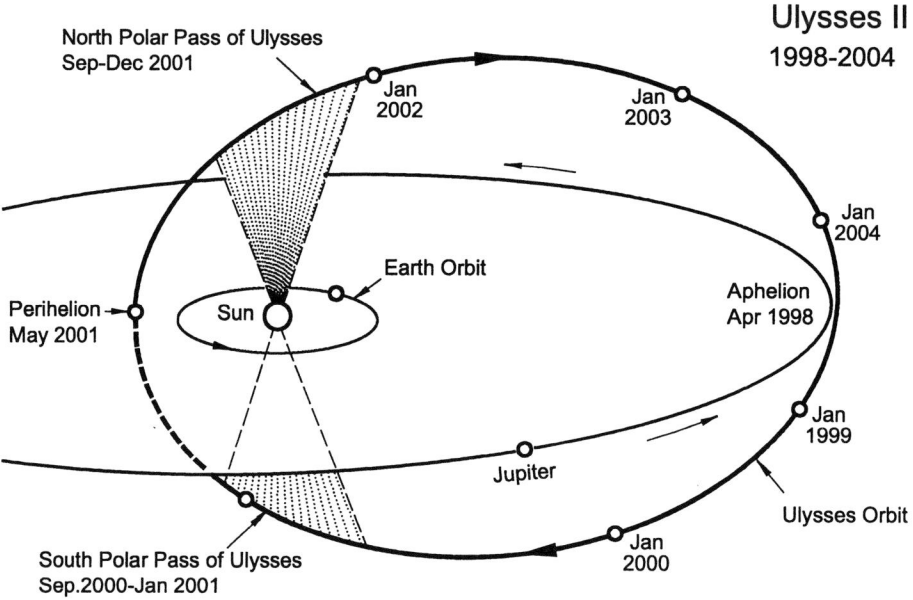

Fig. 15.3. *The Ulysses trajectory in the second cycle, again perpendicular to the ecliptic and above both of Sun's poles.*

The probe's north-polar pass occured in 2001, from September 3 to December 12. This period was used for the realization of special observations, namely, to triangulate accurate positions for the brightest extragalactic gamma-ray sources. Although the Ulysses orbital plane was not exactly
perpendicular to the ecliptic, $i \sim 70°$, it was enough to catch both the Sun's
polar regions.

As expected, Ulysses will continue operation until returning to the vicinity of Jupiter in late 2004. However, by then the slowly declining power
output of the probe's plutonium radioisotope generators will require periodic switching off of several of the instruments.

Some elements of Ulysses orbit are as follows:

Earth–Jupiter segment: from 16 October 16, 1990 to December 30, 2001

Semi-major axis	$a = 8.99150 \ a.u.$
Eccentricity	$e = 0.889166$
Inclination	$i = 1°.990719$

Final out-of-ecliptic orbit: from March 19, 1992 to December 2001

Semi-major axis	$a = 3.3730033 \ a.u.$
Eccentricity	$e = 0.60306$
Inclination	$i = 72°.128$

4 Pioneers: Out of the Solar System

In the early 1970s two probes were launched for the first time with a final aim to cross the entire Solar system to the last planet, Pluto, and then to become the first human-made objects to leave the Solar system forever.

The first, 250 kg Pioneer 10, was launched on March 3, 1972 with an initial velocity of $14.3 \ km \ s^{-1}$. In July of the same year the spacecraft entered the asteroid belt, leaving it in mid-February 1973. Then, entering with a heliocentric velocity of $10.6 \ km \ s^{-1}$ into Jupiter's sphere of action, Pioneer 10 passed a minimum distance of 130 000 km from Jupiter's atmosphere. Exactly a minute later, the spacecraft passed at a distance of 18 000 *km* from Jupiter's nearest but very small satellite Amalthea, and in 17 minutes found itself in the zone of action of the first Galilean satellite Io, imaging for the first time its surface and atmosphere and discovering a surprisingly large number of active volcanoes. Then, speeding up in Jupiter's gravitational field and making a sharp turn, Pioneer 10 left Jupiter's sphere of gravity with a hyperbolic velocity of $22.1 \ km \ s^{-1}$, and continued its free flight towards the outer regions of the Solar system without meeting any other planet. In 1979, Pioneer 10 crossed the Uranus orbit, and in 1987 Pluto's orbit, and with a velocity of $11.3 \ kms^{-1}$ this spacecraft left the Solar system; the trajectory of this flight is shown in Fig. 15.4.

A year later, on April 6, 1973 the next spacecraft, Pioneer 11, was launched on an analogous heliocentric trajectory. A year later, in April 1974, Pioneer 11 passed at a close distance, 42 600 km, and with a high velocity, $48 \ km \ s^{-1}$, the outer boundaries of Jupiter's atmosphere, and, undergoing a much sharper turn than in the case of Pioneer 10, it turned to Saturn. Four and a half years after the encounter with Jupiter and six and a half years after its launch from the Earth, on September 1, 1979, Pioneer 11 reached Saturn, at a distance of 20 000 km from the upper layers of its atmosphere. It was remarkable that, while crossing Saturn's rings, the probe succeeded in avoiding any damage. Then, the spacecraft passed Saturn's largest satellite Titan at the distance of 335 000 km, and left for the outer regions of

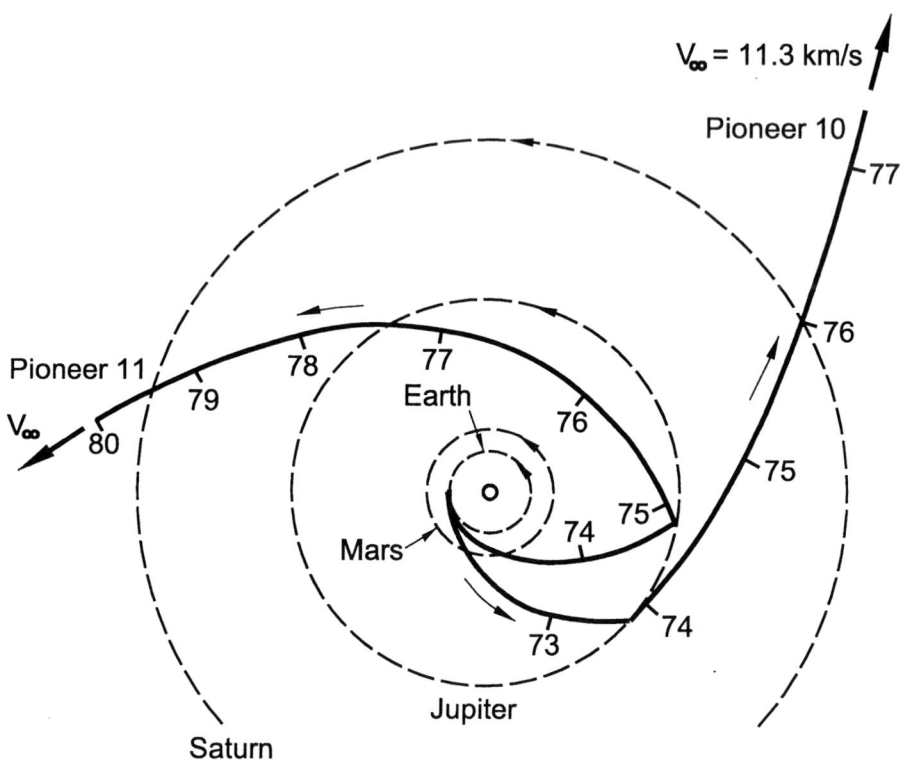

Fig. 15.4. *Trajectories of interplanetary flights of Pioneer 10 and Pioneer 11 as projected on the plane of the ecliptic.*

the Solar system, crossing sequentially the orbits of the remaining planets. Pioneer 11 left the Solar system at the end of 1988 in a direction opposite to the direction of Pioneer 10.

The flight of Pioneer 10 posed at least two puzzles. The first was the possibility of the existence of a Trans-Neptunian object at 56 a.u. – a distance, from which radio signals took 7 hours and 47 minutes to reach the Earth. Since apparently Pioneer 10 had experienced gravitational deflection in December 1992, the unknown celestial body may be from the so-called Kuiper Belt. The second puzzle was the tiny anomalous acceleration towards the Sun – about 10 billion times smaller than the acceleration from Earth's gravity – apparently detected in the motion of Pioneer 10, as well as of Pioneer 11 and even Ulysses.

Until Pioneer 10 reaches a distance of about 1.5 parsec – in about 126 000 years from now – it will still be dominated by the gravitation field of the

Sun. Later, Pioneer 10 will journey through the Galaxy, influenced by the field of the nearby stars.

As to Pioneer 11, the last communication from it was received in November 1995, shortly before the Earth's motion carried it out of view of the spacecraft antenna. Pioneer 11 may pass near one of the nearby stars in about 4 million years.

5 Voyagers' Mission

In mid-1977 two probes, Voyager 1 and Voyager 2, were sent on a long tour. Voyager 2 was launched first, on August 20, 1977. Two weeks later, on September 5, 1977, Voyager 1 was launched on a faster, shorter trajectory. Both Voyagers were identical probes. Each was equipped with instruments to conduct up to 10 different experiments – magnetometers, plasma detectors, infrared and ultraviolet sensors, cosmic-ray receivers, and, an important part – color television cameras as well as radio communication systems. Both Voyagers were traveling too far from the Sun to use solar panels. Instead they were equipped with power sources, radioisotope thermoelectric generators, which convert the heat produced from the natural radioactive decay of plutonium to electricity. The probes were programmed to be oriented in a "safe" state, i.e. the instruments were directed away from the Sun, with the radio dish oriented constantly towards the Earth, ready to receive commands.

In the process of preparation of the Voyagers project, over 10 000 trajectories were initially considered allowing close flyby of the giant planets, Jupiter and its satellites, Saturn and its satellites, first of all Titan, the largest natural satellite in the Solar system; in addition, the chosen flight path for Voyager 2 also preserved the option to continue on to Uranus and Neptune.

During the flights of both Voyagers, the so-called "gravity assist" technique was used, first demonstrated via the Mariner 10 spacecraft on the Venus–Mercury mission in 1973–74. One can outline the perturbative maneuver, i.e., first, the strong change in the direction of velocity of a spacecraft and, second, the increase in heliocentric velocity of a spacecraft as a result of its flyby of a planet. The perturbative maneuver during the flight of *Voyager 2* was performed curiously successfully, thus revealing in general the practical possibilities of this powerful tool of celestial mechanics. As a result, the Voyager 2 flight time, for example, to Neptune was reduced from 30 to 12 years.

Thus two probes were launched, *Voyager 1* and *Voyager 2*, with the same aim, namely, to perform an Earth–Jupiter–Saturn–Uranus–Neptune tour with close flybys of each of those planets. Instead of waiting for the rare opportunity of a perfect configuration of the planets, the chosen path was based on essential modification of the trajectories after the flyby of the first planet, in a manner to have the final velocity directed towards the next planet. The most obvious solution, i.e. the realization of corrections of direction of the probes by means of jet propulsion, was excluded for a simple reason, namely, the required amount of fuel for realization of all three maneuvers would have been extremely large. However, at the flybys of the first three planets – Jupiter, Saturn and Uranus, very minor corrections should be enough; these could be performed by miniature bursts of jet propulsion during the intervals between flybys.

The main aim of both *Voyagers* was the transmission to the Earth of images of these planets and their satellites obtained at very small distances, as well as the investigation of physical conditions in the interplanetary medium along the path of the flight.

Up to Jupiter the flight of both Voyagers was successful, in a classical scheme, spending from 1.5 to 2 years on this voyage. Also successful for both spacecraft was the first perturbative maneuver in Jupiter's sphere of action, after which they continued their flight in new trajectories in the direction of Saturn.

Fig. 15.5 gives an image of the trajectories of flight of both *Voyagers* as well as of the configuration of all four planets at the moment of their meetings with the spacecraft. Because of the non-monotonous changes in the trajectories after each encounter with a planet, the complete trajectories from Earth to Neptune cannot be represented by a smooth curve – an ellipse or a parabola has to be drawn.

Voyager 1 made its close encounter with Jupiter on March 5, 1979, and Voyager 2 followed with its closest approach occurring on July 9, 1979. The first spacecraft, Voyager 1, flew within 206 700 km of the planet's clouds, and Voyager 2 passed within 570 000 km. The trajectory of Voyager 1 around Jupiter and its transit near all Galilean satellites is shown in Fig. 15.6. The passages by Io, Ganymede and Callisto were realized at quite near distances.

Although astronomers had studied Jupiter for centuries, they were surprised by many of Voyager's discoveries, which we will mention in Section 7.

6   The Grand Tour of Voyager 2

Voyager 2 reached Saturn on August 25, 1981, i.e. four years after its launch from the Earth, and, successfully realizing the perturbative maneuver in Saturn's sphere of action, it took the route towards Uranus. As for Voyager 1, though it reached Saturn rather earlier, on November 12, 1980, its perturbative maneuver was not successful, and hence its further flight became aimless.

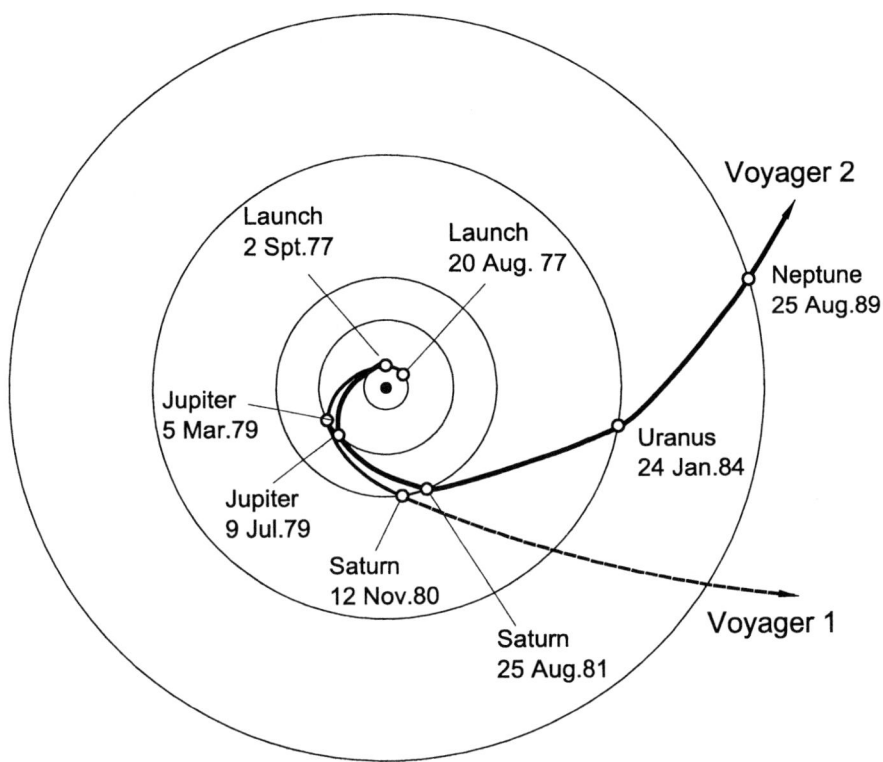

Fig. 15.5. *A heliocentric view of the Voyagers' trajectories. The perturbative maneuver was realized in the case of Voyager 2 in the spheres of action of Jupiter, Saturn and Uranus, after which Voyager 2 reached Neptune; in the case of Voyager 1 the maneuver was performed only once near Jupiter.*

Voyager 2 reached Uranus on January 24, 1986, i.e. nearly 8.5 years after its launch, and here, near Uranus, the third and last perturbative maneuver was successfully realized. Three and a half years later *Voyager 2* finally reached Neptune on August 25, 1989, i.e. nearly 12 years after its launch from the Earth.

The brilliantly performed tour of *Voyager 2* was remarkable not only for the fact that three properly realized perturbation maneuvers were executed but also for the perfect functioning of all its on-board facilities, and first of all, the system of stabilization and orientation. The requirements posed for *Voyager* grew with distance. Commands, traveling at the speed of light, reach the spacecraft with increasing delays; the correction of a new error can take too long for it to be performed successfully. E.g. close to Jupiter, the

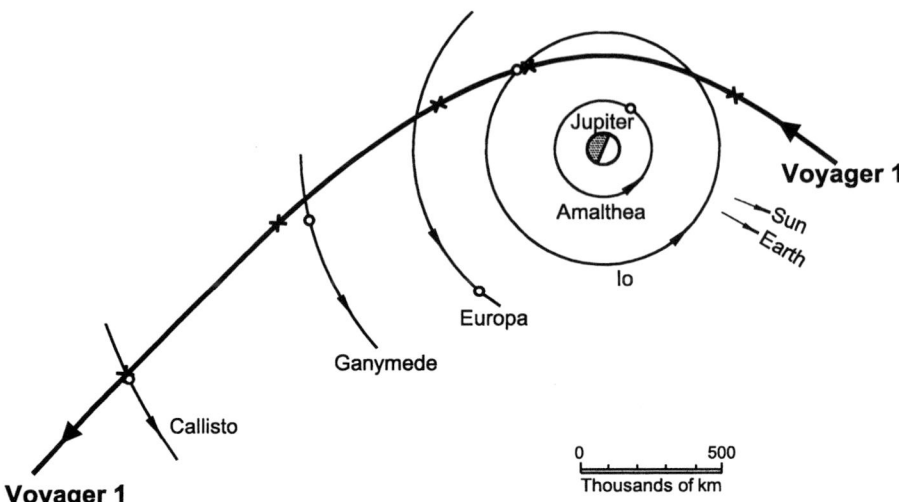

Fig. 15.6. *The trajectory of Voyager 1 around Jupiter with a transit near the small satellite Amalthea and all four Galilean satellites, Io, Europa, Ganymede and Callisto. In the case of Io, Ganymede and Callisto the Voyager's passage was realized at very near distances.*

Voyagers were less than about one light-hour away, but by the time *Voyager 2* had reached Neptune, that number was increased to four light-hours.

In the first solo planetary flyby, Voyager 2 made its closest approach to Uranus within 81 500 kilometers of the planet's cloud tops. Uranus is distinguished by the fact that it is tipped on its side. Its unusual position is thought to be the result of a collision with a planet-size body in the Solar system's history. Radiation belts at Uranus were found to be of an intensity similar to those at Saturn.

Voyager 2 found 10 new moons around Uranus, bringing their total number to 15 instead of 5 known (the present number is 20). Most of the new moons are small, with the largest measuring about 150 km down to the smallest at about 30–50 kilometers.

When Voyager 2 flew within 5000 km of Neptune on 25 August, 1989, three and a half years after leaving Uranus, the planet was the most distant member of the Solar system from the Sun. Even though Neptune receives only three percent as much sunlight as Jupiter does, it remains a dynamic planet and surprisingly showed several large, dark spots reminiscent of Jupiter's hurricane-like storms.

The strongest winds in any planet were measured on Neptune – up to 2000 kilometers per hour.

Triton, the largest of the moons of Neptune, was shown to be not only the most intriguing satellite of the Neptunian system, but one of the most interesting in the Solar system. An extremely thin atmosphere extends about 800 km above Titan's surface.

Six new moons were found around Neptune during Voyager 2's voyage, all small and close to Neptune's equatorial plane. Voyager 2 also solved many of the questions about Neptune's rings.

Both Voyagers capabilities in fact were much greater than expected when they left the Earth: their two-year planned mission became four, and their five-year lifetimes stretched to 12 years and more.

Eventually, between them, the Voyagers would explore all the giant outer planets of the Solar system, 48 of their moons, and unique systems of rings, atmosphere conditions, magnetic fields, radiation belts etc.

In 1997, Voyager 1 and Voyager 2 marked their 20th anniversary in space, and Pioneers their 25th, as they continue their lonely journeys. *Voyager 2*, after crossing Pluto's orbit, must leave the Solar system, moving sometime within the heliopause, where the strength of the magnetic field of the Sun is of the same order as that of the Galaxy. *Voyager 2* is planned to continue its measurements and transmit data to the Earth up to 2017, i.e. up to the exhaustion of its energy resources. Later, it will continue its uncontrolled flight within the interstellar medium towards the neighboring stars.

In Fig. 15.7, the trajectories of both Pioneers and both Voyagers up to the outer boundaries of Solar system are shown. Pioneer 11 is moving the most directly towards the boundary of the heliopause, but the Voyagers are moving faster. Pioneer 10 is leaving the Solar system towards Taurus, Pioneer 11 towards Scutum, Voyager 1 towards Ophiuchus, and Voyager 2 towards Telescopium.

Pioneer 11 is headed directly toward the heliopause.

Had the Voyagers mission ended after the Jupiter and Saturn flybys alone, the results obtained would have been enough to stand as historic. In fact, over the years, the Voyagers have sent to Earth information that has revolutionized the science of planetary astronomy in general.

7   The Avalanche of Discoveries of Voyager 2

We mentioned some of the discoveries made by the Voyagers. Let us devote some more attention to some of them, which posed a inconceivable number of unexpected problems and puzzles for planetary astronomers, geologists, volcanologists, specialists on planetary atmospheres, magnetospheres, etc. In fact, with the Voyagers a new epoch of studying the planets and their satellites began.

Thus the first encounter of Voyager 2 took place with Jupiter. The flight

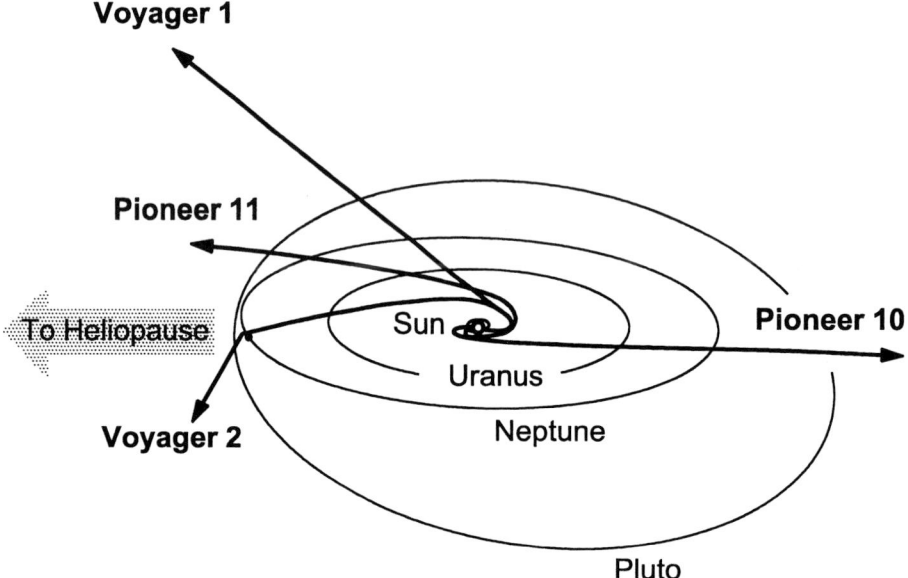

Fig. 15.7. *Paths of the two Pioneers and two Voyagers up to the outer boundaries of the Solar system. Pioneer 10 is leaving the Solar system in the opposite direction as compared with the directions of the other three probes.*

trajectory and the time of the launch were chosen in such a way as to have the nearest flyby of Jupiter, as well to trace the largest possible number of satellites, first of all the Galilean ones; the trajectory of this flight is shown in Fig. 15.8. The first flyby, from a considerable distance, was that of Amalthea; however, for unknown reasons it was not captured by the cameras.

The first passage at a close distance, 40 000 km, was realized with Io, the nearest to Jupiter among the Galilean satellites. The discovery of active volcanism on the satellite Io was apparently the greatest unexpected discovery, since it was the first time that active volcanism had been observed on another body in the Solar system. Nearly ten active volcanoes were found erupting rocks up to the altitude of 480 km. The speeds of the explosions were also measured, up to 440 $m\,s^{-1}$, and lava streams were detected around the volcanoes. Io's volcanoes are explained via the heating of the satellite by tidal pumping of the central planet, Jupiter, as well as by the nearest large satellites, Europa and Ganymede.

In fact, Voyager fixed for the first time an active volcano on other celestial bodies – on a planet's satellite. So, the Earth is not only celestial body-planet with volcano activity.

Voyager 2's encounter with Europa, though from a large distance, 241 000 km, turned to be fruitful; for the first time the exact diameter of this Jovian satellite was determined. Its next meeting was with Ganymede and Callisto, both on rather near distances. Both satellites are nearly of one and the same size, ∼ 2 500 km in radius, both are covered by ice and rocks and possess a complex structure of surfaces.

Voyager 2 photographed Jupiter's ring from a distance of 1.5 million km, which appeared very thin and narrow, completely different from Saturn's well-known ring. Jupiter's ring glows from the eclipsed sun's strongly refracted twilight. Jupiter's Great Red Spot was revealed as a complex storm moving counterclockwise.

The next encounter of Voyager 2 was with Saturn on November 8, 1980, four days before Voyager 1 reached Saturn and passed it at 124 200 km distance, i.e. at about one Saturnian diameter distance, crossing the trajectories of all satellites. The schematic diagram of the system of Saturn's six satellites along with the trajectories of both Voyagers is shown in Fig. 15.8. During this flight high-resolution color images of all six satellites were obtained with various cameras and at various distances. Titan was imaged from 4 000 km distance, Tethus and Enceladus from more than 200 000 km, while the probes passed reasonably close to Mimas, Dione and Rhea.

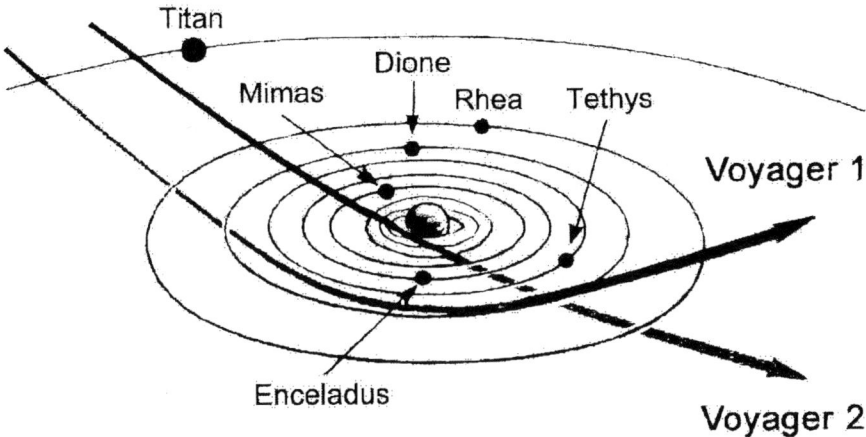

Fig. 15.8. *A schematic diagram of the trajectories of Voyager 1 and Voyager 2 through Saturn's six satellites.*

An enormous number of discoveries were made during the flights of the

Voyagers. First of all, this includes the discovery of over 1000 discrete rings of Saturn, of 100–150 m thickness each.

A giant crater was discovered on Saturn's nearest satellite Mimas, making an enormous circular scar 145 km across – more than one-third the satellite's diameter (400 km); this crater has a rim 9 km above its floor (Plate XI).

Actually a new class of satellite was discovered. Voyager 2 recorded high-resolution images of Saturn's moons Dione, Rhea and Tethus, with features in some cases as small as 1.5–2 km. Of special attention is Saturn's largest satellite Titan (see the next section). Dione's surface is similar to that of the Moon, with thousands of craters, the largest nearly 100 km in diameter, and a well-developed central peak. A similar structure was observed for the other satellite – Rhea.

Nearly four and a half years after leaving Saturn, on January 24, 1986, Voyager 2 arrived at Herschel's planet, Uranus. This was the first spacecraft to have visited Uranus; its distance from the Earth means that radio signals take in 2 h 45 min. to reach the Earth. Uranus is distinguished by the fact that it is tipped on its side.

Voyager 2 first encounter was with Uranus' nearest and smallest satellite Miranda, entirely composed of stone according to the obtained images; an extremely impressive Voyager 2 image of this terribly disfigured body-satellite is shown in Plate V.

Twelve minutes later Voyager 2 passed Uranus 81 000 km above the cloud upper layers. The early results, indicating that the Uranus atmosphere is composed of hydrogen and methane, with helium nearly 15%, i.e. comparable with those of the Sun, Jupiter, and Saturn, were confirmed. Uranus' atmosphere is very hot, about 480° C, as of Venus.

For the first time, Voyager's cameras fixed clear-cut images of all nine rings, discovering the faint 10th ring. The nine rings were discovered earlier, on March 10, 1977, via stellar occultations at ground-based astronomical observations. The diameter of the outer ring is 100 000 km, twice the length of Uranus' diameter.

In Plate IX, an extremely impressive image of Uranus with its first four rings, obtained in August 14, 1994, with the help of the Hubble Space Telescope, is shown.

Now about Uranus' five satellites. Unlike the Voyagers' visits to Jupiter and Saturn, when encounters with the satellites were spread out over several days, everything happened very quickly at Uranus. All Uranus' known – at that period – five satellites, Miranda, Ariel, Umbriel, Titania and, farthest, Oberon, judging by Voyager's images, are covered by ice of various thickness, which is why all these satellites are called "Ice Worlds".

Voyager 2 also discovered ten new small satellites, earlier unknown, with diameters 30–100 km, rotating in the immediate vicinity of Uranus, in Miranda's orbit.

When Voyager 2 passed within 4900 km of Neptune on August 25, 1989, three and a half years after leaving Uranus, Neptune was the most distant planet from the Sun. A dynamic deep blue atmosphere was fixed on the whole surface of this planet, with small white clouds of methane, and a large spot of about the size of the Earth and similar to Jupiter's Great Red Spot.

Six new moons were found around Neptune, all small and close to Neptune's equatorial plane. Voyager 2 also solved many of the questions about Neptune's four rings. Thus, at least four planets of the Solar system possess rings – Jupiter, Saturn, Uranus and Neptune.

Voyager 2 arrived at Neptune via a flyby trajectory shown in Fig. 15.9.

Voyager 2 passed 40 000 km from Neptune's largest satellite, Triton (Fig. 15.9), which is larger than the Earth's Moon and the only satellite in the Solar system with retrograde orbital motion around the central planet. The panoramic pictures taken by Voyager 2's camera are extremely impressive; the terrain of this satellite, fringed Triton, is geologically young with deformations of icy, frozen nitrogen. An extremely thin atmosphere expends about 800 km above Triton's surface.

The second satellite, Nereid, is 15 times farther from Neptune than Triton (that is why its location is not shown in Fig. 15.9). Voyager 2 passed Nereid at 4.7 million km, so that the spacecraft's images only revealed its size – about 340 km across, one and half times larger than known before. By the way, Nereid has the highest known eccentricity, $e = 0.75$, among all satellites of the Solar system, a value more usual for comets.

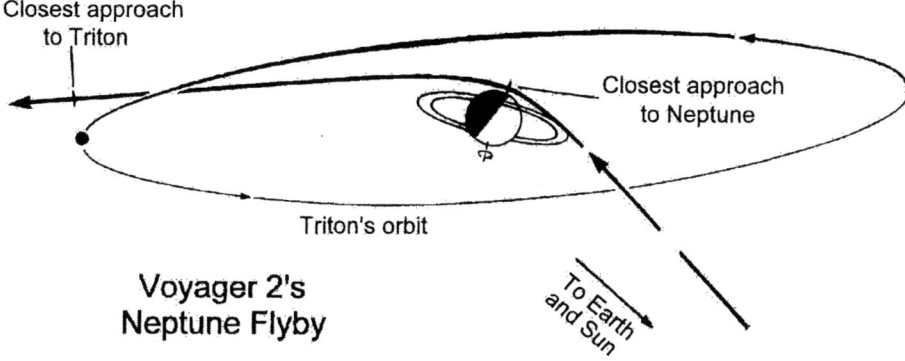

Fig. 15.9. *Voyager 2's flyby trajectory around Neptune and its transit near the satellite Triton.*

With the final image of Triton from 4.9 million km distance, Voyager 2 completed its impressive tour of the outer planets of the Solar system. If an optimistic schedule holds, Voyager 2 will continue to send back data for another 25 years, i.e. up to 2017, having journeyed to the edge of the Solar system and into interstellar space.

In Fig. 15.6, a panorama – view traces are shown for both Pioneers and both Voyagers, leaving the Solar system for ever.

At present, both Voyagers and both Pioneers are very far outside of the Solar system as they cruise toward interstellar space. Even if their cameras were active, they probably could not detect the presence of comets around them in this space, but precise tracking of their trajectories may reveal the combined gravitational perturbations of the distant local masses.

8   Magellan: Venus Radar Mapping

The atmosphere of Venus is completely opaque for optical and even infrared wavelengths, and this circumstance makes absolutely impossible the photography of its surface from the spacecraft orbiting around this enigmatic planet. The only possibility of the mapping of its surface is radar ranging in wavelengths for which its atmosphere is transparent, that is, in the centimeter wavelengths. The Magellan space probe has been destined just for this problem.

Magellan was launched on April 25, 1989, and via a long trajectory looping 1.5 times around the Sun, on August 10, 1990, i.e. in 15 months, arrived at Venus, in an elliptical polar 3.26-hour orbit around Venus. The orbit had $85°.5$ inclination to the planet's equator, varying in altitude between 294 and 8 450 km. For 37 minutes of each orbit, when the probe was closest to Venus, the spacecraft bounced short radio pulses off the planet's surface. The 12.6 cm wavelength beam penetrated through the thick mantle of clouds. On each circuit Magellan covered a strip of ground 20 to 25 km wide and some 15 000 km long. The resulting echoes from these long, narrow strips were recorded on board and transmitted to Earth, enabling radar-brightness maps of Venus' surface with a resolution 120 m across to be created. As the spacecraft passed overhead, a second radar system repeatedly determined the distance to the surface to an accuracy of 10 m, thus storing data for a global topographical map.

As a result, Venus' surface was mapped completely, including the discovery of mountain ranges, giant valleys of a few km in depth, 200–300 km in width and over 1000 km in length. Mountains with peaks up to 10 km high were detected as well. The resources of the Magellan probe are supposed to enable the mapping of Venus repeatedly for decades.

9   Galileo Probe onto Jupiter

The idea of the Galileo project – *Jupiter Orbiter Probe* – goes back to 1977, the year both Voyagers were launched.

Galileo consists of two parts, the *Orbiter* itself, a spacecraft for the realization of the flight from Earth to Jupiter, and the *Probe*, to enter Jupiter's atmosphere. The creation of both parts of this project, the Orbiter by Jet Propulsion Laboratory and the Probe by NASA's Ames Research Center, was completed in early 1989, and on October 18, 1989, Galileo was launched.

Galileo's mission consisted of two parts, first, the interplanetary flight Earth–Jupiter and, second, the separation of the Probe from Orbiter and its descent with the help of a parachute into Jupiter's atmosphere.

Galileo's highest degree original trajectory to Jupiter is shown in Fig. 15.10. After launch and passing into a heliocentric trajectory with a velocity of 39 $km\,s^{-1}$ – the third and the final magnitude, including also the component of Earth's circular velocity – 30 $km\,s^{-1}$ on its orbit around the Sun, both the Orbiter and the Probe, as one unit, continued their flight in the direction of Venus. After a gravitational maneuver in Venus' gravitation field, on February 10, 1990, Galileo continued its flight via an orbit that would cross the Earth's orbit relatively soon, and after nearly eight months, on December 8, 1990, found itself in the Earth's gravitational field – this was the first meeting with the Earth since launch.

Gaining the first (1) additional acceleration in the Earth's gravitational field, Galileo entered a new and larger orbit in the direction of the asteroid Gaspra, a 17-kilometer-long body. After a short meeting with this asteroid on December 29, 1991, during which its color images of it were obtained, Galileo continued its flight and in ten months realized its second (2) meeting with the Earth on December 8, 1992, exactly two years after the first (1) meeting, at the very same point as at its first meeting with the Earth (1). So, Galileo completed its about three-year cruise through the inner Solar system.

At the second encounter with the Earth (2), Galileo was accelerated in such a manner that its flight to Jupiter via an extremely large ellipse would cross the asteroid belt nearly at the halfway point. However, before reaching Jupiter, Galileo had an encounter with a second asteroid, Ida, at the distance of about 2 400 km.

It is remarkable that Galileo's Earth–Jupiter flight occurred via a rather complex and at the same time the highest degree interesting trajectory of interplanetary flight, namely, with three gravitation maneuvers, two around the Earth and one around Venus, and with close flybys of two asteroids. This trajectory has nothing in common with Hohmann's trajectory – a pure ellipse touching two circular–elliptic orbits of two planets. But this Earth–Jupiter trajectory, having nothing in common with the trajectories of the

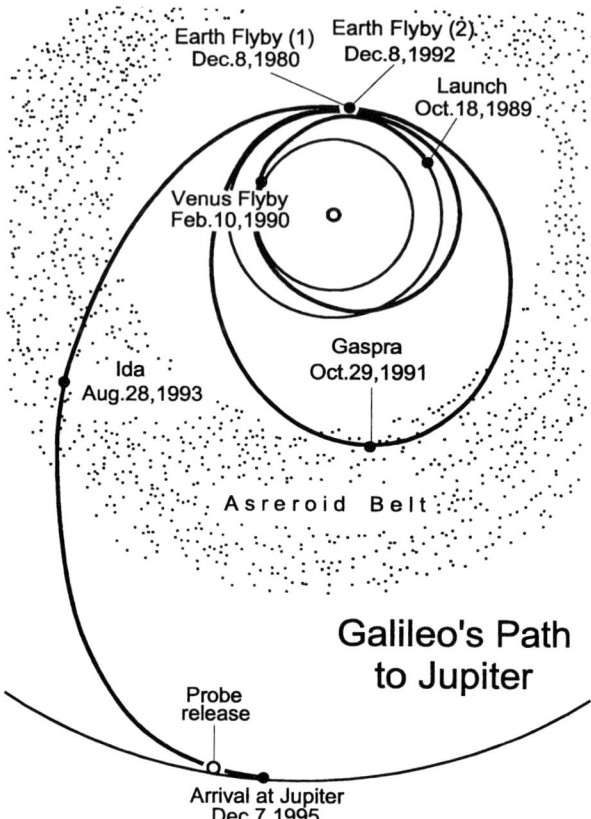

Fig. 15.10. *The trajectory of Galileo's Earth–Jupiter six-year cruise through the inner solar system with nearby transits of two asteroids, Gaspra and Ida. The route around the Earth and Venus was needed to gain enough velocity in the gravitational fields of these planets for the cruise to Jupiter.*

interplanetary flights examined above, particularly in Chapter XII, was the most economic path from the energetic point of view.

Of course, Hohmann's orbits preserve their principal significance. However, side by side with this, Galileo's trajectory should be taken as a model of a new approach as well, namely, when the remote flights are concerned, what turns out to be important in the search for trajectories satisfying at least two demands:

– To realize transits of spacecraft past massive planets at short distances with the aim of acquiring of an additional impulse in the gravitational fields of the planets.

– To pass near many celestial bodies of all types, massive and not massive, with the aim of studying them.

Thus, after six-years duration and more than 3.8 billion km an a route

Earth (0) – Venus – Earth (1) – Gaspra – Earth (2) – Ida, the spacecraft Galileo, with two components, the Orbiter and the Probe, on December 7, 1995, reached its ultimate objective, Jupiter, the largest planet of the Solar system.

Galileo harvested secrets of many worlds.

The first close-up look was at the asteroid, Gaspra. Galileo imaged this asteroid of size 9×18 km from a distance of 5300 km via exceptionally clear and high-resolution color snapshots. The structure of the surface of this asteroid is of extreme interest.

Later, the careful processing of Galileo's observations brought to light an interesting discovery, namely, that Gaspra has its own magnetic field. The observed fact of this was the abrupt change in the orientation of the interplanetary magnetic field recorded when Galileo and Gaspra were closest. As it turned out, Gaspra has a magnetic field strong enough to create a large, elongated bubble in the Solar wind.

The discovery of the existence of a magnetic field in Gaspra, an asteroid-sizes celestial body, is an event of extreme importance bearing in mind that magnetic fields are not an intrinsic property of all Solar-system objects. For example, neither Venus nor the Moon has one; Mars' is barely detectable, if it exists at all.

However, certain meteorites, particularly of iron, can be distantly magnetized. So, Gaspra's putative field would be a good match for scaled-up versions of these.

The history of Gaspra's magnetic field should be included in the scientific problems raised by Galileo's extremely fruitful mission.

Galileo next encountered a second asteroid, Ida, also elongated, of size 56×28 km. Again, high resolution images were obtained from a distance of 10500 km. The remarkable discovery was that this asteroid is not alone: mimicking the major planets, this minor planet–asteroid, as it turned out, has a kilometer-wide satellite (Plate VI). The essence of this discovery is that asteroids also can have companions – satellites, although it still is not clear to what degree the gravitational field of such a minor body as an asteroid can be enough in order to keep any body, even of meteorite mass.

In all cases, the impression from these images, both of Gaspra and Ida, which are exceptionally distinct and incredibly clear, is strong.

Five months before reaching Jupiter, the Probe was released from the Orbiter and was then free to continue its journey separately until its entry into Jupiter's atmosphere. On December 7, 1996, when the Orbiter reached its closest point to the planet, 215000 km above Jupiter's atmosphere, the 340 kg Probe, influenced by 300 times Earth's gravity and high temperature, decelerated from 47 $km\,s^{-1}$ velocity and began its fiery entry into Jupiter's atmosphere. Fig. 15.11 and Table 15.2 illustrate the dynamical and physical characteristics of this final stage of the flight of the Probe.

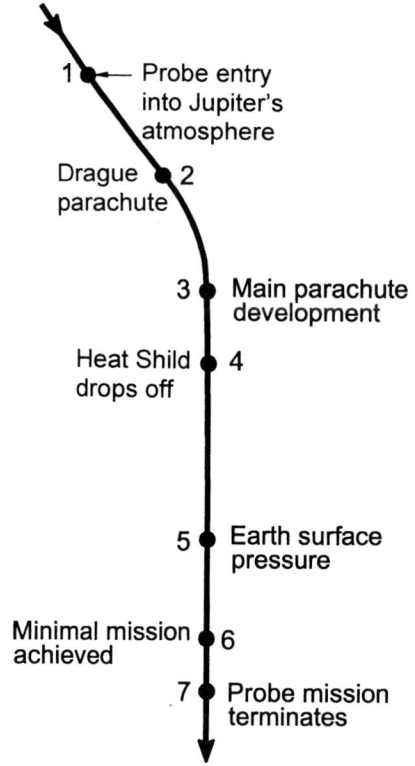

1 — Probe entry into Jupiter's atmosphere

Drague parachute — 2

3 — Main parachute development

Heat Shild drops off — 4

5 — Earth surface pressure

Minimal mission achieved — 6

7 — Probe mission terminates

Fig. 15.11. *The sequence of stages of the Galileo Probe entry into Jupiter's atmosphere, before the parachute opens to let it sink leisurely through the cloud decks.*

Decelerating rapidly, within two minutes Probe's parachute was deployed. A half-minute later the heat shield fell away, and the Orbiter started to captured the data transmitted from the Probe's six instruments.

The duration of transmission was 75 minutes, during which the Probe descended ~ 200 km when the atmospheric pressure jumped from 0.1 to 30 bars, after which, in nearly an hour, the onboard accelerometers were stopped. During this period extremely valuable data on the atmospheric temperature, pressure, composition, etc. was transmitted.

T a b l e   15.2

The main physical characteristics of Jupiter's atmosphere as transmitted by the Galileo Probe during its entry into the atmosphere

S t a g e s	1	2	3	4	5	6	7
Time, min	0	1.88	1.92	2.05	8.33	38	60 to 78
Altitude, km	450	50	48	45	0	−92	−134 to −163
Pressure, bar	$5\,10^{-8}$	0.08	0,09	0.10	1.0	10	20 to 30
Temperature, C	-8	−160	−161	−162	−107	+63	+140 to +190

As follows from Table 15.2, even in the outer layers of Jupiter's atmosphere, within a depth of ∼200 km, negligible as compared with Jupiter's diameter, 142 600 km, the physical and dynamical conditions are striking: high pressures, up to 30 bars, and high temperatures, up to $+200°C$.

Galileo Orbiter started functioning on December 7, 1997. In the next two years it performed 15 close flybys of the Galilean satellites Io, Europa, Ganymede and Callisto. During that period it transmitted information of extreme importance on Jupiter and its satellites, particularly on Io and Europa. In the first case, Io, the Voyagers' discovery is confirmed, namely, the existence of a unique celestial object – a small planet with an extremely volcanic activity. Galileo Orbiter survived its closest brush with Io on February 22, 2000. This flyby, the fourth and last, at the smallest distance from Io, only 198 km (!), concentrated on studies of volcanoes and plasma tori generated along the Io's orbit by volcanic eject.

As for Europa, its views were extremely imaginative and intriguing, suggesting the existence of a subsurface ocean of a liquid water (?) at 100 km depth, half of it, 50 km, in the form of ice (Plate X). Two examples from Galileo's quick-look orbits around Io (right) and Callisto are shown in Fig. 15.12.

Galileo's exceptional scientific efficiency was a result of successful engineering and technological solutions demonstrated by the creators of both the Orbiter and the Probe.

10   Pluto–Kuiper Express

Pluto, the smallest planet of the Solar system – it is only twice as large as the Earth's Moon – has remained enigmatic since its discovery in 1930. Pluto is the only planet in the Solar system not yet viewed close-up by spacecraft, and given the great distance and tiny size, the study of this planet continues to challenge planetary astronomers. Most of the information, key questions, about Pluto has been obtained by astronomers since the late 1970s, more precisely in 1978 when Charon, Pluto's moon, was discovered, but more

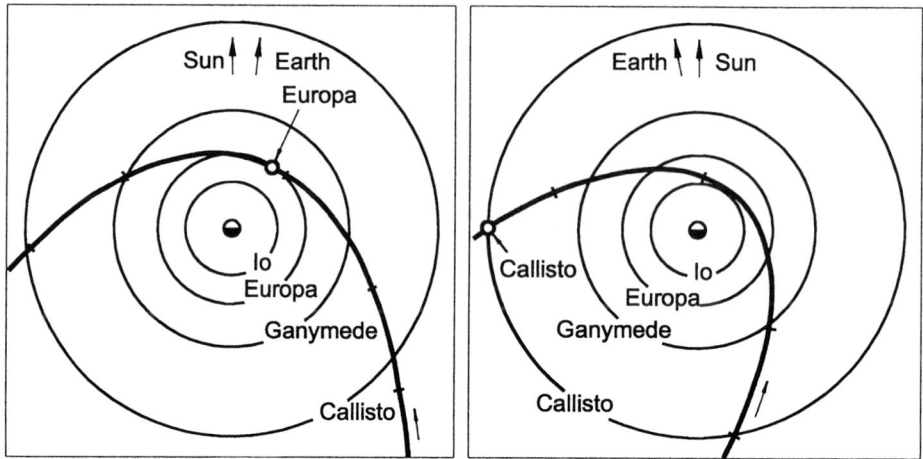

Fig. 15.12. *Examples of Galileo's quick-look orbits around Galilean satellites Io (right) and Callisto.*

effectively due to the clearest image of the planet and its satellite obtained by the Hubble Space Telescope on February 21, 1994, when the planet was 4.4 billion km or 30 a.u. from the Earth – a distance from which radio signals took 4 hours 10 minutes to reach the Earth. The Hubble observations showed that Charon is bluer than Pluto, which means that they have different surface composition and structure. A bright highlight on Pluto suggests that it has a smoothly reflecting surface. A bright area parallel to the equator on Pluto was discovered.

Usually astronomers were acquainted that planets may have satellites, and that the masses of the satellites of the planets of the Solar system are significantly smaller as compared with the mass of the planet itself. In the case of Charon the situation is something unexpected: Charon is only half of the diameter of Pluto; for comparison, Earth's Moon is one-quarter the diameter of the Earth. As a result, for the first time, an idea has arisen, namely, that in the "round dance" of celestial bodies there may be *binary planets* in analogy to binary stars. Thus the Pluto–Charon system resembles a binary planet.

Pluto's parameters have led planetary astronomers to debate whether Pluto has to be considered a major planet or a minor one, or both, a regular planet, or a trans–Neptunian icy object, like those of a few hundred km in size beyond Neptune's orbit, over 300 of which have already been discovered, most of them a few hundred kilometers across.

This problem acquired a special significance against the background of the latest progress in latest the sphere of so-called *Extrasolar Planets*, i.e. the planets discovered around the nearby bright stars; this problem is discussed above, in Section 18 of Chapter XI. In particularly, perhaps in the distant future, progress can be expected when the discovery of satellites around these extrasolar planets turned out to be possible. Pluto–Charon type pairs should be searched for in the extrasolar planetary systems.

The physical and geometrical characteristics of the Pluto–Charon system, in view of their peculiarity are presented in Table 15.3 in comparison with the Earth–Moon system.

As follows from the last two rows of Table 15.3, i.e. from the mass and diameter ratios, there is enough basis for the **binary planet** concept. Moreover, Charon is the only natural satellite in the Solar system with an extremely large inclination of orbital plane of revolution, 65°, to the plane of Pluto's orbit around the Sun (Fig. 15.13).

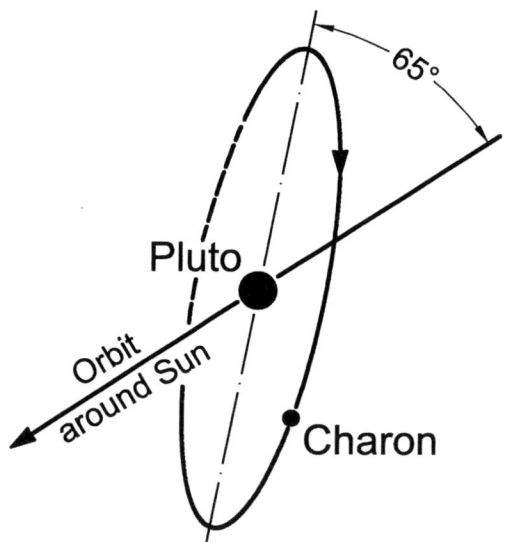

Fig. 15.13. *Space orientation of Pluto's satellite Charon's plane of orbit to Pluto's orbital plane: the angle between them is 65°.*

Now about the Pluto-Kuiper Express Mission. The purpose of this mission is to conduct the first reconnaissance of Pluto and Charon with a low mass flyby spacecraft. This mission will complete the exploration of this last unvisited planet and address fundamental questions about the origin of the Solar system, since the cold Pluto is an oddity among the planets and may provide considerable information about the early periods of planetary evolution.

T a b l e   15.3

Comparative characteristics of Earth–Moon and Pluto–Charon systems

	Earth–Moon System	Pluto–Charon System
Separation	384 000 km	20 000 km
Revolution period	27.32 days	6.39 days
P l a n e t		
Diameter	12 750 km	∼3000 km
Mass (Earth = 1)	1.0	0.0023
Density (water = 1)	5.5	∼1
Rotational period	1 day	6.39 day
S a t e l l i t e		
Diameter	3480 km	∼1300 km
Mass (Earth = 1)	0.01215	0.0002
Density (water = 1)	3.3	∼1
Rotational period	27.32 days	6.39 days
Planet / Satellite mass ratio	81.3	∼10
Planet / Satellite diameter ratio	3.7	2.3

In Fig. 15.14 the computed trajectory of the Pluto–Kuiper Express is shown. According to the preliminary program, the Pluto–Kuiper Express will be launched on December 18, 2004, to arrive the Pluto in about 8 to 12 years, depending, apparently, on the efficiency of acceleration in Jupiter's gravitational field. In one variant, the expected arrival date at Pluto-Charon is December, 2012. The mission can be extended to include possible encounters with one or more objects within the Kuiper Belt.

Beyond Pluto lies the recently discovered Edgeworth–Kuiper Disk of *Ice Dwarfs* or minor planets. Its history may be connected with the formation of Earth's atmosphere and even biosphere. It is difficult to overestimate the extremely scientific importance of the Pluto-Kuiper Express project.

11   Cassini: En Route to Titan

Titan, Saturn's largest satellite, as mentioned above, is considered one of the most intriguing bodies of the Solar system. **Titan's mystery** is expected to unriddle with the help of the Cassini Mission and Huygens Probe. The Huygens Probe is designed to survive a landing on either solid ground or, most probably, a hydrocarbon sea, performing an analysis of the surface and atmosphere.

On October 15, 1997, the *Cassini* spacecraft was launched towards Saturn with 12 scientific instruments on board the *Huygens Probe*. Cassini will

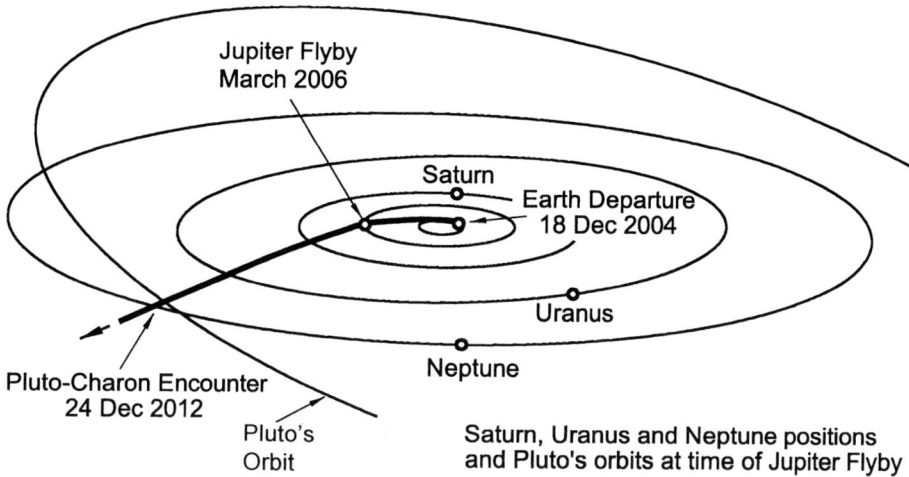

Fig. 15.14. *The favorable trajectory for the Pluto-Kuiper Express for the 2004 launch with a Jupiter gravity assist in 2006 and with expected arrival at the Pluto-Charon system in 2012.*

become the first to orbit Saturn upon arrival in 2004. Then, Cassini will deliver the Huygens Probe to Saturn's largest moon Titan, the largest satellite in the Solar system after Jupiter's Ganymede. Cassini will transmit information about the atmosphere and surface of this satellite – the results of the observations of the Huygens Probe.

Titan is a mysterious celestial body in many respects. Indeed, the existence of a dense atmosphere for Titan is confirmed. It contains methane, associated with more complex organic molecules. Moreover, there are reasons to assume that Titan's surface represents an ocean of liquid methane. According to the Mauna Kea 1999 infrared observations the possibility of methane rains on Titan is not excluded either.

The radio ranging of Titan by Voyager 1 indicated that its atmosphere is much denser than was expected before. Nitrogen is its principal constituent (99%), the atmospheric pressure at ground level is over 1.5 times greater than of Earth, and the temperature is −181° C, hence, many gases, including methane, should be in liquid state. Also, the temperature first decreases with altitude from the surface, then warms again. The presence of hydrogen cyanide is confirmed as well, and the existence of liquid nitrogen, as well as of numerous other gases, both simple and molecular, is expected. The opaque hazes are laced with at least 10 organic components, including propane, acetylene and hydrogen cyanide, as well as argon and molecular hydrogen.

Voyager 1 discovered a thick disk of neutral hydrogen atoms surrounding Saturn and extending from its satellite Rhea's orbit up to Titan and beyond.

The flight trajectory of Cassini is shown in Fig. 15.15. The main idea is based again on an already well-tested means, the gravity-assist technique. This was used four times during the flight, twice to Venus in 1998 and 1999, one to Earth in 1999 and the final one to Jupiter on the last day of 2000. Among those, the Earth flyby is estimated as especially important. During its tour through the asteroid belt, a close flyby of an asteroid, Masursky, is envisaged as well.

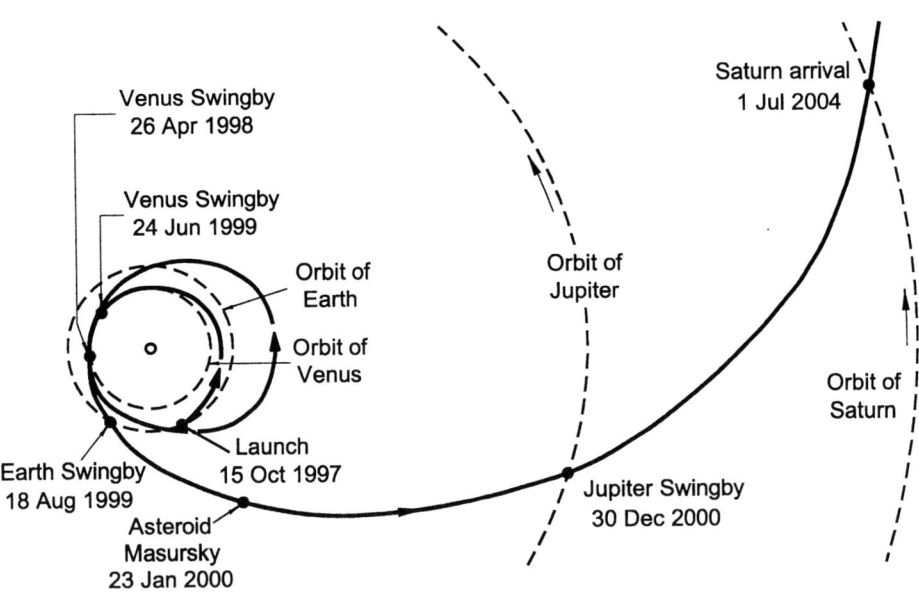

Fig. 15.15. *Cassini's interplanetary Earth–Saturn trajectory. After the launch from the Earth on October 15, 1997, Cassini tested four gravity assists during the cruise to Earth, Venus and Jupiter.*

Cassini's flyby of Jupiter is quite a distant one, with the closest approach at about 140 Jupiter's radii, i.e. 14 million km.

The time schedule of the Cassini flight is of a definite interest, and below we represent the most important events of the mission.

L a u n c h	– October 15, 1997
Venus 1 flyby	– April 26, 1998
Venus 2 flyby	– June 24, 1999
Earth swingby	– August 18, 1999
Flyby of asteroid Masursky	– January 23, 2000
Jupiter flyby	– December 30, 2000
Scientific observations begin	– January 1, 2004
Maneuver to target the Probe to Titan	– September 12, 2004
Huygens Probe separation from spacecraft to go to Titan	– November 8, 2004
Saturn orbit insertion	– July 1, 2004
Huygens encounter with Titan	– January 14, 2005
End of nominal mission, after four-year tour	– June 30, 2008
Launch mass: Cassini Orbiter with Probe	– 5235 kg

In January 2000 the Cassini spacecraft, on the route to the flyby of Jupiter and its arrival at Saturn in 2004, passed the minor planet–asteroid Masursky, just as in the schedule provided above, and took a series of snapshots from 1.6 million km away. This asteroid is about 15–20 km across and probably not of stone.

In the last days of December, 1999, at the end of the 20th century, the Galileo Orbiter, which passed on September 8, 1999, falls inward to make a close pass by the planet on December 29. One day later, the Cassini Probe left for Saturn, giving a final close-up stereo view of the Jovian system as shown in Fig. 15.16. Cassini is studying Jupiter from December 2000 to March 2001, even through its closest approach will be only 9.8 million km. But this is still close enough for Jupiter's enormous gravity to increase the Probe's velocity by 2.2 km per second. In fact, Jupiter's *gravitational sphere of influence* is 48 million km, more than the distance between the orbits of Earth and Venus. During the double encounter Galileo will be inside Jupiter's magnetosphere (465 000 km from the planet), Cassini outside, and the pair will study how changes in the Solar wind affect conditions in near-Jovian space. Cassini will then swing on toward Saturn, arriving there in mid-2004.

As to Galileo Orbiter, it can continue orbiting around Jupiter for a long time. Perhaps arrangement will be made for Galileo to crash into Jupiter's atmosphere or directly on to Europa.

Total cost of Cassini–Huygens mission was $3.5 billion, making its up to 2000, the most expensive scientific space project.

12   The Road to Halley

Nearly 69 km per second – this was the speed of the spacecraft to pass safely through comet Halley. In conditions of such unprecedented velocity three probes, Giotto, Vega 1 and Vega 2, observed comet Halley at the period of its appearance in the neighborhood of the Sun in early 1986.

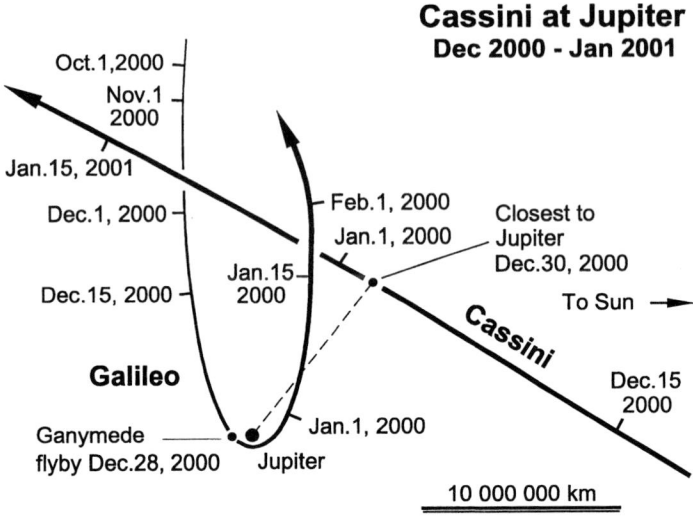

Fig. 15.16. *Path of the Cassini spacecraft through Jupiter's zone in the period from 15 December, 2000, up to 15 January, 2001. Cassini's closest distance from Jupiter, 9.8 million kilometers, will be on 30 December, 2000. Later in 2000 and in early 2001 the Cassini spacecraft joins the Galileo's Orbiter for joint studies.*

Interplanetary spacecraft usually encounter their targets at much slower speeds, but Halley courses through the Solar system in a retrograde direction, with motion opposite to that of the nine planets and most of their natural satellites. Also, the comet's orbital plane is inclined 18° to the Earth's orbital plane, as shown in Fig. 15.17. Consequently, all probes chose to meet the comet when it passed downward through this plane. At perihelion, February 9, 1986, Halley was on the far side of the Sun as seen from the Earth and relatively close to Venus.

In all, five probes were dispatched to visit Halley: Giotto, Vega 1 and Vega 2, and Sakigake and Suisei. The main goal of the probes – to photograph the comet's head – nucleus, and its huge halo of hydrogen in ultraviolet.

Giotto's trajectory, which is quite original, is shown in Fig. 15.18 (R. Farquher, D. Dunham). It is a hexahedral closed curve with six rhythmic knots, in one of which the flyby of comet Halley's head was realized on March 14, 1986, eight and a half months after the launch on July 3, 1985. The Giotto probe obtained the first detailed images of the nucleus of Halley's comet. In doing so, the comet's dust particles sandblasted the probe at about 70 km per second, severely damaging it.

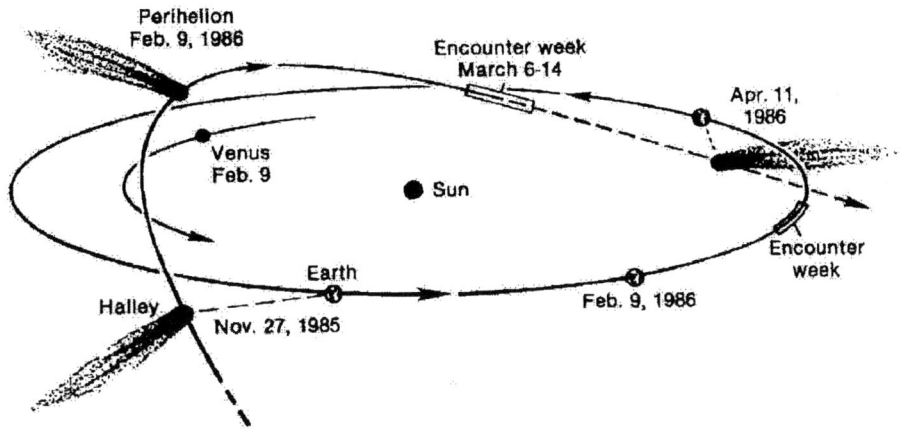

Fig. 15.17. *The trajectory of Halley's comet near the Sun. The comet travels in a retrograde orbit inclined at 18° to the ecliptic. At perihelion on February 9, 1986, Halley was on the far side of the Sun as seen from the Earth.*

The flyby distance from Halley's nucleus was smallest in the case of Giotto, 480 km. Both Vega probes were traveling faster, 80 km per second relative to the comet's nucleus, but they were also staying farther away, at about 9500 km for Vega 1 and 3200 km for Vega 2. For Suisei and Sakigake the distances were 200 000 km and over 3 million km, consequently. All probes met the comet when they passed downward through Earth's orbital plane in early March, 1986. For the first time, the size of the comet's nucleus, about 20 km, was revealed.

13 Lunar Prospector: Ice on the Moon

The spacecraft Lunar Prospector was launched on January 6, 1998 with an aim to survey the Moon's resources and search for water at its poles. Traveling by the Earth–Moon standard trajectory, it was captured by the Moon to an elliptical orbit around the poles as shown in Fig. 15.19. After repeated firings over the next four and a half days (105 hours), the Prospector entered a two-hour low circular orbit just 100 km above the lunar landscape.

The spacecraft carries instruments to measure alpha particles, neutrons, electrons, and gamma rays emanating from the surface. Energetic neutrons are spawned when cosmic-ray particles smash into atoms on the Moon's surface. The neutrons then slow down gradually as they ricochet off the nuclei of other atoms.

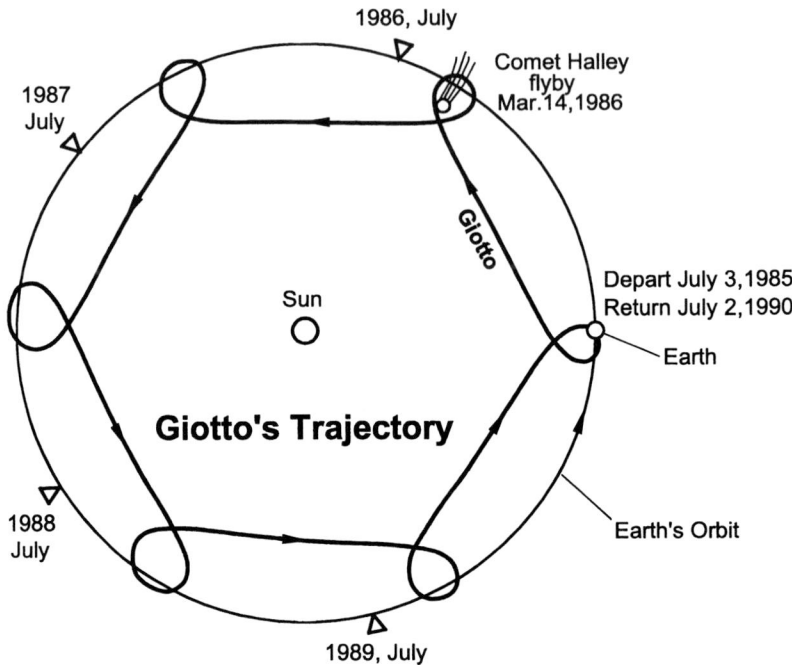

Fig. 15.18. *The trajectory of the Giotto probe, launched from the Earth on July 3, 1985, with a flyby of Halley's head on May 14, 1986, and the return after exactly five years and exactly on the same point of the Earth's orbit around the Sun.*

At each of its numerous polar passages the Prospector detected intermediate energy neutrons dropping in number by a few percent, 3.4% at the north pole, and 2.2% at the south, proving the excess of hydrogen at the Lunar poles.

Water is the likeliest compound to contain that hydrogen, hence, the Lunar Prospector had proved the existence of water in the form of ice at both poles of the Moon. It seems that future residents on the Moon would not have any problems with water... .

The Lunar Prospector, this 160 kg and surprisingly cheap probe, only $63 million dollars, with five instruments, after a one and a half year active and exceptionally productive mission, crashed into an unnamed 51-kilometer-wide crater, directly onto the slope of its terraces, flying at a shallow angle only 1.2 km above its opposite slope, on July 31, 1999. This crater lies only 65 km from the Moon's south pole. To add to the mission's dramatic end: Lunar Prospector carried ashes, in a little polycarbonate vial, from

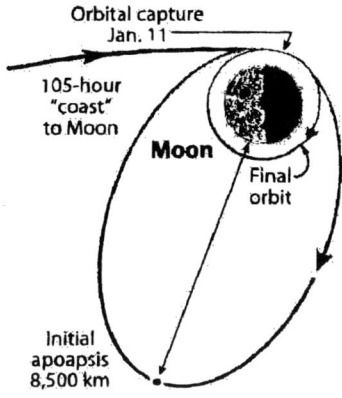

Fig. 15.19. *Lunar Prospector around the Moon. The circular orbit passes both poles of the Moon 100 km above its surface.*

astronomer–geologist Shoemaker's remains. By resolution of the International Astronomical Union, this 51-kilometer target crater on the Moon's south pole, unseen from the Earth, was named after Shoemaker.

14 A Tour to Comet Kopff

Short-period comets are generally easier for probes to follow than long-period ones, as they travel more slowly through the center of the Solar system. From this point of view, the periodic comet Kopff, discovered in 1906, with a 6.5-year period, presents particular interest: this comet has been traced on every return, and astronomers believed that it would be a good choice if a probe could be send to study it from a near distance. Also, it allows a rendezvous two years before perihelion, actually around aphelion, and, hence, represents a unique possibility to carry out continuous observations on its behavior beginning from its minima of activity near aphelion up to the highest activity near perihelion.

The Kopff probe's interplanetary trajectory is also interesting from the point of view of the theory of orbits; its orbit is shown in Fig. 15.20. After the launch on July 5, 1990, the probe went on an elliptic orbit and after four years, on July 12, 1994, arrived at the comet's orbit nearly on its aphelion. During its flyby the probe crossed the asteroid belt, where in passed two asteroids, Namaqua and Lucia, taking a quick look at their surfaces. Four years after leaving Earth the probe entered the orbit around the nucleus of the comet, approaching it to within 10 kilometers at times. But the comet reached perihelion on July 2, 1996, so as it began to develop a coma and tail under the Sun's warming rays, the spacecraft backed off some distance to register the changes.

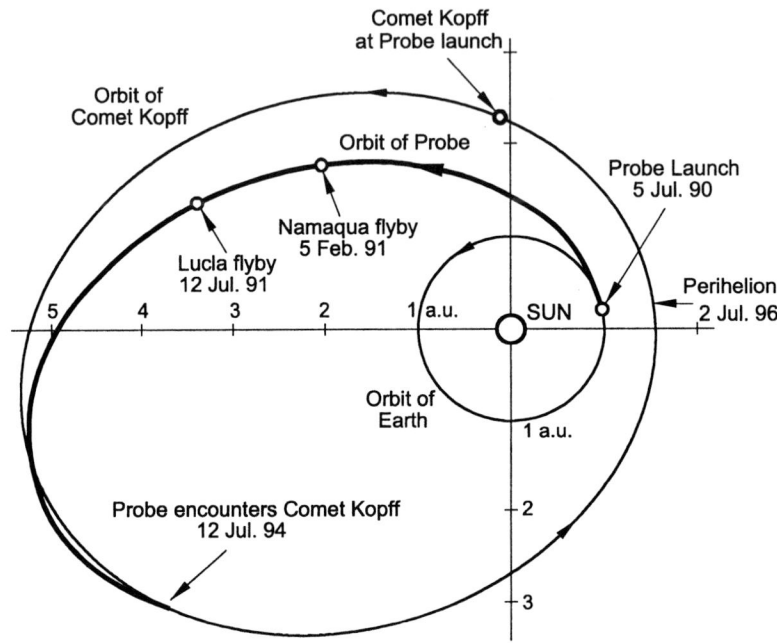

Fig. 15.20. *Trajectory of the comet Kopff probe reaching the comet some two years before perihelion, and staying on the orbit around its nucleus for years for detailed study of the comet's evolution under the action of the Sun's radiation, both corpuscular and thermal.*

The successfully chosen trajectory of the probe, coinciding very well with the comet's orbit, was the crucial aspect of the mission.

The first half of the program was completed by early 1997, giving results of exceptional scientific interest. The probe's study of the comet will be continued for up to six years, that is, during one full cycle of the comet's motion around the Sun.

15   Tempel 2 Rendezvous

We also have an interesting trajectory in the case of the spacecraft CRAF's (Comet Rendezvous Asteroid Flyby) mission to the periodic comet Tempel 2. The latter has an orbital period of 5.26 years, half-major axis $a = 3.0\ a.u.$ and not so large eccentricity of the orbit $e = 0.55$. This trajectory is shown in Fig. 15.21. The spacecraft's passage very close to the asteroid Hestia within the asteroid belt enabled it to get detailed snapshots and mapping of the asteroid's surface.

However, in contrast to the previous case (the comet Kopff probe), CRAF after its launch in February 1993 had its first gravity assist in Venus' gravitational field in August 1993, then, nearly a year later, in June 1994, a second gravity assist in the field of the Earth.

Previous observations revealed that the surface of this short-period comet is covered with a dark, heat-absorbing crust. What is the composition of the surface material? And what are the physical characteristics of the nucleus?

Tempel 2 Rendezvous

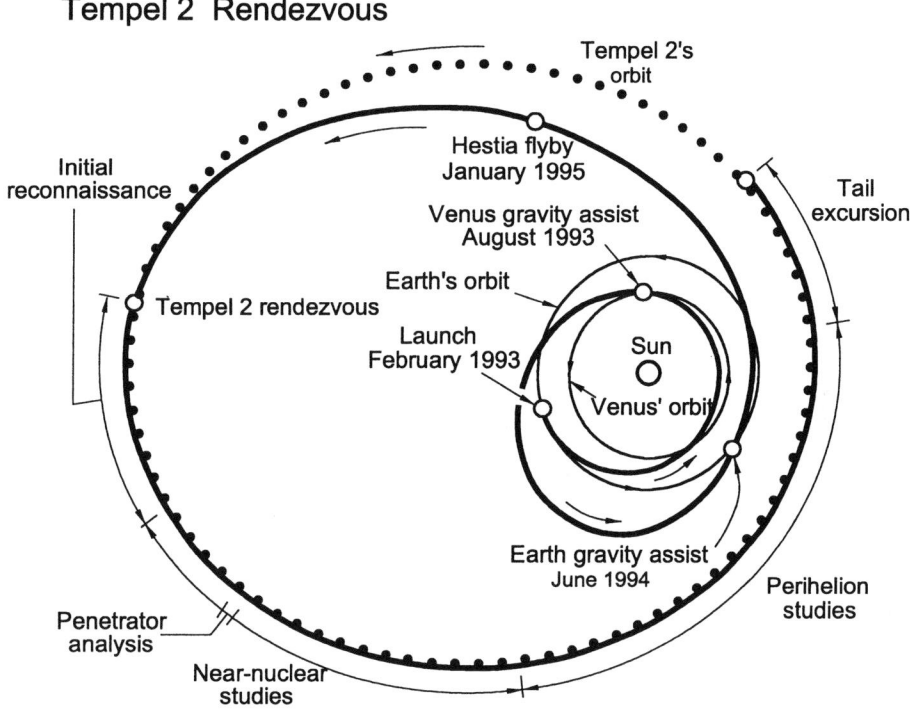

Fig. 15.21. *The trajectory of the CRAF probe around the Sun and the orbit of the Tempel 2 comet. Forty-four months after its launch, the probe arrived at the comet to accompany it through its closest approach to the Sun.*

Note that none of the above-mentioned missions was able to determine, for example, the mass of Halley's comet because of the great flyby speeds. The situation is quite different in the case of Tempel 2. In the above-mentioned data of the orbit of Tempel 2, one can find its orbital velocity of 17.0 $km\,s^{-1}$, exactly four times smaller than in the case of the Halley comet in 1986, namely, about 70 $km\,s^{-1}$. At such a small velocity several slow passages are enough to determine, with the help of CRAF's instruments, the mass of the comet with an accuracy better than 0.1 percent (!). In addition, the entire surface should be mapped up to features only 1 m across. Then, the comet's density could be estimated, revealing whether it is icy or like a well-compressed snowball or as fluffy as freshly fallen snow.... .

Let us turn to Fig. 15.21. During more than half of the orbital period, that is, over three years, CRAF's trajectory coincided completely with the comet's orbit, from its aphelion up to perihelion. Hence, CRAF had a unique possibility to carry out detailed continuous observations of the comet's surface during a long period – three years – recording all variations and events. Particularly at perihelion, when Tempel 2 comes close enough to the Sun to become active, CRAF's instruments would watch the development of gas and dust jets.

Such comet flyby missions could be crucial for the understanding of some of the fundamental problems of the evolution of the Solar system; again, the realization of such missions is largely due to the successful choice of orbit.

16   Messenger: To the Fastest Planet

The fastest planet of the Solar system is Mercury, whit a mean orbital velocity of 47.9 $km\,s^{-1}$. With the shortest orbital cycle, the Mercurian year is equal to 88 Earth days. It also has after Pluto, the largest, eccentricity of its orbit around the Sun: $e = 0.206$. Since the 1970s, when Mariner 10 provided the first close-up look at Mercury, it has been clear that the innermost planet must remain one of the priorities of planetary exploration. It has also been obvious that answers to the key questions about Mercury require detailed study of the planet from its circumplanetary orbit. However, it is also clear that to get a probe to the orbit around Mercury is rather difficult, to some degree even risky and hence will need an exotic design of the probe. The Mercury orbiter *Messenger* was designed to overcome these problems.

The main problem is the acceleration of the spacecraft from the velocity of 30 $km\,s^{-1}$ with which it leaves the Earth, up to Mercury's orbital velocity of 48 $km\,s^{-1}$. To solve this problem with the help of jet engines is practically impossible, since the mass of fuel, hence, of the rocket, would be too large. The only way out is, again, the use of a gravitational maneuver in planets' gravitational fields. As a result, an extremely interesting Earth – Mercury flight trajectory has been developed, which is shown in Fig. 15.22.

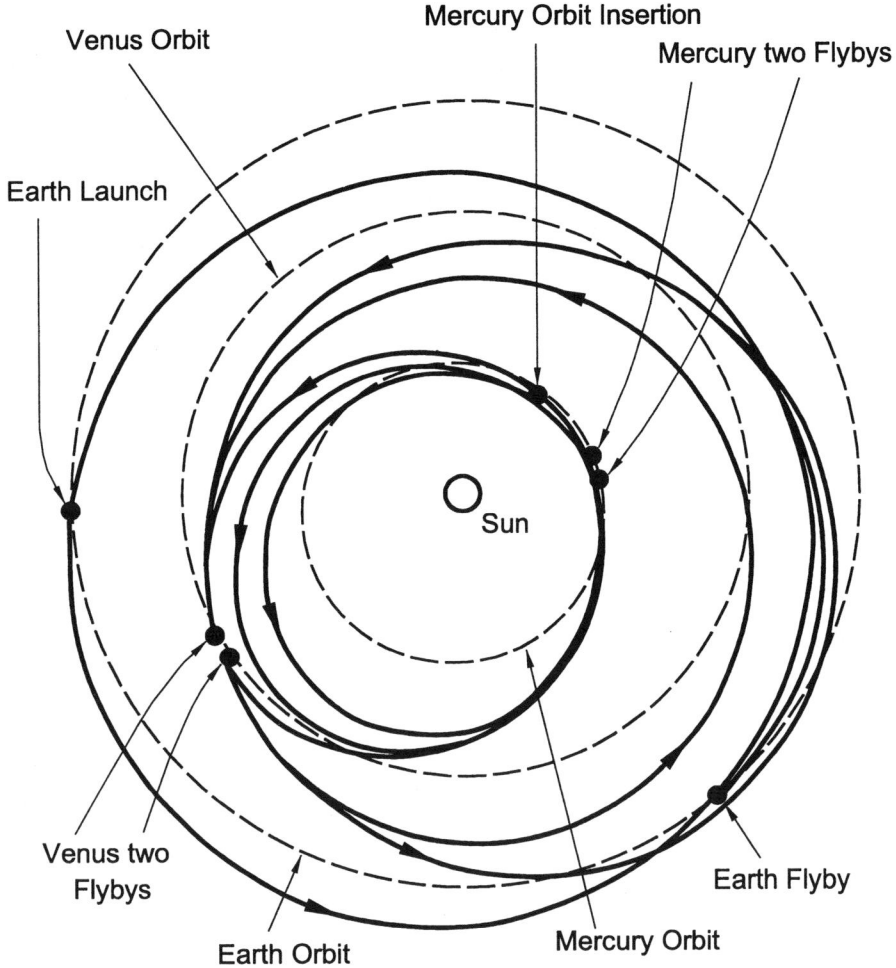

Fig. 15.22. *Trajectory of the Messenger from Earth to Mercury. Messenger will un-dergo, after its launch in January 2004, in total five gravity assists, one by Earth, two by Venus and the last two by Mercury.*

This trajectory is a masterpiece of space missions, chosen from among many thousands of versions, with the use of the five gravity assists, one by Earth, two by Venus and two by Mercury itself.

After the planned launch in January 2004 and two years and nine months of travel, the spacecraft will undergo its first acceleration in November 2006 during the Venus flyby, then the next in April 2007 again during a Venus flyby, then in January 2008 at first flyby of the Mercury and in September of the same year at the second and final flyby of Mercury.

Messenger uses these gravity assists in order to gain the velocity needed to match that of Mercury, before entering into orbit around it. In a gravity assist the spacecraft passes close to a planet, and a tiny amount of planet's angular momentum around the Sun is passed to the spacecraft. The gain in the spacecraft's velocity amounts to hundreds meters per second or more.

During its two flybys of Mercury, Messenger will map in color nearly the whole planet, including most of the part of the planet unseen by Mariner 10 many years ago. The absolute absence of an atmosphere on this planet, in contrast with Venus, must facilitate the realization of this program.

Messenger will be captured by Mercury, as scheduled, in early 2009, i.e. nearly five years after its launch from the Earth. Messenger's computed orbit around Mercury is highly elliptical, 200 km above the surface at its lowest point and over 15 000 km at its highest point, which gives for the eccentricity of that prolonged orbit $e = 0.974$, like we had for Halley's orbit (Table 3.2). The plane of the orbit is inclined 80° to Mercury's rotational axis. This would enable detailed measurements of Mercury's libration, the geology and the composition of a giant impact basin discovered by Mariner 10 and known under the name Caloris.

Messenger's instrumental baggage too is rich: imaging systems, magnetometer, gamma-ray and neutron spectrometers, surface composition, X-ray and plasma spectrometers, as well as, most important, a laser infrared altimeter for the realization of accurate measurements of topography and surface mapping analogously to Venus' Magellan (see section 8 of this chapter).

The planned duration of Messenger's orbital investigations is 12 months, and the expected end of Messenger's activity around Mercury is in 2010.

17   Vikings – Global Surveyor – Pathfinder: A Grand Run to Mars

Certainly, the number one aim among all the planets of the Solar system is Mars, firstly because Mars is considered to be more similar to the Earth than any other planet, and also because Mars is the only planet among the nine planets of Solar system on which humans can in principle stay for a long period, although this red planet is too far away to be a hospitable place to visit even for unmanned spacecraft. Therefore, there is nothing surprising that the biggest number of interplanetary missions have been sent to Mars.

The first significant event was due to Mariner 9, which was launched on May 30, 1971. After a half-year-long interplanetary cruise, in November, 1971 Mariner 9 slipped into orbit around Mars ready to carry out an intense three-month study and mapping of the red planet. Mariner 9 revealed a tremendous variety of surface features, such as dry channels, same of them of giant size (Plate VII), volcanoes and rift zones, which indicate Mars' recent geologic activity, winds of over 100 m per second, i.e. more than half the Martian speed of sound, which probably transport large volumes of surface material over great distances, as well as eroding the rocks and mountains. At the same time, Mariner 9 posed new problems which had to be addressed in future missions.

As a result, a grand program named Viking was launched, a spacecraft consisting of two parts, the Orbiter which had to be constantly in the Martian orbit, and the Lander, a special laboratory which had to land on the Mars to perform chemical-biological analysis of Mars' soil, and transmit the results via the Orbiter to the Earth. The Orbiter meanwhile had to photograph the Martian surface via various cameras.

Two identical probes, Viking 1 and Viking 2, were created, both with an Orbiter and a Lander. Both Vikings were launched practically simultaneously. This vast, extremely complex and billion-dollar project was realized during 15 years by a team of 750 scientists, planetologists, engineers and many other specialists.

Viking 1 was launched on August 20, 1975, and Viking 2 twenty days later, on September 9, 1975. Their Earth–Moon flight trajectory is shown in Fig. 15.23; in fact it was a typical Hohmann orbit, i.e. an ellipse coming into contact with orbits both of the Earth and Mars, but of slightly deformed form. The duration of Viking 1's interplanetary flight was 305 days, and Viking 2 traveled two months longer, for 360 days. During both flights as many as four midcourse corrections were carried out to adjust the spacecraft's rendezvous with Mars. Both Vikings were without solar panels; their power source was independent of sunlight, running on the decay of radioactive plutonium.

Arriving at Mars, Viking was captured by the planet on a highly elliptical orbit with a major semiaxis nearly of 40 000 km, as shown in Fig. 15.24. Traveling by this path, the spacecraft completed one revolution in 24.6 hours i.e. during a Martian day. Then both the Orbiter and the Lander underwent substantial testing. The Lander entered the Martian atmosphere at about 250 km, where it lost 90 percent of its kinetic energy. At about six kilometers' altitude the parachute was deployed, acting up to 1.5 km when the probe was slowed down to 100 m per second. Three descent engines then realized the landing with a velocity of 2.5 m per second, three–four times smaller

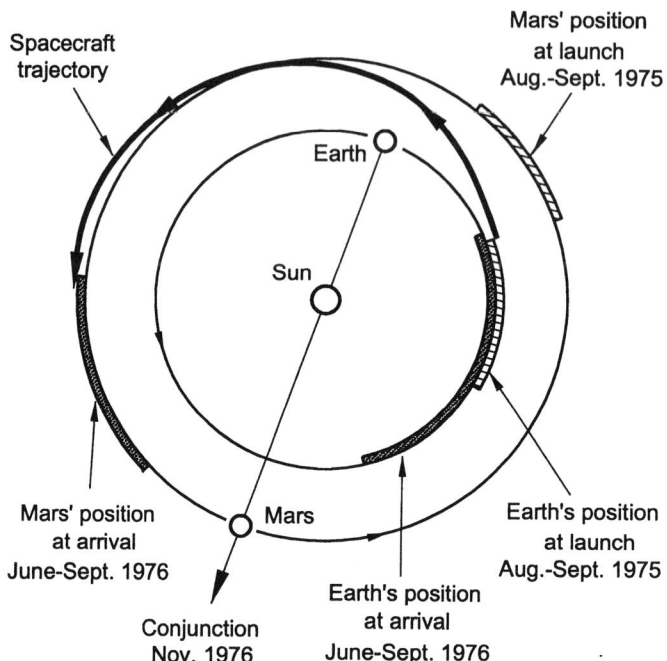

Fig. 15.23. *The Earth–Mars 500 million km flight trajectory of both Vikings. The positions both of the Earth and Mars during the missions are shown, namely, at launch from the Earth and on arrival at Mars. In the same period, the Sun was on the Mars-Earth line, and communication with the Vikings was interrupted for two months.*

than for probes landing on the Earth.

Note a remarkable fact in Fig. 15.22, namely, a long time interval, more than three hours, for the realization of separation of the Lander from the Orbiter and the landing on the given point on the Martian surface.

The main goal of both Landers of both Vikings was the realization of special biological experiments with the aim of finding an answer to the question: is there life on the Mars? To do this, both Viking Landers carried a special device, a Gas Chromotograph Mass Spectrometer, as well as a special mechanical three-meter-long push to probe a sample of the Martian soil. Two cycles of analysis were performed. In both cases no trace of organic molecules were detected.

On the assumption that the high ultraviolet flux from the Sun as well the cosmic rays might have destroyed organic remnants on the Martian surface, Viking's arm removed a piece of rock to get the protected soil beneath it.

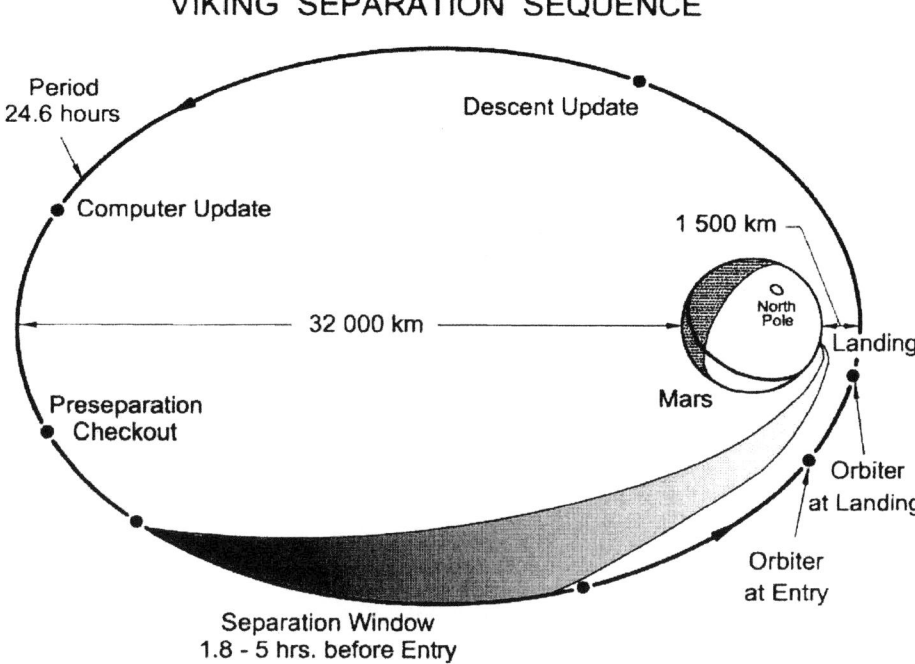

Fig. 15.24. *The orbit of the Viking around the Mars at the moment of its arrival to the red planet. Viking is captured on an highly elliptical orbit with period 24.6 hours, i.e. equal to a Martian day. The lowest point of the orbit is directly at the landing point on the Martian surface.*

Yet again nothing was found. Some biologists, however, did not consider this negative result as decisive for a far reaching conclusion about the existence or absence of life on Mars.

Among the important and intriguing discoveries of the Vikings were the Martian dry rivers; one such image is shown in Plate VII. They were probably a result of liquid flows. The number and variety of those dry rivers posed the question, of where the water on Mars can have gone. The idea is as follows: in view of the existence of very strong dust and wind activity upon the Martian surface, these dry rivers had to be filled up quite soon, during hundreds or at least thousands of years. So, the age of these dry rivers cannot be of a cosmogonical scale. The dry rivers should be considered as one of the largest Martian enigmas.

In one of its orbitings, in February 1977, Viking 1 passed less than 100 kilometers from the Martian inner moon, Phobos. As is often the case in space explorations, while the Vikings answered many questions, they also posed unexpected puzzles. Thus, at the first meeting with Phobos, its most prominent feature, a giant circular crater of 10 kilometers across, was discovered. Phobos itself is only 27 km long. This crater, later named Stickney (Plate IV), is definitely a result of a strong collision with a body at least of five km size, which might essentially change Phobos' initial orbit.

The possibility of the changes of Phobos' orbit was suspected earlier, based, however, on the influence of Martian tidal forces (Sharpless). However, as the later analysis indicates, the orbit of Phobos is gradually shrinking and its original orbit was about twice the size of its present one, and Phobos will reach the Martian surface perhaps in 100 million years (Smith, Born).

The Viking 2 Orbiter flew within 23 km of Deimos, the Martian outer satellite, taking its spectacular close-up pictures with a high linear resolution – up to two–three meters. The images of both satellites, Phobos and Deimos, enable the following conclusion to be drawn:

i. Both satellites are irregularly shaped, but approximately ellipsoidal, some 1.4 times longer than their width. Phobos is about 27 km long, while Deimos is only half that size.

ii. Each satellite rotates synchronized with its orbital motion, so that it keeps its longest axis pointed towards Mars. Thus, both satellites are tidally locked, as is Earth's Moon. Phobos moves around Mars at 9400 km in 7 hours 39 minutes, and Deimos at 23 500 km in 30 hours 18 minutes.

iii. Both have heavily cratered surfaces, but Deimos appeared much smoother and had nothing similar to Phobos' structure. Phobos' surface is scored by numerous long parallel grooves 100 to 200 m wide and 20 m deep (Plate IV). These grooves may or may not be related to the impact that caused the Stickney crater.

The large impact that formed Stickney must also have heated parts of the interior of Phobos. It turns out that most of Phobos' surface, including Stickney and grooves, is very old, by some estimations, probably more than three billion years. However, in view of Phobos' orbital changes, it should be concluded that Stickney's impact is not so old, perhaps 100 million years.

As to Deimos, its composition is still uncertain and seems to be essentially different from that of Phobos.

In all, with the help of both Vikings, 55 000 images of Mars and its satellites, Phobos and Deimos, were obtained.

Ten years later, in 1996, practically simultaneously, two new missions were sent off to the red planet, the Mars Global Surveyor (MGS) and Mars

Pathfinder. The first was launched on November 7, 1996, the second, on December 2, 1996. The first one, MGS, arrived at Mars on September 12, 1997, the second one, Pathfinder, landed and deployed the rover Sojourner.

The scientific program of both these missions in essence was the same as that of the Vikings, but was essentially broadened and more accurate. Under special attention was the sensitivity of the cameras. For example, the cameras of MGC might obtain images of a linear resolution up to one and half meters from a distance of three hundred kilometers.

One of the largest discoveries of Mariner 9 was the Olympus Mons, whose peak was 26.4 km above the "sea level" of Mars. Surprisingly, neither Voyager ever caught this peak. Later, excellent images of Olympus Mons were obtained by Mars Global Surveyor. Now Olympus Mons is accepted as both the highest and largest mountain on Solar system planets by its parameters; its diameter at the base is nearly 600 kilometers. The candela of volcanic origin on the Mons peak is 75 kilometers in diameter.

MGS obtained excellent images with a high degree of linear resolution of the Martian surface, as well of its satellites, Phobos and Deimos; one such image is shown in Plate IV.

MGS's cameras captured, from an altitude of about 350 kilometers, surface structures strongly eroded by strong winds (Plate VIII), discovered isolated rocks of a few meters size on the Martian surface, resolved features as small as 1.4 meters, and obtained so-called "fisheye" views. With the help of MGS' Laser Altimeter was discovered perhaps the largest among the known canyons in the Solar system, the so-called Elysium Rise landscape, a vast volcanic region. The depth of this canyon is three and half kilometers (!); for comparison, the depth of the largest canyon on the Earth's, the Grand Canyon, Arizona, is only one kilometer.

The global temperature map for Mars' surface was obtained. The results included cold polar caps of temperature 175°C, and a relatively warm equatorial region, 10°C.

Mars Pathfinder. On July 4, 1997, the 360 kg Lander of Mars Pathfinder landed on the red planet, near its target zone in Ares Vallis, a plain strewn with a variety of rocks and soils deposited by catastrophic floods early in the planet's history. On the next day a small, six-wheeled, 11.5 kg rover, the Sojourner, rolled down onto the Martian surface and began measuring the chemical makeup of the surrounding terrain. Meanwhile the Lander took pictures of its landing site. The landscape looks as rocky and tortured as the top of a Colorado mountain peak.

Pathfinder became the first spacecraft to enter the Martian atmosphere without prior orbiting of the planet, and thumped down within its ellipse-shaped 100-by-200 km target zone. The next day the rover commenced its study of the composition of the rocks and soils using the Alpha Proton X-ray Spectrometer.

Then, the problem with Martian water, where has all the water gone? Pathfinder discovered visible evidence that water once flowed freely there. It appears that some 1 to 3 billion years ago this region of Mars experienced a strong pulse of near-surface volcanic heat, which melted the ice trapped in underground reservoirs. The fast-moving water carried, then dropped, rocks and sand along the route, shaping the gullies and terraces we see today.

The first two rocks studied by Sojourner were very different; one was similar to a quartz-rich andesite, the other of granite. The initial analysis shows that not one catastrophic event, but many, had occurred.

That was not all. Pathfinder realized many other measurements, of temperature, pressure, dust content in the Martian atmosphere, the magnetic properties of airborne dust, the dust drift through the atmosphere and many, many other investigations.

Even a brief account of the scientific results obtained with the help of these five missions to Mars, Mariner 9, both Vikings, Mars Global Surveyor and Mars Pathfinder, can fill volumes.

18   Mission to Europa

The biggest mystery among Saturn's satellites is Titan. The biggest mystery among Jupiter's satellites is Europa. To decipher Titan's mystery Cassini was sent in 1997. The deciphering of Europa's mystery is the goal of another special mission, the Europa Orbiter, to be launched in 2003.

The main route, Earth–Jupiter, for the Europa probe is standard except for one detail. It is necessary to organize a transit from the Earth's orbital plane to the orbital plane of Jupiter, since the latter is tilted more than that of the Earth. The spacecraft traveling from Earth's plane to Jupiter's has to "climb a hill" to reach the giant planet. Of course, the spacecraft can accomplish this operation at any time after its launch from the Earth, but the best place "to climb" is the intersection point of two orbital planes of two planets, i.e. via the so-called "Broken Plane Maneuver" point, nearly halfway to Jupiter. This operation needs energy equivalent to a change in the velocity of the spacecraft by 227 meters per second, i.e. rather small.

Arriving at Jupiter is only the first stage (I) of this operation.

In the second stage (II), Orbiter has to be captured by Jupiter.

After the planned launch of Europa Orbiter in November 2003, the spacecraft will enter the orbit around Jupiter and start a Galileo-like tour of three of the four largest Jovian satellites – Io, Europa and Ganymede. Using gravity assists from these three satellites, the spacecraft will be captured around Jupiter – this stage (II) is shown in Fig. 15.25 – realizing a slowdown of the orbital period roughly from 200 days to under two weeks. This will take about 1–1.5 years to accomplish. At the end of this stage (II) the "Europa

Endgame" will start.

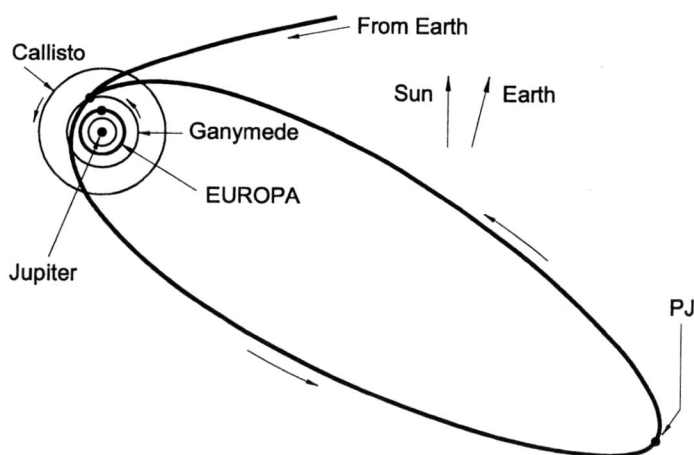

Fig. 15.25. *Europa Orbiter captured around Jupiter. The first and the largest elliptic orbit with the large axis of 21 million km is shown. The Orbiter's smallest velocity in the perijove PJ will be 90 meters per second. After a series of orbitings around Jupiter via decreasing ellipses Orbiter will be captured into the orbit around Europa using gravity assists from all three inner satellites, as well as one outer – Callisto.*

The "Europa Endgame" (stage III) includes the capture of the Orbiter around Europa. Again using gravity assists, in this case by Europa, the spacecraft's engines further "pump down" the spacecraft orbit from a two-week period to just over a four-day period, i.e. equal to the period of Europa around Jupiter, 3.55 days. This is done by "stepping down" resonances, starting at a 1:3 resonance, i.e. one spacecraft orbit around Jupiter per three Europa orbits, to a 5:6 resonance.

This stage takes 3–4 months to complete. Once the spacecraft is in 5:6 resonance, Jupiter's own gravity will be used to pull the spacecraft into a large elliptical orbit around Europa; this situation is shown in Fig. 15.26. A special maneuver will be used to circularize the Orbiter's orbit at 200 km altitude around Europa. From this moment the normal scientific phase of the Europa Orbiter's activity will start, i.e. the detailed observations of the thick icy surface of this striking celestial body.

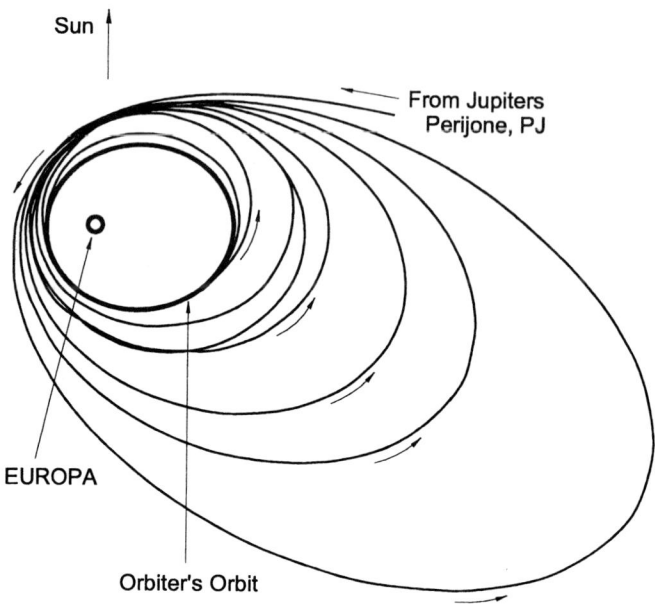

Fig. 15.26. *The Europa Orbiter, captured by Jupiter's satellite Europa into an initial elliptic orbit with a major semiaxis of 5 million km; in the subsequent orbitings with smaller ellipses, the finally will enter Europa's orbit. At that moment the scientific observations of Europa's surface will begin.*

19   Mission to Eros

Eros has been considered a minor planet despite its extremely small size, 7 km, according to the old ground based observations, much smaller as compared even with medium-size asteroids. Its somewhat elliptic orbit with an eccentricity $e = 0.223$ and semimajor axis $a = 1.458AU$ is slightly smaller than the Martian orbit ($a = 1.52AU$). It is clear that with such an orbit and orbital period $P = 642^d$ around the Sun should present definite interest for its detailed study by means of a special scientific probe. As a result, a Near Earth Asteroid Rendezvous mission, later called NEAR-Shoemaker, was launched, carrying, besides various types of cameras, a number of scientific instruments for the detailed study of this celestial body from a very close distance.

The NEAR probe was launched on February 17, 1996. Its trajectory presents definite interest from many points of view, therefore we present it in Fig. 15.27. The final aim has been to organize the capture of NEAR on an orbit around Eros with possibly a closer orbit, perhaps up to 50 km.

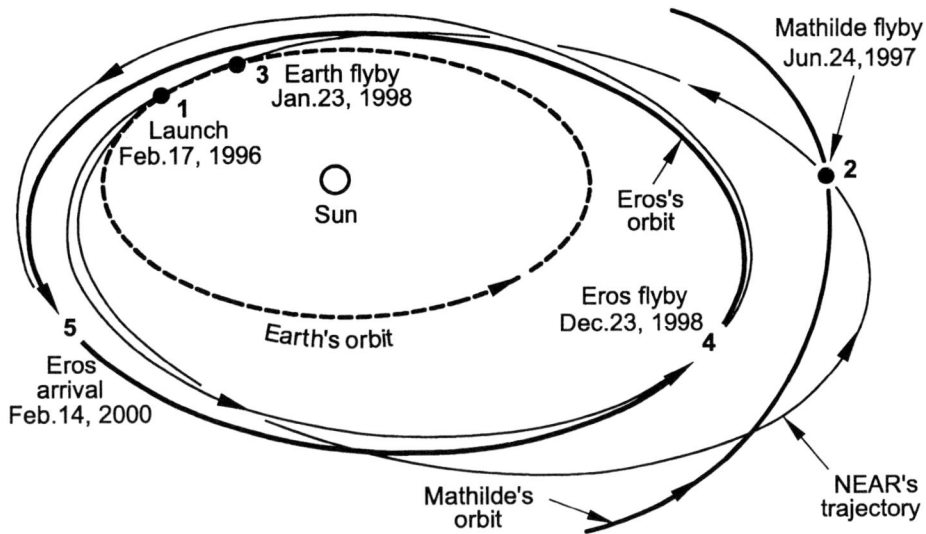

Fig. 15.27. *The flight trajectory of NEAR-Shoemaker from the Earth to Eros by a sequence of very prolonged ellipses. The capture of the spacecraft around Eros was realized on February 14, 2000, after the four-year flight.*

After one and half years, on June 27, 1997, NEAR passed within 1200 kilometers of the asteroid–minor-planet Mathilde, a very dark object, and took more than 500 images during a 25-minute passage at nearly 10 km per second. Mathilde was a colorless, heavily cratered black body of sizes $57 \times 53 \times 50$ km, and an albedo of only 4 percent. Mathilde is one of slowest-spinning asteroids known, requiring 17.4 days to complete one rotation. Apparently, something must have happened to cause such an extremely slow axial rotation.

NEAR did not discover a satellite around Mathilde, so the extremely slow spin of this asteroid remains a mystery.

However, another scenario is conceivable as well, namely, for a long time, say, of the order of a billion years, Mathilde was a binary system and had its own satellite. Then, due to their tidal interaction, Mathilde lost its rotational moment up to the present rate, then relatively recently it also lost its satellite under the action of close transit of another asteroid or a more or less solid body.

The illuminated hemisphere of Mathilde is dominated by two huge impact craters 30 and 20 km across, and two or three others of that size are evident. For a typical impact velocity of 5 km per second, a 3-km asteroid would be needed to cause a 30-km-wide crater on Mathilde. The energy of such an impact should be equivalent to that of 600 billion tons of TNT. It is therefore surprising that Mathilde has withstood several such impacts without being scarred into fragments or totally demolished. This situation concerns, as we shall see later, Eros as well. This suggests that the present surfaces of these asteroids are at least two billion years old.

By the trajectory perturbation method, Mathilde's mass was estimated; it was about 10^{20} g and, by known sizes, the asteroid's density turned out to be less than 1.5 g per cubic cm. So, Mathilde should have a porous interior or, with very little probability, an icy internal composition. More likely, Mathilde is a "rubble pile", whose interior has been pulverized by a long history of severe collisions. It is interesting to note that the Martian moons Phobos and Deimos also have low densities, 1.8 g per cubic cm.

Exactly a half year later, on January 23, 1998, NEAR tested a flyby near the Earth. At the end of the same year, on December 23, 1998, NEAR had its first meeting with Eros, and from a distance of 4100 km the spacecraft's cameras took more than 1100 images of the asteroid. Simultaneously, the recording of the observations by on-board instruments were performed, i.e. of a near-infrared spectrometer and magnetometer. NEAR passed near Eros at a velocity of 1 km per second.

Most of the images were taken specially with the purpose of searching for a natural satellite of Eros. None was seen. The attempt to determine the asteroid's mass by the methods of celestial mechanics, namely, by the tiny deflection of the probe by gravitation perturbation, "looks promising". However, the main goal, to be captured by Eros at this meeting, does not turn out well. At the first flyby in late 1998 the engine misfired, and NEAR did not slip into orbit around Eros as originally planned; an extra trip around the Sun, as shown in Fig. 15.27, was needed to catch up with Eros once again. That happened only a year and two months later, on February 14, 2000, and in this case the spacecraft was captured by Eros on a 370 × 323 kilometer polar orbit around Eros. This operation should be considered a success from the point of view of the organization of interplanetary flights.

The panorama of the interlacing orbits shown in Fig. 15.27, namely, with two natural orbits of the Earth and of Eros, and circulating flight trajectories, looks spectacular.

The spacecraft recorded thousands of images and spectra of its target, Eros, which appeared to be not just a piece of rock.

At this rendezvous for the first time the size of Eros was measured: it looked like a cylinder of length 33 km and diameter 13 km, essentially larger than previously estimated. Also, Eros' surface was covered by many craters, which means that much of its surface must be quite old. This suggests that this asteroid was not ejected toward Earth's vicinity by a relatively recent collision within the asteroid belt. Many giant boulders, and craters definitely not of volcanic origin, dot the surface, one of them of a striking 5.5-km-wide circular form, i.e. of half the asteroid's diameter.

During this rapprochement, dynamists used the parameters of the probe's orbit around Eros to deduce its total mass and then its mean density. The mass appeared to be 5×10^{18} g, the density about 2.4 g per cubic centimeter, typical for Earth's crystal rocks.

The following maneuvers were realized by NEAR before its capture by Eros in early February, 2000:

Rendezvous Maneuver I,	3 Feb. 2000,	at distance 8300 km
Rendezvous Maneuver II,	8 Feb. 2000,	at distance 4500 km
Orbit Insertion Maneuver,	14 Feb. 2000,	at distance 350 km

By early April, 2000, the spacecraft NEAR had started the circling around Eros with a decreasing size of orbit. The time-dependent schedule of this period is presented in Table 15.4. The sequence of the transfer orbits around Eros up to the smallest orbit of 50 km radius is illustrated in Fig. 15.25.

T a b l e   15.4

The schedule of maneuvers for th e transfer of NEAR spacecraft from interplanetary flight trajectory into the smallest orbit with a size of 50 km around Eros.

Orbit's correction Maneuver	Date	Orbit km	Period days	Duration days
3	2 April, 2000	210×100	6.6	9
4	11 April, 2000	101×99	3.4	11
5	22 April, 2000	101×50	2.3	8
6	30 April, 2000	51×49	1.2	68

NEAR was the first spacecraft to orbit an asteroid.

Both the data in Table 15.4 and Fig. 15.28, as well as the above-mentioned results, enable one to draw some conclusions. First, the possibility for a spacecraft to be captured on a close circular orbit not around an ordinary planet but around an asteroid of extremely small mass was shown. Second, the possibility to realize maneuvers, during a month, around an asteroid with progressively decreasing sizes of orbit up to the preliminary programmed smallest orbit was confirmed. Third, the spacecraft was shown to be successfully kept during a long period, more than two months, on the smallest circular orbit around an asteroid. Fourth, the classic method of

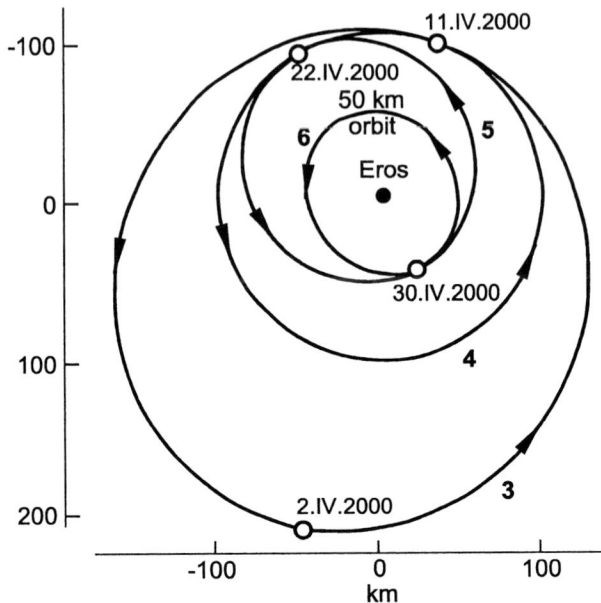

Fig. 15.28. *The sequence of the transfer of spacecraft NEAR around the asteroid Eros up to the smallest orbit with 59 km radius around the Eros.*

celestial mechanics, namely, to obtain the mass of an unknown planet by the perturbations of the system "Planet–Planet", was successfully applied, for the first time, to the system "Asteroid–Probe" for the determination of the mass of the asteroid.

For the first time in the history of interplanetary flights the perturbation maneuver has been applied twice, at flybys of two asteroids, Mathilde and Eros, and as a result, the masses and densities of those medium-size asteroids have been discovered. For comparison, the gravitational attraction between the asteroid and the probe is 10^{20}–10^{23} times weaker than that of the attraction of the Earth and the Moon.

Thus the perturbations undergone by space probes can be efficient tools for the accurate tracing of any variation of the gravitational field on the route.

During two months the NEAR probe realized a wide program of scientific observations and measurements, including laser rangefinder measurements to map the surface, X-ray and gamma–ray spectrometric measurements for the determination of Eros' composition and physical properties of its surface minerals, as well as radio transmitter observations aiming to determine the characteristics of NEAR's orbit and to obtain the asteroid's true mass.

The final part of NEAR's mission was even more spectacular. The project leaders managed the non-planned landing of the probe on the asteroid. February 12, 2001 became a history-making day: NEAR's probe left its 35 km circular orbit and landed on the surface of Eros. The last picture of EROS' surface was obtained from an altitude 120 m. This was the first landing of a space probe on a surface of an asteroid.

NEAR-Shoemaker mission, cost $212 million, should be considered as one of most successful interplanetary experiments.

20 Binary Asteroids

Thus, 10–20 km size asteroids possess a sufficient gravitational field to hold a space probe in a closed orbit as a natural satellite. Then one may ask whether there are binary asteroids. The ground-based observations confirmed such a phenomenon.

The history of binary asteroids begins, apparently, in 1993 when the Galileo spacecraft, on a flyby through the asteroid belt, fixed a companion-satellite, only 1.5 km across, around an elongated, 56 km by 26 km asteroid, *Ida*, of 2.5 h orbital period (Plate VI).

The next case involved the minor planet 1994 AW1, an Earth-approacher no more than about 1 km wide. Astronomers monitored its brightness during its 1994 flyby of the Earth and found small variations showing that it rotates in 2.5 h. But the light curve contained a surprise: the asteroid also dimmed regularly every 11.2 hours, suggesting that it was being eclipsed by a satellite of about half its size, orbiting rather closely in a 22.4 h period. There were apparently two dimmings per orbit, when the large body occulted the small one, and vice versa.

A similar discovery was made for another small near-Earth asteroid, 1991 VH. It not only varied in brightness with a 2.6 h period but also showed eclipses every 33 h caused by an object of 40 percent of its size.

When monitoring asteroids' light curves, a signature of a satellite eclipsing the 1-km-wide Earth-crosser 3671 Dionysus was noticed. In addition to its 2.7-hour rotation period, Dionysus showed eclipses every 28 hours.

For the first time, ground-based astronomers directly imaged a satellite-companion orbiting an asteroid, the minor planet 45 Eugenia. Its satellite was first seen in a series of 16 images taken during a 10-day span in November, 1998 using an adaptive optical system of 3.6 m the Canada–France–Hawaii Telescope.

Eugenia itself is about 215 km in diameter; its moon is about 13-km in diameter and orbits Eugenia in 4.7 days at a distance of 1190 km. These data determined Eugenia's density of about 1.2 g per cubic cm. So, its interior may consist, perhaps, mostly of ice, making it a long-tied comet.

For two decades radars have been used to image dozens of minor planets, among them the main-belt asteroid 216 Kleopatra. It was discovered that Kleopatra is shaped like a dumbbell, with a handle that looks substantially narrower than the two bodies when seen pole-on, but not when seen from within the equatorial plane. Kleopatra's overall size is 217 by 90 km. It seemed as if it had a void within the central "handle", which would mean that two big lobes are orbiting with a period of 5.4 hours. Judging from the radar images, the probability that Kleopatra is a binary system is too high, if assuming the size of the main body of $\sim$120 km and $\sim$80 km for its satellite.

Satellites around two other asteroids, 762 Pulcova and 90 Antiope, were discovered using the above-mentioned 3.6 m CFHT telescope in the first case, and 9.8 m Keck II telescope, both in Hawaii, for the second one.

Pulcova's companion is 4 magnitudes dimmer than its 140-kilometer-wide parent, suggesting a diameter of perhaps 20 km. It orbits every 4.0 days at a mean distance of about 800 km.

Antiope's two components have nearly the same brightness, each of roughly 80 km across. They are separated by just 170 km and revolve around each other in 16.5 hours.

Until now dynamists had been assuming that the asteroid satellites are most likely captured impact debris, but Antiope's near-twins cannot be explained so easily.

Radar observations reveal that the small asteroid 2000 DP$_{107}$ also might be a binary. The two components are about 800 and 300 m in diameter, and orbit each other in about 1.8 days at a distance of only 2.6 kilometers. One more asteroid, FG$_3$, needs confirmation of its binary nature.

The data obtained by early 2001 about binary asteroids are summarized in Table 15.5.

So, the existence of binary asteroids is a confirmed fact. As to the interpretations, they may be various with far reaching consequences. According to the current idea, many asteroids are piles only loosely held together by gravity. Even a soft collision between asteroids could result in many large pieces being expelled, while two pieces of nearly same trajectory could end up orbiting each other. Alternatively, the disruption could be induced via tidal forces at, say, near–Earth passage.

To the point, the Earth itself bears evidence that asteroids sometimes approach the Earth in pairs. Among Earth's 28 largest impact craters, there are doublets, i.e. as if formed by the simultaneous impact of two bodies. It is remarkable that this fact had never been noticed before.

T a b l e   15.5

Binary asteroids discovered by ground-based and space probe's observations. In the eighth column, the computed radii of the orbits are given, in the last column computed eccentricities of the orbits.

Binary asteroid Name	Year of discovery	Asteroid diameter D km	Satellite diameter D_s km	Orbiting Period P hours	Mass of Binary M g	Orbit's radius a, km Obsr.	Orbit's radius a, km Comp.	Eccentric. of orbit e
I d a	1993	52x26	1.5	2.5	$6.9\,10^{19}$	-	22.0	-
1994 AW1	1994	1	-	22.4	$1.3\,10^{15}$	-	2.3	-
1996 FG$_3$	1996	1	-	16.15	$1.3\,10^{15}$	-	1.9	-
1991 VH	1997	1	-	23	$1.3\,10^{15}$	-	3.1	-
Dionysus	1997	1	-	28	$1.3\,10^{15}$	-	2.8	-
Eugenia	1999	215	13	113	$6.5\,10^{21}$	1190	1120	0.01
Kleopatra	2000	120	80	5.4	$1.6\,10^{21}$	-	100	-
Pulcova	2000	140	20	96	$3.6\,10^{21}$	800	890	0.06
Antiope	2000	80	80	16.5	$1.3\,10^{21}$	170	200	0.08
2000 DP$_{107}$	2000	0.8	0.3	43.2	$0.7\,10^{15}$	2.6	3.8	0.15

The data presented in Table 15.5 give an opportunity to carry out some quantitative analysis as well. In particularly, there is a possibility to predict the sizes of satellite's orbits and compare them with observations.

From Eq. (66) of Chapter III we have for the dependence of the semimajor axis a on the orbital period P and on the system's total mass M in the CGS system:

$$a = 1.19 \cdot 10^{-3}\, P^{2/3}\, M^{1/3} \text{ cm,} \qquad (15.1)$$

where P is in seconds, M in grams and a in centimeters. As to the total mass of the main spherical asteroid of the binary, we have, ignoring the satellite mass (with an exception for Antiope with components of equal sizes): $M = 0.523\, D^3 \gamma$, where D is the asteroid's diameter (third column in Table 15.5), γ the density of asteroid's body.

With the help of Eq. (15.1) and known values of P (fifth column) the calculated semimajor axes of these binary asteroids were obtained, assuming $\gamma = 2.5$ g per cubic centimeter for all asteroids, except Eugenia for which we adopted the above-mentioned value $\gamma = 1.2$ g per cubic centimeter. The results are presented in the eighth column of Table 15.5.

For the last four binary asteroids in Table 15.5 the observed distances between the components are known (seventh column). Comparing both these numbers, observed and computed, we can see a surprisingly good accordance between these magnitudes, especially in the case of Eugenia. This means that the parameters adopted for these binaries, particularly for the masses (sixth column), are not far from being quite similar.

As to the circular (orbital) velocities, they are very small, from 0.2 m per second (Dionysus) up to 20 m per second (Eugenia), much smaller as compared with the planetary orbital velocities, i.e. tens of km per second.

The orbital period P is independent of the eccentricity of the elliptical orbit. However, the observed distance between the components at the fixed phase of motion depends on the orbit's eccentricity. Then, some differences between the observed and calculated intercomponent distances, which seems to be real at least in the case of the last binary, 2000 DP_{107}, may be attributed to the effect of the ellipticity of the orbit, the numerical values of which are not difficult to find; they are presented in the last column of Table 15.5.

Thus, the gravity of even a one-km celestial body can hold a body, as it turned out, even in the conditions of the asteroid belt. This conclusion should be evaluated as of particular importance from the point of view of dynamics and the early evolution of the Solar system.

21   Chandra's Orbit

On June 23, 1999, NASA launched an X-ray observatory, named Chandra in honour of Subrahmanyan Chandrasekhar who had died several years before. Chandra was the most powerful X-ray observatory ever raised into the space. Therefore, the problem of the exploitation of this multi-mirror 14-meter-long, more than four and half tons, 1.5 billion dollar cost, superpower observatory with unique productivity, acquired extreme importance. To solve this problem without long exposure times to record X-ray spectra is impossible. However, unlike ground-based telescopes which can quickly switch from one target to another, space telescopes slew over the sky far more slowly, e.g. there is not enough time to slew to another target in the opposite celestial hemisphere, and then return to the original target within a single observing operation.

Chandra needs long enough continuous exposures, of the order of many hours, for registration, say, of X-ray spectra of variable X-ray binaries, and hence has to avoid Earth's blocking, as in the case, for example, of the Hubble Space Telescope.

The only solution to this problem, i.e. performance of long continuous observation times, can be possible only at extremely elongated elliptic orbits around the Earth. For Chandra such an orbit was found, shown in Fig. 15.29.

Chandra's orbit is a highly elongated ellipse, with apogee, the farthest

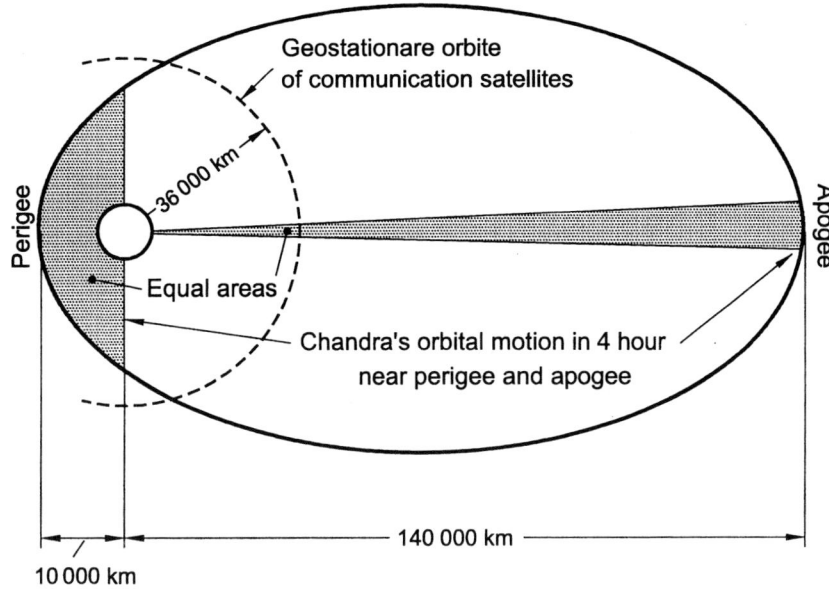

Fig. 15.29. *Chandra X-ray observatory's orbit, a highly elongated ellipse, with a farthest distance from the Earth of 140 000 km and a period of an orbital revolution of 64.2 hours. Thanks to Kepler's second law of orbital motion, the observatory will spend most of each orbit near the apogee.*

distance from the Earth, of 140 000 km, much beyond the geostationary circular orbit of communication satellites of radius 36 000 km, and nearly one-third of the way to the Moon. Its perigee is 10 000 km, much farther than that of the Hubble telescope, 600 km. The eccentricity of this ellipse is big, $e = 0.81$, and also quite large is the orbital period, 64.2 hours. At the apogee the Earth appears as a disk less than 5° in diameter. From such a large distance, in apogee, Chandra may realize continuous observations during many hours, four, five and more. In Fig 15.29, the shaded areas, near apogee and perigee, indicate the equal surfaces, according to Kepler's Second law, covered during 4 hours of flight of the spacecraft through the zones of apogee and perigee.

Chandra was taken to the orbit with the help of the Space Shuttle Columbia, with a maximum altitude of 600 km, but in this case it was 283 kilometer only. So, the problem was then to transfer Columbia from a low circular orbit into the high and elongated elliptic orbit, the nearest distance of which is only 10 000 km. This was an extremely difficult and complex problem, which was solved by the astronauts of Columbia with stunning success in perfect correlation with the ground-based flight controllers and thus piloting the Shuttle to the proper attitude for Chandra's deployment.

This transit from Columbia's initial low circular orbit into Chandra's high elliptic orbit is not shown in Fig. 15.29.

The large distance of Chandra's orbit from the Earth makes this orbit less sensitive to the various types of perturbations, and, hence, much more stable during a long period.

This brief story of Chandra may be perceived as an another impressive example of the successfully application of the theory of orbits for the solution of purely practical problems of space research.

22   Genesis Mission: Search of the Origin

According to current knowledge, the Solar System was formed 4.6 billion years ago at the gravitational collapse of the solar nebula, a cloud left by a previous generation of stars. As time went, the condensation of the matter led to the formation of the Sun and other celestial bodies, i.e. planets and their moons, comets, asteroids etc.

However, many key problems concerning the formation process of the planetary system and the properties of the planets still remain unclear, and many mysteries remains unanswered. For example:

– How this transformation from solar nebula to planets took place?

– Why did some planets, like Venus, develop thick and high-temperature atmospheres?

– How did the Earth become hospitable to life?

Partial answers from previous studies suggest that planets, moons, and even asteroids are significantly different in composition. But without knowing the properties of the original solar nebula, no complete answers are possible.

The Sun which contains over 99% of all material in the Solar system can help to find the answer to many questions. While its interior has been modified by nuclear reactions, the outer layers of the Sun are composed essentially of the same material as the original solar nebula. So, the comparison of the Sun's composition with the known planetary composition may yield significant answers.

The goal of the Genesis mission is to measure the isotopic composition of elements, first of all, of oxygen, silicon, nitrogen, and noble gases. These data will enable to better evaluate the isotopic variations in meteorites, comets, lunar samples, and hence to obtain improved measures of Solar elemental abundances.

The Genesis spacecraft will be placed into the orbit around Lagrangian point L_1, of the Earth–Sun system where the gravity of both bodies is balanced; the distance of the point L_1 from the Earth is nearly one million

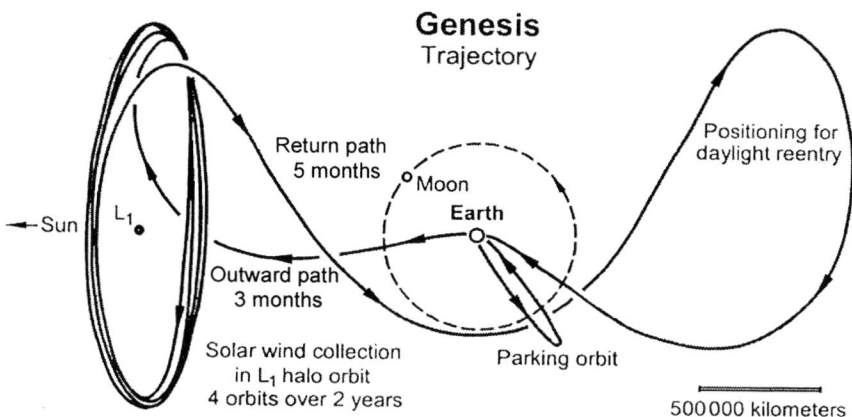

Fig. 15.30. *The Genesis spacecraft's flight path from the Earth up to the circular trajectory around the libration point L_1 of the Earth–Sun system. The spacecraft's orbital plane around the point L_1 is perpendicular to the line Earth–Sun. After collecting Solar wind samples for two years (loops at left), the spacecraft will dispatch a capsule to Earth.*

kilometer (Section 14, Chapter IX). Its remarkable flight trajectory from the Earth up to the orbit, nearly circular, is shown in Fig. 15.30. Once in orbit, Genesis will operate its collector arrays and begin collecting particles of the Solar wind which will imbed themselves in specially designed high purity waters. After two years, the sample collectors will be returned to the Earth inside the Science Canister. The samples will be stored and cataloged under ultra-pure conditions.

The schedule of the mission is as follows. It was launched on August 8, 2001. In about three months flights the Genesis spacecraft comes into an orbit around the point L_1 along the Earth–Sun line. Then, the probe performs four orbits within two years, to return to the Earth in 2004.

23 Solar Probe Trajectory

We conclude this chapter devoted to interplanetary missions with the same type of flights as was considered at the start of this chapter, namely, flights to the Sun.

The Solar Probe, JGA, is tentatively scheduled to leave in 2007 in the direction of the Sun with at least two successive close passages near the Sun.

The trajectory of this flight is shown in Fig. 15.31. Again, as in the previous case (Fig. 15.1), the spacecraft will perform a perturbative maneuver within Jupiter's sphere of action after which it will be directed to the Sun. The location of the Earth is fixed at two successive perihelions, i.e. in 2010 and 2015.

About a year after launch, the Solar probe will encounter Jupiter for a gravity assist. Actually, the spacecraft will undergo a retro-gravity assist which will almost stop the spacecraft in its tracks, and sling it up over Jupiter's North pole, to literally fall back towards the Sun. This maneuver will propel the Solar probe towards a passage near the north pole, equator and the south pole of the Sun. This will be the first perihelion (nearest to Sun) passage of the Solar probe timed to occur during "Solar Max", when the sunspot activity is at its highest level during the 11-year sunspot cycle.

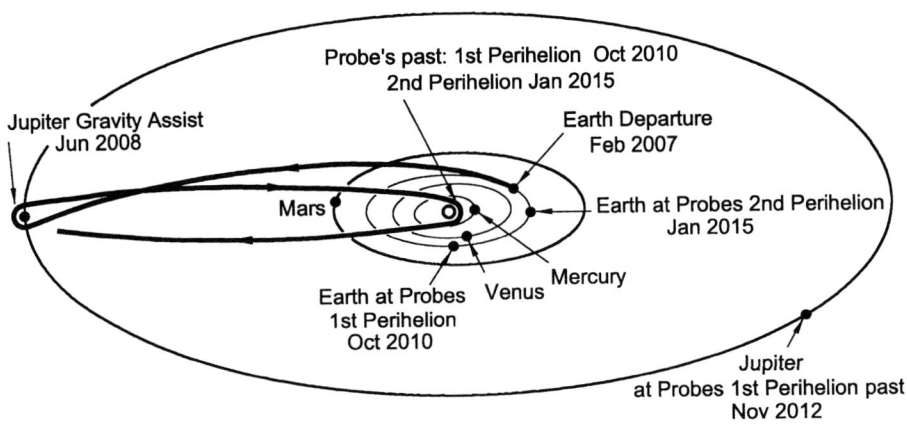

Fig. 15.31. *Planned Solar probe flight trajectory to the Sun with at least two perihelion passages. At the formation of initial orbit, the Probe will be accelerated in Jupiter's sphere of action. The Earth's locations at the Probe's first and second perihelion transits are shown as well.*

A planned second perihelion passage will occur during the Solar minimum phase of the cycle

So, the sequence of the main stages of the Solar probe flight is as follows:

Earth departure	February 2007
Jupiter gravity assist	June 2008
First perihelion	October 2010
Second perihelion	January 2015

Thus, the intervals between perihelion passages yield four years and three months.

The main essence of the JGA flight is the possibility to carry out observations of the Sun from very near distances, especially at the period of Solar maximum.

24   Overview of the Solar System

Thus, in the epoch of interplanetary flights, with triumphal missions of Pioneers, Voyagers, Galileo, the no less remarkable tours of Ulysses, Magellan, NEAR's landing on Eros, Lunar Prospector, and many other spacecraft and missions, the striking images from the Hubble Space Telescope, the flights to nearly all planets and their satellites, to all corners of the Solar system and even beyond, resulted in an infinite number of discoveries opening entirely new worlds, new phenomena, new concepts. If, decades ago, astronomers were aware, say, of only five satellites around Uranus, now there are twenty; if previously Saturn was known as the only planet having a ring, now at least four planets are know to possess rings of various forms, sizes and structure. Saturn itself is now possess thousands of rings and not just the three that were known for three centuries. As a result, our knowledge about the composition of the Solar system and its components has changed drastically. This occurred under the convincing conduction of Celestial Mechanics. Actually we are witnesses of a revolution in planetary astronomy.

Below, in three tables, the data are presented for the main members of the Solar system, i.e. the planets and their satellites.

PHYSICAL PARAMETERS OF THE SUN AND PLANETS

N a m e	M a s s g	Equatorial Radius km	Density g cm^{-3}	Axial Rotation Period	Obliquity (*) degrees	Surface Gravity cm s^{-2}	Escape Velocity km s^{-1}
S u n	$1.989 \cdot 10^{33}$	696 000	1.41	25.4　d	7.25	27398	617
Mercury	$3.302 \cdot 10^{26}$	2 439	5.43	58.646 d	0	363	4.43
Venus	$4.868 \cdot 10^{27}$	6 051	5.20	43.018 d	177.33	860	10.36
Earth	$5.974 \cdot 10^{27}$	6 378	5.52	23.934 h	23.45	982	11.19
Mars	$6.418 \cdot 10^{26}$	3 396	3.93	24.623 h	25.19	374	5.03
Jupiter	$1.899 \cdot 10^{30}$	71 492	1.33	9.925 h	3.08	2590	59.54
Saturn	$5.685 \cdot 10^{29}$	60 268	0.69	10.656 h	26.73	1130	35.49
Uranus	$8.683 \cdot 10^{28}$	25 559	1.32	17.24　h	97.92	1040	21.33
Neptune	$1.024 \cdot 10^{29}$	24 764	1.54	16.11　h	28.80	1400	23.61
Pluto	$1.32　\cdot 10^{25}$	1 170	2.0	6.387 h	119.6	650 ?	1.25

(*) Obliquity is the angle between a planet's equatorial plane and the plane of its orbit; values exceeding 90° (Venus, Uranus, Pluto) mean that the planet has retrograde rotation. Uranus' rotational axis, particularly, lies nearly in its orbital plane. The planet's rotational axis is exactly perpendicular to its orbital plane in the case of Mercury (0°) and nearly perpendicular in the case of Jupiter (3°).

ORBITAL PARAMETERS OF PLANETS

N a m e	Period years	Semimajor Axis AU	Mean orb. Velocity km s^{-1}	Orbit's Eccentricity e	Inclination of Orbit's Plane degrees
Mercury	0.2408	0.38710	47.89	0.205631	7.0048
Venus	0.6152	0.72333	35.03	0.006773	3.3947
Earth	1.0000	1.00000	29.79	0.016710	0.0000
Mars	1.8807	1.52366	24.13	0.093412	1.8506
Jupiter	11.856	5.20336	13.06	0.048393	1.3053
Saturn	29.424	9.53707	9.64	0.054151	2.4845
Uranus	83.747	19.1913	6.81	0.047168	0.7699
Neptune	63.723	30.0690	5.43	0.008586	1.7692
Pluto	248.02	39.4817	4.74	0.248808	17.1417

SATELLITES OF THE PLANETS OF SOLAR SYSTEM

Numbers are the diameters of satellites in kilometers

I n n e r P l a n e t s			
V e n u s	**M e r c u r y**	**E a r t h**	**M a r s**
none	none	Moon 3476	Phobos 26.8×18.6
			Deimos 15.6×10.2

O u t e r P l a n e t s			
J u p i t e r	**S a t u r n**	**U r a n u s**	**N e p t u n e**
Metis 60×34	Pan 20	Cordelia 26	Naiad 58
Adrastea 20×14	Atlas 36×28	Ophelia 32	Thalassa 80
Amalthea 250×128	Prometheus 148×68	Bianca 44	Despina 150
Thebe 116×84	Pandora 110×62	Cressida 66	Galatea 160
Io 3642	Epimetheus 138×106	Desdemona 58	Larissa 208×178
Europa 3130	Janus 198×152	Juliet 84	Proteus 436×402
Ganymede 5268	Mimas 398	Portia 110	Triton 2706
Callisto 4806	Enceladus 498	Rosalind 58	Nereid 340
Leda 10	Tethus 1058	Belinda 68	
Himalia 170	Telesto 30×16	Puck 154	
Lysithea 24	Calypso 30×16	Miranda 480×466	**P l u t o**
Elara 80	Dione 1120	Ariel 1159	
Ananke 20	Helene 32	Umbriel 1170	Charon 1250
Carme 30	Rhea 1528	Titania 1580	
Pasiphae 36	Titan 5150	Oberon 1520	
Sinope 28	Hyperion 370×226	Caliban 60	
	Iapetus 1440	Sycorax 120	
	Phoebe 230×210	Stephano 20	
		Prospero 20	
		Setebos 20	

Epilogue

This brief review of certain interplanetary missions aimed to illustrate the ever increasing role of Celestial Mechanics, particularly of the Theory of Orbits in the ongoing study of the Solar system. The unprecedented success of some of the missions described is clearly due to the developed accurate theory of trajectories and orbits, including the perturbation methods.

Two requirements of the trajectories have to be outlined here. First, any flight should be accomplished with rapprochements with possibly larger number' of planets, satellites, asteroids etc. Second, the acceleration effects in the gravitational fields of the planets and satellites should be used most efficiently. In the first case, the scientific outcome of the missions is broadened essentially. In the second case, the durations of the flights can be reduced drastically, up to three and more times. In some cases, for example flights to Mercury, the perturbation maneuver is the only way to be captured around the planet.

It appears almost usual: all remote flights were realized with rapprochements with asteroids. The probability to have rendezvous with asteroids can be estimated without difficulty: over 20 000 asteroids are already cataloged, one-tenth of the estimated number, discovered during the past 200 years since of the discovery of the first asteroid, Ceres, in 1801.

Much more difficult is the problem with the realization of gravity assists in the field of a given planet, since the slightest inaccuracy can be fatal, as happened with Voyager 1. Nevertheless, in many cases multiple gravitating maneuvers are the only way out of the situation, as in the case of the Messenger: on the Earth–Mercury route it needs a record number of gravity assists, five, in the gravitational fields of three planets.

Among the already realized projects mentioned above, the Galileo Earth–Jupiter mission should indeed be considered as a good choice from all points of view, with the realization of two gravity assists, passage by two asteroids, capture by Jupiter, entry of a probe with scientific instruments into Jupiter's atmosphere, and finally, transmission during two years of thousands of color images of all Galilean satellites and Jupiter itself.

The contemporary space flights represent a spectacular gallery of interplanetary trajectories and, at the same time, mark the triumph of Celestial Mechanics, the epoch of merging of the theory and practice, as presumably hardly was dreamt by the founders of Celestial Mechanics two centuries ago.

Chapter XVI

Canonical Equations of Celestial Mechanics

1 Statement of the Problem

The entire history of celestial mechanics is nothing other than a history of continuous development of new methods, tools, principles, transformations, etc., without which no solution of any fundamental problem is imaginable. Good examples are the geometrical concepts, which have led finally to the discovery of new integration variables – the eccentric anomaly E in the case of an elliptic orbit, or the function F in the case of a hyperbolic orbit. Those quantities enabled the integration of corresponding differential equations as well as the determination of the equations of motion. Another example would be the **general operation**, a remarkable method in perturbation theory, which enables one to rewrite Euler's set of differential equations for osculating elements of the orbit, the solution of which by successive approximations leads to the discovery of the desired dependence between elements of the orbit and time.

The appearance more than one and a half centuries ago of the theory of **canonical equations**, the Lagrange equations in **generalized** coordinates and the Hamiltonian equations in **canonical** coordinates should undoubtedly be considered among the highlights of celestial mechanics. The role of canonical formalism, owing to its strictness and comprehensiveness, obviously concerns not only celestial mechanics; at present it is one of the most important branches of theoretical mechanics. The theory of canonical equations obviously covers the systems that consist of an arbitrary number of material points under the action of conservative forces. A particular case of such systems is those containing two bodies interacting by Newtonian gravitation; here both the Lagrange and the Hamilton equations lead to rather elegant treatments of the two-body problem with well-known results – the Keplerian relationships.

Note, however, that the mathematical treatment of the theory of canonical equations is far from simple. In particular, it concerns the Hamiltonian canonical equations. Though, insofar as here the question is about the principal novelty introduced by the Hamiltonian equations into celestial mechanics, the mathematical difficulties cannot be considered fatal, they are as an obstacle, at least in principle, to the further development of the theory. Convincing examples are the KAM theorem with consequences concerning the global stability of systems that could be associated with the planetary systems, as well the stability of Lagrangian triangle solutions. Such results were obtained as a result of deep analysis of Hamiltonian systems and canonical equations, by means of overcoming extreme mathematical difficulties.

The present chapter is devoted to the canonical systems – to the derivation and analysis both of Lagrangian equations in generalized coordinates and of Hamiltonian equations in canonical variables. Obviously we will confine ourselves to outlining only the basics of the general theory and consideration of a few examples of its application. The results obtained here, of course, do not add new essential knowledge to the given particular physical problem, but they demonstrate the efficiency of this approach.

2   Generalized Coordinates

Consider a system consisting of k material points of masses m_1, m_2, ..., m_k and performing some smooth motion. We assume that the motion of these material points takes place not independently of each other but rather with their mutual interaction by some force. In this particular case it can be the Newtonian force of gravity. However, in general, the behavior of the system will be determined by the conditions that are called **constraints**. These liaisons may be expressed in terms of equations including the coordinates of all points x_i, y_i, z_i, as well as the time t. In the general case these equations can be written in the form

$$
\begin{aligned}
f_1\left(t; x_1, y_1, z_1, \ldots, x_k, y_k, z_k\right) &= 0, \\
f_2\left(t; x_1, y_1, z_1, \ldots, x_k, y_k, z_k\right) &= 0, \\
&\ \ \vdots \\
f_l\left(t; x_1, y_1, z_1, \ldots, x_k, y_k, z_k\right) &= 0.
\end{aligned}
\tag{1}
$$

The position of any point i in the system is determined by three coordinates x_i, y_i, z_i, the number of the points is k, therefore the total number of variables, i.e. coordinates, will be $3k$. However, the number of constraints is l; this is also the number of equations in (1). Obviously, the number of these equations l will be less than the total number of variable coordinates, i.e. $3k$. Therefore, in the limit when $l = 3k$, we shall have in Eq. (1) as many equations as the number of unknown quantities x_i, y_i, and z_i, which can be obtained by solving the dynamical problem.

Thus, the number of equations in Eq. (1) is less than the three variables x_i, y_i, and z_i. However, all these variables are connected with each other in Eq. (1). Therefore, we can express these variables as functions of new, completely independent parameters with total number equal to $3k - l$. Moreover, these new parameters can be chosen arbitrarily, i.e. it is not compulsory to have, say, coordinates; it is sufficient that they be bearers of definite and simple mechanical or geometrical meanings – velocity, momentum, etc. Lagrange was the first to mention such a possibility.

Thus, the location of the system will be determined by n independent parameters, i.e.

$$q_1, q_2, \ldots, q_n,$$

where $n = 3k - l$. In this case the coordinates x_i, y_i, z_i can be expressed as functions of these parameters and t. As a result we will have the following relationship for the dependence of coordinates x_i, y_i, z_i on the parameter q_i

$$
\begin{aligned}
x_i &= x_i \left(q_1, q_2, \ldots, q_n; t\right), \\
y_i &= y_i \left(q_1, q_2, \ldots, q_n; t\right), \\
z_i &= z_i \left(q_1, q_2, \ldots, q_n; t\right),
\end{aligned}
\tag{2}
$$

where $i = 1, 2, \ldots, k$.

These parameters, $q_1, q_2, \ldots, q_n$, uniquely determine the coordinates x_i, y_i, z_i and, hence, the position of the system at a given moment of time t. These parameters are called **generalized coordinates**. Lagrange formulated the problem thus: it is necessary to derive the differential equations for the generalized coordinates q_j.

3 Lagrange Equations

We start to form the set of differential equations of motion for a material point m_i in the system of n material points of masses $m_1, m_2, \ldots, m_n$ and coordinates $x_1, y_1, z_1, x_2, y_2, z_2, \ldots, x_n, y_n, z_n$. In the most general case, in which no limitations are imposed on the nature of the forces acting between the material points $m_1, m_2, \ldots, m_n$, these equations can be written, for the mass m_i, in the form

$$m_i \frac{d^2 x_i}{dt^2} = X_i,$$

$$m_i \frac{d^2 y_i}{dt^2} = Y_i, \qquad (3)$$

$$m_i \frac{d^2 z_i}{dt^2} = Z_i,$$

where X_i, Y_i, and Z_i are the sums of the components of the forces acting along coordinate axes x, y, z.

Our aim is to derive differential equations of motion for the material point m_i, analogously to Eq. (3) and represented through the new variables q_j. Let us derive first of all the auxiliary expressions.

Denote by T the kinetic energy of the system. As a matter of fact, up to now we knew only one way to represent of the kinetic energy of the system, namely, through the coordinates of the point x_i, y_i, z_i and the time t

$$T = \frac{1}{2} \sum m_i \left[\left(\frac{dx_i}{dt} \right)^2 + \left(\frac{dy_i}{dt} \right)^2 + \left(\frac{dz_i}{dt} \right)^2 \right]. \qquad (4)$$

We aim now to represent T through the new variables q_i and their derivatives with respect to time $\dot{q}_i$. We have from Eq. (2)

$$\dot{x}_i = \frac{dx_i}{dt} = \frac{\partial x_i}{\partial q_1} \dot{q}_1 + \frac{\partial x_i}{\partial q_2} \dot{q}_2 + \cdots + \frac{\partial x_i}{\partial q_n} \dot{q}_n + \frac{\partial x_i}{\partial t},$$

$$\dot{y}_i = \frac{dy_i}{dt} = \frac{\partial y_i}{\partial q_1} \dot{q}_1 + \frac{\partial y_i}{\partial q_2} \dot{q}_2 + \cdots + \frac{\partial y_i}{\partial q_n} \dot{q}_n + \frac{\partial y_i}{\partial t}, \qquad (5)$$

$$\dot{z}_i = \frac{dz_i}{dt} = \frac{\partial z_i}{\partial q_1} \dot{q}_1 + \frac{\partial z_i}{\partial q_2} \dot{q}_2 + \cdots + \frac{\partial z_i}{\partial q_n} \dot{q}_n + \frac{\partial z_i}{\partial t},$$

where

$$\dot{q}_j = \frac{dq_j}{dt}, \qquad j = 1, 2, \ldots n, \qquad (6)$$

is none other than the velocity of the variation of the parameter q_j with time. Correspondingly the magnitudes $\dot{q}_1, \dot{q}_2, \ldots, \dot{q}_n$ are called the **generalized velocities** of the system.

Now, with the help of Eq. (5) we can express T through the generalized coordinates q_j, generalized velocity $\dot{q}_j$ and time t. Differentiating Eq. (4) by $\dot{q}_j$, we have

$$\frac{\partial T}{\partial \dot{q}_j} = \sum m_i \left(\dot{x}_i \frac{\partial \dot{x}_i}{\partial \dot{q}_j} + \dot{y}_i \frac{\partial \dot{y}_i}{\partial \dot{q}_j} + \dot{z}_i \frac{\partial \dot{z}_i}{\partial \dot{q}_j} \right), \qquad (7)$$

where
$$\dot{x}_i = \frac{dx_i}{dt}, \quad \dot{y}_i = \frac{dy_i}{dt}, \quad \dot{z}_i = \frac{dz_i}{dt}.$$

By differentiating now $\dot{x}_i$, $\dot{y}_i$, $\dot{z}_i$ from Eq. (5) by $\dot{q}_j$, we obtain the following remarkable relationships:

$$\frac{\partial \dot{x}_i}{\partial \dot{q}_j} = \frac{\partial x_i}{\partial q_j},$$

$$\frac{\partial \dot{y}_i}{\partial \dot{q}_j} = \frac{\partial y_i}{\partial q_j}, \tag{8}$$

$$\frac{\partial \dot{z}_i}{\partial \dot{q}_j} = \frac{\partial z_i}{\partial q_j}.$$

Taking into account Eq. (8), we have from Eq. (7) finally

$$\frac{\partial T}{\partial \dot{q}_j} = \sum_{i=1}^{k} m_i \left(\dot{x}_i \frac{\partial x_i}{\partial q_j} + \dot{y}_i \frac{\partial y_i}{\partial q_j} + \dot{z}_i \frac{\partial z_i}{\partial q_j} \right). \tag{9}$$

Analogously we obtain, differentiating Eq. (4) now by q_j

$$\frac{\partial T}{\partial q_j} = \sum_{i=1}^{k} m_i \left(\dot{x}_i \frac{\partial \dot{x}_i}{\partial q_j} + \dot{y}_i \frac{\partial \dot{y}_i}{\partial q_j} + \dot{z}_i \frac{\partial \dot{z}_i}{\partial q_j} \right). \tag{10}$$

It is necessary to obtain one more expression for the derivatives $\dot{x}_i$, $\dot{y}_i$ and $\dot{z}_i$ with respect to q_j, that enter into these equations. We have from the first line of Eq. (5)

$$\frac{\partial \dot{x}_i}{\partial q_j} = \frac{\partial^2 x_i}{\partial q_1 \partial q_j} \dot{q}_1 + \frac{\partial^2 x_i}{\partial q_2 \partial q_j} \dot{q}_2 + \cdots + \frac{\partial^2 x_i}{\partial q_n \partial q_j} \dot{q}_n + \frac{\partial^2 x_i}{\partial t \, \partial q_j}. \tag{11}$$

On the other hand, by differentiating, first, the right-hand side of Eq. (2) by q_j, then obtaining its full derivative with respect to t, we have (bearing in mind also that the operation of partial differentiation does not depend on the order of differentiation)

$$\frac{d}{dt} \left(\frac{\partial x_i}{\partial q_j} \right) = \frac{\partial^2 x_i}{\partial q_1 \partial q_j} \dot{q}_1 + \frac{\partial^2 x_i}{\partial q_2 \partial q_j} \dot{q}_2 + \cdots + \frac{\partial^2 x_i}{\partial q_n \partial q_j} \dot{q}_n + \frac{\partial^2 x_i}{\partial t \, \partial q_j}. \tag{12}$$

The right-hand sides of Eq. (11) and (12) are identical. Therefore we have

$$\frac{\partial \dot{x}_i}{\partial q_j} = \frac{d}{dt} \left(\frac{\partial x_i}{\partial q_j} \right). \tag{13}$$

Analogously, we obtain for $\dot{y}_i$ and $\dot{z}_i$

$$\frac{\partial \dot{y}_i}{\partial q_j} = \frac{d}{dt} \left(\frac{\partial y_i}{\partial q_j} \right),$$

$$\frac{\partial \dot{z}_i}{\partial q_j} = \frac{d}{dt} \left(\frac{\partial z_i}{\partial q_j} \right). \tag{14}$$

Then we have from Eq. (10) finally, considering Eqs. (13) and (14)

$$\frac{\partial T}{\partial q_j} = \sum_{i=1}^{k} m_i \left[\dot{x}_i \frac{d}{dt} \left(\frac{\partial x_i}{\partial q_j} \right) + \dot{y}_i \frac{d}{dt} \left(\frac{\partial y_i}{\partial q_j} \right) + \dot{z}_i \frac{d}{dt} \left(\frac{\partial z_i}{\partial q_j} \right) \right]. \tag{15}$$

Thus, we have completed the preparatory part of our problem. Let us turn now to the main problem, to the derivation of differential equations of motion of the material point m_i in terms of their dependence on the new variables q_j.

After multiplication of Eqs. (3) by $\partial x_i/\partial q_j$, $\partial y_i/\partial q_j$, and $\partial z_i/\partial q_j$, correspondingly, and summing the results, we obtain

$$\sum_{i=1}^{k} m_i \left(\frac{d^2 x_i}{dt^2} \frac{\partial x_i}{\partial q_j} + \frac{d^2 y_i}{dt^2} \frac{\partial y_i}{\partial q_j} + \frac{d^2 z_i}{dt^2} \frac{\partial z_i}{\partial q_j} \right)$$
$$= \sum_{i=1}^{k} \left(X_i \frac{\partial x_i}{\partial q_j} + Y_i \frac{\partial y_i}{\partial q_j} + Z_i \frac{\partial z_i}{\partial q_j} \right), \tag{16}$$

where the index j takes values from 1 to n. The expression within the parentheses in Eq. (16) can be rewritten in the following manner:

$$\frac{d^2 x_i}{dt^2} \frac{\partial x_i}{\partial q_j} + \frac{d^2 y_i}{dt^2} \frac{\partial y_i}{\partial q_j} + \frac{d^2 z_i}{dt^2} \frac{\partial z_i}{\partial q_j} = \frac{d}{dt} \left(\dot{x}_i \frac{\partial x_i}{\partial q_j} + \dot{y}_i \frac{\partial y_i}{\partial q_j} + \dot{z}_i \frac{\partial z_i}{\partial q_j} \right)$$
$$- \left[\dot{x}_i \frac{d}{dt} \left(\frac{\partial x_i}{\partial q_j} \right) + \dot{y}_i \frac{d}{dt} \left(\frac{\partial y_i}{\partial q_j} \right) + \dot{z}_i \frac{d}{dt} \left(\frac{\partial z_i}{\partial q_j} \right) \right]. \tag{17}$$

Then Eq. (16) takes the form

$$\frac{d}{dt} \sum_{i=1}^{k} \left(\dot{x}_i \frac{\partial x_i}{\partial q_j} + \dot{y}_i \frac{\partial y_i}{\partial q_j} + \dot{z}_i \frac{\partial z_i}{\partial q_j} \right)$$
$$- \sum_{i=1}^{k} m_i \left[\dot{x}_i \frac{d}{dt} \left(\frac{\partial x_i}{\partial q_j} \right) + \dot{y}_i \frac{d}{dt} \left(\frac{\partial y_i}{\partial q_j} \right) + \dot{z}_i \frac{d}{dt} \left(\frac{\partial z_i}{\partial q_j} \right) \right] \tag{18}$$
$$= \sum_{i=1}^{k} \left(X_i \frac{\partial x_i}{\partial q_j} + Y_i \frac{\partial y_i}{\partial q_j} + Z_i \frac{\partial z_i}{\partial q_j} \right).$$

Comparing this expression with the one derived above, i.e. with Eqs. (9) and (15), it is not difficult to see that the first summation on the left-hand side of Eq. (18) coincides with Eq. (9) and the second term of Eq. (18) with Eq. (15). Therefore we can write, instead of Eq. (18)

$$\frac{d}{dt}\left(\frac{\partial T}{\partial \dot{q}_j}\right) - \frac{\partial T}{\partial q_j} = R_j, \qquad j = 1, 2, \ldots, n, \tag{19}$$

where we have denoted

$$R_j = \sum_{i=1}^{k}\left(X_i \frac{\partial x_i}{\partial q_j} + Y_i \frac{\partial y_i}{\partial q_j} + Z_i \frac{\partial z_i}{\partial q_j}\right). \tag{20}$$

Eq. (19) represents the generalized coordinates, or the parameters $q_1, q_2, \ldots, q_n$ as time functions, and equations of this type are known as **Lagrange equations**. Lagrange equations, as follows from Eqs. (19) and (6), represent themselves as a set of n differential equations of second order. Eq. (19) represents the general form of Lagrange equations, when no limitations are imposed on the nature of the function R_j, i.e. the relation of the components of acting force X_i, Y_i, and Z_i.

When applied to real systems with k bodies interacting according to the law of Newtonian gravity, the Lagrange equations can be considerably simplified due to the fact that the Newtonian law has a potential, i.e. the components of X_i, Y_i, and Z_i can be represented through a force function U in the following manner:

$$X_i = \frac{\partial U}{\partial x_i}, \quad Y_i = \frac{\partial U}{\partial y_i}, \quad Z_i = \frac{\partial U}{\partial z_i}. \tag{21}$$

In this case T depends on $t, q_1, q_2, \ldots, q_n$, as well as on $\dot{q}_1, \dot{q}_2, \ldots \dot{q}_n$, and U depends only on $t, q_1, q_2, \ldots, q_n$.

Substituting the values of X_i, Y_i, and Z_i from Eq. (21) into Eq. (20), we obtain for the function R_j

$$R_j = \sum_{i=1}^{k}\left(\frac{\partial U}{\partial x_i}\frac{\partial x_i}{\partial q_j} + \frac{\partial U}{\partial y_i}\frac{\partial y_i}{\partial q_j} + \frac{\partial U}{\partial z_i}\frac{\partial z_i}{\partial q_j}\right). \tag{22}$$

The right-hand side of this expression represents the full derivative of the function U on q_j. Therefore we have

$$R_j = \frac{\partial U}{\partial q_j}, \qquad j = 1, 2, \ldots, n. \tag{23}$$

Taking into account Eq. (23), the Lagrange equations can be written in the form

$$\frac{d}{dt}\left(\frac{\partial T}{\partial \dot{q}_j}\right) - \frac{\partial(U+T)}{\partial q_j} = 0, \qquad j = 1, 2, \ldots, n. \qquad (24)$$

The Lagrange equations, even in relatively simple form, can be solved only if the kinetic energy T and potential function U are expressed as a function of time t, generalized variables $q_1, q_2, \ldots, q_n$ and their derivatives $\dot{q}_1, \dot{q}_2, \ldots, \dot{q}$.

Let us introduce the notation

$$L = T + U. \qquad (25)$$

Taking into consideration that the function U does not depend on $\dot{q}_1, \dot{q}_2, \ldots, \dot{q}_n$, i.e.

$$\frac{\partial U}{\partial \dot{q}_j} \equiv 0,$$

Eq. (24) finally can be written in the following symmetrical and elegant form:

$$\frac{d}{dt}\left(\frac{\partial L}{\partial \dot{q}_j}\right) - \frac{\partial L}{\partial q_j} = 0, \qquad (26)$$

where $L = L(t, q_k, \dot{q}_k)$ is called a **Lagrange function** or **Lagrangian**.

Eq. (26) determines fully the motion of the system, i.e. its state at any moment of time. In the Lagrange equations, both the generalized coordinates q_j and the generalized velocities $\dot{q}_j$ are used. These equations, or more precisely, their integration over independent coordinates, offer the shortest means of the solution of the central problem of mechanics – the problem of motion of the system. Indeed, while using the Lagrange equations we will have the least number of differential equations, it being equal to the number of degrees of freedom of the system.

As a matter of fact, the Lagrange equations (of the second kind and in the system of conservative forces) follow directly from the d'Alembert principle, i.e. as a result of introduction of generalized coordinates in the d'Alembert principle:

$$\sum\left(X_i - m_i \frac{d^2 x_i}{dt^2}\right),$$

$$\sum\left(Y_i - m_i \frac{d^2 y_i}{dt^2}\right),$$

$$\sum\left(Z_i - m_i \frac{d^2 z_i}{dt^2}\right).$$

The d'Alembert principle states that the external forces X, Y, and Z and the forces of inertia $m\ddot{x}$, $\ddot{y}$, and $\ddot{z}$ must always be in equilibrium. Strictly speaking, this principle represents nothing other than Newton's third law: the action is equal to its reaction. However, the d'Alembert principle's merit is in the fact of the serious involvement of this principle in the dynamics, thus reducing the problem of dynamics to a simple problem of statics. Below we will examine several examples of application of Lagrange equations to the problems of celestial mechanics.

4 Canonical Equations of Free Motion

Consider the free motion of a point mass m with coordinates x, y, z. In this case $n = 3$ and therefore

$$q_1 = x, \qquad q_2 = y \qquad q_3 = z,$$
$$\dot{q}_1 = \dot{x}, \qquad \dot{q}_2 = \dot{y}, \qquad \dot{q}_3 = \dot{z}.$$

For the kinetic energy of this point we have

$$T = \frac{1}{2} m \left(\dot{x}^2 + \dot{y}^2 + \dot{z}^2 \right). \tag{27}$$

We have also for the components of the acting force X, Y, and Z, represented through the potential function U (taking into account also the sign)

$$X = -\frac{\partial U}{\partial x}, \qquad Y = -\frac{\partial U}{\partial y}, \qquad Z = -\frac{\partial U}{\partial z}.$$

We shall confine ourselves to representing the computations only for the coordinate x. The sequence is as follows.

We have for the derivatives of the function U

$$\frac{\partial U}{\partial x} = \frac{\partial U}{\partial q_1} = -X,$$
$$\frac{\partial U}{\partial \dot{x}} = 0, \tag{28}$$
$$\frac{d}{dt} \left(\frac{\partial U}{\partial \dot{x}} \right) = 0,$$

and for derivatives of the function T

$$\frac{\partial T}{\partial x} = \frac{\partial T}{\partial q_1} = 0,$$
$$\frac{\partial T}{\partial \dot{x}} = m\dot{x}, \tag{29}$$
$$\frac{d}{dt} \left(\frac{\partial T}{\partial \dot{x}} \right) = m\ddot{x}.$$

For the coordinate x we have from the Lagrange equations Eq. (26), if $L = T - U$

$$\frac{d}{dt}\left(\frac{\partial(T-U)}{\partial\dot{x}}\right) + \frac{\partial(T-U)}{\partial x}$$

$$= \frac{d}{dt}\left(\frac{\partial T}{\partial\dot{x}}\right) - \frac{d}{dt}\left(\frac{\partial U}{\partial\dot{x}}\right) - \frac{\partial T}{\partial x} + \frac{\partial U}{\partial x},$$

or, by substituting the values of corresponding derivatives from Eqs. (28) and (29), we arrive at the following well-known expression, i.e. the differential equation of motion in the usual Newtonian form. This trivial result of course does not contain anything new; it merely illustrates the technique of application of Lagrange equations to analysis of the motion of a point mass.

5   Equations of Motion in Spherical Coordinates

Lagrange equations can also be used for the transformation of equations of motion into different coordinate systems. As an example, we shall examine here the transformation of Eq. (3) into spherical coordinates.

The Cartesian coordinates x, y, z of the point mass m are related to its spherical coordinates r, φ and θ through the following formulas:

$$\begin{aligned}
x &= r\,\cos\varphi\,\cos\theta, \\
y &= r\,\cos\varphi\,\sin\theta, \\
z &= r\,\sin\varphi,
\end{aligned} \tag{30}$$

where r is the radius-vector of the point m, φ is the latitude, and θ is the longitude. By differentiating Eq. (30) we obtain for the derivatives $\dot{x}$, $\dot{y}$, and $\dot{z}$

$$\begin{aligned}
\dot{x} &= \dot{r}\,\cos\varphi\,\cos\theta - r\dot{\varphi}\,\sin\varphi\,\cos\theta - r\dot{\theta}\,\cos\varphi\,\sin\theta, \\
\dot{y} &= \dot{r}\,\cos\varphi\,\sin\theta - r\dot{\varphi}\,\sin\varphi\,\sin\theta + r\dot{\theta}\,\cos\varphi\,\cos\theta, \\
\dot{z} &= \dot{r}\,\sin\varphi + r\dot{\varphi}\,\cos\varphi.
\end{aligned} \tag{31}$$

On substituting these values of $\dot{x}$, $\dot{y}$, and $\dot{z}$ into Eq. (27) we obtain for the kinetic energy T of the point mass m

$$T = \frac{1}{2}\,m\left(\dot{r}^2 + r^2\dot{\varphi}^2 + r^2\dot{\theta}^2\,\cos^2\varphi\right). \tag{32}$$

For the potential energy we have in the most common case

$$U = U\left(r, \varphi, \theta\right) \tag{33}$$

Thus we have for the generalized coordinates q_1, q_2, q_3 and velocities $\dot{q}_1$, $\dot{q}_1$, and $\dot{q}_3$

$$q_1 = r, \qquad q_2 = \varphi \qquad q_3 = \theta,$$
$$\dot{q}_1 = \dot{r}, \qquad \dot{q}_2 = \dot{\varphi}, \qquad \dot{q}_3 = \dot{\theta}. \tag{34}$$

From Eqs. (31)–(33) we have for the first coordinate $(j = 1)$

$$\frac{\partial T}{\partial q_1} = \frac{\partial T}{\partial r} = mr\dot{\varphi}^2 + mr\dot{\theta}^2 \cos^2\varphi,$$

$$\frac{\partial T}{\partial \dot{q}_1} = \frac{\partial T}{\partial \dot{r}} = m\dot{r},$$

$$\frac{d}{dt}\left(\frac{\partial T}{\partial \dot{q}_1}\right) = m\ddot{r}, \tag{35}$$

$$\frac{\partial U}{\partial q_1} = \frac{\partial U}{\partial r}.$$

By substituting all these into the Lagrange equations (26) with $j = 1$, we obtain the first differential equation of the motion:

$$\frac{d^2 r}{dt^2} - r\left(\frac{d\varphi}{dr}\right)^2 - r\,\cos^2\varphi\left(\frac{d\theta}{dt}\right)^2 = \frac{1}{m}\frac{\partial U}{\partial r}. \tag{36}$$

Analogously one can obtain for the second coordinate q_2

$$\frac{\partial T}{\partial q_2} = \frac{\partial T}{\partial \varphi} = -mr^2\dot{\theta}^2 \cos\varphi \, \sin\varphi,$$

$$\frac{\partial T}{\partial \dot{q}_2} = \frac{\partial T}{\partial \dot{\varphi}} = mr^2\dot{\varphi},$$

$$\frac{d}{dt}\left(\frac{\partial T}{\partial \dot{q}_2}\right) = 2mr\dot{r}\dot{\varphi} + mr^2\ddot{\varphi}, \tag{37}$$

$$\frac{\partial U}{\partial q_2} = \frac{\partial U}{\partial \varphi}.$$

For the second differential equation of the motion $(j = 2)$ we have

$$\frac{d}{dt}\left(r^2\frac{d\varphi}{dt}\right) + r^2 \, \cos\varphi \, \sin\varphi \left(\frac{d\theta}{dt}\right)^2 = \frac{1}{m}\frac{\partial U}{\partial \varphi}. \tag{38}$$

Finally, we have for the third coordinate q_3

$$\frac{\partial T}{\partial q_3} = \frac{\partial T}{\partial \theta} = 0,$$

$$\frac{\partial T}{\partial \dot{q}_3} = \frac{\partial T}{\partial \dot{\theta}} = mr^2\dot{\theta} \, \cos^2 \varphi,$$

$$\frac{d}{dt}\left(\frac{\partial T}{\partial \dot{q}_3}\right) = 2mr\dot{r}\dot{\theta} \, \cos^2 \theta - 2mr^2\dot{\varphi}\dot{\theta} \, \sin \varphi \, \cos \varphi + mr^2\ddot{\theta} \, \cos^2 \varphi, \tag{39}$$

$$\frac{\partial U}{\partial q_3} = \frac{\partial U}{\partial \theta},$$

and then the third differential equation of the motion will be written in the following form ($j = 3$):

$$\frac{d}{dt}\left(r^2 \, \cos^2 \varphi \, \frac{d\theta}{dt}\right) = \frac{1}{m}\frac{\partial U}{\partial \theta}. \tag{40}$$

6   Derivation of Kepler's First Law from the Lagrange Equations

Consider the motion of a point mass m (a planet) in a plane with polar coordinates r and θ with respect to the central body of mass M (Sun). In this case we have $\varphi = 0$, $d\varphi/dt = 0$ and hence

$$x = r \, \cos \varphi,$$

$$y = r \, \sin \varphi,$$

Accordingly we have

$$U = U \, (r, \, 0, \, 0) = k^2 \, \frac{mM}{r},$$

$$\frac{\partial U}{\partial r} = F_r = -k^2 \, \frac{mM}{r^2} = -mg,$$

$$\frac{\partial U}{\partial \varphi} = 0,$$

$$\frac{\partial U}{\partial \theta} = 0.$$

With these data we obtain from Eq. (36) for the first equation of the motion

$$m\frac{d^2r}{dt^2} = mr\left(\frac{d\theta}{dt}\right)^2 - mg. \tag{41}$$

Here the first term on the right-hand side is centrifugal force, the second term, mg, is the gravitational force, and the term on the left-hand side is the resulting force. As a matter of fact, Eq. (41) is one of the differential equations of the motion containing two unknown variables, r and θ. In order to find them, it is necessary to have the second equation of motion.

For the second equation, by substituting into Eq. (40) it $\varphi = 0$ and $\partial U/\partial\theta = 0$, we have

$$\frac{d}{dt}\left(r^2\frac{d\theta}{dt}\right) = 0. \tag{42}$$

From that

$$r^2\frac{d\theta}{dt} = c. \tag{43}$$

This is Kepler's second law, i.e. the constancy of sectorial velocity of motion of the point m with respect to the central body.

The joint solution of systems (41) and (43) can be carried out by a known scheme, i.e. first, the derivation of the equation of the orbit (Chapter III), and then of the equation of the motion through various orbits (Chapter IV), etc.

7   Transformation with Respect to the Rotating Frame of Coordinates

As the last example of application of the Lagrange equations, we will now examine the problem of transformation of the differential equations of motion written for the stationary i.e. rest system of coordinates ξ, η, ζ:

$$m\ddot{\xi} = F_\xi, \quad m\ddot{\eta} = F_\eta, \quad m\ddot{\zeta} = F_\zeta, \tag{44}$$

in relation to the rotational system of coordinates x, y, z. Let the axis $0Z$ in the new rotating system coincide with the axis $O\zeta$ of the old (rest) system, and $v = v(t)$ be the angle between the axes OX and $O\xi$; here $v(t)$ is a known function of time. Then we shall have for the transition from the old system ξ, η, ζ to the new system x, y, z

$$\xi = x\,\cos v - y\,\sin v,$$

$$\eta = x\,\sin v + y\,\cos v, \tag{45}$$

$$\zeta = z.$$

After differentiating, we obtain

$$\dot{\xi} = \dot{x}\,\cos v - \dot{y}\,\sin v - \dot{v}\,(x\,\sin v + y\,\sin v),$$

$$\dot{\eta} = \dot{x}\,\sin v + \dot{y}\,\cos v + \dot{v}\,(x\,\cos v - y\,\cos v)\,,$$

$$\dot{\zeta} = \dot{z}\,.$$

The expression for the kinetic energy T in this case takes the following form:

$$T = \frac{1}{2}m\left(\dot{\zeta}^2 + \dot{\eta}^2 + \dot{\xi}^2\right)$$

$$= \frac{1}{2}m\left[\dot{x}^2 + \dot{y}^2 + \dot{z}^2 + 2\dot{v}(x\dot{y} - y\dot{x}) + \dot{v}^2\left(x^2 + y^2\right)\right]\,.$$

(46)

Taking as generalized coordinates x, y, z, i.e. assuming that

$$q_1 = x\,,\quad q_2 = y\,,\quad q_3 = z\,,$$

we shall have for the partial derivatives of kinetic energy T along the three coordinates x, y, z

$$\frac{\partial T}{\partial x} = m\dot{v}\,(\dot{y} + \dot{v}x)\,,\qquad \frac{\partial T}{\partial \dot{x}} = m\,(\dot{x} - \dot{v}y)\,,$$

$$\frac{\partial T}{\partial y} = m\dot{v}\,(-\dot{x} + \dot{v}y)\,,\qquad \frac{\partial T}{\partial \dot{y}} = m\,(\dot{y} + \dot{v}x)\,,$$

$$\frac{\partial T}{\partial z} = 0\,,\qquad \frac{\partial T}{\partial \dot{z}} = m\dot{z}\,.$$

By substituting these derivatives into the Lagrange equations (26) we obtain the following three differential equations of motion of the point m in the rotational coordinates x, y, z:

$$\ddot{x} - 2\dot{v}\dot{y} - \dot{v}^2 x - \ddot{v}y = \frac{1}{m}X\,,$$

$$\ddot{y} + 2\dot{v}\dot{x} - \dot{v}^2 y + \ddot{v}x = \frac{1}{m}Y\,,$$

(47)

$$\ddot{z} = \frac{1}{m}Z\,.$$

where X, Y, and Z are now the projections of the acting forces on to the motionless axes x, y, z.

8   Derivation of the Jacobi Integral from the Lagrange Equations

In the restricted three-body problem we often have a situation in which the rotation of the coordinate system around the axis OZ takes place uniformly, i.e. with constant angular velocity n. Then we have

$$v = nt, \qquad \dot{v} = \frac{d(nt)}{dt} = n, \qquad \ddot{v} = 0.$$

In this case Eq. (47) takes the following form:

$$\ddot{x} - 2n\dot{y} - n^2 x = \frac{1}{m} X,$$

$$\ddot{y} + 2n\dot{x} - n^2 y = \frac{1}{m} Y, \tag{48}$$

$$\ddot{z} = \frac{1}{m} Z.$$

These are nothing other than the well-known differential equations of the restricted three-body problem for the motion of an infinitely small mass m in the field of two bodies of finite masses, μ and $1 - \mu$. In this case X, Y, and Z are the sums of the components of the actions of the finite mass on the point m. Assuming the existence of a force function U in the form

$$U = \frac{1}{2} mn^2 \left(x^2 + y^2\right) + \frac{1 - \mu}{r_1} + \frac{\mu}{r_2},$$

we can reduce Eq. (48) to a more compact form:

$$\frac{d^2 x}{dt^2} - 2n \frac{dy}{dt} = \frac{1}{m} \frac{\partial U}{\partial x},$$

$$\frac{d^2 y}{dt^2} + 2n \frac{dx}{dt} = \frac{1}{m} \frac{\partial U}{\partial y}, \tag{49}$$

$$\frac{d^2 z}{dt^2} = \frac{1}{m} \frac{\partial U}{\partial z}.$$

This system in its turn can be reduced to the well-known Jacobi integral

$$V^2 = \left(\frac{dx}{dt}\right)^2 + \left(\frac{dy}{dt}\right)^2 + \left(\frac{dz}{dt}\right)^2 = 2U - C, \tag{50}$$

where V is the total velocity of the point m, and C is the constant of integration.

The examples examined above reveal the great possibilities of the Lagrange equations. The only disadvantage of these equations, apparently, is the fact that they are of second order. By increasing somehow the number of unknown functions, one can reduce the degree of the equations, i.e. replace Eq. (26) by equivalent differential equations of the first order. This problem was solved by Hamilton in a perfect way.

9   Canonical Equations

Let us start from the Lagrange differential equations of motion (26) for a system with n bodies:

$$\frac{d}{dt}\left(\frac{\partial L}{\partial \dot{q}_j}\right) - \frac{\partial L}{\partial q_j} = 0 \, . \tag{51}$$

when

$$L = T + U \, , \tag{52}$$

where T is the kinetic energy, and U is the potential energy (with opposite sign). In other words, the acting forces R_j in the Lagrange equations (19), written for the general case, are the partial derivatives of the function U:

$$R_j = \frac{\partial U}{\partial q_j} \, .$$

Let us introduce, instead of derivatives $\dot{q}_j$ of the generalized coordinates q_j, new variables p_j in the following manner:

$$p_j = \frac{\partial L}{\partial \dot{q}_j} \, . \tag{53}$$

If the potential function U does not depend on the velocity $\dot{q}_j$, we shall have from Eq. (53), taking into account Eq. (52)

$$p_j = \frac{\partial T}{\partial \dot{q}_j} \, . \tag{54}$$

Then Eq. (51) will be written in the form

$$\frac{dp_j}{dt} - \frac{\partial L}{\partial q_j} = 0 \, . \tag{55}$$

In order to reveal the content of the new variable p_j, it is enough to apply Eq. (27) for T, which is the kinetic energy of a free point. Bearing in mind that, in this case, $\dot{x} = \dot{q}_1$, $\dot{y} = \dot{q}_2$, and $\dot{z} = \dot{q}_3$, we shall have for p_j from Eqs. (54) and (27)

$$p_j = m\dot{q}_j \,, \tag{56}$$

i.e. p_j are linear functions of the quantities of $\dot{q}_j$. At the same time the p_j form a system of independent variables.

Insofar as $\dot{q}_j$ are the velocity components of the point m, then from Eq. (56) we can understand the physical content of this new variable p_j; it corresponds to the momentum, or more precisely, to the components of the latter. Therefore, p is called the **generalized momentum** or simply the **momentum**.

Thus, instead of n variables q_1, q_2, ..., q_n, we now have $2n$ variables q_j and p_j, i.e.

$$\begin{aligned} q_1, q_2, \ldots, q_n \,, \\ p_1, p_2, \ldots, p_n \,. \end{aligned} \tag{57}$$

These new variables q_j and p_j are called **canonical**.

Recall that, in the case of the Lagrange equations, the variables were q_j and $\dot{q}_j$, i.e.

$$\begin{aligned} q_1, q_2, \ldots, q_n \,, \\ \dot{q}_1, \dot{q}_2, \ldots, \dot{q}_n \,. \end{aligned}$$

Our aim is to derive the differential equations determining the canonical variables. The solution of this problem is reached by means of the Hamiltonian equations.

10 Hamiltonian Equations

The Lagrange equations deal with the Lagrangians, i.e. , the Lagrange function L, the kinetic potential

$$L = T + U \,. \tag{58}$$

The Hamilton equations are written for the Hamiltonian H given by the expression

$$H = T - U \,, \tag{59}$$

which is related to the Lagrange function through the expression

$$H = 2T - L \,. \tag{60}$$

The Hamiltonian function represents the complete energy of the system if the latter is a conservative one, i.e. when H is represented as a sum of kinetic and potential energies.

Our aim is to derive the differential equations of motion that determine the variables q and p. It appears that a rather simple and symmetrical form of these equations can be reached by introducing the Hamiltonian function.

The Lagrange function, as was shown above, is represented through the quantities q_j, $\dot{q}_j$, and t, i.e. $L = L(q_j, \dot{q}_j, t)$, while the Hamiltonian function is represented via p_j, q_j and t, i.e. $H = H(p_j, q_j, t)$. The kinetic energy of the system, T, depends, as is not difficult to show, on p_j and q_j. Indeed, if q_j are the generalized coordinates, and $\dot{q}_j$ the generalized velocities, then the kinetic energy T_j and momentum p_j of the point m_j will be $T_j = 1/2 m_j \dot{q}_j^2$ and $p_j = m_j \dot{q}_j$. Hence, $T_j = 1/2 \dot{q}_j p_j$, and for the complete kinetic energy of the system T we will have

$$T = \frac{1}{2} \sum_{j=1}^{k} \dot{q}_j p_j \,. \tag{61}$$

Therefore we can write for the Hamiltonian function H from Eq. (60)

$$H(p_j, q_j, t) = \sum_{j=1}^{k} \dot{q}_j p_j - L(q_j, \dot{q}_j, t) \,. \tag{62}$$

After differentiation of the function H, i.e. modifying p_j and q_j, we obtain

$$dH = \sum \frac{\partial H}{\partial p_j} dp_j + \sum \frac{\partial H}{\partial q_j} dq_j \,. \tag{63}$$

On the other hand, we have from Eq. (62)

$$dH = d\left(\sum \dot{q}_j p_j - L \right) = \sum \dot{q}_j dp_j + \sum p_j d\dot{q}_j - \sum \frac{\partial L}{\partial q_j} dq_j - \sum \frac{\partial L}{\partial \dot{q}_j} d\dot{q}_j$$

or, taking into account Eq. (53),

$$dH = \sum \dot{q}_j dp_j = -\sum \frac{\partial L}{\partial q_1} dq_j \,. \tag{64}$$

By comparison of the right-hand sides of Eqs. (63) and (64), we obtain

$$\frac{\partial L}{\partial q_j} = -\frac{\partial H}{\partial q_j} \,, \qquad \dot{q}_j = \frac{\partial H}{\partial p_j} \,. \tag{65}$$

Comparing now the first of these equations with Eq. (55), we obtain finally

$$\frac{dp_j}{dt} = -\frac{\partial H}{\partial q_j} \,,$$
$$\frac{dq_j}{dt} = \frac{\partial H}{\partial p_j} \,. \tag{66}$$

Eq. (66) are called **canonical equations**, or **Hamiltonian equations**, the function

$$H = H(p_j, q_j, t)$$

being the **Hamiltonian** of the system (66). Below we will discuss examples of application of the Hamiltonian equations in several problems of celestial mechanics.

11 The Time-Independent Hamiltonian

In cases in which the Hamiltonian is explicitly independent of time, i.e. when $H = H(p_j, q_j, 0)$, Eq. (66) always has an integral in the form $H = constant$. It can be demonstrated easily: multiply the first equation in (66) by dq_j/dt, and the second one by dp_j/dt, and take the sum:

$$\sum \frac{\partial H}{\partial q_j} \frac{dq_j}{dt} + \sum \frac{\partial H}{\partial p_j} \frac{dp_j}{dt} = 0. \tag{67}$$

Insofar as H depends only on p_j and q_j, the left-hand side of Eq. (67) must be a complete derivative of the function H with respect to t, so that

$$\frac{dH}{dt} = 0,$$

whence

$$H = constant.$$

In this case in which H does not depend explicitly on time, this integral is the **integral of energy**, i.e. H represents the total energy of the system.

12 Canonical Transformations

Let us replace the canonical variables q and p in Eq. (66) by new variables Q and P determined by the equations

$$\begin{aligned} Q_j &= Q_i(t; q_1, q_2, \ldots, q_k; p_1, p_2, \ldots, p_k), \\ P_j &= P_i(t; q_1, q_2, \ldots, q_k; p_1, p_2, \ldots, p_k). \end{aligned} \tag{68}$$

If these equations can be represented in canonical form, i.e.

$$\frac{dQ_i}{dt} = \frac{\partial K}{\partial P_i}, \qquad \frac{dP_i}{dt} = -\frac{\partial K}{\partial Q_i}, \tag{69}$$

then the transformation of the form (68) is called canonical; here K is a new characteristic function, which is defined by the formula

$$K = H + \frac{\partial W}{\partial t}. \tag{70}$$

The proof of this important theorem has no direct relation to the subsequent discussion, therefore we omit it. Note only that, under such a transformation, the following remarkable condition called the **canonical condition** is implemented:

$$\sum p_i dq_i - \sum P_i dQ_i = d'W, \tag{71}$$

where $d'W$ is the complete differential of some function W with respect to all variables, besides the time. Just this condition transforms Eq. (66) into Eq. (69) where K is given by Eq. (70), and H and W are expressed through the new variables Q_i and P_i

13   The Hamilton–Jacobi Equation

The universality and power of the canonical representation of equations of motion, unfortunately, do not always correspond to the efforts made for the solution of the equations. From this point of view one cannot claim that the canonical form of the equations makes the solution of a concrete problem of celestial mechanics easier; presumably the usual methods are more often much more efficient and reachable. However, during the investigation of these equations an important theorem was proved by Hamilton and Jacobi, which constituted an essential step forward.

Consider the canonical system

$$\frac{dq_i}{dt} = \frac{\partial H}{\partial p_i}, \qquad \frac{dp_i}{dt} = -\frac{\partial H}{\partial q_i}. \tag{72}$$

By introducing new variables Q_i and P_i satisfying the condition of canonicity (71), we obtain

$$\frac{dQ_i}{dt} = \frac{\partial K}{\partial P_i}, \qquad \frac{dP_i}{dt} = -\frac{\partial K}{\partial Q_i} \tag{73}$$

where the Hamiltonian is defined from Eq. (70) as

$$K = \frac{\partial W}{\partial t} + H(t; q_1, q_2, \ldots, q_k; p_1, p_2, \ldots, p_k). \tag{74}$$

Eq. (72) can be solved if the function W can be obtained, so that $K = 0$. In this case we shall have from Eq. (73)

$$Q_i = \alpha_i, \quad P_i = -\beta_i, \qquad (75)$$

where α_i and β_i are constants of integration. On the other hand, by expanding the differential $d'W$ under the condition of canonicity (71) and equating the terms in the two parts of this equation, we obtain

$$p_i = \frac{\partial W}{\partial q_i} \qquad P_i = -\frac{\partial W}{\partial Q_i}. \qquad (76)$$

Replacing now the quantity Q_i in W by α_i in accordance with Eq. (75), we have the function W in the form

$$W = W(t; q_1, q_2, \ldots, q_k; \alpha_1, \alpha_2, \ldots, \alpha_k).$$

which satisfies the relationships

$$\frac{\partial W}{\partial q_i} = p_i, \qquad \frac{\partial W}{\partial \alpha_i} = \beta_i, \qquad (77)$$

and enables us to express q_i and p_i through t and arbitrary constants α_i and β_i (the total number of which is $2k$). Then the general solution of Eq. (72) is reduced to the derivation of the function W satisfying the equation $K = 0$ which can be written in the form, in view of Eqs. (74) and (76), of

$$\frac{\partial W}{\partial t} + H\left(t; q_1, q_2, \ldots, q_k; \frac{\partial W}{\partial q_1}, \frac{\partial W}{\partial q_2}, \ldots, \frac{\partial W}{\partial q_k}\right) = 0. \qquad (78)$$

This is the Hamilton–Jacobi equation, a first-order in partial derivatives and with independent variables $q_1, q_2, \ldots, q_k$, the total number of which is k, plus time t.

14 Solution of Canonical Systems. The Hamilton–Jacobi Theorem

The only unknown quantity in the Hamilton–Jacobi equation (78) is the function W, which must contain: a) unknown variables q of total number k; b) arbitrary constants α of total number k; and c) time t. In the most general case in which the Hamiltonian H also depends on time, obtaining the general integral of the canonical system is reduced to the Hamilton–Jacobi theorem, which we present here without proof.

The Hamilton–Jacobi theorem states that: if $W(t; q_1, \ldots, q_k;$ $\alpha_1, \ldots, \alpha_k)$ is the complete integral of the Hamilton–Jacobi equation (78), then the general integral of the canonical system (77) is given by the following equations:

$$\frac{\partial W}{\partial \alpha_1} = \beta_1, \qquad \frac{\partial W}{\partial q_1} = p_1,$$

$$\frac{\partial W}{\partial \alpha_2} = \beta_2, \qquad \frac{\partial W}{\partial q_2} = p_2, \qquad (79)$$

$$\frac{\partial W}{\partial \alpha_k} = \beta_k, \qquad \frac{\partial W}{\partial q_k} = p_k.$$

Here $\beta_1, \beta_2, \ldots, \beta_k$ are k new arbitrary constants.

If the Hamiltonian H does not depend explicitly on time t, then the equation in partial derivatives (78) is simplified. Indeed, by substituting into the equation

$$\frac{\partial W}{\partial t} + H = 0$$

the function

$$W = -\alpha_1 t + V(q_1, q_2, \ldots, q_k; \alpha_1, \alpha_2, \ldots, \alpha_k),$$

where α_1 is an arbitrary constant, we obtain

$$\frac{\partial W}{\partial t} = -\alpha_1. \qquad (80)$$

Then instead of Eq. (78) we obtain the Hamilton–Jacobi equation in the form

$$H\left(q_1, q_2, \ldots, q_k; \frac{\partial W}{\partial q_1}, \frac{\partial W}{\partial q_2}, \ldots, \frac{\partial W}{\partial q_k}\right) = \alpha_1. \qquad (81)$$

On the other hand, we have

$$\frac{\partial W}{\partial \alpha_1} = -t + \frac{\partial V}{\partial \alpha_1} = \beta_1. \qquad (82)$$

Then the solution of Eq. (81) can be written in a form in accord with the Hamilton–Jacobi theorem (79)

$$\frac{\partial V}{\partial \alpha_1} = t + \beta,$$

$$\frac{\partial V}{\partial \alpha_2} = \beta_2, \qquad (83)$$

$$\frac{\partial V}{\partial \alpha_k} = \beta_k,$$

and we also have

$$\frac{\partial V}{\partial q_1} = p_1 \,,$$

$$\frac{\partial V}{\partial q_2} = p_2 \,, \tag{84}$$

$$\cdot \quad \cdot \quad \cdot$$

$$\frac{\partial V}{\partial q_k} = p_k \,,$$

or in more general form

$$\frac{\partial V}{\partial \alpha_1} = t + \beta \,, \qquad \frac{\partial V}{\partial \alpha_i} = \beta_i \,, \qquad \frac{\partial V}{\partial q_j} = p_j \,, \tag{85}$$

where $i = 2, 3, \ldots, k$ and $j = 1, 2, \ldots, k$.

15   The Canonical Form of the Equation of Motion in the Central Field

Consider the motion of point mass m in the gravitational field of a central body in spherical coordinates r, θ, φ (see section 5 of this chapter). In this case the expression for kinetic energy is given by Eq. (32):

$$T = \frac{1}{2} m \left(\dot{r}^2 + r^2 \dot{\varphi}^2 + r^2 \dot{\theta}^2 \cos^2 \varphi \right), \tag{86}$$

and the Hamiltonian H is determined by Eq. (59):

$$H = T - V \,. \tag{87}$$

We assume as before that $\dot{q}_1 = \dot{r}$, $\dot{q}_2 = \dot{\varphi}$ and $\dot{q}_3 = \dot{\theta}$. From these data we obtain for the canonical variables p_r, p_φ and p_θ

$$p_\gamma = \frac{\partial T}{\partial \dot{q}_1} = \frac{\partial T}{\partial \dot{r}} = m\dot{r} \,,$$

$$p_r = \frac{\partial T}{\partial \dot{q}_2} = \frac{\partial T}{\partial \dot{\varphi}} = mr^2 \dot{\varphi} \,, \tag{88}$$

$$p_\theta = \frac{\partial T}{\partial \dot{q}_3} = \frac{\partial T}{\partial \dot{\theta}} = m\dot{\theta} r^2 \cos^2 \varphi \,.$$

From that we obtain for $\dot{r}$, $\dot{\varphi}$ and $\dot{\theta}$

$$\dot{r} = \frac{p_r}{m},$$

$$\dot{\varphi} = \frac{p_\varphi}{mr^2}, \tag{89}$$

$$\dot{\theta} = \frac{p_\varphi^2}{mr^2 \cos^2 \varphi},$$

and also

$$r^2 \dot{\varphi}^2 = \frac{p_\varphi^2}{m^2 r^2},$$

$$r^2 \dot{\theta}^2 \cos^2 \varphi = \frac{p_\theta^2}{m^2 r^2 \cos^2 \theta}. \tag{90}$$

Finally, we assume for the potential function

$$U = \frac{K}{r} = k^2 \frac{1+m}{r}. \tag{91}$$

From Eqs. (68), (69) and (73) we obtain for the Hamiltonian function H, taking into account also Eqs. (71) and (72)

$$H(p_r, p_\varphi, p_\theta) = \frac{1}{2m} \left(p_r^2 + \frac{p_\varphi^2}{r^2} + \frac{p_\theta^2}{r^2 \cos^2 \varphi} \right) - \frac{K}{r}. \tag{92}$$

From Eqs. (74) and (71) we obtain for the equation of the motion in Hamiltonian form, Eq. (66)

$$\frac{\partial H}{\partial p_r} = \dot{q}_1 = \dot{r} = \frac{p_r}{m},$$

$$\frac{\partial H}{\partial p_\varphi} = \dot{q}_2 = \dot{\varphi} = \frac{p_\varphi}{mr^2}, \tag{93}$$

$$\frac{\partial H}{\partial p_\theta} = \dot{q}_3 = \dot{\theta} = \frac{p_\theta}{mr^2 \cos^2 \varphi}.$$

and, using the relationships (70) and (71)

$$-\frac{\partial H}{\partial r} = \dot{p}_r = \frac{p_\varphi^2}{mr^3} + \frac{p_\theta^2}{mr^3 \cos^2 \varphi} - \frac{K}{r^2},$$

$$-\frac{\partial H}{\partial \varphi} = \dot{p}_\varphi = -\frac{p_\theta^2 \sin \varphi}{mr^2 \cos^3 \varphi}, \tag{94}$$

$$-\frac{\partial H}{\partial \theta} = \dot{p}_\theta = 0.$$

If we move to the flat problem, i.e. to the case of motion within the plane around the central body, then we will have for the equation of motion in Hamiltonian form, adopting also $\theta = 0$ and $p_\theta = 0$:

$$
\begin{aligned}
\frac{\partial H}{\partial \rho_r} &= \frac{p_r}{m}, \\
\frac{\partial H}{\partial p_\varphi} &= \frac{p_\varphi}{mr^2}, \\
-\frac{\partial H}{\partial r} &= \frac{p_\varphi^2}{mr^3} - \frac{K}{r^2}, \\
-\frac{\partial H}{\partial \varphi} &= 0.
\end{aligned}
\tag{95}
$$

The relationships presented here can be used, in particular, for the derivation of the canonical elements of elliptic motion.

16 Canonical Elements for Elliptic Motion

To obtain the canonical elements for elliptic motion it is necessary to have the expression for the Hamilton–Jacobi equation in spherical coordinates. We have from Eq. (78)

$$
\frac{\partial W}{\partial t} + \frac{1}{2} \left[\left(\frac{\partial W}{\partial r} \right)^2 + \left(\frac{\partial W}{\partial \varphi} \right)^2 \frac{1}{r^2} + \frac{1}{r^2 \cos^2 \varphi} \left(\frac{\partial W}{\partial \theta} \right)^2 \right] - \frac{k_1^2}{r} = 0, \tag{96}
$$

where $k_1^2 = k^2(1 + m)$.

Eq. (96) contains explicitly neither t nor θ. Therefore its solution can be represented in the form (see the previous section)

$$
W = \alpha_1 t + \alpha_2 \theta + W_1. \tag{97}
$$

On substituting this expression into Eq. (96), we obtain the following differential equation for W_1:

$$
\left(\frac{\partial W_1}{\partial r} \right)^2 + \frac{1}{r^2} \left(\frac{\partial W_1}{\partial \varphi} \right)^2 + \frac{\alpha_2^2}{r^2 \cos^2 \varphi} = \frac{2k_1^2}{r} - 2\alpha_1. \tag{98}
$$

The variables r and φ in this equation can be separated easily, if any solution of this equation can be found. Obviously, by introducing one more constant α_3 the desired solution can be represented in the form

$$
\left(\frac{\partial W_1}{\partial \varphi} \right)^2 + \alpha_2^2 \sec^2 \varphi = \alpha_3^2, \tag{99}
$$

which gives the second equation for r:

$$\left(\frac{\partial W_1}{\partial r}\right)^2 + \frac{\alpha_3^2}{r^2} = \frac{2k_1^2}{r} - 2\alpha_1 \, . \tag{100}$$

By integrating Eqs. (99) and (100) within arbitrary limits, we obtain finally for the sought function W

$$W = \alpha_1 t + \alpha_2 \theta + \int_0^\varphi (\alpha_3^2 - \alpha_2^2 \sec^2 \varphi)^{1/2}$$
$$+ \int_{r_0}^r \left(-2\alpha_1 + \frac{2k_1^2}{r} - \frac{\alpha_3^2}{r^2}\right)^{1/2} dr \, . \tag{101}$$

However, the general solution is given, in accord with the Hamilton–Jacobi theorem, by the equalities

$$\beta_1 = \frac{\partial W}{\partial \alpha_1}, \quad \beta_2 = \frac{\partial W}{\partial \alpha_2}, \quad \beta_3 = \frac{\partial W}{\partial \alpha_3} \, . \tag{102}$$

With the help of these relationships we can express the coordinates as functions of t and six canonical elements α_1, α_2, α_3 and β_1, β_2, β_3. Our final aim is the determination of the relation between canonical elements α_1, α_2, α_3 and β_1, β_2, β_3, on the one hand, and the usual Keplerian elements of elliptic motion on the other.

Substituting W from Eq. (101) into the first relationship in Eq. (102) and taking a derivative by α_1, we obtain

$$\beta_1 = t - \int_{r_0}^r \left(-2\alpha_1 + \frac{2k_1^2}{r} - \frac{\alpha_3^2}{r^2}\right)^{-1/2} dr \, . \tag{103}$$

At the moment of passage of the planet through perihelion we have $t = T$ and $r = r_0$. Therefore we can write from Eq. (103)

$$\beta_1 = T \, , \tag{104}$$

as well as

$$2\alpha_1 r^2 - 2k_1^2 r + \alpha_3^2 = 0 \, . \tag{105}$$

The roots of this equation are

$$r_0 = \frac{k_1^2}{2\alpha_1} - \frac{1}{2}\left(\frac{k_1^4}{\alpha_1^2} - 2\frac{\alpha_3^2}{\alpha_1}\right)^{1/2} \, ,$$
$$r_1 = \frac{k_1^2}{2\alpha_1} + \frac{1}{2}\left(\frac{k_1^4}{\alpha_1^2} - 2\frac{\alpha_3^2}{\alpha_1}\right)^{1/2} \, . \tag{106}$$

However, r_0 and r_1 are the minimal and maximal values of the radius-vector and in the case of elliptic motion are represented through Keplerian elements a and e – semimajor axis and eccentricity:

$$r_0 = a(1 - e), \quad r_1 = a(1 + e). \tag{107}$$

From that we find

$$r_0 + r_1 = 2a, \quad r_0 r_1 = a^2(1 - e^2). \tag{108}$$

However, for the same combinations $r_0 + r_1$ and $r_0 r_1$ we have from Eq. (106) in canonical elements

$$r_0 + r_1 = \frac{k_1^2}{\alpha_1}, \quad r_0 r_1 = \frac{\alpha_3^2}{2\alpha_1}. \tag{109}$$

Therefore we obtain finally from Eqs. (108) and (109) for the canonical elements α_1 and α_3, represented through Keplerian elements:

$$\alpha_1 = \frac{k_1^2}{2a}, \tag{110}$$

$$\alpha_3 = k_1[a\left(1 - e^2\right)]^{1/2} = k_1\sqrt{p}. \tag{111}$$

The second relationship in Eq. (102) gives, after the substitution of the value of W from Eq. (101)

$$\beta_2 = \theta - \alpha_2 \int_0^\varphi \sec^2 \varphi \left(\alpha_3^2 - \alpha_2^2 \sec^2 \varphi\right)^{-1/2} d\varphi. \tag{112}$$

When $\varphi = 0$, the planet is located at one of the nodes of its orbit, hence $\theta = \Omega$, where Ω is the longitude of the ascending knot, i.e. it is a Keplerian element. Therefore we have from Eq. (112)

$$\beta_2 = \Omega. \tag{113}$$

As for the equation
$$\alpha_3^2 - \alpha_2^2 \sec^2 \varphi = 0,$$

presumably it can be fulfilled if $|\varphi| = i$, where i is the inclination of the orbital plane to the plane of the ecliptic. Then we obtain, taking into account also Eq. (111)

$$\alpha_2 = \alpha_3 \cos i = k_1 \sqrt{p} \cos i. \tag{114}$$

Finally, from the last relationship in Eq. (102), we obtain

$$\beta_3 = \alpha_3 \int_0^\varphi \left(\alpha_3^2 - \alpha_2^2 \sec^2 \varphi \right)^{-1/2} d\varphi$$

$$-\alpha_3 \int_{r_0}^r \frac{1}{r^2} \left(-2\alpha_1 + \frac{2k_1^2}{r} - \frac{\alpha_3^2}{r^2} \right)^{-1/2} dr \,. \tag{115}$$

Let us introduce the argument of longitude $u = \omega + v$, as we had in the case of Keplerian motion. Then, from the corresponding spherical triangle, we can write

$$\sin \varphi = \sin i \sin u \,.$$

At the moment of passage of the planet through perihelion we have $r = r_0$, $v = 0$, and therefore $u = \omega$, where ω is the longitude or angular distance of perihelion from the line of nodes. Then we obtain from (115) $\beta_3 = \omega = \pi - \Omega$.

Thus, we have obtained all six canonical elements α_1, α_2, α_3 and β_1, β_2, β_3, represented through six Keplerian elements a, e, i, ω, Ω, and T for elliptic motion. They are

$$\alpha_1 = k^2 M / 2a \,, \qquad \beta_1 = T \,,$$

$$\alpha_2 = k \sqrt{M} \left(1 - e^2 \right)^{1/2} a^{1/2} \cos i \,, \qquad \beta_2 = \Omega \,, \tag{116}$$

$$\alpha_3 = k \sqrt{M} \left(1 - e^2 \right)^{1/2} a^{1/2} \,, \qquad \beta_3 = \omega = \pi - \Omega \,.$$

The contrary is also possible, i.e. the representation of Keplerian elements through canonical ones:

$$a = k^2 M / 2\alpha_1 \,, \qquad T = \beta_1 \,,$$

$$e^2 = 1 - 2\alpha_1 \alpha_3^2 / k^4 M^2 \,, \qquad \Omega = \beta_2 \,,$$

$$\cos i = \alpha_2 / \alpha_3 \,, \qquad \omega = \beta_3 \,.$$

The canonical elements given by Eq. (116) were derived originally by Jacobi in 1842 and are known as **Jacobi canonical elements**.

17   Delaunay Canonical Elements

The set of Lagrange differential equations, Eq. (56) (see Chapter X) for the variation of Keplerian elements a, e, $\ldots$, T in terms of their dependence on time, can also be presented through the Jacobi canonical elements, i.e. using the relationships (116) between a, e, i, ω, Ω, T and α_1, α_2, α_3, β_1, β_2, β_3 and in accord with the equations

$$\frac{d\alpha_k}{dt} = \frac{\partial R}{\partial \beta_k}, \quad \frac{d\beta_k}{dt} = -\frac{\partial R}{\partial \alpha_k}, \quad i = 1, 2, 3.$$

One of the Jacobi elements, α_1, enters into the perturbation function R explicitly, through a, as well as indirectly, through the mean longitude λ. To remove this defect, Delaunay introduced new elements slightly differing from Eq. (116), namely

$$L = k\sqrt{M}\,a^{1/2}, \qquad l = n(t - T),$$

$$G = k\sqrt{M}\left(1 - e^2\right)^{1/2} a^{1/2} \qquad g = \pi - \Omega, \tag{117}$$

$$H = k\sqrt{M}\left(1 - e^2\right)^{1/2} a^{1/2}\cos i \qquad h = \Omega.$$

In contrast to β_1, here l is the mean anomaly. The elements L, G, and H are measures of sectorial velocity, and l, g, h are the angular variables; these elements, L, G, H, l, g, and h are called **Delaunay canonical elements**; they were first used in 1867 in the theory developed by Delaunay for the motion of the Moon.

The differential equations of unperturbed motion in terms of the Delaunay elements have the form

$$\begin{aligned}
\frac{dL}{dt} &= \frac{\partial R^*}{\partial l}, & \frac{dl}{dt} &= -\frac{\partial R^*}{\partial L}, \\
\frac{dG}{dt} &= \frac{\partial R^*}{\partial g}, & \frac{dg}{dt} &= -\frac{\partial R^*}{\partial G}, \\
\frac{dH}{dt} &= \frac{\partial R^*}{\partial h}, & \frac{dh}{dt} &= -\frac{\partial R^*}{\partial H},
\end{aligned} \tag{118}$$

where

$$R^* = \frac{k^4 M^2}{2L^2} + R,$$

Here R is the perturbation function.

Delaunay elements are used for the interpretation of motions only in elliptic orbits. It follows already from the fact that the elements G and H must be nonzero and positive, that it is possible only if $0 < e < 1$, i.e. only for motions via elliptic orbits.

Besides Jacobi and Delaunay elements, other systems of canonical transformations have also been introduced into celestial mechanics, and used with various degrees of success for the solution of certain problems; one can mention the two systems of canonical transformations of Poincaré, systems of Levi-Civita, Hill, De Sitter, etc. Delaunay elements have also been used successfully in the problems of stellar dynamics. One such problem will be discussed in the last section of this chapter.

18   Canonical Equations and Flights to Planets with Small Traction

Interplanetary flights both to the outer and to the inner planets, as discussed above in Chapter XIII, can be realized in the Keplerian trajectories. These trajectories in all their diversity require constrained conditions, i.e. to be performed under the gravitational action of a central body (the Sun) and with given starting conditions concerning the direction and the magnitude of the initial velocity. No other forces influence the spaceship during the flight.

However, interplanetary flights can also be realized by a small traction acting during the flight. In this case both the magnitude of the force and the direction of its action are changing continuously during the motion. Among the factors determining the schedule and the regime of work of the ionic engine are the character and aim of the flight, i.e. rapid flights, slow flights with capture by landing on a planet or nearby flight. Of course, one must also take into account the required duration of the flight.

The interplanetary flight of a spaceship equipped with an engine of low traction can be realized with any preliminary chosen trajectory. However, in contrast to Keplerian motion, controlled only by the gravity of central body, in this case one should take into consideration also the contribution of the action of the ionic engine. As a result, the problem becomes much more complicated. As a matter of fact, here we have a three-body problem: Sun–Spaceship–Perturbation caused by small traction.

Continued action of a small traction can lead to essential modification of the Keplerian orbit. Moreover, the integral effect of perturbation caused by traction of ionic engines can be comparable with the gravitational action of the of central body (the Sun). Therefore, the question really reduces to the three-body problem, so that for its correct solution it is necessary to have the exact equations of motion.

The exact equations of motion in our case, i.e. for flight with small traction and in the field of a central body, can be derived by means of canonical equations, in particular, the Lagrange equations (Ehricke, 1962). Insofar as one of these forces acting on the ship, namely, the small traction, is not a conservative force, i.e. has no initial potential, the Lagrange equations written for the general case, Eq. (19), should be used:

$$\frac{d}{dt}\left(\frac{\partial T_k}{\partial \dot{r}_i}\right) - \frac{\partial T_k}{\partial x_i} = F_{x_i}, \qquad (119)$$

where $\dot{q}_i$ is replaced by $\dot{x}_i$, q_i is replaced by x_i, T is the specific energy of the ship, F_{xi} is the component of the main vector of the forces acting on the spaceship and related to the mass of the spaceship. The nature of these forces in our case is clear: one of them – gravitation of the central body which has potential, and other one – the force of small traction, is not of potential type. Therefore we can write for F_{xi}

$$F_{x_i} = \frac{\partial U}{\partial x_i} + f_{x_i}, \qquad (120)$$

where $f = F/m$, F is the force of small traction, and m is the spaceship's mass.

The spaceship's trajectory is determined by the coordinates of its position x_i and by the component of the velocity vector $\dot{x}_i$ along the coordinate line. In the case of a two-dimensional system we have four variables: x_1, x_2, $\dot{x}_1$, and $\dot{x}_2$. This is the minimal number of constraints imposed on the trajectory. The requirements concerning the optimization of the flight pose additional conditions which determine the program or schedule of the traction.

Applying the Lagrange equations (119) and using the polar coordinates r and η, we can write for two-dimensional motion

$$x_1 = r, \quad x_2 = \eta,$$

$$\dot{x}_1 = \dot{r}, \quad \dot{x}_2 = \dot{\eta},$$

$$x = r\cos\eta,$$

$$y = r\sin\eta,$$

$$\dot{x} = \dot{r}\cos\eta - r\dot{\eta}\sin\eta,$$

$$\dot{y} = \dot{r}\sin\eta + r\dot{\eta}\cos\eta.$$

For the ship's specific kinetic energy we have

$$T_k = \frac{1}{2}\left(\dot{x}^2 + \dot{y}^2\right) = \frac{1}{2}\left(\dot{r}^2 + r^2\dot{\eta}^2\right).$$

For the Lagrange equations for $i = 1$, we have

$$\frac{\partial T_k}{\partial r} = 0 + \frac{1}{2} 2r\dot\eta^2 = r\dot\eta^2 \,,$$

$$\frac{\partial T_k}{\partial \dot r} = \frac{1}{2} 2\dot r = \dot r \,,$$

$$\frac{d}{dt}\left(\frac{\partial T_k}{\partial \dot r}\right) = \ddot r \,.$$

For the acting force $F_{x_1} = F_r$ from Eq. (120), taking for potential function $U = K/r$ we have

$$F_r = \frac{\partial U}{\partial r} + f_r = -\frac{K}{r^2} + f_r \,.$$

From these relationships we derive the first equation of motion:

$$\ddot r - r\dot\eta^2 + \frac{K}{r^2} - f_r = 0 \,. \tag{121}$$

Or, using the expression for instantaneous value of the specific kinetic moment

$$C = r^2\dot\eta \,, \tag{122}$$

we can rewrite Eq. (121) in slightly different form:

$$\ddot r - \frac{C^2}{r^3} + \frac{K}{r^2} - f_r = 0 \,. \tag{123}$$

At $i = 2$ we find analogously that

$$\frac{\partial T_k}{\partial \eta} = 0 \,,$$

$$\frac{1}{r}\frac{\partial T_k}{\partial \dot\eta} = \frac{1}{2}\frac{\partial}{\partial \dot\eta}(r\dot\eta^2) = r\dot\eta = \frac{C}{r} \,,$$

$$\frac{1}{r}\frac{d}{dt}\left(\frac{\partial T_k}{\partial \dot\eta}\right) = \frac{1}{r}\frac{d}{dt}(r^2\dot\eta) = 2\dot r\dot\eta + r\ddot\eta = \frac{\dot C}{r} \,,$$

$$F_\eta = \frac{\partial U}{\partial \eta} + f_\eta = f_\eta \,.$$

Correspondingly, we find for the second equation of motion

$$r\ddot\eta + 2\dot r\dot\eta - f_\eta = 0 \,, \tag{124}$$

or, taking a derivative from (122),

$$\dot{C} - r f_n = 0 \,. \tag{125}$$

In these equations f_r and f_η are the radial and transversal components of the acceleration of the traction, respectively.

The integration of Eqs. (123) and (125) gives the radial and transversal components of the velocity, $\dot{r}$ and $r\dot{\eta}$, as well as the value of C/r. The spaceship's coordinates r and η, are obtained via integration of the following obvious equations:

$$\dot{r} - V_r = 0 \,,$$
$$C - r^2 \dot{\eta} = 0 \,. \tag{126}$$

In the limiting case in which $f_r = f_\eta = 0$, i.e. when the small traction is absent, Eqs. (123) and (125) are reduced to the equations of the Keplerian orbit

$$\ddot{r} - \frac{C^2}{r^3} + \frac{K}{r^2} = 0 \,, \tag{127}$$

$$r\ddot{\eta} + 2\dot{r}\dot{\eta} = \dot{V_a} + \frac{V_a V_r}{r} = 0 \,, \tag{128}$$

where $V_a = r\dot{\eta}$ is the transversal component of velocity.

Thus, the appearance of the functions f_r and f_η, which are acceleration components, caused by the traction of ionic engines, in the equations of the spaceship's motion (123) and (125), introduce a new feature into the type of the trajectory and the flight parameters. In the general case f_r and f_η depend on the type of traction – radial, transversal, tangential or intermediate, as well on the absolute magnitude of acceleration of the traction f, on the inclination angle θ of the trajectory (the angle between the velocity vector of the trajectory and the local horizon), and on the angle φ between the traction acceleration vector and transversal direction. All these quantities can have different values in various intervals of the flight and at various moments of time. Therefore in general one can write for f_r and f_η

$$f_r = f_r(f, \, \theta, \, \varphi, \, t) \,,$$
$$f_\eta = f_\eta(f, \, \theta, \, \varphi, \, t) \,. \tag{129}$$

This is the time schedule of the traction with respect to all coordinates and for the whole duration of the flight.

All these are related to the mechanics of flight of the space probe under the action of a small traction. There are also certain criteria that often appear to be crucial for the choice of the type of the flight and the trajectory. For example, the criteria characterizing the dynamic conditions at the points of the start and the destination, i.e. the boundary conditions. Or the criteria characterizing the aim of the flight and the main requirements: e.g. the minimal duration of the flight for a given trajectory and fixed parameters of the ionic engine – its weight, size, energetic resources, system of traction regulation, etc. Often the condition of least flight duration is the most crucial.

Thus, the final version and structure of the flight are chosen, i.e. the structure of the functions in Eq. (129) is known. Then we start the integration of Eqs. (123) and (125) by numerical methods. The most important parameter, i.e. the duration of the flight, is estimated via integration to the end, i.e. after the derivation of the complete trajectory of the flight.

As an example, in Fig. 14.18 (Chap. XIV) the schemes of rapid and slow flights with small traction from the Earth to Mars (below) and to Venus (top) are shown: the trajectories are calculated with the help of Eqs. (123) and (125). The parameters of the trajectory under different regimes of ionic engines are also shown. On each trajectory, both the directions and the magnitudes of active acceleration by the small traction are shown. Hyperbolic trajectories are denoted by BE; in this case the acceleration of small traction is directed tangentially to the trajectory; however, with one difference, i.e. in the case of the flight to Mars the tangent of the traction is directed in the direction of motion of the space probe, whereas in the case of the flight to Venus it is directed opposite to the motion (trajectories E both schemes). The duration of the flight to Mars is five months at $f_p = 110\mu g_0$ ($1\mu g_0 = 10^{-6} g_0$, where g_0 is the acceleration due the gravity at the Earth's surface; hence, $110\mu g_0 = 1.10 \cdot 10^{-4} g_0$). The duration of the flight to Venus is three months. In both cases the traction decreases monotonously during the flight, reaching zero at the destination.

Trajectories finished by landing on the planet are denoted by C. Trajectories $C-1$, $C-2$, and $C-3$ differ from each other by their magnitude of acceleration due by traction. Thus, e.g. in the case of Mars the flight duration with the trajectory $C-1$ is six months with initial acceleration $f_p = 140 \mu g_0$, and one month in the case of trajectory $C-3$ with $f_p = 6000 \mu g_0$. Recall that the duration in the case of flight to Mars in a Hohmann trajectory is 260 days (see Section 3, Chapter XI), i.e. about nine months. The efficiency in the case of $C-1$ is about 30%.

Durations of flights are less in the case of Venus. Thus, the duration of the flight by the trajectory $C-1$ is four months at $f_p = 150 \mu g_0$, and one month by $C-3$ at $f_p = 3650 \mu g_0$. The duration of flight to Venus in a Hohmann trajectory is 146 days, i.e. about five months. The efficiency is about 20%.

Note that the efficiency using small traction is far higher in the case of flights to the outer planets than with flights to the inner planets, and that the efficiency increases rapidly on moving away from the Sun. Qualitatively this effect can be represented by the ratio of two accelerations, f_p/F_r, of the small traction (f_p) and Newtonian gravity (F_r); this ratio increases strongly with increasing distance of the probe from the Sun, according to the law

$$\frac{f_p}{F_r} \sim r^2 .$$

Therefore a space probe with an ionic engine that has appeared in the zone of the most distant planets, Uranus, Neptune, and Pluto, can realize maneuvers under the action of small traction. The external boundary of the Solar system is the zone of high efficiency of small traction.

Up to now we have mentioned only one way to escape from the Solar system, namely, the perturbation maneuver in the sphere of action of a massive planet. However, in this case it is necessary first to direct the probe into the sphere of action of a given planet, which generally speaking, requires engines with strong traction, and hence, a considerable amount of energy.

Now we see that it is possible to leave the Solar system if the space probe is equipped with ionic engines of rather long-duration action and with small energy resources. The considered examples, i.e. interplanetary flights with small traction, once more demonstrate the efficiency of the application of canonical systems, particularly of the Lagrange equations for the derivation of the exact equations of motion of a space probe.

19 The Artificial Satellite in the Gravitational Field of a Flattened Planet

The motion of an artificial satellite around a flattened planet is a problem that is within the scope of perturbation theory. However, it can be solved much more easily by means of the Hamilton–Jacobi equation. To a first approximation, the Hamiltonian H_0 should be identified with the potential μ/r, as in the case of undisturbed motion, and solution of the Hamilton–Jacobi equation leads to the ordinary Keplerian ellipse. The disturbed Hamiltonian H_1 itself gives new canonical equations, which determine the changes in time of the previous canonical constants obtained in the first approximation.

However, it appears (Sterne, 1958) that one can use also the undisturbed Hamiltonian H_0, including the main part of the flattening effects and leading to a solvable Hamilton–Jacobi equation with separated variables. Let us choose the Hamilton function H for which the exact solution can be obtained in the following form

$$H = \frac{1}{2}\left(p_r^2 + \frac{p_\lambda^2}{r^2 \cos^2 \delta} + \frac{p_\delta^2}{r^2}\right) + V_1(r) + \frac{1}{r}^2 V_2(\delta), \qquad (130)$$

where r, δ, λ are the satellite's spherical coordinates relative to the central planet (the Earth); p_r, p_δ, and p_λ are the moments associated with r, δ, λ; and V_1 and V_2 are arbitrary functions of the radius-vector r and declination δ, respectively.

The Hamilton–Jacobi equation with a Hamiltonian in the form (130) is an equation with separated variables, and its solution can be represented in the following form

$$W = \int_{r_0}^{r} \frac{L}{r} dr + \int_{0}^{\delta} M \, d\delta + \alpha_3 \lambda - \alpha_1 t \,, \qquad (131)$$

where

$$
\begin{aligned}
L^2 &= 2r^2\alpha_1 - 2r^2 V_1(r) - \alpha_2^2 \,, \\
M^2 &= \alpha_2^2 - \alpha_3^2 \sec^2 \delta - 2V_2(\delta) \,.
\end{aligned}
\qquad (132)
$$

The canonical constants α_1, α_2, and α_3 (per unit mass of the satellite) are the measures of the total energy (α_1), the total moment of quantity of motion in the orbit at $V_2 = 0$ (α_2), and the component of the moment of the quantity of motion directed along the axis (α_3), respectively.

The canonical solution in our case has the following form (see Section 15)

$$t - \beta_1 = \frac{\partial V}{\partial \alpha_1} = J(\alpha_1, \alpha_2, r)\,,$$

$$-\beta_2 = \frac{\partial V}{\partial \alpha_2} = G(\alpha_2, \alpha_3, \delta) - j(\alpha_1, \alpha_2, r)\,,$$

$$\beta_3 = -\lambda - \frac{\partial V}{\partial \alpha_3} = -\lambda + g(\alpha_2, \alpha_3, \delta)\,, \qquad (133)$$

$$p_r = \dot{r} = \frac{1}{r} L(\alpha_1, \alpha_2, r)\,,$$

$$p_\delta = r^2 \dot{\delta} = M(\alpha_2, \alpha_3, \beta)\,,$$

$$p_\lambda = r^2 \dot{\lambda} \cos \delta = \alpha_3\,,$$

where through J, G, j and g there are denoted the following four integrals

$$
J = \int_{r_0}^{r} \frac{r}{L} dr, \qquad j = \int_{r_0}^{r} \frac{\alpha_2}{rL} dr\,,
$$
$$
G = \int_{0}^{\delta} \frac{\alpha_2}{M} d\delta, \qquad g = \int_{0}^{\delta} \frac{\alpha_3}{M} \sec^2 \delta \, d\delta\,,
\qquad (134)
$$

and the canonical constants β_1, β_2, and β_3 are, correspondingly, the time T of the passage of a satellite through the perihelion (β_1), the longitude of perihelion from the line of nodes ω at $V_2 = 0$ (β_2) and the angular distance or longitude of the ascending node Ω on the equator (β_3). Now, if we choose the Hamiltonian H_0 in the form

$$H_0 = \frac{1}{2}\left(p_r^2 + \frac{p_\lambda^2}{r^2 \cos^2 \delta} + \frac{p_\delta^2}{r^2}\right) - \frac{\mu}{r}\left[1 + \frac{B}{r^2}\left(1 - \frac{3}{2}\sin^2 i\right)\right.$$
$$\left. - \frac{3B}{ra\,(1 - e^2)}\left(\sin^2 \delta - \frac{1}{2}\sin^2 i\right)\right], \tag{135}$$

then it will be of the same form as Eq. (130). In particular, bearing in mind that Eq. (135) is a good approximation for the gravitational potential of the Earth $(B = J_2 R^2/2$, where R is the Earth's equatorial radius), insofar as the ignored terms $(J_3, J_4,$ etc.$)$ one about 10^3 times smaller than the term in J_2. The constant i in Eq. (135) is the inclination of the orbital plane to the plane of the ecliptic.

The disturbed Hamiltonian H_1 is now determined as

$$H_1 = H_0 - H\,,$$

and has the following form

$$H_1 = 3\mu B\left(\sin^2 \delta - \frac{1}{2}\sin^2 i\right)\left(\frac{1}{r}^3 - \frac{1}{ar^2\,(1 - e^2)}\right), \tag{136}$$

where $B \approx 0.000546$ for the Earth. This Hamiltonian determines the following canonical equations relative to the old canonical constants

$$\dot\alpha_i = \frac{\partial H_1}{\partial \beta_i}\,, \qquad \dot\beta_i = \frac{\partial H_1}{\partial \alpha_i}\,, \qquad i = 1,\,2,\,3.$$

On comparing Eq. (135) with Eq. (130), we find

$$V_1(r) = -\frac{\mu}{r} - \frac{\mu B}{r^3}\left(1 - \frac{3}{2}\sin^2 i\right)\,,$$
$$V_2(\delta) = \frac{3\mu B}{a\,(1 - e^2)}\left(\sin^2 \delta - \frac{1}{2}\sin^2 i\right)\,, \tag{137}$$

and, hence, in the case of the exact undisturbed solution of Eq. (133) we have for L and M

$$L^2 = 2r^2\alpha_1 + 2\mu r - \alpha_2^2 + \frac{2\mu B}{r}\left(1 - \frac{3}{2}\sin^2 i\right)\,,$$
$$M^2 = \alpha_2^2 - \alpha_3^2\sec^2 \delta - \frac{6\mu B}{p}\sin^2 \delta + \frac{3\mu B}{p}\sin^2 i\,, \tag{138}$$

where $p = a(1 - e^2)$.

The next step is the computation of four integrals (134) using also Eq. (138). They are elliptic ones, and are computed, first, by expansion by the series and, then, via integration by parts. As a result we obtain for the first integral

$$
\begin{aligned}
J = \frac{a}{(-2\alpha_1)^{-1/2}} & \left\{ E \left(1 + \frac{c}{2a} \right) - e \sin E + \frac{3c^2}{8a^2} \frac{q}{(1 - e^2)^{1/2}} \right. \\
& \left. + \frac{5c^3}{16a^3} \left[\frac{e}{1 - e^2} \frac{\sin E}{1 - e \sin E} + \frac{q}{(1 - e^2)^{3/2}} \right] + \dots \right\} ,
\end{aligned}
\tag{139}
$$

where

$$
\tan \frac{q}{2} = \left(\frac{1 + e}{1 - e} \right)^{1/2} \tan \frac{E}{2} .
$$

In the same manner we find for the second integral $j(w)$

$$
j(w) = \frac{\Delta j}{2\pi} \left[w - \left(\frac{k^2}{4} + \frac{k^4}{8} \right) \sin w + \frac{3}{128} k^4 \sin 2w + \dots \right] ,
\tag{140}
$$

where

$$
\sin^2 w = \frac{(r_1 - c)(r - r_0)}{(r - c)(r - r_0)} ,
$$

$$
k^2 = \frac{c(r_1 - r_0)}{r_0(r_1 - c)} ,
$$

$$
\frac{\Delta j}{2\pi} = \frac{\alpha_2}{(-2a_1)^{1/2}} \frac{1 + (1/4)k^2 + (9/64)k^4 + \dots}{[r_0(r_1 - c)]^{1/2}} ,
$$

and k^2 is a quantity of the order of $3 \cdot 10^{-5}$.

The expression for the third integral $G(u)$ has the following form:

$$
\begin{aligned}
G(u) = \frac{\Delta G}{2\pi} & \left\{ u - \frac{1}{8} \left[\left(\frac{x_1}{x_2} \right)^2 + \frac{1}{2} \left(\frac{x_1}{x_2} \right)^4 \right] \sin 2u \right. \\
& \left. + \frac{3}{256} \left(\frac{x_1}{x_2} \right)^4 \sin 4u + \dots \right\} ,
\end{aligned}
\tag{141}
$$

where $x_2 = \sin \delta$, $x_1 = \sin u$, and also

$$
\frac{\Delta G}{2\pi} = \frac{x_1 \alpha_2}{(\alpha_2' - \alpha_3^2)^{1/2}} \left[1 + \frac{1}{4} \left(\frac{x_1}{x_2} \right)^2 + \frac{9}{64} \left(\frac{x_1}{x_2} \right)^4 + \frac{25}{256} \left(\frac{x_1}{x_2} \right)^6 + \dots \right] ,
$$

where

$$\alpha_2' = \alpha_2^2 + \frac{3\mu B}{p} \sin^2 i.$$

Finally, we have for the fourth integral $g(u)$

$$g(u) = \frac{\alpha_3}{|\alpha_3|} \left[\arctan \left(1 - x_1^2\right)^{1/2} \tan u - fu \right.$$
$$\left. + \frac{3}{32} \varepsilon^2 x_1^2 \left(1 - x_1^2\right)^{1/2} \sin 2u + \ldots \right], \tag{142}$$

where

$$f = \frac{\varepsilon}{2} \left(1 - x_1^2\right)^{1/2} \left[1 - 9\left(1 - \frac{x_1^2}{2}\right)\right],$$

$$\varepsilon = 1 / \left(x_2^2 - 1\right).$$

Now we can present the undisturbed solution by the following equations, which at the same time give the sequence of computation of coordinates r, δ and λ:

$$n_r(t - \beta_1) = E - e \sin E, \tag{143}$$

$$r = a(1 - e \cos E), \tag{144}$$

$$\tan \frac{w}{2} = \left(\frac{1 + e - c/a}{1 - e - c/a}\right)^{1/2} \tan \frac{E}{2}, \tag{145}$$

$$j(w) = \frac{\Delta j}{2\pi} \left[w - \left(\frac{k^2}{4} + \frac{k^4}{8}\right) \sin w + \frac{3}{128} k^4 \sin 2w + \ldots\right], \tag{146}$$

$$G(u) = j(w) - \beta_2, \tag{147}$$

$$G(u) = \frac{\Delta G}{2\pi} \left\{ u - \frac{1}{8} \left[\left(\frac{x_1}{x_2}\right)^2 + \frac{1}{2} \left(\frac{x_1}{x_2}\right)^4\right] \sin 2u \right.$$
$$\left. + \frac{3}{256} \left(\frac{x_1}{x_2}\right)^4 \sin 4u + \ldots \right\}, \tag{148}$$

$$\sin \delta = x_1 \sin u, \tag{149}$$

$$g(u) = \pm \left[\arctan \left(1 - x_1^2\right)^{1/2} \tan u + fu \right.$$
$$\left. + \frac{3}{32} \varepsilon^2 x_1^2 \left(1 - x_1^2\right)^{1/2} \sin 2u + \ldots \right], \tag{150}$$

$$\lambda = g(u) - \beta_3, \tag{151}$$

where n_r is the mean angular motion of the satellite, given by

$$n_r = \frac{(-2a_1)^{1/2}}{a} \frac{1}{1 + c/2a + 3/2(c/2a)^2(1 - e^2)^{-1/2}} . \qquad (152)$$

The sequence of the computations of coordinates of the satellite r, δ, λ with the help of Eqs. (143)–(152) for the given moment of time is as follows.

For the given moment of time t we obtain E from Eq. (143) by the method of successive approximations, after which we obtain the first coordinate – the distance r, from Eq. (144). At the same time we obtain w from Eq. (145), after which $j(w)$ from Eq. (146) and $G(u)$ from Eq. (147). Solving now Eq. (148) by the known left-hand side, we obtain the value of u by the method of successive approximations. Having u, we obtain readily the second coordinate of the satellite – δ from Eq. (149). Then we obtain from the known value of u the numerical value of the integral $g(u)$ with the help of Eq. (150), after which the third coordinate λ from Eq. (151). This completes the determination of the coordinates of an artificial satellite of the Earth: distance r, declination δ and right ascension λ for a given moment of time t.

The method described does not involve perturbations, but, via only the modified mean motion n_r, enables one to compute the position (ephemerides) of the satellite almost as easily as in the case of ellipse in the two-body problem. The example just considered demonstrates again the vast applicability of the Hamilton–Jacobi theorem for solution of similar problems.

20 The Distribution of Binary Stars by Eccentricities

Let us consider one more example of the application of canonic transformations, this one from stellar dynamics. It is associated with the debate between Jeans and Ambartzumian in 1930s concerning the interpretation of the observed distribution of binary stars in terms of the eccentricities of their orbits.

Already in the early 1930s it was obtained that, among the binary stars with known orbits, the number of observed binaries with eccentricities smaller than e is proportional to e^2. On the other hand, Jeans showed in 1935 that, with the statistically balanced distribution (Boltzmann), as the number density of stars dN in the phase space is given by an exponential law with respect to the satellite's energy E, i.e.

$$dN \propto e^{-E/\sigma} ,$$

where σ is the parameter of the Boltzmann distribution, so the same dependence should be fulfilled, i.e. dN must be proportional to e^2. From that the conclusion was drawn that, at least in the case of the binary systems, the state of the most probable distribution has already been reached, which in its turn leads readily to the so-called long cosmological time scale, i.e 10^{13} years.

Ambartzumian (1937), however, proved that exactly the same distribution $dN \propto e^2$ must be satisfied not only when the distribution of density in phase space is Boltzmannian but also when the density is an arbitrary function $f(E)$ of energy E. This proof for our purposes, is certainly of interest, first of all because the technique of canonical transformations is used for it.

Let the three coordinates of the position of the satellite of a binary system x, y, z and the three components of its momentum p_x, p_y, and p_z with respect to the main star $(m_\star)$ compose the phase space, so that one has for the number density of the binaries

$$dN = f(E)\, dx\, dy\, dz\, dp_x\, dp_y\, dp_z \,. \tag{153}$$

We will now perform a canonical transformation in the phase space from the variables x, y, z, p_x, p_y, and p_z to Delaunay canonical elements L, G, H, l, g, and h according to Eq. (117):

$$
\begin{aligned}
L &= k\sqrt{M}a^{1/2}, & l &= n(t - T), \\
G &= k\sqrt{M}(1 - e^2)^{1/2}a^{1/2}, & g &= \omega = \pi - \Omega, \\
H &= k\sqrt{M}(1 - e^2)^{1/2}a^{1/2}\cos i, & h &= \Omega,
\end{aligned}
\tag{154}
$$

where a and e are the major semiaxis and the eccentricity of the orbit of the second component of the binary system, respectively, i is the inclination of the orbit's plane, $n(t - T)$ is the mean anomaly, ω is the angular distance of perihelion from the node, Ω is the longitude of the ascending node, and $M = m + m_\star$ is the total mass of the binary system. Under canonical transformations the volume element of phase space is conserved or, in other words, the Jacobian of transformation equals unity:

$$dx\, dy\, dz\, dp_x\, dp_y\, dp_z = dL\, dG\, dH\, dl\, dg\, dh \,.$$

We also have for the kinetic energy of the satellite of mass m

$$E = -\frac{mV^2}{2} = \frac{m}{2}\frac{k^2 M}{a} = -\frac{k^4 m M^2}{2L^2} \,.$$

Hence, the energy depends only on the element L, i.e. $f(E) = f(L)$. Then the number of pairs dN, when L is within the interval between L and $L+dL$, and G is larger than some magnitude G_0, will be

$$dN = 8\pi^3 f(L)dL \int_{G_0}^{L} dG \int_{0}^{G} dH \,, \tag{155}$$

bearing in mind that L, g and h are varying independently from each other within the interval between 0 and 2π, H takes the values from 0 (at $i = 90°$) to G (at $i = 0$), and G varies, according to its definition, between 0 (when $e = 1$) to L (when $e = 0$). Therefore, from Eq. (155), we have

$$dN = 4\pi^3 f(L)\,(L^2 - G_0^2)\,dL\,,$$

and from (154):

$$L^2 - G_0^2 = k^2 M a e_0^2 = L^2 e_0^2\,,$$

where e_0 is the eccentricity corresponding to the orbit with given L and $G = G_0$.

Thus, the number of stars with $e < e_0$, i.e. with $G > G_0$, and for L within the limits L and $L + dL$, is equal to

$$dN\,(e < e_0) = 4\pi^3 f\,(L)\,L^2 e_0^2\,dL\,.$$

From that we find for the number of all orbits with $e < e_0$

$$N(e < e_0) = 4\pi^3 e_0^2 \int_0^\infty f(L)L^2\,dL\,. \tag{156}$$

The integral on right-hand side is a constant number, and therefore we have finally

$$N\,(e < e_0) \propto e_0^2\,. \tag{157}$$

Thus, if the density in the phase space is an arbitrary function of L, i.e. of the total energy and only it, then the total **number of binary stars with eccentricities smaller than e_0 is proportional to e_0^2.**

Thus, even if the observed number $N(e_0)$ were proportional to e_0^2, one could never conclude that the phase density must be proportional to $\exp(-E/\sigma)$, i.e. that equipartition in energy had been reached. The result, Eq. (157), claims that one must always have $N(e_0) \sim e_0^2$, at any distribution of energy, under the sole condition that the phase density does not depend on other elements. Therefore, from the observed dependence $N(e_0) \propto e_0^2$ one can draw make no unique conclusions concerning the equipartition and the time scale of stellar systems. Together with other considerations, in particular, the absence of a dissociative equilibrium between binary systems and single stars, this supports the so-called short cosmogonical scale, i.e. 10^{10} years.

We will not go into the details of the comparison of the law (157) with observational data, the dissociative equilibrium, relaxation time scales, etc. and related evolutionary aspects of the stellar systems. Our aim was to outline that one of the most interesting results in stellar dynamics, Eq. (157), was obtained by successful application of canonical transformations. On the other hand, this fact indicates the exceptional possibilities and productivity of this technique of celestial mechanics.

SELECTED ASTRONOMICAL CONSTANTS

Gauss constant, k	0.017 202 098 95	(1809)
	0.017 202 098 950 000	(2000)
Astronomical unit, AU	$1.495\ 978\ 706\ 65 \times 10^{13}$ cm	
Speed of light, c	$2.997\ 924\ 585 \times 10^{10}$ cm s^{-1}	
Gravitational constant, G	$6.672\ 59\ 10^{-8}$ dyn cm^2 g^{-2}	
Mass of the Sun	$1.989\ 15 \times 10^{33}$ g	
Mass of the Earth	$5.974\ 25 \times 10^{27}$ g	
Solar constant	1 368 W m^{-2}	
Sun–Jupiter mass ratio	1047.3486	
Earth–Moon mass ratio	81.30059	
Equatorial radius of the Earth	6378.136 km	
Obliquity of the ecliptic (J2000)	23°26'21.412"	
Earth sidereal day	23 h 56 m 4.09054 s	
Sidereal year	365.25636 d	
Semimajor axis of the Earth's orbit	1.000000032 AU	
Parsec, pc	206 264.806 AU	$= 3.0856785 \times 10^{18}$ cm
		$= 3.261633$ light year
Age of the Solar system	4.55×10^9 yrs	
Age of the Galaxy	$10\text{–}15 \times 10^9$ yrs	

SELECTED DATA FOR THE SUN

Diameter		1 391 020 km	109 Earth's diameter
Mass		1.9895×10^{33} g	333 000 Earth's mass
Density:	center	151.3 g cm^{-3}	
	mean	1.409 g cm^{-3}	
Temperature:	center	15 557 000 K	
	photosphere	5780 K	
	chromosphere	10 000 – 20 000 K	
	reversing layer	100 000 – 300 000 K	
	corona	2 000 000 – 3 000 000 K	
Rotation period		25. 4 d (sidereal)	
Escape velocity		617.7 km s^{-1}	
Solar constant (at Earth)		1368 w m^{-2}	

FROM THE CHRONICLE OF SPACE FLIGHTS

Spacecraft		Launch	Target	Notes
Sputnik	USSR	1957 Oct.4	Earth	First satellite
Vostok	USSR	1961 Apr.12	Earth	First man in space
Lunar Orbiter 4	USA	1967 May 4	Moon	Photographic mapping
Venera 4	USSR	1967 Jun.12	Venus	Atmospheric probe
Lunar Orbiter 5	USA	1967 Aug.1	Moon	Photographing mapping
OSO 4	USA	1967 Oct.18	Sun	Solar observatory

Table, continued

Luna 14	USSR	1968 Apr.7	Moon	Mapped lunar gravity field
Apollo 8	USA	1968 Dec.21	Moon	Manned lunar orbiter
OSO 5	USA	1969 Jan.22	Sun	Solar observatory
Apollo 10	USA	1969 May 18	Moon	Lunar orbit test-precursor
Apollo 11	USA	1969 Jul.16	Moon	First manned lunar landing
Apollo 12	USA	1969 Nov.14	Moon	Manned lunar landing
Apollo 13	USA	1970 Apr.11	Moon	Lunar landing-crew returned
Venera 7	USSR	1970 Aug.17	Venus	First successful lander:Dec.15,1970
Uhuru	USA	1970 Dec.12	Earth	First X-ray observatory in space
Apollo 14	USA	1971 Jan.31	Moon	Manned lunar landing
Apollo 15	USA	1971 Jul.26	Moon	Manned lunar landing
Pioneer 10	USA	1972 Mar.3	Jupiter	Jupiter flyby on Dec.3, 1973
Apollo 16	USA	1972 Apr.16	Moon	Manned lunar landing
Apollo 17	USA	1972 Dec.7	Moon	Manned lunar landing
Luna 21	USSR	1973 Jan.8	Moon	Unmanned lunar rover
Pioneer 11	USA	1973 Apr.6	Jupiter	Jupiter flyby on Dec.4, 1974
Mariner 10	USA	1973 Nov.3	Venus	Flew by Venus, Feb.5, 1974
Soyuz 13/Orion 2	USSR	1973 Dec.18	Earth	First manned space observatory
Venera 9	USSR	1975 Jun.8	Venus	Landed Oct.22, 1975
OSO 8	USA	1975 Jun.21	Sun	Solar observatory in Earth orbit
Viking 1 lander	USA	1975 Aug.20	Mars	Landed July 20, 1976
Viking 2	USA	1975 Sept.9	Mars	Orbited Mars Aug.7, 1976
Viking 2 lander	USA	1975 Sept.9	Mars	Landed Sept.3, 1976
Voyager 2	USA	1977 Aug.20	Jupiter	Flew by Jupiter July 9, 1979
Voyager 1	USA	1977 Sept.5	Jupiter	Flew by Jupiter Mar.5, 1979
IUE	USA	1978 Jan.26	Earth	Ultraviolet observatory
Venera 14	USSR	1981 Nov.4	Venus	Landed Mar.5, 1982
IRAS	USA	1981 Jan.26	Earth	Infrared observatory
Vega 1	USSR	1984 Dec.15	Halley	Flew by comet Halley at 8890 km
Giotto	ESA	1985 July 2	Halley	Flew by comet Halley at 596 km
Magellan	USA	1989 May 5	Venus	Orbited Venus: radar mapper
Hipparcos	USA	1989 Aug.8	Earth	Astrometric observatory
Galileo Orbiter	USA	1989 Oct.18	Jupiter	Flew by Venus: Feb.10, 1990
				Flew by asteroid Gaspra: Oct.29,1991
				Flew by Earth: Dec.8,1992
				Flew by asteroid Ida: Aug.28,1993
				Orbited Jupiter: Dec. 7, 1995
Galileo Probe	USA	1989 Oct.18	Jupiter	Entered Jupiter's atmosphere
HST	USA/ESA	1990 Apr.25	Earth	2.4 m Hubble Space Telescope
Ulysses	USA/ESA	1990 Oct.6	Sun	Injected into Solar polar orbit
Compton	USA	1991 April	Earth	Gamma ray observatory
SOHO	ESA	1995 Dec.2	Sun	Solar observatory
NEAR	USA	1996 Feb.17	Eros	Orbited asteroid Mathilde
				Landed asteroid Eros Febr.12, 2001
Mars Surveyor	USA	1996 Nov.7	Mars	Orbited Mars: Sept. 12, 1997
Mars Pathfinder	USA	1996 Dec.2	Mars	Landed Mars: deployed rover
Cassini	USA/ESA	1998 Jan.6	Saturn	En route to Saturn
Lunar Prospector	USA	1998 Jan.6	Moon	Orbited Moon
Chandra	USA	1999 Jul.9	Earth	X-ray observatory
MAP	USA	2001 Jun.30	L_2 point	Microwave Anisotropy Probe
Genesis	USA	2001 Aug.8	L_1 point	Solar wind probes

References

The Classical Period

Ambartzumian V.A. 1937, On the Statistics of Binary Stars. *Astron. Zh.* **14**, 207.

Arenstorf R.F. 1963, Periodic Solution of Restricted Three-body problem Representing Analytic Continuations of Keplerian Elliptical Motions. *Amer. J. Math.* **85**, 27.

Broucke R. 1962, *Recherches d'orbites périodiques dans le probléme restrein plans (system Terre-Lune).* Université de Louvain.

Bruns H. 1887, Über die Integrale des Vielkörper-Problems. *Acta Math.* **11**, 25.

Chandrasekhar S. 1960, *Principles of Dynamics of Stellar Systems.* Dover, New York.

Chandrasekhar S. 1961, *Hydrodynamic and Hydromagnetic Stability.* Clarendon Press, Oxford.

Darwin G.H. 1897, Periodic Orbits. *Acta Math.* **21**, 99.

Darwin G.H. 1910, On Certain Families of Periodic Orbits, *Mon. Not. Roy. Astr. Soc.* **70**, 604.

Deprit A. and Henrard J. 1965, Symmetric Double Asymptotic Orbits in the Restricted Three-body Problem. *Astron. J.* **70**, 271.

Egorov V.A. 1965, *Problems of the Motion of the Moon.* Nauka, Moscow, (in Russian)

Farquhar R.W. 1972, A Halo-orbit Lunar Station. *Astronaut. Aeronaut.* **10**, No. 6.

Hill G.W. 1878, Research in the Lunar Theory. *Amer. J. Math.* I. Works, Vol. I, 284.

Hohmann W. 1925, *Die Erreichbarkeit des Himmelskörpers.* Munich, Oldenburg.

Kislik M.D. 1964, Sphere of Influence of Large Planets and the Moon. *Space Investig.* **2**, No. 6, (in Russian).

Kordylewski K. 1961, Photographische Untersuchungen des Libration-Punktes L im System Erde-Mond. *Acta Astron.* **11**, 165.

Message P.J. 1958, The Search for Asymmetric Periodic Orbits in the Restricted Problem of Three Bodies. *Astron. J.* **63**, 443.

Moulton F.R. 1920, *Periodic Orbits,* Carnegie Inst., Washington.

Poincaré H. 1910, *Leçons de mécanique céleste.* Tômes I–III. Paris.

Rabe E. and Schanzle A. 1962, Periodic Librations About the Triangular Solutions of the Restricted Earth-Moon Problem and their Orbital Stabilities. *Astron. J.* **67**, 732.

Roy A.E. 1978, *Orbital Motion,* Adam Hilger, Bristol. 396.

Strömgrén E. 1934, Symmetrische und unsymmetrische librationsähnliche Bahnen in Problem Restreint mit asymptotisch-peridischen Bahnen als Grenzbahnen. *Copenhagen Obs. Publ.* **No 97**.

Sundman K.F. 1912, Mémoire sur le probléme des trois corps. *Acta Math.* **36**, 105.

Szebehely V. 1967, *Theory of Orbits.* Academic Press, New York.

Tisserand F. 1889-96, *Traité de Mecanique Céleste,* Tômes I (1889), II (1891), III (1894), IV (1896). Paris.

Later Studies

Anderson Jr.A.D. 1989, *Introduction to Flight.* McGraw-Hill, New York.

Arnold V.I. 1989, *Mathematical Methods of Classical Mechanics.* Springer-Verlag, Berlin.

Arnold V.I. 1989, *Huygens and Barrow, Newton and Hooke.* Nauka, Moscow (in Russian).

Brumberg V.A. and Brumberg E.V. 1999, *Celestial Mechanics at High Eccentricities*, Gordon and Breach, New York.

Celnikier L.M. 1993, *Basics of Space Flights.* Editions Frontiéres, New York.

Encyclopedia of the Solar System, 1999, eds. P. R. Weissman, L.-A. McFadden and T.V. Johnson, Academic Press.

Farinella P. *et al.*, 1994, Asteroids Falling into the Sun. *Nature* **371**, 314.

Fernandez J. 1994, Dynamics of Comets: Recent Developments and New Challenges. *IAU Symp.* **160**, p. 223, Kluwer Acad. Publ., Dordrecht.

Ferraz-Mello S. (Ed.), 1992, Chaos, Resonance and Collective Dynamical Phenomena in the Solar System. *IAU Symp.* **152**, Kluwer Acad. Publ., Dordrecht

Fitzsimmons A. (Ed.), 2001, *Minor Bodies in the Outer Solar System.* Springer-Verlag, Berlin.

Fridman A.M. and Gorkavy N.N. 1999, *Physics of Planetary Rings*, Springer-Verlag, Berlin.

Ford E.B., Seager S., Turner E.L. 2001, Characterization of Extrasolar Terrestial Planets from Diurnal Photometric Variability. *Nature* **412**, 885.

Gurzadyan G.A. 1996, *Theory of Interplanetary Flights.* Gordon and Breach, New York.

Gurzadyan G.A. 1997, *Physics and Dynamics of Planetary Nebulae.* Springer-Verlag, Berlin.

Gurzadyan G.A. 1998, Common Chromospheres - Roundchroms - as Means for the Study of Binary Systems. *Mon. Not. Roy. Astr. Soc.* **290**, 607.

Gurzadyan G.A. 2000, Binary Globular Clusters. *New Astronomy* **5/6**, 349.

Gurzadyan V.G. 2000, Astronomy and the Fall of Babylon. *Sky & Telescope* **100**, 40.

Gurzadyan V.G. and Pfenniger D. (Eds.), 1994, *Ergodic Concepts in Stellar Dynamics.* Springer-Verlag, Berlin.

Gurzadyan V.G. and Ruffini R. (Eds.), 2000, *The Chaotic Universe.* World Scientific, New York.

Harland D.M. 2000, *Jupiter Odyssey: The Story of NASA's Galileo Mission.* Praxis/Springer-Verlag, New York.

Hills J.G. and Leonard P.J.T. 1995, Earth-crossing Asteroids: The Last Days Before Earth Impact. *Astron.J.* **109**, 401.

Jakosky B.M., Hunderson B.G. and Mellon M.T. 1993, Chaotic Obliquity and the Nature of the Martian Climate. *Bull Amer. Astron. Soc.* **25**, 1041.

Klavatter J.J. 1989, Rotation of Hyperion I. Observations. *Astron. J.* **97**, 570.

Laskar J. 1994, The Stability of the Solar System. In: V.G. Gurzadyan, D. Pfenniger (Eds.) *Ergodic Concepts in Stellar Dynamics*. Springer-Verlag, Berlin.

Laskar J. 1997, Large Scale Chaos and the Spacing of the Inner Planets. *Astron. Astrophys.* **317**, L75.

Morbidelli A. 2002, *Modern Celestial Mechanics*. Taylor and Francis, London.

NASA homepage for space missions: http://spacescience.nasa.gov/ missions/

Sussman G.J. 1992, Chaotic Evolution of the Solar System. *Science* **257**, 56.

Tremaine S. 1994, Is the Solar System Stable? *The George Darwin Lecture*, Roy. Astron. Soc., London.

Touma J., Wisdom J. 1993, The Chaotic Obliquity of Mars. *Science* **259**, 1294.

Valtonen M.J. (Ed.), 1992, *The Few Body Problem*. Kluwer Acad. Publ., Dordrecht.

Verger F. *et al*, 2000, *The Cambridge Encyclopedia of Space*, Cambridge University Press, Cambridge.

Veverka J. *et al*, 2001, The Landing of the NEAR-Shoemaker Spacecraft on Asteroid 433 Eros. *Nature* **433**, 390.

Weissman P. 1994, Comet Shoemaker-Levy 9: Events after the Events. *Nature* **372**, 404.

Wiessel W.E. 1989, *Space Flight Dynamics*. McGraw-Hill, New York.

Wright J.L. 1992, *Space Sailing*. Gordon and Breach, New York.

Index